Metallabenzenes

Metallabenzenes

An Expert View

Edited by
L. James Wright
University of Auckland
New Zealand

WILEY

This edition first published 2017

Registered Office(s)
John Wiley & Sons, Inc., 111 River Street, Hoboken, NJ 07030, USA
John Wiley & Sons Ltd, The Atrium, Southern Gate, Chichester, West Sussex, PO19 8SQ, UK

Editorial Office
9600 Garsington Road, Oxford, OX4 2DQ, UK

For details of our global editorial offices, customer services, and more information about Wiley products visit us at www.wiley.com.

Wiley also publishes its books in a variety of electronic formats and by print-on-demand. Some content that appears in standard print versions of this book may not be available in other formats.

Library of Congress Cataloging-in-Publication Data

Names: Wright, L. James (Leonard James), 1954- editor.
Title: Metallabenzenes : an expert view / edited by L. James Wright, University of Auckland, New Zealand.
Description: Hoboken, NJ : Wiley, 2017. | Includes bibliographical references and index. |
Identifiers: LCCN 2017010554 (print) | LCCN 2017012640 (ebook) | ISBN 9781119068099 (pdf) | ISBN 9781119068082 (epub) | ISBN 9781119068068 (cloth)
Subjects: LCSH: Benzene--Derivatives. | Aromatic compounds.
Classification: LCC QD341.H9 (ebook) | LCC QD341.H9 M51278 2017 (print) | DDC 547/.611--dc23
LC record available at https://lccn.loc.gov/2017010554

Cover image: L. James Wright and Benjamin J. Frogley
Cover design by Wiley

Set in 10/12pt WarnockPro by SPi Global, Chennai, India

10 9 8 7 6 5 4 3 2 1

Contents

List of Contributors *xi*
Preface *xiii*

1 **Metallabenzenes and Fused-Ring Metallabenzenes of Osmium, Ruthenium and Iridium: Syntheses, Properties and Reactions** *1*
Benjamin J. Frogley, Warren R. Roper and L. James Wright

1.1 Introduction *1*
1.2 Syntheses and Properties of Metallabenzenes with Methylthiolate Substituents *2*
1.2.1 Osmabenzenes *2*
1.2.2 Iridabenzenes *7*
1.3 Syntheses and Properties of Fused-Ring Metallabenzenes *10*
1.3.1 Osmabenzofurans *11*
1.3.2 Ruthenabenzofurans *14*
1.3.3 Iridabenzofurans *17*
1.3.4 Iridabenzothiophenes *19*
1.3.5 Iridabenzothiazolium Cations *21*
1.4 Reactions of Metallabenzenes and Metallabenzenoids *24*
1.4.1 Electrophilic Aromatic Substitution Reactions *24*
1.4.2 Rearrangement to Cyclopentadienyl Complexes *34*
1.4.3 Nucleophilic Aromatic Substitution Reactions *38*
1.5 Concluding Remarks *43*
References *43*

2 **The First Iridabenzenes: Syntheses, Properties, and Reactions** *51*
John R. Bleeke

2.1 Introduction *51*
2.2 Basic Theory *52*
2.3 Discovery of the First Stable Metallabenzenes *54*
2.4 Synthesis of Iridabenzene *54*
2.5 Valence Bond Structures and Electron Counting for Iridabenzene *56*
2.6 The Tris(trimethylphosphine) Reaction System *57*
2.7 Structure and Spectroscopy of Iridabenzene 3 *58*

2.8 Chemical Reactivity of Iridabenzene 3 *60*
2.8.1 Ligand Substitution *60*
2.8.2 Oxidative Addition/Oxidation *63*
2.8.3 Cycloaddition *65*
2.8.4 Theoretical Study of Cycloaddition *68*
2.8.5 Electrophilic Addition *69*
2.8.6 Coordination to $Mo(CO)_3$ *71*
2.9 Iridaphenol *73*
2.10 Synthesis and Spectroscopy of Iridapyrylium *78*
2.11 Valence Bond Structures and Electron Counting for Iridapyrylium *80*
2.12 Chemical Reactivity of Iridapyrylium 37 *80*
2.12.1 Ligand Addition *80*
2.12.2 Cycloaddition *81*
2.13 Comparison of Iridabenzene 3 and Iridapyrylium 37 *83*
2.14 Synthesis and Spectroscopy of Iridathiabenzene *83*
2.15 Structure of Iridathiabenzene 50 *85*
2.16 Chemical Reactivity of Iridathiabenzene 50 *86*
2.16.1 Ligand Addition and Cycloaddition *86*
2.16.2 Coordination to Other Metals *86*
2.17 Comparison of Iridathiabenzene 50 and Iridapyrylium 37 *89*
2.18 Synthesis and Structure of a Neutral Iridathiabenzene *89*
2.19 Spectroscopy of Neutral Iridathiabenzene 56 *90*
2.20 Chemical Reactivity of Neutral Iridathiabenzene 56 *91*
2.21 Related Metal-Coordinated Metallabenzenes *94*
2.22 Aromaticity *99*
2.22.1 Electronic Structure *99*
2.22.2 Structural (Geometric) Features *99*
2.22.3 Magnetic Properties *100*
2.22.4 Stability *100*
2.22.5 Chemical Reactivity *101*
2.22.6 Conclusion *101*
2.23 Final Word *102*
References *102*

3 Metallabenzenes and Valence Isomers via the Nucleophilic 3-Vinylcyclopropene Route *109*
Michael M. Haley

3.1 Project Origin and Inspiration (A Nod to Binger, Bleeke, Grubbs, Hughes, and Roper) *109*
3.2 Ligand Synthesis (An Exercise in Over-Engineering) *111*
3.3 Iridabenzenes and Valence Isomers (Success after Six Long Years) *114*
3.3.1 Iridabenzene/Iridabenzvalene Synthesis *114*
3.3.2 Mechanisms of Iridabenzvalene Isomerization and Iridabenzene Rearrangement *120*
3.3.3 Iridabenzene/Iridabenzvalene Spectroscopic Properties *123*

3.3.4 Iridabenzene/Iridabenzvalene Solid-State Structures *125*
3.4 Platinabenzenes (How You Get Your Chemistry on a Beer Coaster) *126*
3.4.1 "Irrational" Platinabenzene Synthesis *126*
3.4.2 "Rational" Platinabenzene Synthesis *129*
3.4.3 Platinabenzene Spectroscopic Properties and Solid-State Structures *130*
3.5 Odds and Sods (Ones that Got Away) *132*
3.5.1 16-Electron, High Oxidation State Iridabenzenes *133*
3.5.2 Hammett Plot of Iridabenzvalene Isomerization *135*
3.5.3 Rhodabenzvalene and Putative Dewar Rhodabenzene *137*
3.6 Conclusion (So Long, and Thanks for All the Fish) *139*
3.7 Acknowledgements *140*
References *140*

4 Iridabenzenes and Iridanaphthalenes with Supporting Tris(pyrazolyl)borate Ligands *145*
Margarita Paneque and Nuria Rendón

4.1 Introduction *145*
4.2 Synthetic Routes to Iridaaromatic Derivatives with Supporting Tris(pyrazolyl)borate Ligands *146*
4.2.1 Oxidatively Induced Ring Contraction *146*
4.2.2 Synthesis of Iridaaromatics by Ring Expansion Reactions *148*
4.2.2.1 Protonation of a Vinylidene-Iridacyclopentene *149*
4.2.2.2 Preparation of Tp^{Me2}-Iridaaromatics by Reaction of Olefins with Iridacyclopentadienes *153*
4.2.2.2.1 Iridabenzenes *153*
4.2.2.2.2 Iridanaphthalenes *157*
4.2.2.3 Other Procedures for Ring Expansion Reactions *161*
4.2.2.3.1 Reaction of a Tp^{Me2}-Iridaindene with Bis(trimethylsilyl)acetylene *161*
4.2.2.3.2 Coupling of Two Molecules of Acetylene and a Further C_1 Fragment in a TpIr Precursor *164*
4.3 Reactivity of Iridaaromatics with Supporting Tris(pyrazolyl)borate ligands *166*
4.4 Structural Data for Iridaaromatics with Supporting Tris(pyrazolyl)borate Ligands *171*
4.5 Spectroscopic Data for Iridaaromatics with Supporting Tris(Pyrazolyl)Borate Ligands *175*
4.6 Conclusions *177*
References *178*

5 Chemistry of Metallabenzynes and Rhenabenzenes *185*
Guochen Jia

5.1 Introduction *185*
5.2 Chemistry of Metallabenzynes *186*
5.2.1 Routes to Construct Metallabenzyne Rings *186*

5.2.1.1 Reactions of Terminal Alkynes with $OsCl_2(PPh_3)_3$ *186*
5.2.1.2 Reactions of Allenylcarbene and Vinylidene Complexes with Acetylides or Alkynes *186*
5.2.1.3 Oxidative Addition Reactions of Vinyl Carbyne Complexes *188*
5.2.1.4 Isomerization of Alkyne-Carbene Complexes *189*
5.2.2 Chemical Properties of Metallabenzynes *189*
5.2.2.1 Reactions Not Involving Carbons of the Metallabenzyne Ring *189*
5.2.2.2 Electrophilic Substitution Reactions *191*
5.2.2.3 Nucleophilic Addition Reactions *193*
5.2.2.4 Migratory Insertion Reactions *193*
5.2.3 Structural Properties of Metallabenzynes *202*
5.2.4 Ring Strain of Metallabenzynes *203*
5.2.5 Electronic Structures of Metallabenzynes *204*
5.2.5.1 Aromatic Properties of Metallabenzynes *205*
5.3 Chemistry of Rhenabenzenes *207*
5.3.1 Synthesis of Rhenabenzenes *207*
5.3.2 Structural and Aromatic Properties of Rhenabenzenes *208*
5.3.3 Rearrangement of Low-valent Rhenabenzene to η^5-Cyclopentadienyl Complexes *209*
5.4 Summary *215*
References *216*

6 Metallabenzenoid Compounds Bearing Phosphonium Substituents *219*
Hong Zhang and Haiping Xia

6.1 Synthesis *220*
6.1.1 [5+1] Synthesis *220*
6.1.1.1 Syntheses Starting from $HC{\equiv}CCH(OH)C{\equiv}CH$ *220*
6.1.1.2 Syntheses Starting from $HC{\equiv}CCH(OH)CH{=}CH_2$ and $HC{\equiv}CCCH_3(OH)CH{=}CH_2$ *223*
6.1.1.3 Syntheses Starting from $HC{\equiv}CCH(OH)Ph$ *223*
6.1.1.4 Synthesis Starting from $HC{\equiv}CCH(OH)Et$ *225*
6.1.2 [4+2] Synthesis *226*
6.1.3 [3+3] Synthesis *228*
6.2 Structure and Bonding *229*
6.3 Reactions *232*
6.3.1 Ligand Substitution Reactions *232*
6.3.2 Nucleophilic Addition and Nucleophilic Aromatic Substitution Reactions *233*
6.3.3 Cycloaddition Reactions *239*
6.3.4 Formation of Unsaturated Organic Rings *240*
6.4 Physical Properties *243*
6.4.1 Electrochemical Properties *244*
6.4.2 Optical Properties *247*
6.4.3 Other Properties *249*
6.5 Polycyclic Metallabenzenoid Compounds Bearing Phosphonium Substituents *249*

6.5.1 Synthesis by Means of S_NAr Reactions *251*
6.5.2 Synthesis by Means of Electrophilic Cyclization Reactions *255*
6.6 Future Prospects *261*
References *262*

7 Theoretical Studies of Metallabenzenes: From Bonding Situation to Reactivity *267*
Israel Fernández and Gernot Frenking

7.1 Introduction *267*
7.2 Structure and Bonding Situation *268*
7.2.1 Bonding Situation and Molecular Orbitals *268*
7.2.2 Aromaticity of Metallabenzenes *272*
7.2.2.1 [4*n*+2]-rule *273*
7.2.2.2 Absolute Hardness *273*
7.2.2.3 Magnetic Descriptors *274*
7.2.2.4 Energetic Descriptors *277*
7.3 Computational Studies on Synthetic Pathways towards Metallabenzenes *279*
7.4 Computational Studies on the Reactivity of Metallabenzenes *286*
7.4.1 Electrophilic Aromatic Substitution Reactions *286*
7.4.2 Nucleophilic Aromatic Substitution Reactions *288*
7.4.3 Addition and Cycloaddition Reactions *292*
7.4.4 Rearrangement Reactions *294*
7.5 Concluding Remarks and Outlook *296*
7.6 Acknowledgements *297*
References *298*

Index *305*

List of Contributors

John R. Bleeke
Department of Chemistry
Washington University
USA

Israel Fernández
Departamento de Química Orgánica I
Facultad de Ciencias Químicas
Universidad Complutense de Madrid
Spain

Gernot Frenking
Fachberich Chemie
Philipps-Universität Marburg
Marburg
Germany and Donostia International
Physics Center (DIPC)
Spain

Benjamin J. Frogley
School of Chemical Sciences
University of Auckland
New Zealand

Michael M. Haley
Department of Chemistry &
Biochemistry and Materials Science
Institute
University of Oregon
USA

Guochen Jia
Department of Chemistry
The Hong Kong University of Science
and Technology
P. R. China

Margarita Paneque
Instituto de Investigaciones
Químicas (IIQ)
Departamento de Química Inorgánica
and Centro de Innovación en Química
Avanzada (ORFEO-CINQA)
Consejo Superior de Investigaciones
Científicas (CSIC) and Universidad
de Sevilla
Spain

Nuria Rendón
Instituto de Investigaciones
Químicas (IIQ)
Departamento de Química Inorgánica
and Centro de Innovación en Química
Avanzada (ORFEO-CINQA)
Consejo Superior de Investigaciones
Científicas (CSIC) and Universidad
de Sevilla
Spain

Warren R. Roper
School of Chemical Sciences
University of Auckland
New Zealand

L. James Wright
School of Chemical Sciences
University of Auckland
New Zealand

Haiping Xia
Department of Chemistry
Xiamen University
P. R. China

Hong Zhang
Department of Chemistry
Xiamen University
P. R. China

Preface

Professor Warren R. Roper FRS
(Photo taken in 1989, seven years after his group isolated the first metallabenzene). Source: Image provided courtesy of Professor Warren R. Roper FRS.

Metallabenzenes are a fascinating class of compounds which can be viewed as analogues of benzene in which one of the CH groups has formally been replaced by an isolobal transition metal fragment. Although benzene itself was discovered by Michael Faraday in 1825, it was not until 1982 that the first metallabenzene, an osmabenzene, was isolated by Warren R. Roper and his group at the University of Auckland, New Zealand. Since that time, interest in these compounds has steadily grown, and today it has become a flourishing area of chemistry with new developments and discoveries regularly appearing in the literature. A diverse range of synthetic approaches to these compounds have been developed. The more common of these involve the direct addition to the metal of groups that contain the five carbon atoms that eventually become the metallabenzene ring atoms, the insertion of a carbon atom from an adjacent ligand into the M–C bond of a metallacyclopentadiene, and the ring contraction of metallacycloheptatrienes.

As might be expected, the presence of the transition metal in the metallabenzene ring has a profound effect on the properties of these compounds. Theoretical studies confirm that metallabenzenes are indeed aromatic compounds with aromatic stabilization energies between *ca.* 20 and 80% that of benzene. However, the delocalized π-system is more complex than that of benzene. Instead of the six π-electrons present in benzene, reported metallabenzenes have either eight or ten π-electrons, depending on whether the *p*-orbitals of ligands such as chloride participate in the π-bonding scheme of the metallacyclic ring. Some of the occupied metallabenzene π-orbitals have Möbius character and so the standard Hückel (4n + 2) π-electron rule cannot be used as a criterion for aromaticity.

The presence of the metal adds an important dimension to the reaction chemistry of metallabenzenes. Reactions that are not observed (e.g. cycloadditions) or are not even possible for benzene (e.g. coupling of the two Cα atoms to form a cyclopentadienyl ligand) have been reported. At the same time other reactions that are characteristic for benzene such as electrophilic aromatic substitution and π-coordination to metal fragments such as $Mo(CO)_3$ have also been observed. The metal significantly influences the spectral and structural properties of metallabenzenes. For example, in the ^1H and ^{13}C NMR spectra the resonances of the Cα (metal-bound) atoms and attached protons appear at very low field values. These resonances are almost invariably observed in between the values observed for related carbene and σ-vinyl complexes and are consistent with partial multiple bond character between the metal and the Cα atoms. The shifts of the remaining three carbon atoms and accompanying protons are in the normal ranges found for benzene derivatives. Structurally, metallabenzenes also display some distinctive features. Although the five-ring carbon atoms are always approximately coplanar, the metal is sometimes found significantly displaced from this plane, while in other cases it sits within this plane. Theoretical studies have shown that both electronic and steric effects are responsible for the location of the metal relative to the five-carbon plane. Unlike the situation for benzene, the energy profile associated with moving the metal out of the five-carbon plane is very shallow since this movement considerably decreases the π-antibonding interactions associated with the ring.

Although major advances have been made, the study of metallabenzenes is still very much in the early stages of development and it can be expected that many important new developments await discovery. The compounds and the reactions they undergo not only are of intrinsic interest but also do much to broaden our understanding of aromaticity. Furthermore, it can be anticipated that, owing to the special properties some of these species exhibit, future applications may be found in areas such as photoelectronics, molecular magnets, conducting polymers, fluorescent molecular probes, and new materials.

The field of metallabenzene chemistry is in the unusual situation that almost all the major synthetic, reaction chemistry, spectroscopic and structural studies have thus far come from just six research groups around the world. Accordingly, this book is arranged so that the work of each of these groups is covered in the form of personal perspectives in the first six chapters. Our own work (Warren R. Roper, L. James Wright and co-workers) from New Zealand, which includes the syntheses of osma-, ruthena- and iridabenzenes (including the first metallabenzene), the electrophilic aromatic substitution and nucleophilic aromatic substitution of hydrogen reactions they undergo, and the synthesis and reactions of fused-ring metallabenzenes, is covered in the first chapter. The research of John R. Bleeke's group (USA) is presented in Chapter 2, and includes, amongst other things, the synthesis and chemistry of the first iridabenzenes as well as heteroatom-substituted analogues such as iridathiabenzenes and iridapyrylium. The chapter ends with a short summary of metal-coordinated metallabenzenes. Michael M. Haley and co-workers (USA) in Chapter 3 describe the development of new synthetic routes to iridabenzenes and the first platinabenzenes, as well as studies of the mechanisms of synthesis and decomposition. Margarita Paneque and Nuria Rendón (Spain) summarize in Chapter 4 their group's contributions that include the development of a diverse range of metallaaromatics (including the first metallanaphthalene) which display unique chemistry and utilize supporting tris(pyrazolyl)borate ligands. The work of Guochen Jia and his group (Hong Kong), which has led to many new metallabenzenes incorporating osmium and rhenium as well as the related metallabenzyne species, is covered in Chapter 5. Haiping Xia and Hong Zhang (China) describe, in Chapter 6, the many significant contributions they and their co-workers have made to the field, largely through investigations into osma- and ruthenabenzenes containing one or more triphenylphosphonium ring substituents. These include new synthetic routes, reaction chemistry, bonding interactions, and the formation of fused-ring derivatives. Important computational investigations into the nature of metallabenzenes and the reactions of these compounds have been made by a number of different groups. In the final chapter (Chapter 7), this work is summarized by Israel Fernández and Gernot Frenking (Spain and Germany), who also highlight their own major contributions to this field.

L. James Wright
Auckland
May 2017

1

Metallabenzenes and Fused-Ring Metallabenzenes of Osmium, Ruthenium and Iridium: Syntheses, Properties and Reactions

*Benjamin J. Frogley, Warren R. Roper and L. James Wright**

School of Chemical Sciences, University of Auckland, Auckland, New Zealand

1.1 Introduction

The origin of metallabenzenes can be traced back almost two centuries to the discovery of benzene by Michael Faraday in 1825 [1]. He managed to separate benzene from the oily liquid obtained as a by-product during the manufacture of an "illuminating gas" by the destructive distillation of fish or whale oil. He correctly described a wide range of its properties and identified the formula as two proportions of carbon to one proportion of dihydrogen gas – thus describing it by the name "bicarburet of hydrogen". A few years later, in 1833, it was also isolated by the German chemist Eilhard Mitscherlich by the distillation of benzoic acid from gum benzoin. Mitscherlich correctly noted that it was identical to Faraday's bicarburet of hydrogen and gave it the name "benzin", from which the common name benzene is derived [2].

The molecular structure eluded chemists for many years. It was not until 1865 that German scientist Friedrich August Kekulé proposed the six-membered cyclohexatriene ring structure with alternating single and double bonds which subsequently led to the development of the concept of aromaticity [3, 4]. These advances revolutionised organic chemistry and began a flood of research into this exciting new area of so-called aromatic chemistry. Benzene is now considered the archetypical aromatic compound, and it is often used as the yardstick against which other species are compared with regard to aromatic character. While a precise definition of "aromaticity" remains somewhat nebulous, properties associated with benzene that have been classically used to characterise aromaticity include planarity, bond length equalisation, π-electron delocalisation, aromatic stabilisation energy, diamagnetic ring currents and electrophilic substitution, rather than addition, reactions. In more recent times, determinations of aromatic stabilisation energies by computational methods have been used to obtain more tangible measures of aromaticity.

Heteroaromatic species could be considered the next generation of aromatic compounds to be discovered. Amongst this large class of compounds, there are many

*Corresponding author: lj.wright@auckland.ac.nz

Metallabenzenes: An Expert View, First Edition. Edited by L. James Wright.

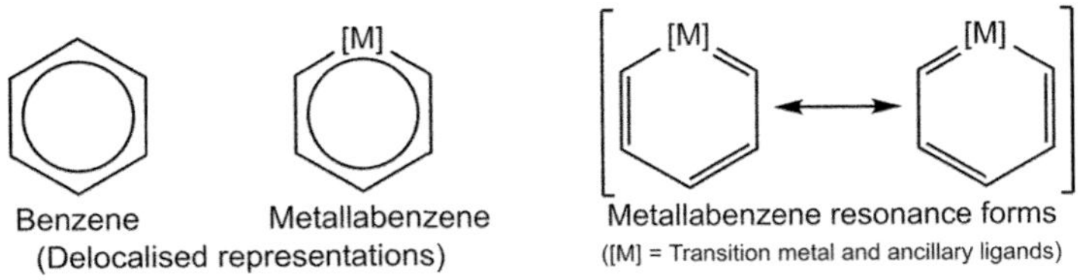

Chart 1.1 Metallabenzene delocalised representation and contributing resonance forms.

six-membered heterocycles that can be thought of as benzene analogues in which one CH unit of benzene has been formally replaced by an appropriate heteroatom. Pyridine, with the heteroatom nitrogen, was one of the earliest examples, and Scottish scientist Thomas Anderson is credited with the first report of this compound in 1849 [5]. Since then, related benzene analogues incorporating a wide array of main group heteroatoms have been isolated and these include, but are not limited to, phosphorus [6, 7], arsenic [7, 8], silicon [9], antimony [10], bismuth [10], germanium [11] and tin [12].

Metallabenzenes, which are perhaps the third generation of related aromatic compounds, have arrived comparatively recently in this timeline. The notion of formally replacing one CH unit of benzene with an appropriate transition metal (and its ancillary ligands) was proposed theoretically in 1979 by Thorn and Hoffmann [13] and it was only a short three years later before the first metallabenzene, an osmabenzene, was synthesised and characterised by Warren Roper and co-workers in New Zealand [14]. The aromatic character of this and other metallabenzenes (Chart 1.1) has now been thoroughly established through a range of different computational methods, and these are discussed in detail in Chapter 7 of this book. From these beginnings a new class of aromatic compounds, the metallabenzenes, was born.

In this chapter we provide a personal perspective on the contributions our group has made to this field, including studies of the syntheses, properties and reaction chemistry of osma-, ruthena- and iridabenzenes as well as related fused-ring derivatives.

1.2 Syntheses and Properties of Metallabenzenes with Methylthiolate Substituents

1.2.1 Osmabenzenes

In the 1970s and early 1980s, there was an acceleration of research efforts focused on the organometallic chemistry of transition metals, particularly on species where a transition metal is multiply bonded to a carbon donor ligand. We had been working in this area for some time and had developed a number of new carbene [15, 16], carbyne [17, 18] and thiocarbonyl [19–21] complexes of second- and third-row transition metals. We were aware of the 1979 theoretical paper by Thorn and Hoffmann that briefly describes the possibility of metallabenzenes as stable species [13]. Therefore, a few years later, when we were exploring the coordination of ethyne at the osmium centre of the zero-valent complex $Os(CS)(CO)(PPh_3)_3$, it did not take long for us to realise the CS ligand and two ethyne molecules had cyclised at the metal centre to produce the first metallabenzene, the osmabenzene, $Os(C_5H_4\{S\text{-}1\})(CO)(PPh_3)_2$ (**1**) (Scheme 1.1). In this

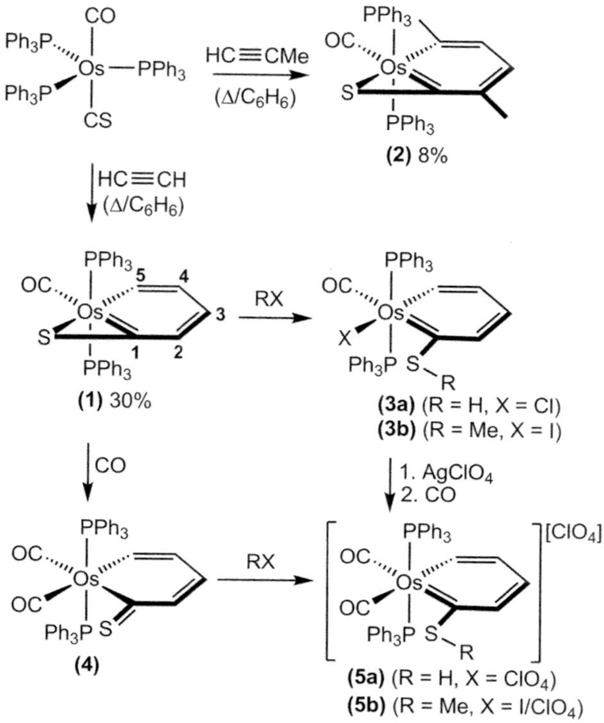

Scheme 1.1 Preparation of osmium complexes **1–5**.

compound the six-membered metallacyclic ring comprises the thiocarbonyl carbon atom, the four carbons of the two ethyne molecules and the osmium atom. The sulfur atom is also coordinated to the osmium metal centre, generating a secondary three-membered Os–C–S osmathiirene ring [14]. Therefore, the osmabenzene **1** could also formally be considered an osmabenzothiirene [22].

To synthesise the osmabenzene **1**, a solution of $Os(CS)(CO)(PPh_3)$ in benzene or toluene was treated with a slow stream of ethyne at 70°C for 20 min. Dark-brown crystals of pure **1** were formed in around 30% yield following purification by recrystallisation from *n*-hexane and column chromatography of the solid obtained [23].

This reaction can be considered a formal [1+2+2] cyclisation at the osmium centre of two molecules of ethyne and the carbon of the CS ligand. The most likely mechanism has been determined computationally using the model complex $Os(CS)(CO)(PH_3)_3$ [24]. The adduct **1A** (Scheme 1.2) is formed by coordination of the first molecule of ethyne after phosphine dissociation. The thiocarbonyl and ethyne ligands then combine to give the osmacyclobutenethione **1B**. The propensity of ligands such as CS to engage in cyclisation and migratory insertion reactions has proven to be invaluable in the synthesis of a number of metallaaromatic compounds. Coordination of the second molecule of ethyne, to give **1C**, and subsequent insertion of both carbon atoms into the four-membered ring gives the osmacyclohexadienethione **1D**. Finally, coordination of the sulfur atom to osmium results in aromatisation of the six-membered ring and formation of the osmabenzene **1**.

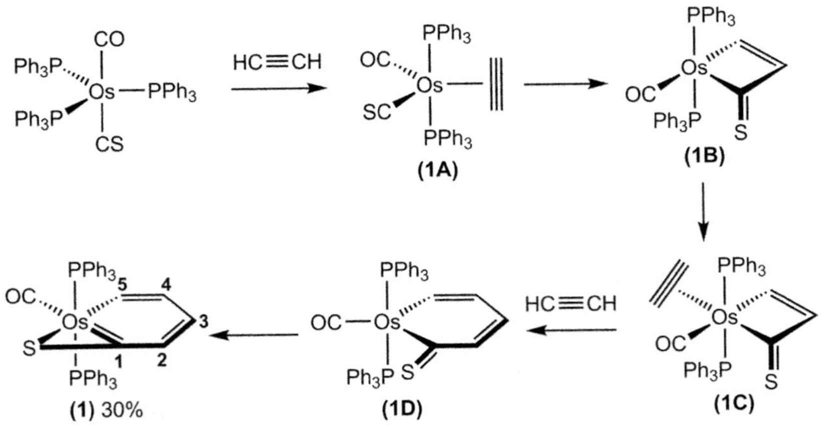

Scheme 1.2 Proposed mechanism for the synthesis of the osmabenzene **1** based on computational studies using PH_3 model compounds.

This cyclisation reaction is not limited to ethyne, and later we found the related dimethyl-substituted osmabenzene $Os(C_5H_2\{S\text{-}1\}\{Me\text{-}2\}\{Me\text{-}4\})(CO)(PPh_3)_2$ (**2**) (Scheme 1.1) is formed as dark-brown crystals when $Os(CS)(CO)(PPh_3)_3$ is treated with propyne, albeit in the low yield of 8%. The major product from this reaction is the complex $OsH(C{\equiv}CMe)(CS)(CO)(PPh_3)_2$ which arises from the simple C–H oxidative addition of propyne. Fortunately, this can be easily separated from the metallabenzene **2** by column chromatography and isolated in 23% yield [25].

Our original report of the first metallabenzenes also included several derivatives of **1** which could be prepared through reactions in which the osmium–sulfur bond was cleaved. The sulfur atom in **1** is nucleophilic and readily undergoes protonation with hydrochloric acid or alkylation with methyl iodide to give the neutral osmabenzenethiol **3a** or the methylthiolate-substituted osmabenzene **3b**, respectively (Scheme 1.1). The sulfur atom in **1** is also displaced from osmium on treatment with carbon monoxide. The resulting osmacyclohexadienethione, **4**, does not have the same π-bond delocalisation about the six-membered ring that is present in **1**, but this can be returned by protonation or alkylation of the thione sulfur. Thus, treatment of **4** with perchloric acid or methyl iodide followed by crystallisation in the presence of sodium perchlorate gives the corresponding osmabenzenes **5a** or **5b**, respectively, which are the cationic analogues of **3a** and **3b** (Scheme 1.1) [14]. **5a** or **5b** can be prepared by an alternative route starting from **3a** or **3b**, respectively, as indicated in Scheme 1.1.

A key question that had to be addressed in the original paper describing the osmabenzene **1** was whether it was best described as a metallabenzene with delocalised π-bonding or, alternatively, as an osmacyclohexatriene with localised double bonds. Key information that strongly supported a delocalised π-system was provided by the single crystal X-ray structure determination (see Figure 1.1). The structure of **1**, and later **2** (Figure 1.2), showed a planar six-membered metallacyclic ring with similar carbon–carbon bond lengths that were midway between standard sp^2 carbon–carbon single (1.46 Å) and double (1.34 Å) bonds. Importantly, the two osmium–carbon bonds were essentially equal in length and midway between those of typical single and double osmium–carbon bonds [26]. Since ring planarity and bond length equalisation are both

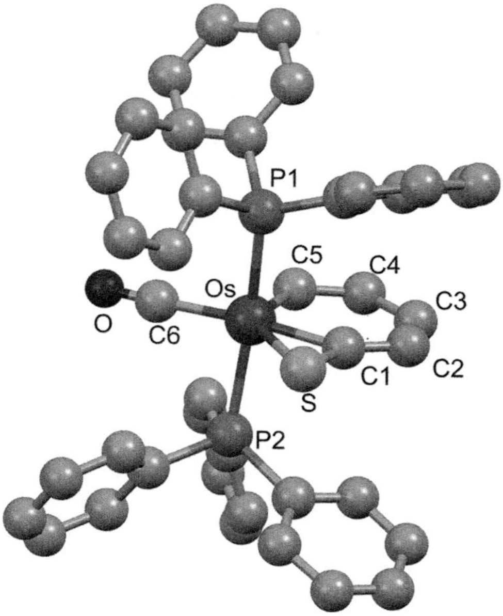

Figure 1.1 Molecular structure of osmabenzene **1**. Hydrogen atoms have been omitted for clarity. Selected distances [Å]: Os–C1 2.00(1), Os–C5 2.00(1), Os–S 2.474(3), Os–C6 1.92(1), C1–C2 1.36(2), C2–C3 1.38(2), C3–C4 1.42(2), C4–C5 1.39(2).)

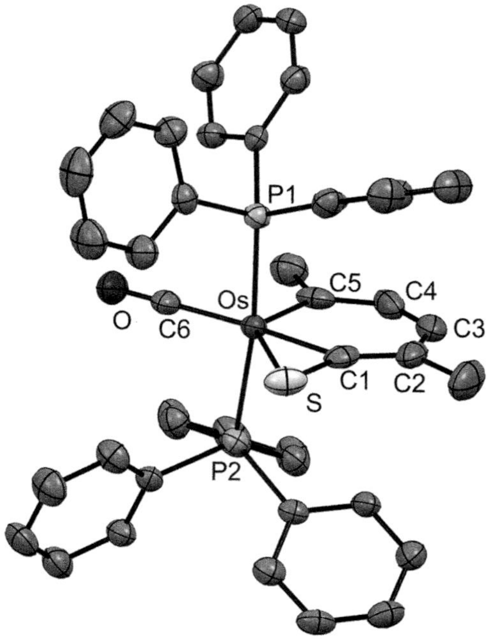

Figure 1.2 Molecular structure of osmabenzene **2** showing 50% probability thermal ellipsoids. Hydrogen atoms have been omitted for clarity. Selected distances [Å]: Os–C1 2.027(4), Os–C5 2.022(5), Os–S 2.4990(12), Os–C6 1.908(4), C1–C2 1.365(6), C2–C3 1.400(7), C3–C4 1.385(7), C4–C5 1.415(6). (*See color plate section for the color representation of this figure.*)

classic indicators of aromatic character, the structural data (together with the low field chemical shifts of the ring protons and carbon atoms in the ^1H and ^{13}C NMR spectra) [23] strongly supported a metallabenzene formulation for **1**.

Regarding the η^2-C(S) moiety in both **1** and **2**, the Os–S bond lengths (2.474(3) and 2.4990(12) Å, respectively) are slightly longer than normal Os–S single bonds, while the C–S bond lengths, (1.66(1) and 1.689(6) Å, respectively) are slightly shorter than standard C–S single bonds. These observations suggest that in both complexes the sulfur atom is not strongly bound to the metal centre [14, 25].

An X-ray crystal structure determination has also been obtained for the osmabenzene $Os(C_5H_4\{SMe\text{-}1\})Cl(CO)(PPh_3)_2$, an analogue of **3b** which has a chloride rather than iodide ligand. The structural parameters associated with the osmabenzene ring in this complex display the same key features found for **1** and **2**, i.e. near bond length equalisation and ring planarity (Os–C1 2.109(3), Os–C5 2.027(3), C1–C2 1.412(4), C2–C3 1.371(4), C3–C4 1.393(5), C4–C5 1.368(4) Å) [27].

The NMR spectroscopic data obtained for the osmabenzenes **1** and **2** provide important indications that a delocalised representation of the π-bonding within the metallacyclic ring is appropriate. In the ^1H NMR spectrum of **1**, H5, which is attached to a metal-bound carbon atom (see Scheme 1.1 for numbering system), is found at 13.95 ppm [23]. This notable down-field chemical shift is consistent with transition metal–carbon multiple bonding character and approaches, but does not quite meet the down-field shifts typically observed for related osmium carbene protons (*ca.* 18 ppm) [27]. H3 and H4 in **2** are found in the ^1H NMR spectrum at the "benzene-like" chemical shifts of 7.59 and 6.65 ppm, respectively.

The metal-bound carbon atoms of **2**, C1 and C5, are found at 257.43 and 220.07 ppm, respectively in the ^{13}C NMR spectrum. These considerable down-field chemical shifts are also consistent with osmium–carbon multiple bonding character. The remaining ring carbon atoms that are remote from the metal centre are observed at positions similar to those found in benzene and benzene derivatives (119.89 (C2), 147.98 (C3) and 129.22 (C4) ppm) [25]. These distinctive chemical shifts for the ring protons and carbon atoms are not limited to the osmabenzenes with the η^2-CS moiety, as is evidenced by the very similar spectra found for the methylthiolate-substituted osmabenzene $Os(C_5H_4\{SMe\text{-}1\})Cl(CO)(PPh_3)_2$ (^1H NMR: 6.65 (H2), 7.07 (H3), 6.57 (H4), 13.27 (H5) ppm, ^{13}C NMR: 237.4 (C1), 121.6 (C2), 145.8 (C3), 123.8 (C4), 211.0 (C5) ppm) [27].

Further general features are apparent in the NMR spectra of these and related osmabenzenes. Many of the resonances in the ^1H and ^{13}C NMR spectra are split into fine triplets due to coupling to the two phosphorus atoms of the mutually *trans* triphenylphosphine ligands. In most cases, coupling to the two phosphorus atoms is observed for carbons C1 and C5 in the ^{13}C NMR spectra, and $^2J_{CP}$ is usually in the order of 5–10 Hz. When present, the H1 and/or H5 protons will often also display coupling to phosphorus in the ^1H NMR spectra and this is usually in the order of 1–3 Hz. This phenomenon can sometimes also be noted in the atoms more remote from the metal, although this is much less common. In addition, the ring protons on the majority of related metallabenzenes and metallabenzenoids usually display long-range proton–proton coupling across three or more bonds in a manner similar to the *meta*-coupling often observed in aromatic organic compounds.

1.2.2 Iridabenzenes

The success encountered in synthesising these osmabenzenes was in large part due to the remarkable propensity of the thiocarbonyl ligand to undergo cyclisation and migratory insertion reactions. A natural progression was to then extend this work and prepare iridabenzenes using the same strategy. At the time this work was carried out, a number of iridabenzenes had already been reported using alternative routes, and these are discussed in Chapters 2, 3 and 4 of this book. However, it was reasoned that if the thiocarbonyl route was successfully applied to iridium it would give iridabenzenes with very different ancillary ligands and ring substituents to those that had been prepared previously.

It had already been shown that the iridacyclopentadiene complex $[Ir(C_4H_4)(NCMe)(CO)(PPh_3)_2][OTf]$ (^-OTf or triflate = $^-O_3SCF_3$) could be prepared by a cyclisation reaction involving two ethyne molecules at the iridium centre of the cation $[Ir(NCMe)(CO)(PPh_3)_2][OTf]$ [28]. This iridium cation is in turn obtained from Vaska's complex, $IrCl(CO)(PPh_3)_2$ [29], through treatment with silver triflate in acetonitrile solvent [30]. Although the CO ligand in this complex does not insert into the iridacyclopentadiene ring, we reasoned that the thiocarbonyl analogue might undergo this rearrangement to give an iridabenzene that is closely related to **1**.

We had already developed a route to the thiocarbonyl analogue of Vaska's compound, $IrCl(CS)(PPh_3)_2$ [20, 31], and so the cation, $[Ir(CS)(NCMe)(PPh_3)_2][OTf]$ (**6**), was readily accessible through treatment of this complex with AgOTf/NCMe. When a dichloromethane solution of this orange species was treated with ethyne under ambient conditions, the cationic iridacyclopentadiene complex, $[Ir(C_4H_4)(CS)(NCMe)(PPh_3)_2][OTf]$ (**7**), formed spontaneously (Scheme 1.3). This complex could be isolated as a yellow solid if the temperature was kept low (*ca.* 0°C) and precipitation carried out rapidly by addition of *n*-hexane [32]. The barrier to rearrangement of this cationic iridacyclopentadiene to the corresponding iridabenzene, through migratory insertion of the CS ligand, is relatively low. Even at 20°C, after 1 h in solution some rearrangement of **7** to the cationic iridabenzene $[Ir(C_5H_4\{S\text{-}1\})(NCMe)(PPh_3)_2][OTf]$ (**9**) can easily be detected by ^1H NMR spectroscopy. Complete conversion of **7** to the iridabenzene **9** occurs on heating **7** under reflux in dichloromethane (*ca.* 40°C) for 16 h (Scheme 1.3). In practice, if the iridabenzene is the target product, all these transformations can be carried out in one pot starting from **6** to give **9** in yields of over 90%. It is noteworthy that if chloride is added to **7** the neutral iridacyclopentadiene complex $Ir(C_4H_4)Cl(CS)(PPh_3)_2$ (**8**) is obtained, and this complex is significantly more resistant to migratory insertion of the CS ligand (Scheme 1.3). However, **8** can be converted into the iridabenzene, $Ir(C_5H_4\{SMe\text{-}1\})Cl_2(PPh_3)_2$ (**13**), by treatment with methyl triflate in a refluxing 1, 2-dichloroethane solution followed by the addition of lithium chloride (Scheme 1.3) [32–34].

The acetonitrile ligand in **9** is fairly labile and the neutral iridabenzene $Ir(C_5H_4\{S\text{-}1\})Cl(PPh_3)_2$ (**10**) can easily be prepared by the addition of chloride (Scheme 1.3) [33]. The iridabenzenes **9** and **10** can both be considered analogues of the original osmabenzene **1**.

Isolation of the intermediate **7** indicates that the reaction mechanism for the formation of iridabenzene **9** is different from that determined by computational studies for the formation of the closely related osmabenzene **1**. Formation of **9** involves cyclisation of two ethyne molecules at iridium to give the isolated intermediate iridacyclopentadiene

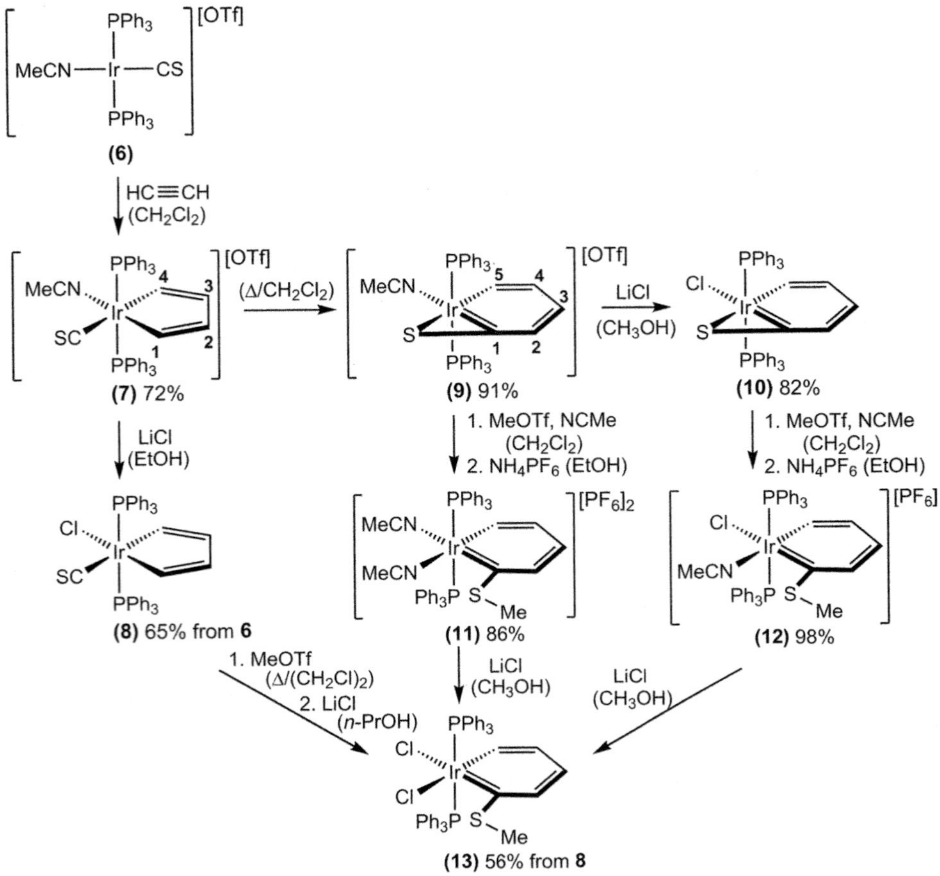

Scheme 1.3 Synthesis of the iridacyclopentadienes **7–8** and the iridabenzenes **9–13**.

7, followed by insertion of CS into an Ir–C bond to give the iridabenzene **9** [32]. In contrast, calculations predict that formation of the osmabenzene **1** occurs via cyclisation of one ethyne and CS to form an osmacyclobutenethione intermediate (**1B**; Scheme 1.2) which subsequently inserts both carbon atoms of a second ethyne to form the six-membered metallacyclic ring in **1** [24].

A number of simple derivatives of the iridabenzenes **9** and **10** are readily accessible through alkylation of the sulfur atoms with methyl triflate (MeOTf). When the methylation reaction of **9** is conducted in the presence of a small amount of acetonitrile, the maroon dicationic iridabenzene $[Ir(C_5H_4\{SMe\text{-}1\})(NCMe)_2(PPh_3)_2][PF_6]_2$ (**11**) can be isolated after addition of NH_4PF_6. Methylation of **10** under the same conditions gives the red monocationic iridabenzene $[Ir(C_5H_4\{SMe\text{-}1\})Cl(NCMe)(PPh_3)_2][PF_6]$ (**12**) (Scheme 1.3). The acetonitrile ligands in both complexes are labile, and treatment of either **11** or **12** with lithium chloride yields the neutral dichloride iridabenzene, **13**, as a purple crystalline solid (Scheme 1.3) [34].

The crystal structures of iridabenzenes **9** and **13** (Figure 1.3) have been determined and provide interesting comparisons with the structures of the osmabenzenes **1** and **2**

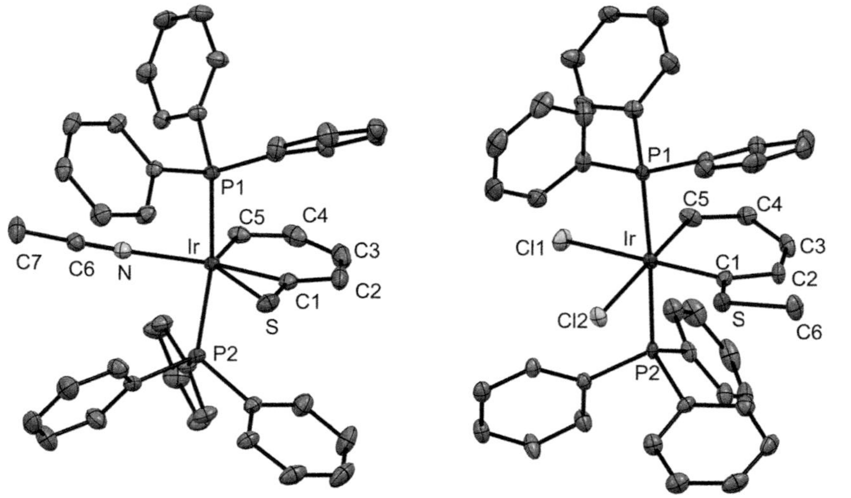

Figure 1.3 Molecular structures of iridabenzenes **9** (left) and **13** (right) showing 50% probability thermal ellipsoids. Hydrogen atoms have been omitted for clarity. Selected distances [Å] for **9**: Ir–C1 1.933(2), Ir–C5 1.989(2), Ir–S 2.6326(6), Ir–N 2.1340(16), C1–C2 1.398(3), C2–C3 1.366(4), C3–C4 1.421(4), C4–C5 1.365(3). Selected distances [Å] for **13**: Ir–C1 1.993(4), Ir–C5 1.992(4), Ir–Cl1 2.4846(10), Ir–Cl2 2.4606(11), C1–C2 1.427(6), C2–C3 1.363(6), C3–C4 1.413(6), C4–C5 1.352(6), S–C6 1.820(5).

(Scheme 1.1) and the iridacyclopentadiene **8**. The bonds within the almost perfectly planar five-membered metallacyclic ring of **8** serve as benchmarks for the lengths of (presumably) essentially localised C–C double bonds and Ir–C(sp^2) single bonds within an iridacyclic ring. The relevant distances in **8** are: iridium–carbon single bonds (Ir–C1 2.051(3), Ir–C4 2.093(3) Å), carbon–carbon double bonds (C1–C2 1.327(4), C3–C4 1.350(5) Å), and carbon–carbon single bonds (C2–C3 1.446(5) Å) [33]. The iridabenzene **9** is also very planar, with the greatest deviation from the Ir, C1–C5, S1 least-squares plane occurring for the iridium metal (0.038 Å). The iridium–carbon bonds (Ir–C1 1.933(2), Ir–C5 1.989(2) Å) are significantly shorter than they are in **8**, as would be expected if there is some multiple bond character in these bonds arising from a delocalised π-bonding system. The remaining carbon–carbon bonds in the IrC_5 ring of **9** display some alternation in their distances but all fall well between the single and double carbon–carbon bond lengths observed in **8** [32, 33]. The structural evidence for the classification of **9** as having a delocalised, aromatic π-bonding system is therefore quite compelling. The bond lengths in the neutral methylthiolate-substituted iridabenzene, **13**, follow the same trends, although in this case the iridium metal lies quite significantly (0.307 Å) out of the least squares plane formed by the iridium metal and carbons C1–C5, although the plane formed by these five carbon atoms is itself very planar[33]. While non-planarity is often associated with a loss of aromaticity in simple carbocyclic compounds, it has been shown computationally that distortions of this type do not adversely affect the aromaticity in metallabenzenes [35].

Interestingly, the sulfur atom of the η^2-C(S) moiety in **9** does not appear to interact with the iridium metal as strongly as it does with the osmium metal in the osmabenzenes **1** and **2**. The Ir–S bond length of 2.6326(6) Å is considerably longer than the Os–S bond lengths in **1** and **2** (2.474(3) and 2.4990(12) Å, respectively), and the Ir–C1–S angle of 93.80(9)° is larger (*cf.* 84.58° and 84.00° in osmabenzenes **1** and **2**, respectively) [14, 25, 33].

In addition to the structural data, the spectroscopic data also support the notion that the iridabenzenes **9–13** are indeed aromatic. In the ^{1}H NMR spectrum of cationic **9**, H5, which is attached to the iridium-bound C5 atom, resonates at 12.47 pm. This is significantly down-field compared to the resonances of the corresponding protons (H1 and H4, which are attached to the metal-bound carbon atoms) in the iridacyclopentadiene complex **7** (7.20 and 6.53 ppm, respectively). Likewise in the ^{13}C NMR spectrum of iridabenzene **9**, the iridium-bound carbons C1 and C5 are found at the low-field positions of 243.9 and 176.9 ppm, respectively, while the iridium-bound carbons C1 and C4 in **7** are found at 152.6 and 136.0 ppm, respectively [32]. The characteristically large down-field chemical shifts that are found for the metal-bound carbon atoms in these iridabenzenes are consistent with there being multiple-bond character in each of the iridium–carbon bonds, further supporting a delocalised π-bonding system [27]. The remaining hydrogen and carbon atoms of **9** that are remote from the iridium metal centre exhibit NMR resonances in regions typical of traditional organic aromatics (^{1}H NMR: 6.24 (H2), 7.50 (H3), 6.81 (H4) ppm, ^{13}C NMR: 117.6 (C2), 153.9 (C3), 127.4 (C4) ppm) [32]. The chemical shifts found in the neutral iridabenzene **10** are very similar (^{1}H NMR: 12.45 (H1), 5.65 (H2), 6.80 (H3) and 6.54 (H4) ppm, ^{13}C NMR: 249.47 (C1), 118.71 (C2), 149.40 (C3), 125.99 (C4), 171.43 (C5) ppm).

The complexes **11–13**, which arise from methylation of the sulfur function in **9** or **10**, display very similar chemical shifts for the metallabenzene ring atoms in the NMR spectra. For example, for the dicationic **11**, H5 is found at 11.59 ppm in the ^{1}H NMR spectrum, while in the ^{13}C NMR spectrum C1 and C5 are found at 224.4 and 180.9 ppm, respectively [32]. In the neutral **13**, H5 is observed at 12.31 ppm (^{1}H NMR spectrum), and C1 and C5 at 229.12 and 198.19 ppm, respectively (^{13}C NMR spectrum). The remaining ring signals are found in the expected aromatic positions [33].

1.3 Syntheses and Properties of Fused-Ring Metallabenzenes

By the start of the 21st century the study of metallabenzenes had expanded into a diverse field featuring a variety of different metal centres and differently substituted carbon skeletons prepared via a wide range of synthetic routes. In spite of this, the sister field of fused-ring metallabenzenes, where the primary six-membered ring is fused with a secondary ring structure, was still very much in its infancy and examples with fused heterocyclic rings were very rare. We therefore directed our attention to compounds of this type.

Metallabenzofurans are now the largest class of the otherwise relatively unexplored fused-ring metallabenzenes. In all the reported examples of metallabenzofurans, the metal occupies a ring junction position and bonds to the oxygen of the fused five-membered furan ring (see Chart 1.2). In principle many other metallabenzofuran isomers are possible, but these have yet to be reported. The three important valence bond structures, as well as a delocalised representation with the general numbering scheme for the known isomer, are given in Chart 1.2.

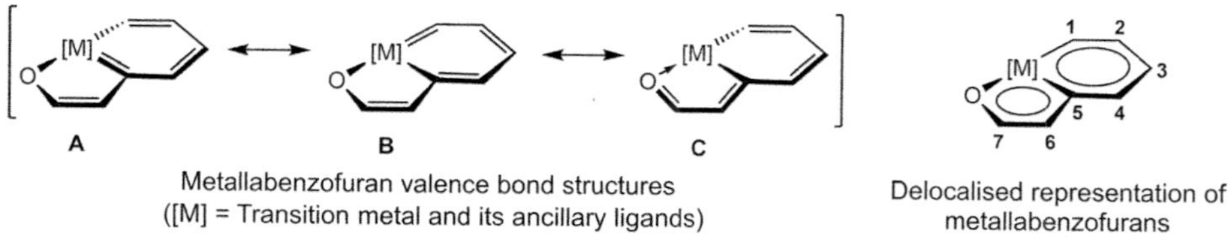

Chart 1.2 General representation and valence bond structures for the single metallabenzofuran isomer that has been reported.

1.3.1 Osmabenzofurans

The synthesis of the first osmabenzene by the reaction between two ethyne molecules and the osmium complex $Os(CS)(CO)(PPh_3)_3$ naturally led us to consider further investigations involving substituted acetylenes. The reaction with propyne to give the osmabenzene **2** has already been noted above. Interestingly, it was found that, with diphenylacetylene, only one molecule adds to $Os(CS)(CO)(PPh_3)_3$, with accompanying loss of both PPh_3 and (surprisingly) CO, to yield the osmacyclopropene complex $Os(CS)(PhC{\equiv}CPh)(PPh_3)_2$ (**14**) (Scheme 1.4) [36, 37]. This complex did not undergo further reaction with diphenylacetylene even in the presence of a large excess of this reagent. However, on treatment of **14** with CO, the coordinated diphenylacetylene could be induced to combine with the thiocarbonyl carbon to give the osmabutenethione, **15**. This very unreactive species bears some similarity to **1B** in Scheme 1.1, which has been proposed as an intermediate in the synthesis of osmabenzene **1**.

Compound **14** is remarkably stable towards further reaction with most alkynes. However, the facile reaction between **14** and CO suggested that **14** might undergo reaction with a suitably activated alkyne and this could lead to a new osmabenzene through formal [1 + 2 + 2] cyclisation of the two alkyne molecules and the thiocarbonyl ligand. One of the activated alkynes chosen for study was methyl propiolate ($HC{\equiv}CCO_2CH_3$). When **14** was heated under reflux in benzene with this alkyne, a reaction ensued but not in the way initially anticipated. The product obtained was the osmabenzofuran

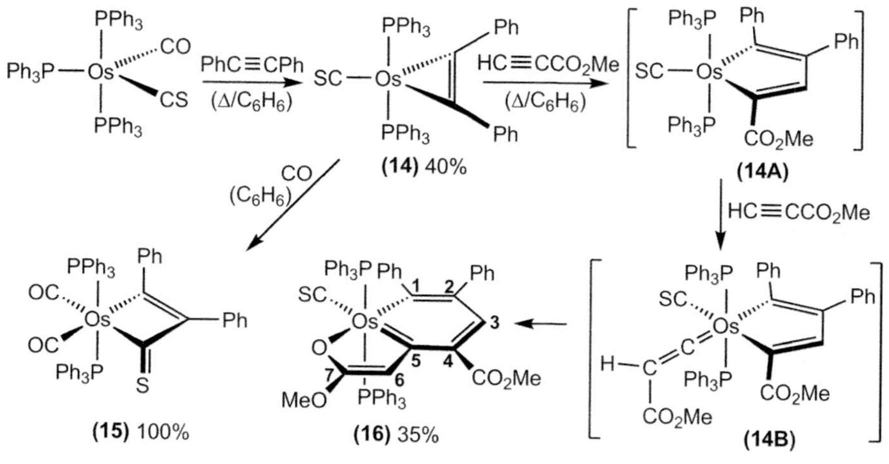

Scheme 1.4 Synthesis of the osmabenzofuran **16**.

$Os(C_7H_2O\{Ph\text{-}1\}\{Ph\text{-}2\}\{CO_2Me\text{-}4\}\{OMe\text{-}7\})(CS)(PPh_3)_2$ (**16**) in 35% isolated yield after chromatography (Scheme 1.4) [38]. Surprisingly, the CS ligand remained intact and was not incorporated into either of the two fused rings. It was proposed that the reaction proceeds via cyclisation of the diphenyl acetylene with a coordinated methyl propiolate to give the intermediate osmacyclopentadiene **14A**. This does not then undergo a migratory insertion reaction with the adjacent CS. Instead an additional methyl propiolate coordinates and rearranges into a vinylidene ligand to give intermediate **14B** (Scheme 1.4). It is well known that coordinated methyl propiolate can rearrange to a vinylidene ligand ($M{=}C{=}CR_2$) [39–41]. After insertion of the vinylidene carbon, coordination of the carbonyl oxygen of the ester substituent forms the fused metallafuran ring in **16** [38].

The osmabenzofuran **16** was found to be relatively unreactive and the extensive ring substitution may be, at least in part, responsible for this. It does, however, undergo protonation and on treatment with aqueous hydrochloric acid the colour changes from blue-green to green. The protonation occurs at C6 and this was confirmed by ^1H NMR spectroscopy (Scheme 1.5). The protonation is readily reversed, and if dilute acid is used only unreacted **16** is collected when attempts are made to isolate the protonated species. However, if gaseous HI (contaminated with a small amount of iodine impurity) is added to a dichloromethane solution of **16,** the green cationic tethered osmabenzene $[Os(C_5H\{Ph\text{-}1\}\{Ph\text{-}2\}\{CO_2Me\text{-}4\}\{CH_2C(O)OMe\text{-}5\})(CS)(PPh_3)_2][I_3]$ (**17**) (Scheme 1.5) can be isolated and structurally characterised. Protonation at C6 saturates this carbon atom and this disrupts the π-delocalisation in the five-membered ring. The five-membered ring can then be viewed as a simple "tethering arm" to the osmabenzene. We have used the term "tethering arm" to distinguish this type of fused ring from those that can support delocalised π-systems [38, 42]. As might be expected, treatment of these protonated osmabenzofurans with base instantly returns the original osmabenzofuran **16**.

An illustration of the chemically robust nature of this osmabenzofuran skeleton is provided by the observation that heating **16** under reflux in a benzene/ethanol mixture containing a small amount of concentrated HCl for 1 h results in the formation of the blue-green trans-esterified osmabenzofuran, $Os(C_7H_2O\{Ph\text{-}1\}\{Ph\text{-}2\}\{CO_2Et\text{-}4\}\{OMe\text{-}7\})(CS)(PPh_3)_2$ (**18**), which can be isolated in 51% yield (Scheme 1.5). It is noteworthy that the methyl ester function on the six-membered metallacyclic ring undergoes *trans*-esterification under these conditions, but the methoxy substituent on the five-membered ring remains intact. This suggests that even under these harsh conditions the osmafuran ring is not opened by cleavage of the Os–O bond to give a pendant methyl ester function [38].

The structures of the osmabenzenoids **16–18** have been determined unambiguously by single crystal X-ray crystallographic studies (Figure 1.4). In all cases the osmium

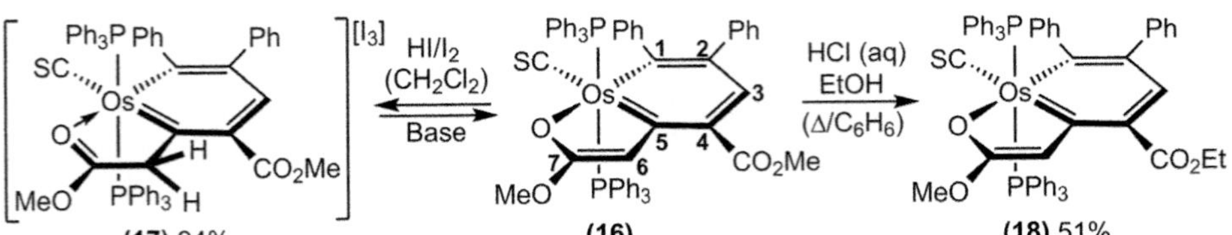

(**17**) 94% (**16**) (**18**) 51%

Scheme 1.5 Preparation of derivatives of **16**.

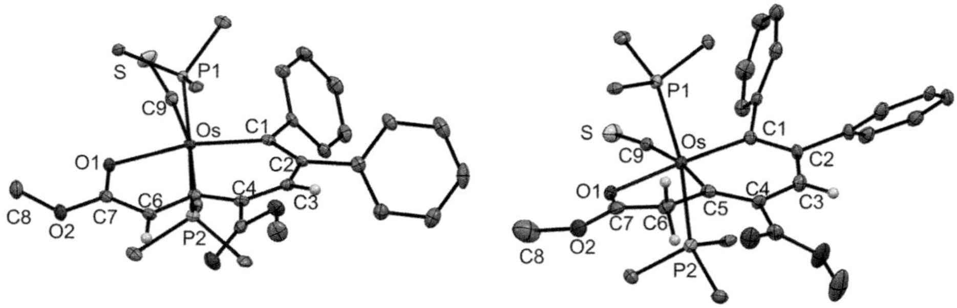

Figure 1.4 Molecular structures of the osmabenzofuran **16** (left) and the cation of the tethered osmabenzene **17** (right) showing 50% probability thermal ellipsoids. Some hydrogen atoms and the phenyl rings of the triphenylphosphine ligands are not shown for clarity. Selected distances [Å] for **16**: Os–C1 2.068(3), Os–C5 2.142(3), Os–O1 2.2178(17), Os–C9 1.846(3), C1–C2 1.376(4), C2–C3 1.442(3), C3–C4 1.359(4), C4–C5 1.436(4), C5–C6 1.371(3), C6–C7 1.416(4), C7–O1 1.250(3). Selected distances [Å] for **17**: Os–C1 1.996(5), Os–C5 2.103(6), Os–O1 2.221(4), Os–C9 1.901(6), C1–C2 1.450(8), C2–C3 1.374(8), C3–C4 1.438(8), C4–C5 1.375(8), C5–C6 1.507(8), C6–C7 1.495(9) C7–O1 1.226(8). (*See color plate section for the color representation of this figure.*)

metal lies at a ring junction position of the two fused rings with the oxygen atom coordinated to osmium. The two fused rings are essentially planar in the osmabenzofurans **15** and **16**. The tethered osmabenzene **17**, on the other hand, displays moderate non-planarity with the mean deviation from the Os, C1–C7, O1 least-squares plane being 0.14 Å. It is perhaps not surprising that the largest deviation from the mean plane was observed for the saturated atom, C6 (0.25 Å) [38].

In the osmabenzofuran **16**, the osmium–carbon bond lengths (Os–C1 2.068(3), Os–C5 2.142(3) Å) are similar to those observed in osmabenzenes [14, 25, 43–45], and the longer Os–C5 distance is most likely due to the *trans* influence of the thiocarbonyl ligand. The carbon–carbon bonds in the two rings display some bond length alternation but all fall midway between the expected distances for single and double bonds and are similar to the C–C distances reported for other metallabenzenes. The structure of **18** is essentially the same as **16** except for the presence of the ethyl rather than methyl group [38].

Protonation of C6 in **16** to give the cationic tethered osmabenzene **17** causes a number of significant changes in the structural parameters. The osmium–carbon bond lengths in **17** are notably shorter (Os–C1 1.996(5), Os–C5 2.103(6) Å), and the bonds to C6 are significantly longer (C5–C6 1.507(8), C6–C7 1.495(9) Å) and are consistent with single bonds. The carbon–carbon bond lengths in the six-membered osmabenzene ring are similar to those found in **16** and **18**. The long C5–C6 and C6–C7 distances in particular provide compelling evidence for the "tethered osmabenzene" formulation of **17** [38].

In the infrared spectrum of **16**, a strong absorbance at 1229 cm^{-1} is assigned to the thiocarbonyl ligand and the two bands at 1695 and 1569 cm^{-1} are assigned to the ester function on C4. Absorbance bands in similar positions are also found in **17** and **18**. The extensive substitution on the osmabenzofuran rings means that there are few signals in the aromatic region of the ^1H NMR spectra of these compounds. The H3 protons are found at 6.99, 8.39 and 6.98 ppm in **16**, **17** and **18**, respectively. It is possible that the lower-field H3 shift for **17** results from the π-system that is now delocalised over only the six atoms of the osmabenzene ring. The single protons on C6 in **16** and **18** are found at 6.51 and 6.49 ppm, respectively, while in the spectrum of the tethered osmabenzene **17**, the two protons on C6 are observed at 2.94 ppm. The large up-field shift is consistent with saturation of C6 and the loss of π-delocalisation in the furan ring. There are significant differences between the ^{13}C NMR spectra of **16** and **17**. For **16**, C1 and C5 are observed at 196.4 and 218.9 ppm, respectively, suggesting multiple bond character in each of the Os–C1 and Os–C5 bonds. The remaining carbon atoms of the two fused rings are found in typical aromatic positions (137.6 (C2), 158.1 (C3), 124.4 (C4), 123.6 (C6), 184.7 (C7) ppm) and are consistent with significant delocalised bonding in the two fused rings. In the ^{13}C NMR spectra of **17**, C1 and C5 are observed at considerably lower field positions (279.2 and 243.5 ppm, respectively), which may suggest increased multiple bonding to osmium with π-delocalisation now only over the six-membered ring. The remaining osmabenzene ring carbon atoms appear in the normal aromatic range (148.3 (C2), 153.8 (C3), 135.7 (C4) ppm). Carbon C6 in **17** is saturated, and accordingly is found in the up-field position of 52.7 ppm. In summary, the combination of the spectral and structural data corroborates the osmabenzene description for this complex [38].

1.3.2 Ruthenabenzofurans

Preparation of the osmabenzofuran complexes focused our attention on the possibility that metallabenzofurans involving other metals can also be prepared. On searching the literature we found that a ruthenium compound which fulfilled the criterion to be considered a ruthenabenzofuran had already been reported (complex **19**; Scheme 1.6), although at the time it was not recognised as such [46, 47]. With the benefit of the knowledge we had obtained from our studies of the osmabenzofurans **16** and **18**, it became clear to us that this alternative formulation was appropriate [42]. The ruthenabenzofuran $Ru(C_7H_3O\{CO_2Me\text{-}2\}\{CO_2Me\text{-}4\}\{OMe\text{-}7\})(CO)(PPh_3)_2$ (**19**) had been prepared by treatment of the ruthenium complex $RuH_2(CO)(PPh_3)_2$ with methyl propiolate in refluxing benzene (Scheme 1.6) [46]. Three molecules of methyl propiolate cyclised at the ruthenium metal centre to form the red-violet product. It was proposed that the

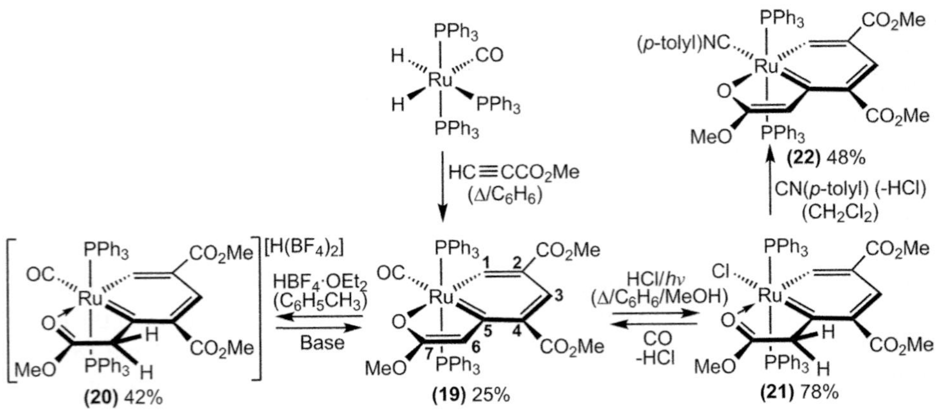

Scheme 1.6 Syntheses of the ruthenabenzene **19** and its derivatives.

mechanism of this reaction first involved the cyclisation of two molecules of methyl propiolate at the ruthenium centre to form a ruthenacyclopentadiene intermediate. A third methyl propiolate then rearranged at the metal centre to form a coordinated vinylidene ligand, the α-carbon atom of which then inserted into an Ru–C bond of the ruthenacyclopentadiene. Coordination of the carbonyl oxygen of the ester function from the vinylidene ligand then formed the fused ruthenafuran ring and at the same time aromatised the six-membered ring, thereby forming the ruthenabenzofuran.

Since we had shown that the furan carbon atom C6 of the osmabenzofuran **16** was susceptible to protonation, the reactivity of ruthenabenzofuran **19** towards acids was investigated. Indeed, it was found that on addition of the anhydrous acid $HBF_4{\cdot}OEt_2$ to a solution of **19** in toluene, protonation at C6 occurred immediately and the dark blue $[Ru(C_5H_2\{CO_2Me\text{-}2\}\{CO_2Me\text{-}4\}\{CH_2C(O)OMe\text{-}5\})(CO)(PPh_3)_2][H(BF_4)_2]$ (**20**) precipitated from solution in 42% isolated yield (Scheme 1.6). The two protons on the saturated carbon atom C6 are observed in the ^1H NMR spectrum at 2.73 ppm. This tethered ruthenabenzene complex, which has been structurally characterised, is stable for several days in solution at ambient temperature. However, on the addition of bases such as triethylamine, the protonation reaction is easily reversed and **19** is returned in excellent yield [42].

A related neutral tethered ruthenabenzene can be obtained if very different reaction conditions are employed. Treatment of a refluxing benzene/methanol solution of the ruthenabenzofuran **19** with trimethylsilyl chloride (which acts as a source of anhydrous hydrochloric acid) while simultaneously irradiating with visible light (to accelerate CO dissociation) over a period of 3–4 h gives the green, tethered ruthenabenzene, $Ru(C_5H_2\{CO_2Me\text{-}2\}\{CO_2Me\text{-}4\}\{CH_2C(O)OMe\text{-}5\})Cl(PPh_3)_2$ (**21**), in 78% yield after purification by column chromatography. Under these conditions C6 is protonated and the carbonyl ligand is replaced by chloride. The two protons on the saturated carbon atom C6 are observed in the ^1H NMR spectrum at 3.19 ppm. The stability of **19** in the presence of HCl, while being irradiated and heated under reflux for hours in a benzene/methanol solution, is remarkable. Although only a few ruthenabenzenes have been reported, it is interesting to note that in some cases they are thermally unstable [48], while other examples show unexpectedly high thermal stability [49–51]. The tethered ruthenabenzene **21** is very stable in solutions containing traces of HCl, but on the addition of bases intractable mixtures of products are formed. Treatment of a dichloromethane solution of **21** with carbon monoxide, on the other hand, results in the replacement of chloride by CO, proton loss from C6 and reformation of the ruthenabenzofuran **19** in good yield. Treatment of **21** with the isoelectronic *p*-tolylisocyanide (CN{*p*-tolyl}) proceeds in a similar manner to give the new ruthenabenzofuran, $Ru(C_7H_3\{CO_2Me\text{-}2\}\{CO_2Me\text{-}4\}\{OMe\text{-}7\})(CN(p\text{-}tolyl))(PPh_3)_2$ (**22**) (Scheme 1.6), which is the *p*-tolylisocyanide analogue of **19** [42].

The NMR spectral data have been reported for the ruthenabenzofurans **19** and **22**, as well as the related tethered ruthenabenzenes **20** and **21**. In addition, the molecular structures of **19**, **20** and **21** have been obtained. This enables useful comparisons to be made between the properties of these ruthenabenzofurans and the closely related tethered ruthenabenzenes.

In the ^1H NMR spectra of the ruthenabenzofurans **19** and **22**, the fused ring protons unsurprisingly appear in similar positions (**19** 11.67 (H1), 7.2–7.5 (H3 obscured by PPh_3 multiplet), 6.06 (H6) ppm [47]; **22** 12.40 (H1), 7.3–7.4 (H3 obscured by PPh_3),

6.17 (H6) ppm). However, the corresponding signals for the metallabenzene ring protons of the related tethered ruthenabenzenes **20** and **21** are both observed at considerably lower-field positions (**20** 14.97 (H1), 8.83 (H3) ppm; **21** 16.45 (H1), 8.76 (H3) ppm) [42]. A similar trend is observed for the chemical shifts of the ring carbon atoms in the ^{13}C NMR spectra. Thus, the signals for the metal-bound carbon atoms C1 and C5 in the tethered ruthenabenzenes **20** and **21** are observed at significantly lower field positions (**20** 290.8 (C1) and 283.6 (C5) ppm, **21** 289.3 (C1) and 287.8 (C5) ppm) than the resonances for the corresponding carbon atoms in the related ruthenabenzofurans **19** and **22** (**19** 232.9 (C1) and 227.0 (C5) ppm, **22** 245.0 (C1) and 232.8 (C5) ppm) [42]. Shifts of the Cα and Hα resonances to lower field values is consistent with increasing π-bonding between the metal and Cα in these closely related compounds and hence may signal increased π-delocalisation over the six-membered metallacyclic rings in the tethered ruthenabenzenes in comparison to the ruthenabenzofurans [22].

There are also some interesting differences in the structural parameters associated with the six-membered metallacyclic rings of the ruthenabenzofurans **19** and **22** compared to those of the tethered ruthenabenzene **20** (Figure 1.5) [42, 46, 47]. While these rings are essentially planar in all three compounds, the Ru–C distances in the tethered ruthenabenzene **20** (Ru–C1 1.933(4), Ru–C5 2.045(5) Å) are significantly shorter than those in the ruthenabenzofurans **19** (Ru–C1 2.004(6), Ru–C5 2.093(5) Å) and **22** (Ru–C1 1.986(3), Ru–C5 2.092(4) Å). In addition, the range of metallabenzene ring C–C distances in **20** show a smaller spread (1.374(7)–1.413(7) Å) compared to the ranges in **19** (1.338(9)–1.450(10) Å) and **22** (1.368(5)–1.475(5) Å). These structural features are also consistent with greater aromatic character for the six-membered metallacyclic ring of the tethered ruthenabenzene **20** compared to the ruthenabenzofurans **19** and **22**.

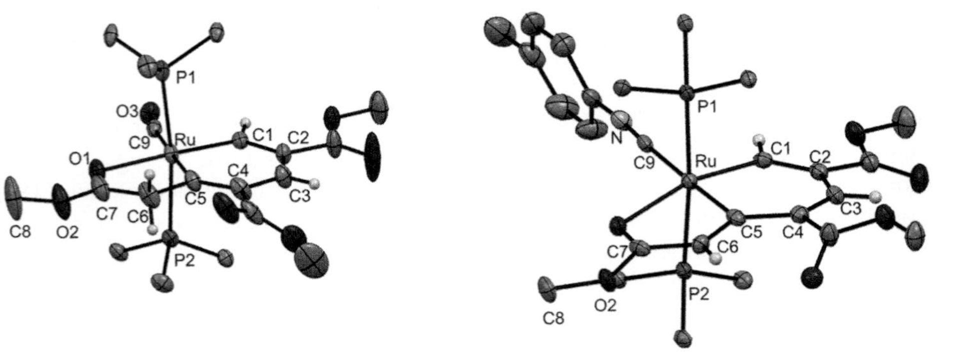

Figure 1.5 Molecular structures of the cation of the tethered ruthenabenzene **20** (left) and the ruthenabenzofuran **22** (right) showing 50% probability thermal ellipsoids. The phenyl rings of the triphenylphosphine ligands and most hydrogen atoms are not shown for clarity. Selected distances [Å] for **20**: Ru–C1 1.933(4), Ru–C5 2.045(5), Ru–O1 2.216(3), Ru–C9 1.941(5), C1–C2 1.404(6), C2–C3 1.383(7), C3–C4 1.413(7), C4–C5 1.374(7), C5–C6 1.518(7), C6–C7 1.504(8), C7–O1 1.231(7). Selected distances [Å] for **22**: Ru–C1 1.986(3), Ru–C5 2.092(4), Ru–O1 2.266(2), Ru–C9 1.954(4), C1–C2 1.368(5), C2–C3 1.437(5), C3–C4 1.376(5), C4–C5 1.475(5), C5–C6 1.376(5), C6–C7 1.437(5), C7–O1 1.242(4).

1.3.3 Iridabenzofurans

The successful isolation of osmabenzofuran and ruthenabenzofuran derivatives led us to consider the synthesis of the related iridabenzofurans. The mechanisms proposed for the formation of the group 8 metallabenzofurans involved, in the last steps, the insertion of a vinylidene ligand (formed by the rearrangement of a coordinated methyl propiolate) into preformed (but not isolated or detected) metallacyclopentadiene intermediates. Since the stable, isolated iridacyclopentadiene $[Ir(C_4H_4)(CO)(NCMe)(PPh_3)_2][O_3SCF_3]$ (**23**) (see Scheme 1.7) was known, it seemed logical that we should investigate the reaction of this compound with methyl propiolate [28]. Attractive features of **23** were that it could be synthesised easily and in high yield through reaction between the iridium(I) cation, $[Ir(NCMe)(CO)(PPh_3)_2][OTf]$, and ethyne, and that the relatively labile acetonitrile ligand should facilitate coordination of the alkyne.

Naturally we were delighted to find that when a mixture of iridacyclopentadiene **23** and an excess of methyl propiolate was heated under reflux in 1, 2-dichloroethane, the bright orange cationic iridabenzofuran $[Ir(C_7H_5O\{OMe\text{-}7\})(CO)(PPh_3)_2][OTf]$ (**24**) (Scheme 1.7) was formed in 88% isolated yield. It seems likely that the mechanism does indeed proceed via the intermediate vinylidene complex **23A**, which undergoes migratory insertion of Cα into the iridacyclopentadiene ring. Coordination of the ester carbonyl oxygen atom to iridium then forms the iridafuran ring. The neutral iridabenzofuran derivative $Ir(C_7H_5O\{OMe\text{-}7\})Cl(PPh_3)_2$ (**25**) may be formed by treating **24** with lithium chloride in refluxing *n*-propanol for several hours. Isolation of the bright red crystalline **25** in very good yield following these reaction conditions is a testament to the stability of the iridabenzofurans **24** and **25** [52].

One of the remarkable features of the iridabenzofurans **24** and **25** is the near absence of substituents on the fused rings. In each case there is only a methoxy substituent at C7 on the iridafuran ring. In all previously reported metallabenzofurans the metallacyclic rings had been quite heavily substituted. This minimal ring substitution, coupled with the high-yielding syntheses of **24** and **25**, paved the way for extended studies of the

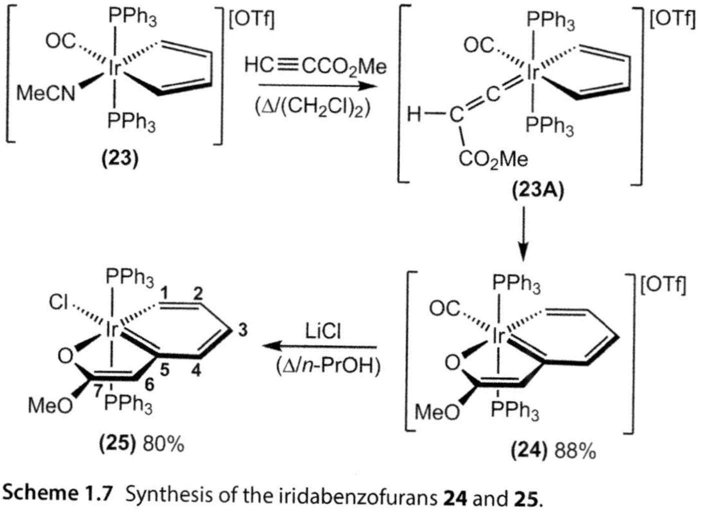

Scheme 1.7 Synthesis of the iridabenzofurans **24** and **25**.

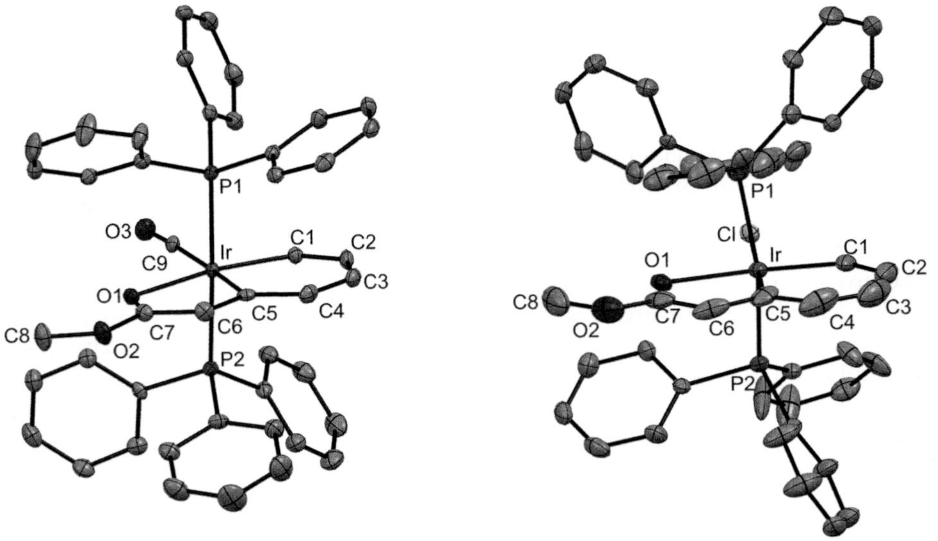

Figure 1.6 Molecular structures of the iridabenzofurans **24** (left) and **25** (right) showing 50% probability thermal ellipsoids. Hydrogen atoms and the counter anion of **24** are not shown for clarity. Selected distances [Å] for **24**: Ir–C1 2.049(3), Ir–C5 2.073(2), Ir–O1 2.2152(19), Ir–C9 1.946(2), C1–C2 1.358(4), C2–C3 1.455(4), C3–C4 1.353(4), C4–C5 1.441(3), C5–C6 1.382(3), C6–C7 1.435(3), C7–O1 1.265(3). Selected distances [Å] for **25**: Ir–C1 1.998(5), Ir–C5 1.987(6), Ir–O1 2.256(3), Ir–Cl 2.4682(13), C1–C2 1.337(7), C2–C3 1.425(9), C3–C4 1.360(9), C4–C5 1.444(8), C5–C6 1.385(8), C6–C7 1.403(8), C7–O1 1.267(6).

reaction chemistry of these species. This was to provide some valuable insights into the chemistry of metallabenzofurans, and this is discussed in Section 1.4.1.

The molecular structures of the iridabenzofurans **24** and **25** were obtained (see Figure 1.6) [52]. In each case the overall geometry about the iridium centre is approximately octahedral with iridium occupying a ring junction position of the essentially planar fused rings. The CO or Cl ligands complete the equatorial plane and the two mutually *trans* triphenylphosphine ligands occupy the axial positions. The iridium–carbon distances in cationic iridabenzofuran **24** (Ir–C1 2.049(3), Ir–C5 2.073(2) Å) are significantly longer than in the neutral derivative **25** (Ir–C1 1.998(5), Ir–C5 1.987(6) Å) but are still within the range of metal–carbon bond lengths observed in iridabenzenes [32, 33, 53–57]. These distances can be compared with the two longer iridium–C(sp^2) single-bond distances found in the cationic iridacyclopentadiene complex [Ir(C_4H_4)(CHCHNEt$_3$)(CO)(PPh$_3$)$_2$][ClO$_4$] (2.114(6), and 2.094(6) Å) [58]. There is a degree of carbon–carbon bond length alternation in the fused rings of both the cationic iridabenzofuran **24** (range of C–C distances 1.353(4)–1.455(4) Å) and in the neutral iridabenzofuran **25** (range of C–C distances 1.337(7)–1.444(8) Å) [52].

The minimal ring substitution in these two complexes means that more detailed information can be obtained from the ^{1}H NMR spectra. The chemical shifts of H1 in **24** (7.70 ppm) and **25** (9.36 ppm) are not found as far down-field as they are in typical iridabenzenes (10–14 ppm), but they are further down-field than the protons on the metal-bound carbon atoms in the iridacyclopentadiene complex **23** (6.75 and 7.30 ppm). The remaining ring protons are observed slightly up-field of typical aromatic compounds (**24** 5.92 (H2), 5.64 (H3), 5.58 (H4), 5.55 (H6) ppm; **25** 5.93 (H2), 5.52 (H3), 4.68 (H4), 4.91 (H6) ppm) [52], and iridabenzenes (6–8 ppm) [34, 59, 60]. Similarly, in the ^{13}C NMR spectra the average chemical shifts of the metal-bound carbon atoms C1 and C5 (**24** 128.20, 197.60 ppm, respectively; **25** 148.61, 193.71 ppm, respectively) [52] are found between those of iridabenzenes (typically 180–250 ppm) [34, 59, 60] and the iridacyclopentadiene **23** (132.00 and 151.84 ppm). The chemical shifts of carbon atoms remote from the iridium are found in characteristic aromatic positions (**24** 123.72 (C2), 140.97 (C3), 125.72 (C4), 121.57 (C6); **25** 121.12 (C2), 141.05 (C3), 121.08 (C4), 114.82 (C6) ppm) [52].

The structural and spectral data are consistent with a π-bonding system that is delocalised over the two fused rings of **24** and **25**. However, the delocalisation over the six-membered metallacyclic rings appears to be less extensive than that in typical iridabenzenes. The data may also point to slightly more delocalisation in neutral **25** than in cationic **24**.

The structural and NMR spectral data collected for all the metallabenzofurans described above suggest that the aromaticity in the primary six-membered metallacyclic ring is somewhat reduced with the addition of a fused furan ring. This would be consistent with the situation for simple organic benzenoids, where computational studies have shown that the aromatic stabilisation energy associated with a six-membered aromatic ring is reduced (compared to benzene) when additional fused rings are present [61].

1.3.4 Iridabenzothiophenes

As is noted above, the cationic iridium(I) thiocarbonyl complex, $[Ir(CS)(NCMe)(PPh_3)_2]$ [OTf], reacts with two molecules of ethyne to give the cationic iridacyclopentadiene complex, $[Ir(C_4H_4)(CS)(NCMe)(PPh_3)_2][OTf]$ (**7**) (see Scheme 1.3). During our studies of the reactions of alkynes with iridium thiocarbonyl complexes, we found that the neutral analogue, $IrI(CS)(PPh_3)_2$, also undergoes reaction with two molecules of ethyne if more vigorous conditions such as heating under reflux in benzene are employed. However, in this case a very different product, $Ir(C_3H_2\{SCHCH\text{-}1\})I(PPh_3)_2$ (**26a**) (Scheme 1.8), is formed. This complex can be viewed as an iridacyclobutadiene with a fused iridathiophene ring. If phenylacetylene or methyl propiolate are used in place of ethyne, the analogous compounds $Ir(C_3H\{SCPhCH\text{-}1\}\{Ph\text{-}3\})I(PPh_3)_2$ (**26b**) or $Ir(C_3H\{SC[CO_2Me]CH\text{-}1\}\{CO_2Me\text{-}2\})I(PPh_3)_2$ (**27**), respectively, are obtained [62].

The iridium centre of the fused-ring iridacyclobutadiene **27** is coordinatively saturated and relatively inert to further insertion of alkynes, most likely due to the tightly bound ancillary ligands. However, it was found that if silver triflate is added to **27** the iodide ligand is precipitated as silver iodide. This opens up access to the metal centre, and on addition of methyl propiolate to the resulting intermediate cationic complex both carbon atoms of the alkyne insert into the iridium–carbon bond of the iridacyclobutadiene ring to form a fused-ring iridabenzene. After addition of lithium chloride the neutral iridabenzothiophene, $Ir(C_5H_2\{SC[CO_2Me]CH\text{-}1\}\{CO_2Me\text{-}2\}\{CO_2Me\text{-}4\})Cl(PPh_3)_2$ (**28**) (Scheme 1.8), can then be isolated as blue-green crystals in good yield [62].

The structure of **28** reveals that the iridium metal occupies a ring junction position of the fused iridabenzene and iridathiophene rings that is remote from the sulfur atom

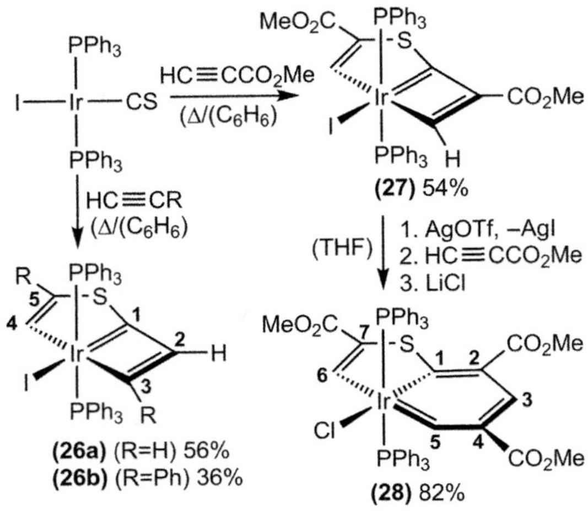

Scheme 1.8 Synthesis of the fused-ring iridacyclobutadienes **26–27** and the iridabenzothiophene **28**.

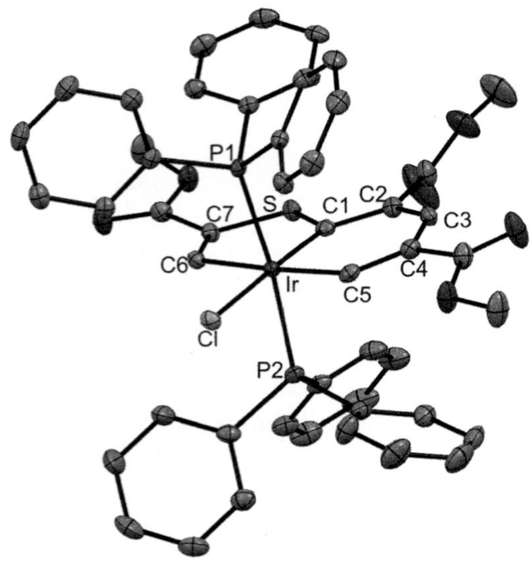

Figure 1.7 Molecular structure of iridabenzothiophene **28** showing 50% probability thermal ellipsoids. Hydrogen atoms are not shown for clarity. Selected distances [Å]: Ir–C1 1.973(3), Ir–C5 2.042(3), Ir–C6 2.106(3), Ir–Cl 2.4685(7), C1–C2 1.449(4), C2–C3 1.361(4), C3–C4 1.434(4), C4–C5 1.360(4), C6–C7 1.341(4), C7–S 1.765(3), S–C1 1.722(3).

(Figure 1.7). The six-membered iridabenzene ring is close to planar and the iridium–carbon bond lengths (Ir–C1 1.973(3), Ir–C5 2.042(3) Å) are similar with distances intermediate between those typical for iridium–carbon single and double bonds. The carbon–carbon bond lengths (C1–C2 1.449(4), C2–C3 1.361(4), C3–C4 1.434(4), C4–C5 1.360(4) Å) are also intermediate in length between those of single and double bonds and show significant bond length alternation, which suggests that the valence bond

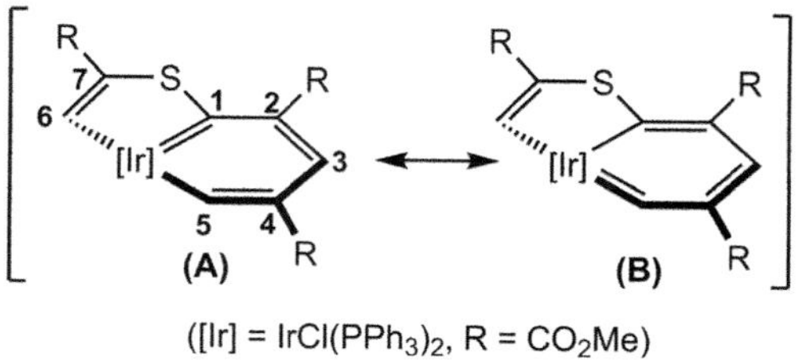

Chart 1.3 Two important valence bond structures of iridabenzothiophene **28**.

structure **A** is more important than **B** (Chart 1.3). Within the five-membered iridathiophene ring, the iridium–carbon bond length, at 2.106(3) Å, is close to an iridium–carbon single bond length, while the carbon–carbon bond length of 1.341(4) Å is more appropriate for a carbon–carbon double bond [62].

The chemical shifts of the iridabenzene carbon and hydrogen atoms in **28** are similar to those reported for other iridabenzenes. The resonance for H5 is observed down-field at 13.25 ppm in the ^{1}H NMR spectrum as a sharp doublet due to a four-bond coupling to proton H3, which is found at 7.63 ppm. The resonances of the metal-bound carbon atoms C1 and C5 are both found significantly down-field (252.7 and 247.5 ppm, respectively) in the ^{13}C NMR spectrum, while the other three ring carbons have typically aromatic chemical shifts (124.0 (C2), 162.0 (C3), 126.8 (C4) ppm). In all cases these chemical shifts fall within the range observed for other iridabenzenes. Within the fused five-membered ring, H6 is found at 10.70 ppm in the ^{1}H NMR spectrum and carbons C6 and C7 at 198.2 and 143.7 ppm, respectively in the ^{13}C NMR spectrum [62]. These resonances are observed somewhat further down-field than would have been expected for a localised bonding situation within the five-membered ring. It is noteworthy that in the ^{13}C NMR spectrum the chemical shifts of the metal-bound carbon atoms of the six-membered metallacyclic ring of **28** are observed further down-field compared to those of the related iridabenzofurans **24** and **25**. Similarly, in the ^{1}H NMR spectrum of **28**, H5 is also found further down-field. These observations may indicate that there is more extensive π-delocalisation in the iridafuran rings of **24** and **25** compared with the iridathiophene ring of **28**. However, appropriate computational studies will be required to provide further information on this point. Complex **28** remains the only example of a metallabenzothiophene to have been reported.

1.3.5 Iridabenzothiazolium Cations

The preferred route to the iridabenzene $Ir(C_5H_4\{SMe\text{-}1\})Cl_2(PPh_3)_2$ (**13**) (see Scheme 1.3) involved methylation of the sulfur atom of the intermediate iridabenzene $[Ir(C_5H_4\{S\text{-}1\})(NCMe)(PPh_3)_2][OTf]$ (**9**) followed by addition of chloride. If triflic acid is added to **9** instead of methyl triflate, the reaction takes a very different course. On treatment of **9** with two equivalents triflic acid (in a dichloromethane solution containing 10 equivalents of acetonitrile), the brown solution immediately turns red. It is proposed that this red intermediate is the thiol-substituted iridabenzene $[Ir(C_5H_4\{SH\text{-}1\})(NCMe)_2(PPh_3)_2][OTf]_2$ (**9A**) (Scheme 1.9). While this intermediate could not be isolated, treatment of the red solution with ammonium hexafluorophosphate (NH_4PF_6) in ethanol enabled the deep-green iridabenzothiazolium cation, $[Ir(C_5H_4\{SC[Me]NH\text{-}1\})(NCMe)(PPh_3)_2]$

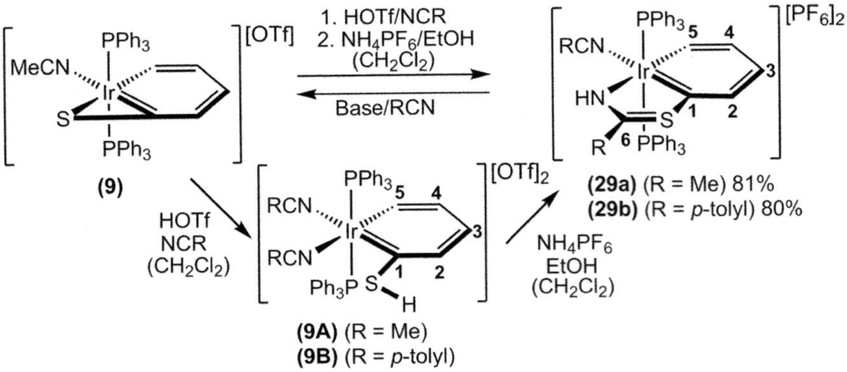

Scheme 1.9 Synthesis of the iridabenzothiazolium cations **29a** and **29b**.

[PF6]$_2$ (**29a**), to be isolated in 81% yield. The analogous complex [Ir(C_5H_4{SC[*p*-tolyl]NH-1})(NC(*p*-tolyl))($(PPh_3)_2$][PF_6]$_2$ (**29b**) can be isolated in 80% yield by employing *p*-tolunitrile rather than acetonitrile (Scheme 1.9) [32]. These iridabenzothiazolium cations can be considered the protonated organometallic analogues of benzothiazoles.

While the intermediates **9A** and **9B** could not be isolated as pure solids, they can be detected by NMR spectroscopy. If protected from air, solutions of the red intermediates generated in situ are stable for days. The ^{1}H NMR spectrum of the intermediate metallabenzenethiol **9A** reveals signals at 11.50 (H5), 7.23 (SH), 7.04 (H4), 6.90 (H3), and 6.89 (H2) ppm, each integrating for one proton, in addition to two singlets integrating for three protons at 2.02 and 2.03 ppm, which correspond to the methyl groups of the two acetonitrile ligands. The structure proposed for intermediate **9A** is essentially the same as that of the iridabenzene [Ir(C_5H_4{SMe-1})($(NCMe)_2$($(PPh_3)_2$][PF_6]$_2$ (**11**) (Scheme 1.3) except that the sulfur is protonated rather than methylated. Indeed, the ^{1}H NMR spectra of **9A** and **11** are very similar. Addition of excess ethanol to the red solution of **9A** causes a swift colour change to deep green and the NMR signals confirm the formation of the iridabenzothiazolium **29a**. Addition of ammonium hexafluorophosphate enables **29a** to be isolated as the crystalline hexafluorophosphate salt. The addition of water instead of ethanol also brought about this conversion. It is not clear exactly what part ethanol or water plays in the conversion of **9A**/**B** to **29a**/**b**, but it is possible these solvents facilitate proton transfer from sulfur to nitrogen [32].

The crystal structure of **29b** has been obtained and the molecular geometry of the iridabenzothiazolium cation is shown in Figure 1.8. The iridium has an approximately octahedral geometry and the two fused rings are nearly planar, with the maximum deviation from the Ir, C1–C6, S, N least squares plane being 0.064(3) Å for nitrogen. The iridium metal lies at a ring junction position, bound to the nitrogen of the fused five-membered ring. The metal–carbon and carbon–carbon bond lengths in the six-membered metallacyclic ring are similar to those observed in other iridabenzenes and do not exhibit any unusual C–C bond length alternation. The carbon–sulfur bond lengths (C1–S 1.754(5), S–C6 1.757(5) Å) and the carbon–nitrogen bond length (1.290(6) Å) are consistent with limited delocalised π-bonding in the five-membered ring and are similar to those reported for organic benzothiazolium ions [32].

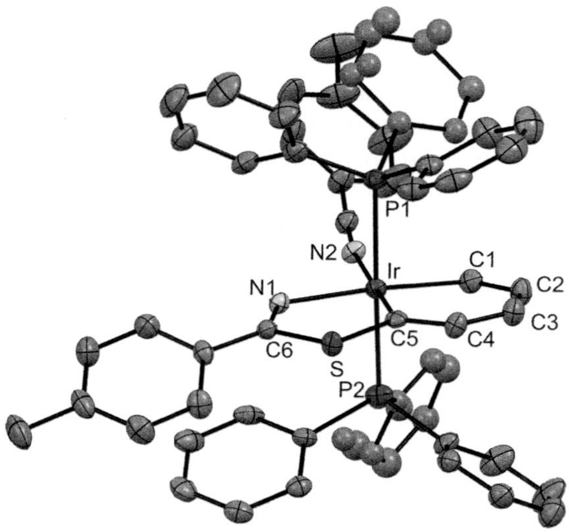

Figure 1.8 Molecular structures of the cation of **29b** showing 50% probability thermal ellipsoids. Hydrogen atoms are not shown for clarity. Selected distances [Å]: Ir–C1 1.953(5), Ir–C5 2.009(5), Ir–N1 2.127(4), Ir–N2 2.093(4), C1–C2 1.396(7), C2–C3 1.385(7), C3–C4 1.402(7), C4–C5 1.360(7), C1–S 1.754(5), S–C6 1.757(5), C6–N1 1.290(6).

The NMR spectra of the two complexes **29a** and **29b** reveal that the π-delocalisation about the six-membered ring is not significantly affected by the addition of the fused thiazolium ring. The aromatic protons in the ^{1}H NMR spectra of **29a** (12.35 (H5), 6.75 (H4), 6.55 (H3), 6.86 (H2) ppm) and **29b** (12.87 (H5), 7.22 (H4), 6.65 (H3), 6.91 (H2) ppm) do not differ significantly from the starting iridabenzene **9**. The NH protons are observed as broad singlets at 10.23 and 10.36 ppm for **29a** and **29b**, respectively, and in the IR spectra the NH groups give rise to broad absorbance bands at 3289 and 3291 cm^{-1} **29a** and **29b**, respectively. Similarly, ^{13}C NMR chemical shifts of the carbon atoms of the iridabenzene rings of **29a** (232.3 (C1), 125.5 (C2), 170.2 (C3), 127.2 (C4), 210.9 (C5) ppm) and **29b** (229.8 (C1), 125.4 (C2), 168.8 (C3), 127.4 (C4), 209.7 (C5) ppm) are typical for iridabenzenes. The resonances for C6 of the thiazolium rings are found at 190.0 ppm in **29a** and at 185.6 ppm in **29b**, each shifted considerably down-field compared to the same carbon atoms in the coordinated ligands prior to annulation [32].

On the basis of the spectroscopic and structural data, the two most important valence bond structures for the fused rings of **29a/b** are given in Chart 1.4. The data suggest there is limited π-delocalisation in the fused five-membered rings.

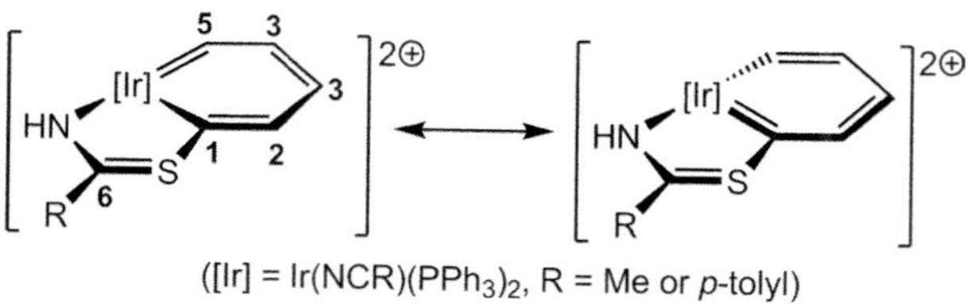

Chart 1.4 Valence bond structures of metallabenzothiazolium cations.

1.4 Reactions of Metallabenzenes and Metallabenzenoids

In addition to developing new synthetic routes to metallabenzenes and studying the spectroscopic and structural features of these unusual compounds, we were also very interested in the chemical reactivity they exhibited. In this respect we were particularly interested in the similarities and differences between metallabenzenes and the organic congeners, the benzenes. Some of the synthetic routes we had developed enabled us to obtain in moderate to very good yields osmabenzenes, iridabenzenes or iridabenzofurans that had only one ring substituent [22, 27, 34]. This placed us in an excellent position to explore the reaction chemistry of these compounds and this became an increasingly important focus of our research.

1.4.1 Electrophilic Aromatic Substitution Reactions

When we first began studies of the reaction chemistry of the osmabenzene $Os(C_5H_4\{SMe\text{-}1\})I(CO)(PPh_3)_2$ (**3b**) (Scheme 1.10) in the late 1990s, it was still very much an open question as to whether metallabenzenes were aromatic compounds. Detailed theoretical studies that were to confirm the validity of an aromatic description came much later. At that time, to obtain information about this fundamentally important point, we explored the reactions of **3b** with electrophiles. We found that **3b** underwent electrophilic substitution reactions – the classically defining reactions of aromatic, benzenoid compounds. Furthermore, the substitution appeared to follow the same "substituent directing rules" observed in benzene chemistry with the electrophile in this case adding *para* to the –SMe ring substituent (Scheme 1.10) [63].

We found that the blue osmabenzene **3b** undergoes ring bromination when treated with bromine in the presence of iron powder (which was added to form the catalyst $FeBr_3$ in situ) at 40°C. The substitution of hydrogen for bromine occurs at the C4 position, *para-* to the SMe group, to furnish the dark-green osmabenzene $Os(C_5H_3\{SMe\text{-}1\}\{Br\text{-}4\})I(CO)(PPh_3)_2$ (**30**) (Scheme 1.10). The iodide ancillary ligand of the osmabenzene has some lability under these conditions and a small amount of exchange for bromide takes place during the reaction. To overcome the problem of forming products with mixed halide ligands the reaction mixture was treated first with silver perchlorate to abstract the metal-bound halide and then with sodium iodide to return the iodide ligand to the metal. Pure **30** was then isolated in 45% yield following column chromatography. It is interesting to note that, although the –SMe group is an *ortho*/*para* director in benzene chemistry, no sign of substitution *ortho* to the –SMe group was found in these or any other related reactions we subsequently carried out with other electrophiles [63]. Much later we carried out computational studies to determine which of the ring atoms was electronically the most susceptible towards attack by

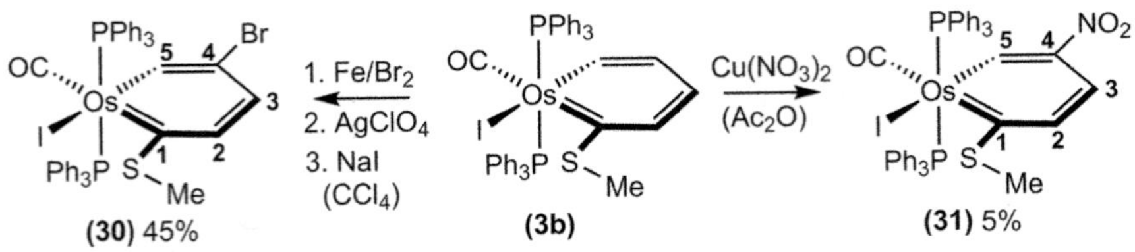

Scheme 1.10 Electrophilic aromatic bromination and nitration reactions of the osmabenzene **3b**.

electrophiles. Indeed, C4 was found to be the preferred site of attack, but this was followed closely by C2 [64]. However, the computational studies did not take into account the influence of steric factors. Presumably our observation of exclusive *para*-substitution is therefore largely a result of the combined steric pressures of the –SMe group and the mutually *trans* PPh_3 ligands in **3b**.

Chlorination occurs at the same ring position, i.e. *trans* to –SMe, and on treatment of **3b** with iodobenzene dichloride ($PhICl_2$) the blue-green product $Os(C_5H_3\{SMe\text{-}1\}\{Cl\text{-}4\})I(CO)(PPh_3)_2$ is formed in 20% yield [63]. Nitration also takes place exclusively at C4 when **3b** is subjected to the fairly mild nitrating conditions of copper nitrate in acetic anhydride (Menke nitration conditions) [65–67]. The mononitroosmabenzene, $Os(C_5H_3\{SMe\text{-}1\}\{NO_2\text{-}4\})I(CO)(PPh_3)_2$ (**31**) (see Scheme 1.13), is isolated in very low yield (5%) as purple crystals following column chromatography [63].

The structural and spectroscopic data indicate the substituted metallabenzene rings in **30** and **31** retain delocalised π-bonding systems. Noteworthy chemical shifts in the ^1H NMR spectra include those of the H5 protons, which are found at 12.61 ppm for the brominated **30** and at 14.00 ppm for the nitrated **31** (*cf.* 12.71 ppm in the unsubstituted **3b**). The down-field chemical shift in **31** arises due to the neighbouring strongly electron-withdrawing nitro group, which is also responsible for the down-field shift of proton H3 (to 7.95 ppm in **31** vs. 6.96 ppm in **3b**) and for carbon C4 in the ^{13}C NMR spectrum of **31** (144.24 ppm *cf.* 123.82 ppm in **3b**). The bromo-substituent in **30**, on the other hand, has far less influence on the chemical shifts of the surrounding atoms, with the only significant difference being an up-field shift to 110.56 ppm for carbon C4. The magnitude and direction of all these shifts are similar to those observed in substituted benzenes. The remaining proton and carbon atoms resonate in positions very similar to the corresponding atoms in the starting osmabenzene **3b**. The crystal structures of **30** and **31** (Figure 1.9) confirm the site of substitution but are otherwise unremarkable compared to other osmabenzenes [63].

We subsequently extended this electrophilic aromatic substitution reaction by bromine to encompass iridabenzenes, osmabenzofurans and iridabenzofurans. We discovered that pyridinium tribromide ($[PyH][Br_3]$) is a convenient solid source of bromine that affords brominated products in good yields and requires minimal purification of the crude products. The purple iridabenzene **13** is exclusively monobrominated with this reagent at the C4 position, the same position (*para* to the –SMe ring substituent) that the closely related osmabenzene is substituted. The dark-purple bromoiridabenzene, $Ir(C_5H_3\{SMe\text{-}1\}\{Br\text{-}4\})Br_2(PPh_3)_2$ (**32**) (see Scheme 1.14), is obtained from this reaction in the remarkably high isolated yield of 87% [33]. The blue osmabenzofuran **16** has only two sites available for ring substitution: C3 in the six-membered iridabenzene ring and C6 of the fused furan ring. The preferred site of bromination was found to be C6, and the heavily substituted osmabenzofuran $Os(C_7HO\{Ph\text{-}1\}\{Ph\text{-}2\}\{CO_2Me\text{-}4\}\{Br\text{-}6\}\{OMe\text{-}7\})(CS)(PPh_3)_2$ (**33**) (Scheme 1.14) could be isolated in good 88% yield as green crystals [38]. No evidence was obtained for bromination at C3 in this compound.

The largely unsubstituted bicyclic ring systems of iridabenzofurans **24** and **25** make them ideal candidates for more in-depth studies of the chemical reactivity of metallabenzofurans. The cationic **24** undergoes exclusive monobromination at the C6 site on the fused furan ring on treatment with pyridinium tribromide to give the substituted iridabenzofuran $[Ir(C_7H_4O\{Br\text{-}6\}\{OMe\text{-}7\})(CO)(PPh_3)_2][OTf]$ (**34**) (Scheme 1.11) as

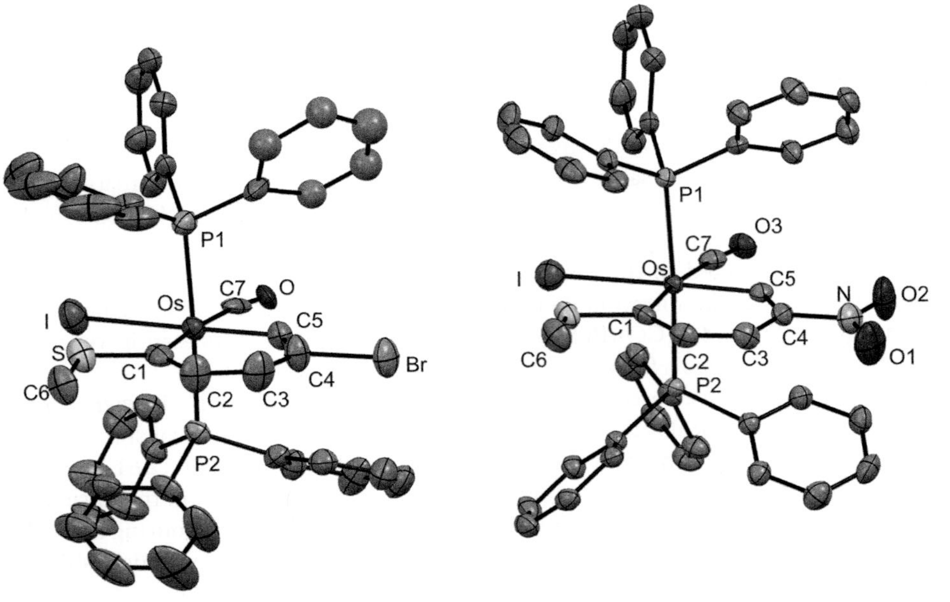

Figure 1.9 Molecular structures of the substituted osmabenzenes **30** (left) and **31** (right) showing 50% probability thermal ellipsoids. Hydrogen atoms are not shown for clarity. Selected distances [Å] for **30**: Os–C1 2.128(9), Os–C5 2.039(9), Os–C7 1.936(13), Os–I 2.8021(9), C1–C2 1.418(13), C2–C3 1.367(14), C3–C4 1.390(14), C4–C5 1.320(13), C1–S 1.702(10), C4–Br 1.968(10). Selected distances [Å] for **31**: Os–C1 2.129(7), Os–C5 2.011(7), Os–C7 1.908(8), Os–I 2.8158(6), C1–C2 1.411(11), C2–C3 1.347(11), C3–C4 1.411(10), C4–C5 1.336(10), C1–S 1.702(7), C4–N 1.476(9). (*See color plate section for the color representation of this figure.*)

red crystals in 70% yield. There was no evidence of further bromination when an excess of the brominating reagent was used, nor was there evidence of any isomers with bromination at a different site. The neutral analogue **25**, however, undergoes exclusive dibromination at C2 and C6 to give the iridabenzofuran $Ir(C_7H_3O\{Br\text{-}2\}\{Br\text{-}6\}\{OMe\text{-}7\})Br(PPh_3)_2$ (**35**) as red crystals in 72% yield. In this case there was no evidence of a monosubstituted product, and even on treatment of **24** with only one equivalent of $[PyH][Br_3]$ a mixture of **35** and unreacted **25** was obtained [52].

In a related electrophilic aromatic substitution reaction, treatment of the iridabenzofuran **25** with mercury trifluoroacetate results in mercuration at the C2, C4 and C6 positions. Following the addition of lithium bromide, the pink trisubstituted iridabenzofuran product $Ir(C_7H_2O\{HgBr\text{-}2\}\{HgBr\text{-}4\}\{HgBr\text{-}6\}\{OMe\text{-}7\})Br(PPh_3)_2$ (**36**) (Scheme 1.12) was isolated in 41% yield. On treatment of **25** with reduced amounts of mercury trifluoroacetate, no mono- or disubstituted products could be isolated. The tri-mercurated compound **36** was the first metallo-substituted metallabenzene derivative to be reported. The carbon–mercury bonds in **36** are easily cleaved on the addition of pyridinium tribromide to give the corresponding tribrominated product $Ir(C_7H_2O\{Br\text{-}2\}\{Br\text{-}4\}\{Br\text{-}6\}\{OMe\text{-}7\})Br(PPh_3)_2$ (**37**) as dark-red crystals in 87% yield [52].

To explain the regioselectivity of all these electrophilic aromatic substitution reactions, a computational study was undertaken. Condensed Fukui functions, derived from DFT calculations, were used as a reactivity index to identify which sites are the

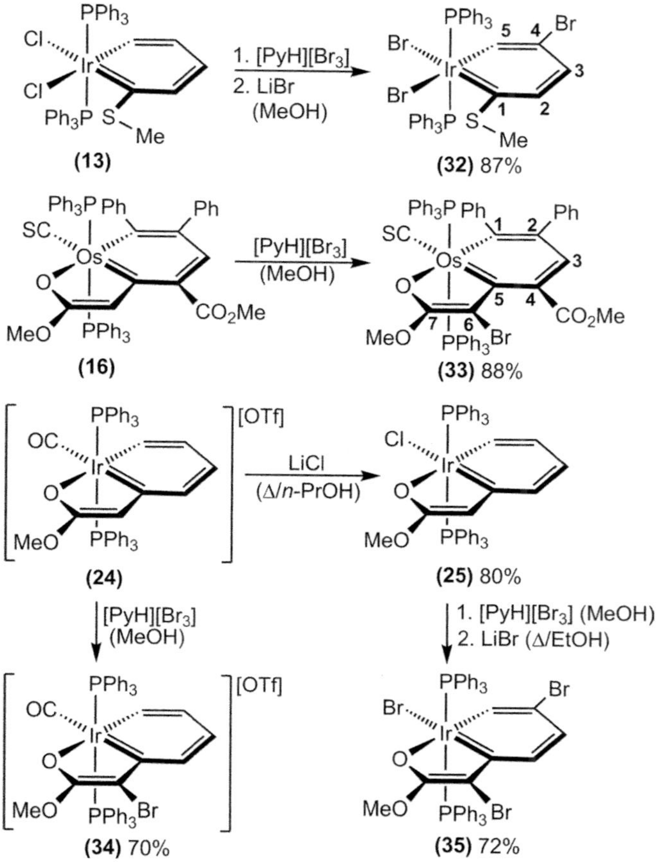

Scheme 1.11 Electrophilic aromatic bromination of the iridabenzene **13**, the osmabenzofuran **16** and the iridabenzofurans **24** and **25**.

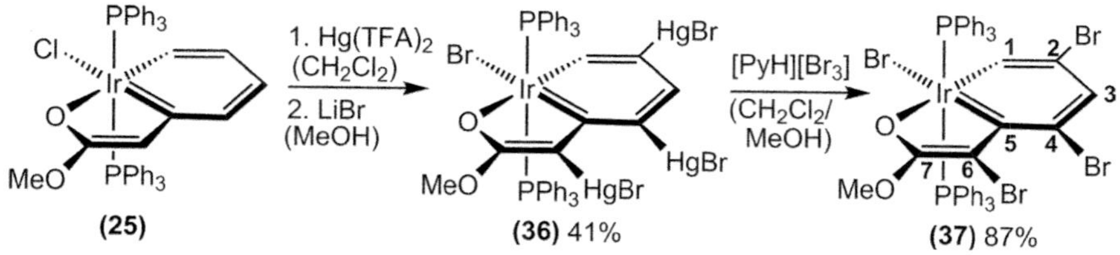

Scheme 1.12 Electrophilic aromatic mercuration of **25** to give the metallo-substituted iridabenzofuran **36** and the cleavage of the C–Hg bonds to give tribrominated iridabenzofuran **37**.

most nucleophilic (i.e. with the largest f_k^- values) and hence the most susceptible to electrophilic attack. It was found that the C1, C2, C4 and C6 sites all possessed large f_k^- values and so electronically these should be the favoured sites for electrophilic substitution. The largest f_k^- value within the fused ring system of the cationic **24** was at C6, consistent with this being the observed site of substitution. The computed f_k^- values for the ring carbons of the monobrominated product **34** are reduced compared to **24**, which may explain why further bromination of **34** is not observed. For the neutral iridabenzofuran analogue **25**, the most electronically favourable site of substitution is again C6, followed by C4 then C2. In this case bromination at both C2 and C6 is observed. On the other hand, mercuration of **25** yields the tri-mercurated product **36** with substitution occurring at C2, C4 and C6, even though the crystal structure of **36** indicates there is significant steric crowding between the substituents at C4 and C6. While the condensed Fukui functions do correctly indicate the preferred sites for electrophilic attack in these compounds, an important limitation of this approach is that it does not take into account steric factors and this limits its value as a predictive tool [52].

In general, we have found that ring bromination of the metallabenzenes or metallabenzofurans which we have studied has minimal effect on the chemical shifts of the remaining unsubstituted ring carbons and protons in the NMR spectra. In addition, aside from the presence of the bromine substituents, there are only small differences in the structures of the brominated products compared to those of the parent compounds. In the ^{13}C NMR spectra the only significant differences between the brominated products and the parent species is an up-field shift of approximately 10–20 ppm for each carbon atom that bears a bromine substituent. In the bromoiridabenzene **32**, this is carbon C4 (104.78 ppm vs. 121.64 ppm in **13**) [33]. Similar up-field shifts can be observed in the bromoiridabenzofurans **34** (104.36 ppm for C6 vs. 121.57 ppm in **24**), **35** (99.95 ppm for C2, 97.35 ppm for C6) and **37** (101.74, 112.90 and 98.33 ppm for C2, C4 and C6, respectively, vs. 121.12, 121.08 and 114.82 ppm, respectively, in **25**). Analogous up-field shifts are observed upon bromination of simple benzenes. The H1 proton of the trimetallo-substituted **36** is shifted down-field to 10.10 ppm (from 9.36 ppm in **25**) in the ^{1}H NMR spectrum, and proton H3 to 5.90 ppm (from 5.52 ppm in **25**). Further analysis of **36** by ^{13}C NMR spectroscopy could not be made due to the very low solubility of this species [52].

The molecular structures of the brominated derivatives **34** and **35** are shown in Figure 1.10, and the structures of the trisubstituted derivatives **36** and **37** are shown in Figure 1.11. The structural parameters associated with the fused rings show only relatively small changes compared to the corresponding parent compounds **24** and **25**. In all cases the structures are consistent with maintenance of the delocalised π-system over the fused rings.

Electrophilic substitution reactions of the iridabenzofurans **24** and **25** have been extended to nitration. Although only very low yields of the corresponding nitroosmabenzene were obtained from **3b**, much higher yields of nitro-substituted iridabenzofurans were obtained from **24** and **25**. On treatment of the cationic iridabenzofuran **24** with copper nitrate in acetic anhydride at 0°C [65–67], substitution occurs only once and exclusively at C2 to give the nitroiridabenzofuran, $[Ir(C_7H_4O\{NO_2\text{-}2\}\{OMe\text{-}7\})(CO)(PPh_3)_2][OTf]$ (**38**) (Scheme 1.13). This product can be isolated in 52% yield as dark orange crystals. There was no evidence of other isomers or further substitution [68]. The regioselectivity of this reaction was initially quite surprising as in the earlier bromination reactions mono-substitution was found to occur exclusively at the C6 position, and the condensed Fukui functions had indicated that C6 was the most favourable site for attack by electrophiles [52]. However,

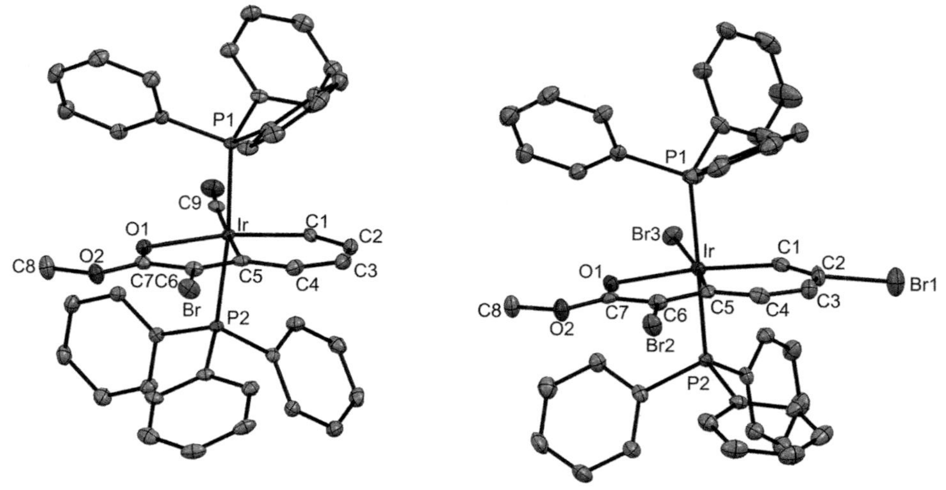

Figure 1.10 Molecular structures of the cation of bromoiridabenzofuran **34** (left) and **35** (right) showing 50% probability thermal ellipsoids. Hydrogen atoms are not shown for clarity. Selected distances [Å] for **34**: Ir–C1 2.019(3), Ir–C5 2.080(3), Ir–O1 2.188(2). Ir-C9 1.948(3), C1-C2 1.350(5), C2-C3 1.451(5), C3-C4 1.350(5), C4-C5 1.443(4), C5-C6 1.353(5), C6-C7 1.445(4), C7-O1 1.261(4), C6-Br 1.909(3). Selected distances [Å] for **35**: Ir-C1 2.023(5), Ir-C5 2.017(5), Ir-O1 2.221(4), Ir-Br3 2.5978(6), C1-C2 1.331(8), C2-C3 1.439(8), C3-C4 1.340(8), C4-C5 1.451(7), C5-C6 1.378(7), C6-C7 1.437(8), C7-O1 1.262(7), C2-Br1 1.948(5), C6-Br2 1.899(5).

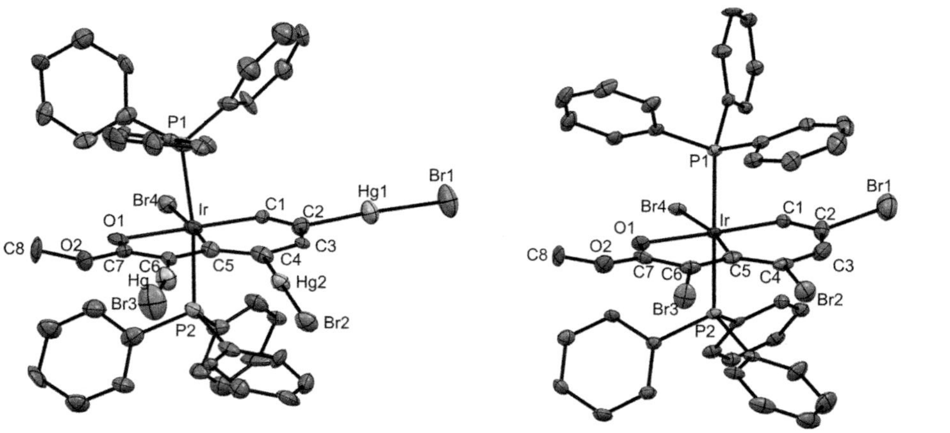

Figure 1.11 Molecular structures of the trisubstituted iridabenzofurans **36** (left) and **37** (right) showing 50% probability thermal ellipsoids. Hydrogen atoms are not shown for clarity. Selected distances [Å] for **36**: Ir-C1 1.99(2), Ir–C5 2.04(2), Ir–O1 2.231(17). Ir-Br4 2.574(3), C1-C2 1.30(3), C2-C3 1.39(3), C3-C4 1.38(3), C4-C5 1.47(3), C5-C6 1.42(3), C6-C7 1.39(3), C7-O1 1.27(3), C2-Hg1 2.13(2), C4-Hg2 2.07(3), C6-Hg3 2.01(2), Hg1-Br1 2.442(3), Hg2-Br2 2.455(3), Hg3-Br3 2.419(3). Selected distances [Å] for **37**: Ir-C1 2.006(8), Ir-C5 2.035(8), Ir-O1 2.159(6), Ir-Br4 2.5746(8), C1-C2 1.306(13), C2-C3 1.397(14), C3-C4 1.352(14), C4-C5 1.453(13), C5-C6 1.369(13), C6-C7 1.450(13), C7-O1 1.243(11), C2-Br1 1.996(9), C4-Br2 1.928(9), C6-Br3 1.912(9).

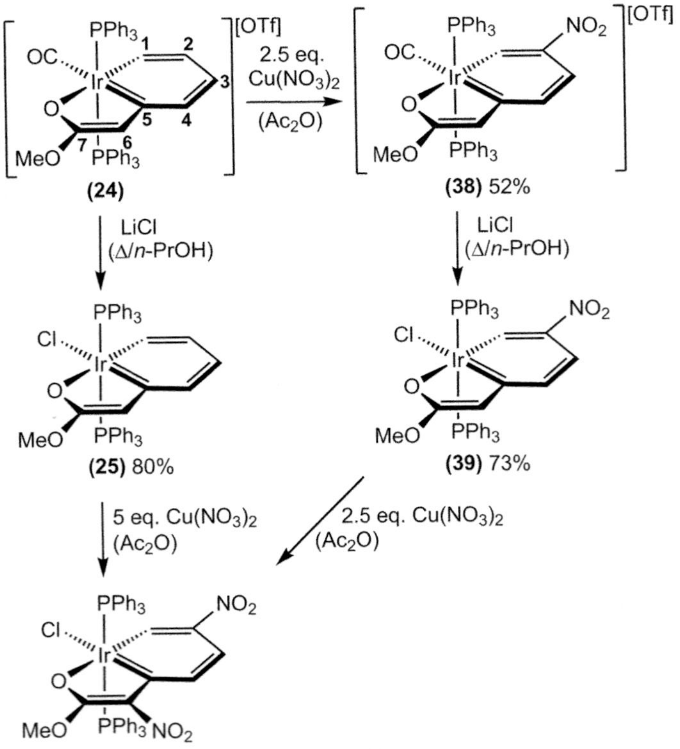

Scheme 1.13 Nitration reactions of the iridabenzofurans **24** and **25**.

it has been reported that nitration of the organic analogue, 2-methylbenzo[*b*]furan, with nitric acid in the presence of tin(IV) chloride gives predominantly 2-methyl-6-nitrobenzo[*b*]furan (which corresponds to nitration at the C2 position of the iridabenzofuran **24**) in addition to lesser amounts of 2-methyl-3-nitrobenzo[*b*]furan (which corresponds to nitration at the C6 position of the iridabenzofuran **24**), with the ratio of these two products being *ca.* 4:1 [69]. In practice, it has been found that the regioselectivity of the nitration of simple benzo[*b*]furans is highly dependent on the nature and position of the substituents on the fused rings, as well as the nitrating reagent employed [70]. Clearly there are a number of subtle but important effects that determine the preferred position of substitution in these nitration reactions. Again, while the f_k^- data successfully indicates the atoms that are susceptible to electrophilic attack in these iridabenzofurans, the method does not enable distinctions to be made reliably between all the potential sites of attack [52, 64].

The neutral analogue **25** can also be nitrated under the same conditions, but in this case the reaction favours disubstitution (at C2 and C6), furnishing the dinitroiridabenzofuran, $Ir(C_7H_3O\{NO_2\text{-}2\}\{NO_2\text{-}6\}\{OMe\text{-}7\})Cl(PPh_3)_2$ (**40**) (Scheme 1.13). When less than two equivalents of copper nitrate are used, some mononitroiridabenzofuran, $Ir(C_7H_4O\{NO_2\text{-}2\}\{OMe\text{-}7\})Cl(PPh_3)_2$ (**39**) can also be isolated from the crude product. However, in practice it is best to prepare this compound by substitution of the carbonyl ligand in **38** with chloride. Subjecting **39** to the same nitrating conditions also furnishes the dinitrated **40**, although of course this is a less convenient route to this compound. The fact that no C6-mononitroiridabenzofuran could be identified as a component of the product mixture when one equivalent of copper nitrate is used suggests that the dinitration reaction proceeds via initial nitration at C2 rather than C6 [68].

In the NMR spectra of **38**, **39** and **40** the electron-withdrawing nitro group causes deshielding of all nearby proton and carbon atoms, giving rise to down-field chemical shifts compared to the related unsubstituted species. In the ^{13}C NMR spectra, the nitro-bearing carbon C2 is found at 144.14, 143.44 and 142.49 ppm in **38–40**, respectively, and the nitro-bearing C6 carbon in **40** at 135.50 ppm. These are comparable to the nitro-bearing carbon atom in nitrobenzene (148.16 ppm) [71] and are found approximately 20 ppm down-field of the related carbon atom of the unsubstituted complexes **24** and **25** (*ca.* 120 ppm). The metal-bound C1 carbon atoms are also shifted 10–20 ppm down-field by the neighbouring nitro group to 139.76, 167.23 and 173.58 ppm in **38–40**, respectively. Likewise, the H1 protons resonate at 9.74, 12.08 and 12.36 ppm in the ^{1}H NMR spectra of **38–40**, respectively (compared to 7.70 and 9.36 ppm in the unsubstituted **24** and **25**, respectively). The remaining proton and carbon atoms also experience slight down-field chemical shifts, diminishing with distance from the nitro group(s).

The regioselectivities of these nitration reactions have been confirmed by X-ray crystal structure determinations of **38–40**, and the molecular structures of **38** and **40** are shown in Figure 1.12. In each of the complexes **38–40**, the nitro substituent at C2 is nearly coplanar with the Ir, C1–C7 and O1 least-squares plane. The second nitro substituent on carbon C6 of complex **40** is much more sterically hindered, and is rotated about 42° from the same plane. The metallacyclic rings of each of the nitrated complexes are close to planar, in some cases even more so than in the unsubstituted analgoues, which is consistent with the computational determination that π-electron-withdrawing substituents favour planarity in metallabenzenes [35]. The metal–carbon and carbon–carbon bond lengths are otherwise very similar to other iridabenzofurans.

Another interesting observation that was noted during the studies of the nitration reactions of iridabenzofurans was that when a halide salt was introduced to the copper nitrate-acetic anhydride nitrating mixture, mixed nitration/halogenation or simple halogenation products were obtained, depending on the conditions used. We were prompted to investigate these reactions because during our initial attempts to nitrate **25** we managed to isolate and characterise a very small amount (*ca.* 2% yield) of the chloro, nitro product **42**. We reasoned that partial decomposition of **25** during the nitration reaction could have released free chloride ions and the presence of these in the nitration mixture could have led to halogenation. We therefore investigated this possibility by deliberately adding halide ion to the nitrating mixture.

When the cationic iridabenzofuran **24** is treated with the standard copper nitrate-acetic anhydride nitrating mixture to which one equivalent of lithium bromide or chloride has been added, halogenation occurs to give **34** or **41** (Scheme 1.14), respectively, as the major products [68]. Substitution occurs at the C6 position, the same position that is brominated when the reaction is carried out with pyridinium tribromide [52]. Minor amounts (<10%) of the mono-nitrated product **38** are also formed in these

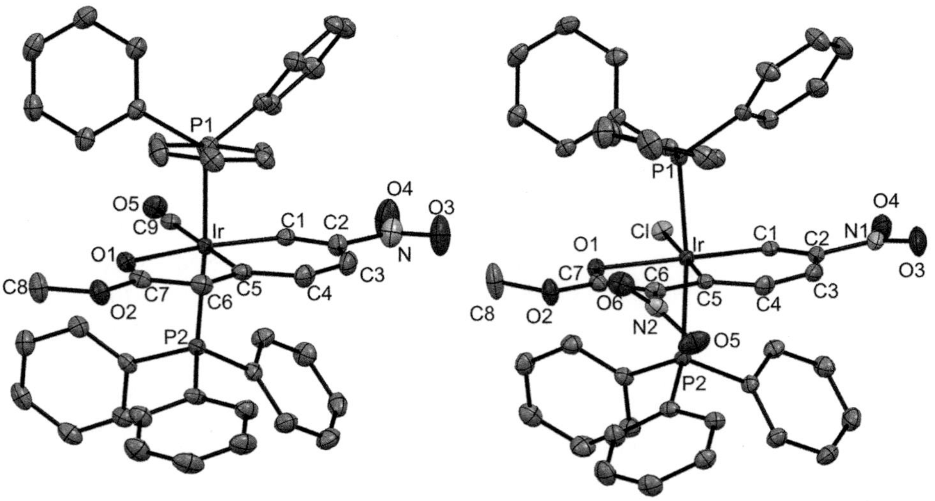

Figure 1.12 Molecular structures of the cation of **38** (left) and **40** (right) showing 50% probability thermal ellipsoids. Hydrogen atoms are not shown for clarity. Selected distances [Å] for **38**: Ir–C1 2.021(4), Ir–C5 2.062(4), Ir–O1 2.186(3), Ir–C9 1.936(4), C1–C2 1.341(6), C2–C3 1.448(6), C3–C4 1.344(6), C4–C5 1.442(5), C5–C6 1.366(6), C6–C7 1.440(6), C7–O1 1.264(5), C2–N 1.488(5). Selected distances [Å] for **40**: Ir–C1 1.963(4), Ir–C5 2.003(4), Ir–O1 2.213(3), Ir–Cl 2.4591(10), C1–C2 1.357(5), C2–C3 1.426(6), C3–C4 1.359(6), C4–C5 1.442(5), C5–C6 1.382(5), C6–C7 1.450(5), C7–O1 1.252(4), C2–N1 1.473(5), C6–N2 1.463(5).

reactions. If two equivalents of lithium bromide are present during nitration of the neutral iridabenzofuran **25**, the dibromoiridabenzofuran **35** is formed (Scheme 1.14) [68]. This is the same product that is obtained by treatment of **24** with pyridinium tribromide [52].

With careful control of the ratio of copper nitrate and lithium chloride, mixed halo, nitro derivatives can be obtained. Thus, if **25** is subjected to nitration with copper nitrate (2.5 equivalents) in the presence of one equivalent of lithium chloride, the C2 site is nitrated while C6 is chlorinated, and the mixed substitution product $Ir(C_7H_3O\{NO_2\text{-}2\}\{Cl\text{-}6\}\{OMe\text{-}7\})Cl(PPh_3)_2$ (**42**) (Scheme 1.14) is obtained as bright red crystals in 47% yield. If a larger amount of copper nitrate (5.0 equivalents) is used in the presence of one equivalent of lithium chloride, trisubstitution occurs with C2 and C4 being nitrated while C6 is chlorinated, to give $Ir(C_7H_2O\{NO_2\text{-}2\}\{NO_2\text{-}4\}\{Cl\text{-}6\}\{OMe\text{-}7\})Cl(PPh_3)_2$ (**43**) in 56% yield as dark red crystals. The former can also be nitrated to give the latter, but in practice the one step process is more convenient and gives better overall yields [68].

The spectroscopic and structural data for the two chloro, nitro-substituted complexes **42** and **43** confirm the formulations and the regioisomers that are formed. In the ^1H NMR spectra, H1 is observed at 12.15 ppm in **42** and 12.55 ppm in **43**, down-field shifts which are similar to those found in the other nitroiridabenzofurans. There are only two further aromatic proton resonances observed for **42** (at 5.30 (H4) and 6.39 (H3) ppm), and only one for **43**, (at 7.11 (H3) ppm). In the ^{13}C NMR spectra of **42** and **43**, C6 which bears the chloro-substituent, is found at 110.56 and 112.06 ppm, respectively. Similar up-field shifts were found for the bromine-bearing carbons (*ca.* 110 ppm) in the

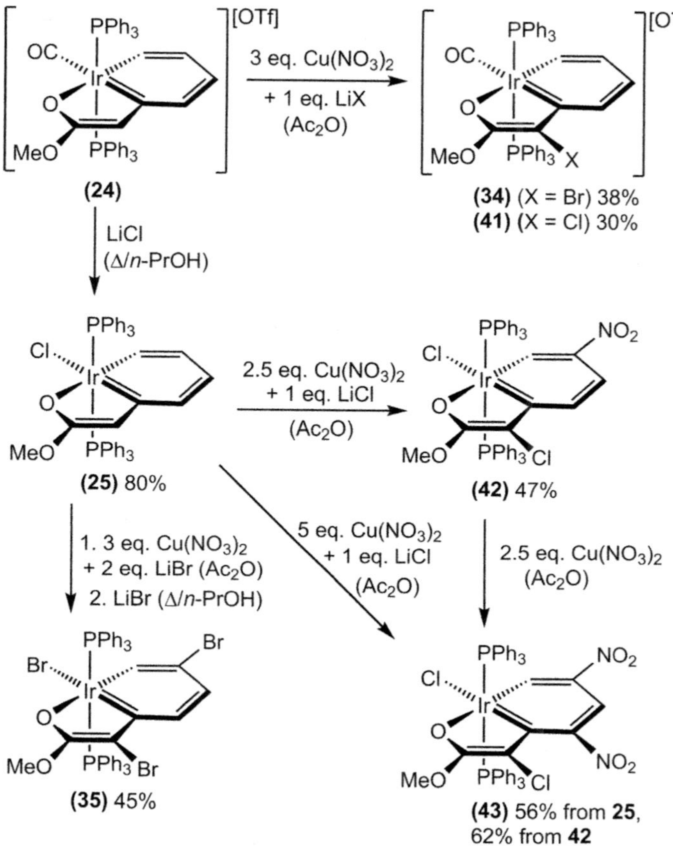

Scheme 1.14 Halogenation and mixed halogenation/nitration products formed by treatment of iridabenzofurans with mixtures of copper nitrate and lithium halides in acetic anhydride.

previously reported iridabenzofurans **34**, **35**, and **37** [52]. As expected, the nitro-bearing carbon atoms resonate in positions similar to the other nitroiridabenzofurans **38–40** (*ca.* 140 ppm), with chemical shifts of 143.36 (C2 in **42**), 143.33 and 143.46 ppm (C2 and C6 in **43**) [68].

Crystal structure determinations of **42** and **43** confirm the substitution patterns in the fused rings. Molecular structures of these compounds are shown in Figure 1.13. Features such as approximately octahedral coordination geometry about iridium, planar fused ring systems and a degree of carbon–carbon bond length alternation are observed in both complexes, just as they are in the other iridabenzofurans. On the basis of the structural and spectroscopic data it does not appear that the π-delocalisation over the fused ring systems is significantly affected by the substituents. In **43** the chloro substituent at C6 and the nitro substituent at C4 are in close proximity and the steric crowding results in the chloro substituent being displaced 0.367 Å above the Ir, C1–C7 O1 least-squares plane and the nitro group rotating out of this plane [68].

Although no direct evidence has been obtained for the mechanism of these reactions that form the mixed halo, nitro iridabenzofuran derivatives, it seems likely that the added chloride reacts with the acetyl nitrate formed in the nitrating mixture [72–74]

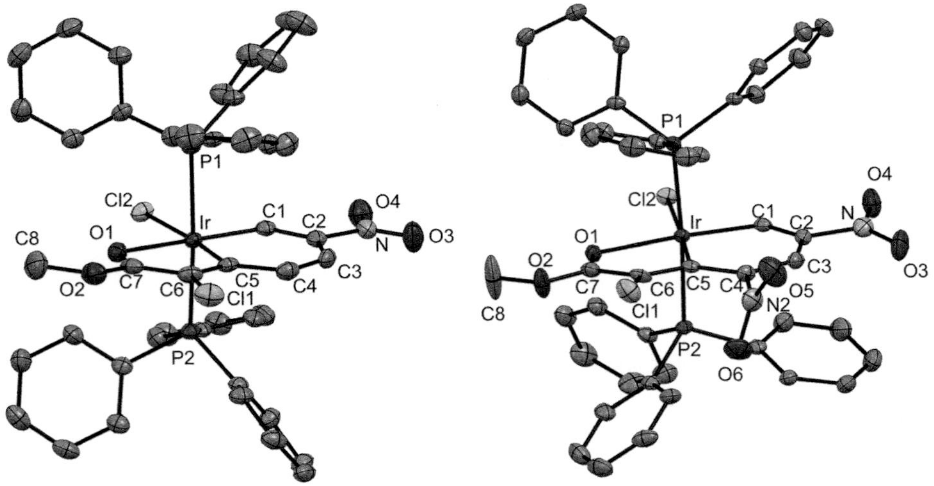

Figure 1.13 Molecular structures of the mixed substitution iridabenzofurans **42** (left) and **43** (right) showing 50% probability thermal ellipsoids. Hydrogen atoms are not shown for clarity. Selected distances [Å] for **42**: Ir–C1 1.965(3), Ir–C5 2.006(3), Ir–O1 2.2122(19), Ir–Cl1 2.4477(9), C1–C2 1.350(4), C2–C3 1.435(4), C3–C4 1.352(4), C4–C5 1.436(4), C5–C6 1.374(4), C6–C7 1.428(4), C7–O1 1.255(4), C2–N 1.475(4), C6–Cl2 1.761(3). Selected distances [Å] for **43**: Ir–C1 1.963(4), Ir–C5 2.017(4), Ir–O1 2.207(3), Ir–Cl1 2.4418(11), C1–C2 1.349(6), C2–C3 1.422(6), C3–C4 1.354(6), C4–C5 1.441(5), C5–C6 1.362(6), C6–C7 1.453(6), C7–O1 1.249(5), C2–N1 1.497(5), C4–N2 1.484(5), C6–Cl2 1.739(4).

to form nitryl chloride and/or chlorine, and one or both of these is the active chlorinating agent [75]. In the case where 2.5 equiv. of $Cu(NO_3)_2$ is used, there is sufficient acetyl nitrate remaining after reaction with the added chloride to introduce only one nitro substituent and ultimately form the observed major product **42** (Scheme 1.14). In the case where 5.0 equiv. of $Cu(NO_3)_2$ is used, there is sufficient acetyl nitrate to effect dinitration in addition to chlorination to give **43** as the major product. Similarly, the dibrominated derivative **35** that is formed on treatment of **25** with $Cu(NO_3)_2$ and LiBr in acetic anhydride (Scheme 1.14) is likely formed through reaction between acetyl nitrate and bromide to form nitryl bromide and/or bromine as the brominating agent [75].

It is worth noting that neither a dichlorinated product nor a mixed bromo/nitro product was detected in the reactions involving copper nitrate, acetic anhydride and lithium halides. However, a mixed bromo/nitro product, $IrCl(C_7H_3O\{NO_2\text{-}2\}\{Br\text{-}6\}\{OMe\text{-}7\})(PPh_3)_2$ (**44**), was obtained on the treatment of **39** with pyridinium tribromide (Scheme 1.15). Bromination of **38**, **40** and **44** did not occur even if up to 5 equiv. of pyridinium tribromide was employed [68].

1.4.2 Rearrangement to Cyclopentadienyl Complexes

One of the important decomposition routes for many metallabenzenes is rearrangement to the corresponding cyclopentadienyl complexes. Clearly, an understanding of the factors that influence this process is important before the rational syntheses of kinetically stable metallabenzenes can be undertaken. Computational studies have

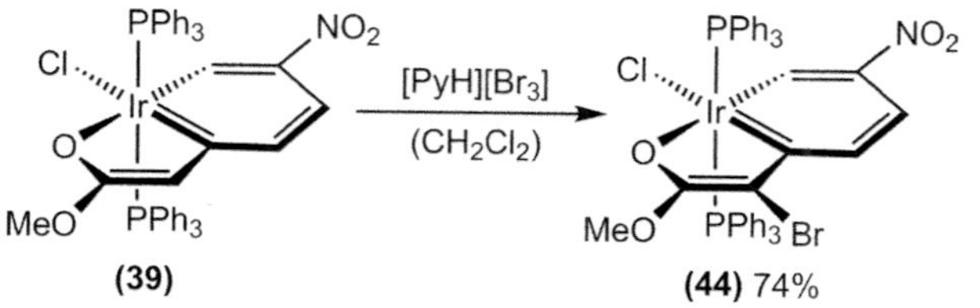

Scheme 1.15 Synthesis of the mixed bromo/nitro iridabenzofuran **44**.

shown that most metallabenzenes are thermodynamically unstable towards this rearrangement and the mechanism of this process has also been studied theoretically [24, 76, 77]. Furthermore, undetected metallabenzenes have been proposed as intermediates in the formation of a number of cyclopentadienyl complexes [42, 62, 78–83]. It is therefore somewhat surprising that well-defined transformations of isolated metallabenzenes into the corresponding cyclopentadienyl complexes have rarely been observed [48, 84–86].

With the isolation of osmabenzene **45** (Scheme 1.16), we had at our disposal the means to synthesise a set of related osmabenzenes that contained different anionic ligands through simple triflate displacement reactions. It was also possible to synthesise a corresponding set of isomeric osmabenzenes via the thermal rearrangement of **3b** in solution (Scheme 1.16). Having these osmabenzenes in hand allowed us to then investigate the influence of different ancillary ligands and the coordination geometry on the relative rates of thermal rearrangement to the corresponding cyclopentadienyl complexes.

The purple osmabenzene $Os(C_5H_4\{SMe\text{-}1\})(cis\text{-}O_3SCF_3)(CO)(PPh_3)_2$ (**45**) is formed in high yield through treatment of osmabenzene **1** with methyl triflate (Scheme 1.16). No crystal structure determination of **45** has been obtained, but the products obtained on addition of coordinating anions are consistent with the geometry depicted in Scheme 1.16. On stirring solutions of **45** under ambient conditions with iodide, chloride, thiocyanate or trifluoroacetate the corresponding blue neutral osmabenzenes $Os(C_5H_4\{SMe\text{-}1\})(cis\text{-}X)(CO)(PPh_3)_2$ (X = I (**3b**), X = Cl (**46a**), X = SCN (**46b**), X = CF_3CO_2 (**46c**)) are formed (Scheme 1.16). In each case, the anionic ligand is

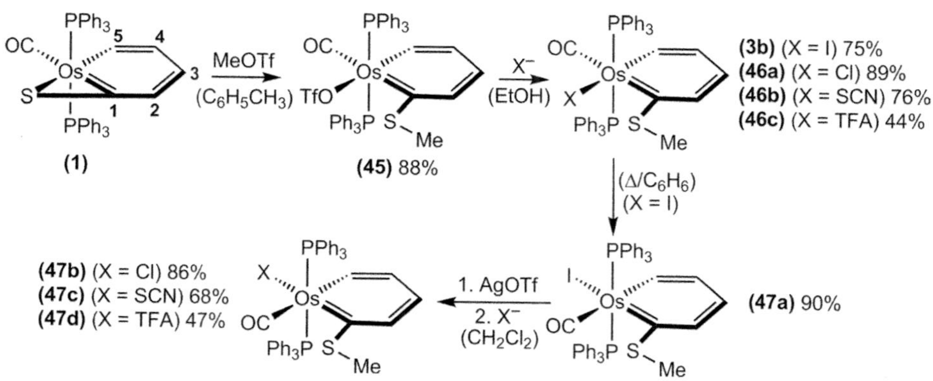

Scheme 1.16 Syntheses of the osmabenzenes **45**, **46a–c** and **47a–d**.

coordinated *cis* to the carbon bearing the methylthiolate group, while the CO ligand is located *trans* to this carbon atom [14, 87, 88].

When the osmabenzene **3b** is heated under reflux in benzene for just 10 min, an isomerisation process occurs in which the positions of the iodide and carbonyl ligands are interchanged, to give the dark-brown osmabenzene, Os(C_5H_4{SMe-1})(*trans*-I)(CO)($(PPh_3)_2$) (**47a**), (Scheme 1.16) in 90% yield. In this isomer the iodide ligand is located *trans* to the carbon bearing the –SMe group and the CO ligand *cis* to this carbon atom. None of the other osmabenzenes **46a-c** underwent this particular thermal isomerism on heating in benzene. As expected, small differences are observed in the ^1H and ^{13}C NMR spectra of **3b** and **47a**, the most notable being the shifts in positions of H5 (12.74 and 12.40 ppm for **3b** and **47a**, respectively) and C1 (244.18 and 237.45 ppm for **3b** and **47a**, respectively) [63, 88]. Isolation of **47a** enabled the corresponding series of isomeric osmabenzenes (Os(C_5H_4{SMe-1})(*trans*-X)(CO)$(PPh_3)_2$ (X = Cl (**47b**), X = SCN (**47c**), X = CF_3CO_2 (**47d**)) to be synthesised through iodide abstraction by Ag^+ followed by the addition of chloride, thiocyanate or trifluoroacetate, respectively (Scheme 1.16) [88].

With both isomers of each member of this family of osmabenzenes with different anionic ligands available to us, we were then able to study the thermal rearrangement of these compounds to the corresponding cyclopentadienyl complexes. It was convenient to study these rearrangements in refluxing toluene solutions because many of the rearrangements occurred at reasonable rates under these conditions. It was found that two osmium cyclopentadienyl products were formed, and these were the cationic bis(triphenylphosphine) cyclopentadienyl complexes **48a–e** and the neutral mono(triphenylphosphine) cyclopentadienyl complexes **49a–c** (Scheme 1.17). The identities and yields of each of the cyclopentadienyl compounds formed were monitored by NMR spectroscopy and mass spectrometry. In most cases the two cyclopentadienyl complexes could be isolated as pure compounds [88].

It was found that the ratio of the two cyclopentadienyl products obtained and the rate at which they formed under these conditions were highly dependent on the nature of the anionic ligand and the relative geometry of this and the adjacent CO ligand with respect to the SMe-substituted osmabenzene ring.

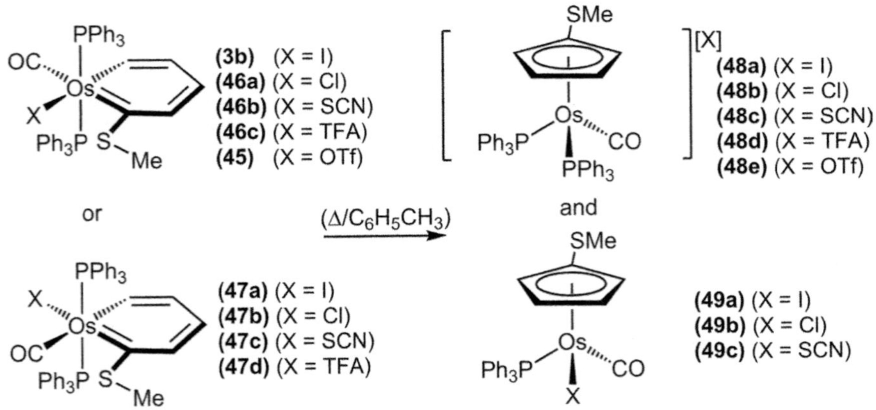

Scheme 1.17 Thermal rearrangement reactions of the osmabenzenes **3b, 45, 46a–c** into the osmium cyclopentadienyl complexes **48a–e** and **49a–c**.

With regard to the blue osmabenzenes **3b**, **45** and **46a–c**, in each case mixtures of the two cyclopentadienyl complexes were obtained. Osmabenzene **3b** gave **48a** and **49a** in a 40:60 ratio, **46a** gave **48b** and **49b** in a 75:25 ratio and **46b** gave **48c** and **49c** in a 50:50 ratio. In each of these three cases, approximately 1 h of heating under reflux was required to fully convert the osmabenzene to the cyclopentadienyl products. In contrast, the analogous trifluoroacetate derivative **46c** gave the cationic cyclopentadienyl complex **48d** as the exclusive product within minutes. The trifluoromethane sulfonate derivative **45** is not soluble in toluene and so thermal rearrangement of this osmabenzene was studied in the lower boiling point solvent 1, 2-dichloroethane. As for **46c**, exclusive rearrangement to the corresponding cationic cyclopentadienyl complex **48e** occurred within minutes.

The brown osmabenzenes **47a** and **47b** showed very similar behaviour to the isomeric compounds **3b** and **46a**, in that mixtures of the two cyclopentadienyl complexes were formed over 1 h and in exactly the same ratios as were obtained from the thermal rearrangements of **3b** and **46a**. It is very likely that the first step in the thermal rearrangements of **3b** and **46a** is isomerisation into **47a** and **47b**. In contrast, the brown osmabenzenes **47c** (SCN *trans* to CSMe) and **47d** (CF_3CO_2 *trans* to CSMe) completely resisted rearrangement into cyclopentadienyl products and were recovered essentially unchanged after heating under reflux in toluene for 1 h. This is in stark contrast to the corresponding isomers (**46b** and **46c**, respectively), both of which readily underwent conversion to cyclopentadienyl products [88].

The molecular structures of two representative examples of the cyclopentadienyl products, **48b** and **49a**, is shown in Figure 1.14.

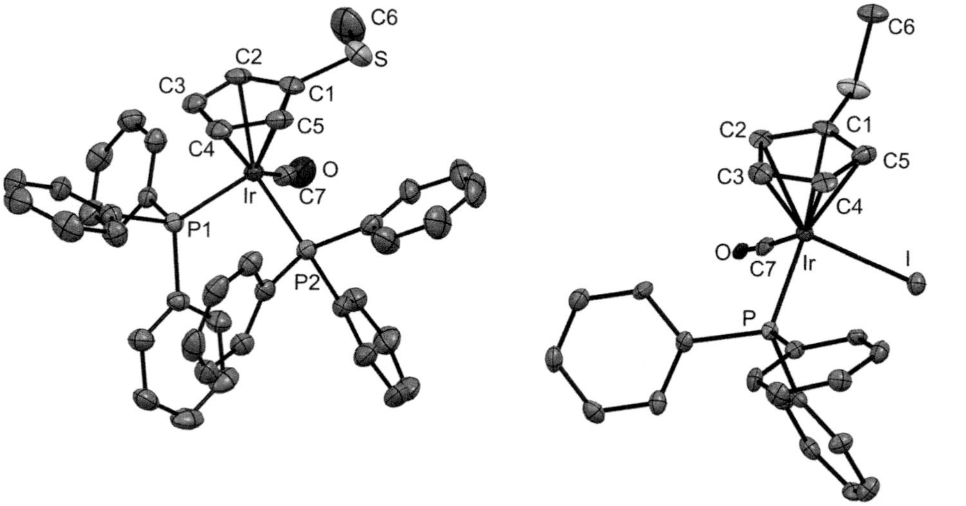

Figure 1.14 Molecular structures of the cation of **48d** (left) and **49a** (right) showing 50% probability thermal ellipsoids. Hydrogen atoms are not shown for clarity. Selected distances [Å] for **48d**: Os–C1 2.327(5), Os–C2 2.275(5), Os–C3 2.243(5), Os–C4 2.266(5), Os–C5 2.292(5), Os–C7 1.857(6), Os–P1 2.3539(14), Os–P2 2.3661(14), C1–C2 1.452(8), C2–C3 1.415(8), C3–C4 1.438(8), C4–C5 1.420(9), C5–C1 1.439(8). Selected distances [Å] for **49a**: Os–C1 2.255(7), Os–C2 2.222(7), Os–C3 2.215(8), Os–C4 2.305(7), Os–C5 2.298(10), Os–C7 1.880(11), Os–P 2.2963(18), Os–I 2.7253(5), C1–C2 1.436(11), C2–C3 1.412(12), C3–C4 1.418(12), C4–C5 1.428(14), C5–C1 1.407(11).

These results show that the rates of thermal rearrangements of metallabenzenes to cyclopentadienyl complexes can be highly sensitive to the nature of the anionic ancillary ligands and to the coordination geometry about the metal. While no direct evidence was obtained for the mechanisms of these rearrangements to cyclopentadienyl products, two important observations were made. First, the cationic and neutral cyclopentadienyl products **48a** and **49a** do not interconvert under the same or similar reaction conditions to those used in the formation of these compounds. These observations strongly suggest that the cyclopentadienyl products **48a–e** and **49a–c** are formed independently via two different pathways. Second, if the osmabenzene **47a** is heated under reflux in toluene in the presence of 10 equivalents of triphenylphosphine, the ratio of the two cyclopentadienyl products formed changes markedly in favour of cationic bis(triphenylphosphine) product **48a** (90%) over neutral mono(triphenylphosphine) product **49a** (10%). This suggests that PPh_3 dissociation is involved in the rate-determining step of the pathway to **49a**. Further, more detailed studies are needed to learn more about the mechanisms of these reactions [88].

It is noted in Section 1.3.2 above that the ruthenabenzofuran $Ru(C_7H_3O\{CO_2Me\text{-}2\}\{CO_2Me\text{-}4\}\{OMe\text{-}7\})(CO)(PPh_3)_2$ (**19**) is remarkably thermally robust. Furthermore, treatment of a refluxing benzene/methanol solution of **19** with trimethylsilyl chloride (an anhydrous HCl source) while simultaneously irradiating with visible light over a period of 4–5 h gives the green, tethered ruthenabenzene, $Ru(C_5H_2\{CO_2Me\text{-}2\}\{CO_2Me\text{-}4\}\{CH_2C(O)OMe\text{-}5\})Cl(PPh_3)_2$ (**21**), in very good yield (78%). The other product that can be isolated in *ca.* 10% yield from this reaction is the ruthenium cyclopentadienyl complex $Ru(\eta^5\text{-}C_5H_2\{CO_2Me\text{-}2\}\{CO_2Me\text{-}4\}\{CH_2CO_2Me\text{-}5\}Cl(CO)(PPh_3)$ (**50**) (Scheme 1.18). This product presumably forms through rearrangement of the tethered ruthenabenzene **21** under the vigorous reaction conditions. It is remarkable that **21** does not decompose at a faster rate under these conditions. The tethering arm and the electron-withdrawing ester functions probably make significant contributions to the stability of **21** [42].

1.4.3 Nucleophilic Aromatic Substitution Reactions

The previous two sections illustrate the important general observation that some reactions which metallabenzenes undergo mirror those classically associated with benzene itself (e.g. electrophilic aromatic substitution), while others have no counterpart in benzene chemistry (e.g. rearrangement to cyclopentadienyl ligands) [22, 34, 59, 60, 89–92].

Nucleophilic aromatic substitution (S_NAr) is a reaction class that is classically associated with benzene and its derivatives. Traditionally, S_NAr proceeds by the addition of a nucleophile to a carbon atom of a suitably electron-deficient (and hence activated

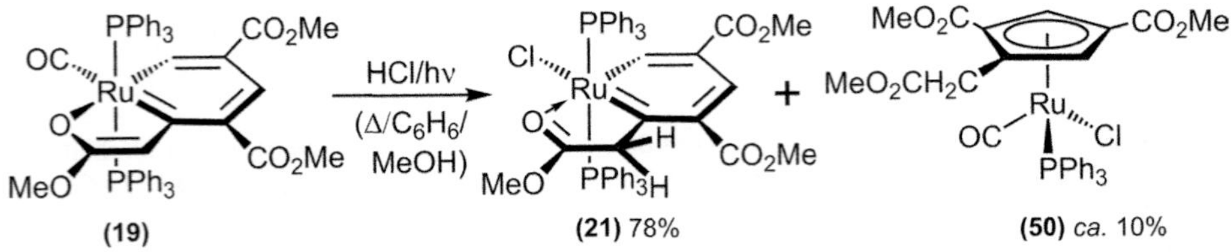

Scheme 1.18 Rearrangement of the ruthenabenzofuran **19** into the cyclopentadienyl complex **50**, as a side product during the synthesis of the tethered ruthenabenzene **21**.

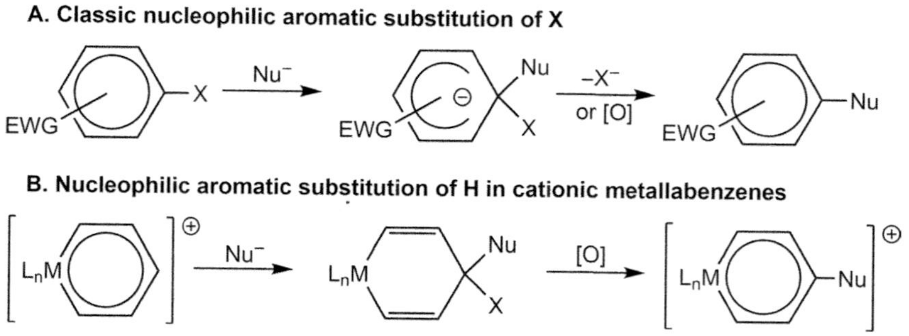

Chart 1.5 Nucleophilic aromatic substitution (S_NAr) of leaving group X (A) or H (B) in benzenes with electron withdrawing groups (EWGs), and the S_NAr of H in cationic metallabenzenes.

benzene) ring. This forms an intermediate σ^X-adduct. If this site of attack is occupied by a good leaving group X, then the substituted benzene product is obtained through loss of X^- (Chart 1.5A). If instead it is occupied by hydrogen (X = H), the elimination of hydride from the σ^H-adduct is much less favourable and an additional process is often required to convert it into the substituted product. This is usually achieved by oxidation of the σ^H-adduct, or by the elimination of HY from it (where Y is a secondary group that is present as part of the nucleophile or as a substituent on the aromatic ring).

During our studies of the chemistry of iridabenzenes and osmabenzenes, we had synthesised a small number of cationic derivatives which had tightly coordinating and chemically robust ancillary ligands. It occurred to us that if these compounds were treated with nucleophiles, attack might be directed towards one of the metallabenzene ring carbon atoms rather than to the metal. If this did happen then the product formed would be very closely related to the σ^X-adducts formed as intermediates in S_NAr reactions of benzenes (Chart 1.5). Oxidation of any metallacyclohexadiene intermediate thus formed could then yield the corresponding substituted metallabenzene.

Two metallabenzenes that seemed to be good candidates for investigating reactions of this type were the red osmabenzene $[Os(C_5H_4\{SMe\text{-}1\})(CO)_2(PPh_3)_2][OTf]$ (**51**) (which can be prepared by treating **1** with methyl triflate and then CO) and the green iridabenzene $[Ir(C_5H_4\{SMe\text{-}1\})(\kappa^2\text{-}S_2CNEt_2)(PPh_3)_2][PF_6]$ (**52**) (which can be simply obtained as the PF_6^- salt from the iridabenzene **13** by replacing the two labile chloride ligands with a bidentate diethyldithiocarbamate ligand) (Scheme 1.19) [64]. Both metallabenzenes are cationic which we reasoned may reduce the electron density in the metallacyclic rings and facilitate attack by negatively charged nucleophiles. In addition, in both complexes there is only one ring substituent, and so steric hindrance to attack by nucleophiles should be minimal.

Indeed, we found that on treatment of either **51** or **52** with sodium borohydride in ethanol the bright colours of the metallabenzenes fade and the colourless neutral metallacyclohexa-1,4-diene complexes $Os(C_5H_3\{SMe\text{-}1\}\{H_2\text{-}3\})(CO)_2(PPh_3)_2$ (**53a**) or $Ir(C_5H_3\{SMe\text{-}1\}\{H_2\text{-}3\})(\kappa^2\text{-}S_2CNEt_2)(PPh_3)_2$ (**54a**), respectively, precipitate from solution (Scheme 1.19). In each case hydride adds to the C3 carbon of the metallabenzene ring, saturating this carbon atom and disrupting the π-electron delocalisation about the metallacyclic ring. These reactions are highly regioselective with the only products detected resulting from nucleophilic addition at C3 [64]. In a related reaction, it had

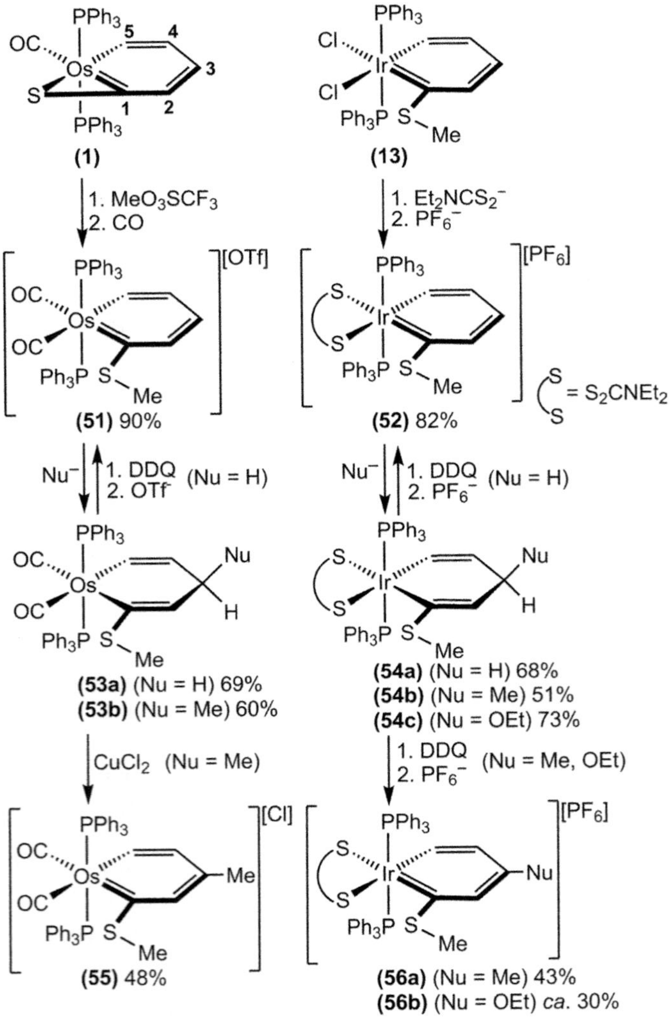

Scheme 1.19 Nucleophilic aromatic substitution reactions of osmabenzenes and iridabenzenes.

previously been reported that hydroxide ion adds reversibly to the same position of a very different iridabenzene, also forming an iridacyclohexadiene [56, 93]. Details of this reaction are provided in Chapter 4.

Loss of the delocalised π-system of the metallabenzene rings of **51** and **52** causes substantial changes to the spectroscopic and structural properties of the resulting complexes. For example, in the ^{1}H NMR spectrum of the osmabenzene **51**, H5 appears at the down-field position of 10.82 ppm while the signal for this same proton moves up-field to 7.08 ppm in the corresponding osmacyclohexa-1,4-diene, **53a**. As expected, the signal for H3 moves considerably up-field (from 7.08 ppm to 2.01 ppm) as C3 becomes saturated in **53a**. The resonances of the ring carbon atoms in the ^{13}C NMR spectra of **51** and **53a** are also significantly different. Loss of the delocalised π-system in **51** causes up-field shifts of close to 100 ppm in the metal-bound carbon atoms

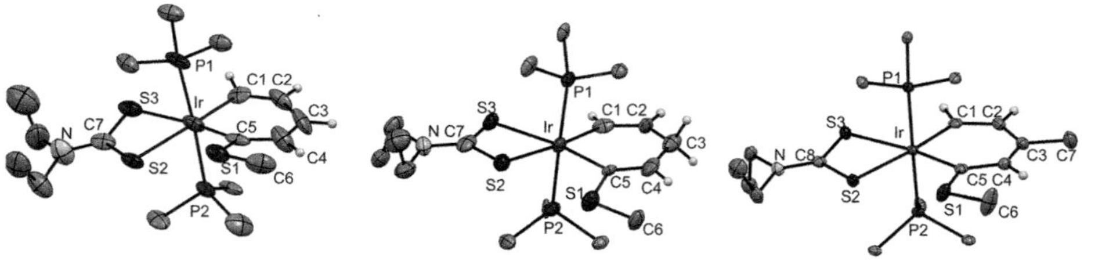

Figure 1.15 Molecular structures of the cation of **52** (left), **54a** (centre) and **56a** (right) showing 50% probability thermal ellipsoids. Phenyl rings on the triphenylphosphine ligands and most hydrogen atoms are not shown for clarity. Selected distances [Å] for **52**: Ir–C1 2.009(16), Ir–C5, 2.009(16), C1–C2, 1.42(2), C2–C3, 1.36(2), C3–C4, 1.44(2), C4–C5, 1.31(2). Selected distances [Å] for **54a**: Ir–C1 2.046(11), Ir–C5 2.073(12), C1–C2 1.321(16), C2–C3 1.460(18), C3–C4 1.465(16), C4–C5 1.340(16). Selected distances [Å] for **56a**: Ir–C1 2.011(3), Ir–C5 2.014(3), C1–C2 1.389(5), C2–C3 1.382(5), C3–C4 1.425(5), C4–C5 1.363(4), C3–C7 1.503(5). (*See color plate section for the color representation of this figure.*)

(from 250.97 (C1) and 199.58 (C5) ppm in **51** to 132.88 (C1) and 123.35 (C5) ppm in **53a**). C3 also shifts from an aromatic position in **51** (154.32 ppm) to a saturated position in **53a** (38.85 ppm). Very similar changes in the chemical shifts of the metallacyclic ring atoms are observed in the NMR spectra of the iridabenzene **52** when it is converted into the iridacyclohexa-1,4-diene **54a** [64].

Crystal structures of the cationic iridabenzene **52** and the corresponding iridacyclohexa-1,4-diene derivative **53a** have been obtained (Figure 1.15). From the structure of **52** it can be seen that the Ir–C and C–C distances within the six-membered metallacyclic ring (Ir–C1 2.009(16), Ir–C5 2.009(16), C1–C2 1.42(2), C2–C3 1.36(2), C3–C4 1.44(2), C4–C5 1.31(2) Å) are consistent with a metallabenzene formulation. In comparison, the corresponding distances within the six-membered metallacyclic ring of **54a** are consistent with a metallacyclohexa-1,4-diene formulation. Thus, the longer iridium–carbon bonds are close to single bond distances (Ir–C1 2.046(11), Ir–C5 2.073(12) Å), the C1–C2 and C4–C5 bond lengths (1.321(16) and 1.340(16) Å, respectively) are consistent with double bonds, and the C2–C3 and C3–C4 distances (1.460(18) and 1.465(16) Å, respectively) are appropriate for single bonds [64].

Other nucleophiles also add to the ring C3 atoms of **51** and **52** to form related metallacyclohexadiene complexes. For example, treatment of **51** or **52** with methyllithium yields the colourless $Os(C_5H_3\{SMe\text{-}1\}\{H\text{-}3\}\{Me\text{-}3\})(CO)_2(PPh_3)_2$ (**53b**) or $Ir(C_5H_3\{SMe\text{-}1\}\{H\text{-}3\}\{Me\text{-}3\})(\kappa^2\text{-}S_2CNEt_2)(PPh_3)_2$ (**54b**), respectively (Scheme 1.19). The iridabenzene cation **52** will also react with sodium ethoxide in ethanol to yield $Ir(C_5H_3\{SMe\text{-}1\}\{H\text{-}3\}\{OEt\text{-}3\})(\kappa^2\text{-}S_2CNEt_2)(PPh_3)_2$ (**53c**). The signals for the metallacyclohexadiene ring atoms in the ^1H and ^{13}C NMR spectra of these complexes are very similar to those observed for **53a** and **54a**. However, in the ^{31}P NMR spectra, significant differences are observed. Unlike the singlet signals that are observed for the two equivalent phosphorus atoms in each of the complexes **53a** and **54a**, four-line second-order spectra are observed for the phosphorus atoms of **53b**, **54b** and **53c**. In these complexes the addition of the methyl or ethoxide groups to C3 removes the horizontal mirror plane that was present in the substrate metallabenzenes, and the two phosphorus atoms become inequivalent. The chemical shift difference between the two phosphorus atoms in each case is of the same order as the coupling constant, which gives rise to the second-order spectra [64].

These highly regioselective nucleophilic additions can also be rationalised by consideration of the condensed Fukui functions for the ring carbon atoms. The computed f_k^+ values for the osmabenzene **51** and iridabenzene **52** indicate that C3 is the most electrophilic atom, followed closely by C5 and C1. Nucleophilic attack should therefore be electronically favourable at each of these three sites. The observed exclusive preference for attack at C3 is most likely due to the C1 and C5 sites being sterically shielded by the metal and its ancillary ligands [64].

These metallacyclohexa-1,4-dienes can be considered intermediates akin to the σ^H-adducts formed during S_NAr reactions of benzenes. The possibility of oxidising these compounds to give the corresponding substituted metallabenzenes was therefore investigated. It was found that treatment with a range of oxidants including 2,3-dichloro-5,6-dicyano-1,4-benzoquinone (DDQ), copper(II) chloride or oxygen results in rearomatisation of the metallacyclohexa-1,4-diene rings and formation of the corresponding metallabenzenes. In the case of **53a** or **54a**, where hydride had been added to C3 of the cationic metallabenzenes **51** or **52**, respectively, treatment with each of these oxidants returned the starting metallabenzenes after addition of a salt containing the appropriate counteranion (Scheme 1.19). In these reactions the highest yields were obtained with DDQ. Oxidation of the methyl-substituted osmacyclohexadiene **53b** gave the methyl-substituted osmabenzene $[Os(C_5H_3\{SMe\text{-}1\}\{Me\text{-}3\})(CO)_2(PPh_3)_2][Cl]$ (**55**) as purple crystals. In this case the best yields were obtained if $CuCl_2$ was used as the oxidant (Scheme 1.19). In a similar manner, oxidation of the iridacyclohexadienes **54b** or **54c** gave the methyl-substituted iridabenzene $[Ir(C_5H_3\{SMe\text{-}1\}\{Me\text{-}3\})(\kappa^2\text{-}S_2CNEt_2)(PPh_3)_2][PF_6]$ (**56a**) or the ethoxy-substituted iridabenzene $[Ir(C_5H_3\{SMe\text{-}1\}\{OEt\text{-}3\})(\kappa^2\text{-}S_2CNEt_2)(PPh_3)_2][PF_6]$ (**56b**), respectively. DDQ was found to be the oxidant that gave the highest yields of the products in these cases [64].

The spectroscopic and structural data obtained for the substituted metallabenzene products closely resemble those found for the starting metallabenzenes **51** and **52**. For osmabenzene **55** the ring methyl substituent is observed as a singlet at 1.74 ppm in the ^1H NMR spectrum. The remaining osmabenzene ring signals observed in the ^1H NMR (5.89 (H2), 6.97 (H4), 10.86 (H5) ppm) and ^{13}C NMR (245.91 (C1), 126.15 (C2), 171.28 (C3), 139.89 (C4), 197.04 (C5) ppm) spectra are almost identical to those of the precursor osmabenzene **51**. Re-establishment of the osmabenzene ring also restores the equatorial mirror plane, and once more only one singlet is observed in the ^{31}P NMR spectrum for the two equivalent phosphorus atoms [64].

Re-establishment of the iridabenzene ring is also evident in the NMR spectra of the methyl-substituted iridabenzene **56a** as well as in the ethoxy-substituted iridabenzene **56b**. In each case the ^1H and ^{13}C NMR spectra of the ring atoms resemble those of the corresponding parent iridabenzene **52**. The bond lengths about the metallacyclic rings in the crystal structures of **56a** (Figure 1.15) (Ir–C1 2.011(3), Ir–C5 2.014(3), C1–C2 1.389(5), C2–C3 1.382(5), C3–C4 1.425(5), C4–C5 1.363(4) Å) and **56b** (Ir–C1 2.022(16), Ir–C5 1.996(13), C1–C2 1.35(2), C2–C3 1.40(2), C3–C4 1.447(19), C4–C5, 1.349(19) Å) also support re-establishment of the delocalised π-systems. In each case the Ir–C and C–C bond distances are appropriate for iridabenzene formulations [64]. Related reactions of metallabenzenes involving intramolecular nucleophilic substitution [51, 94] and *cine*-substitution [95] have also been reported, and these are discussed in Chapter 6.

The sequence of reactions that take the cationic metallabenzenes **51** and **52** to the corresponding metallacyclohexadienes and then to the corresponding substituted cationic metallabenzenes mimics the steps involved in the classic nucleophilic aromatic substitution of hydrogen (S_NAr^H) in benzene chemistry. These reactions therefore provide a further example of where the chemistry of metallabenzenes and benzene overlap.

1.5 Concluding Remarks

Our involvement in both the inception and the subsequent development of the field of metallabenzene chemistry has been a very rewarding experience. From the outset we have been privileged to have worked with some outstanding and highly talented students and it is largely through their hard work and dedication that we have been able to make our particular contributions to this area. In our studies we have been able to develop efficient synthetic routes to metallabenzenes that bear only one ring substituent, synthesise and study a number of interesting fused-ring and tethered metallabenzenes, discover examples of metallabenzene reactions involving electrophilic aromatic substitution and nucleophilic aromatic substitution of hydrogen, as well as study well-defined rearrangements of metallabenzenes to cyclopentadienyl compounds. Many exciting challenges remain in the field of metallabenzene chemistry and we are confident that the field will continue to flourish and capture the imagination of chemists for many years to come.

References

1 Faraday, M. (1825) On new compounds of carbon and hydrogen, and on certain other products obtained during the decomposition of oil by heat. *Philosophical Transactions of the Royal Society of London*, **115**, 440–466.
2 Mitscherlich, E. (1834) Ueber das Benzol und die Säuren der Oel- und Talgarten. *Annalen der Pharmacie*, **9**(1), 39–48.
3 Kekulé, A. (1865) Sur la constitution des substances aromatiques. *Bulletin de la Societe Chimique de France*, **3**, 98–110.
4 Kekulé, F. A. (1865) Note sur quelques produits de substitution de la benzine. *Bulletin De l'Academie Royale De Medecine De Belgique*, **19**, 551–563.
5 Anderson, T. (1851) Ueber die Producte der trocknen Destillation thierischer Materien. *Justus Liebigs Annalen der Chemie*, **80**(1), 44–65.
6 Märkl, G., Lieb, F., Merz , A. (1967) A new synthesis of phosphabenzene derivatives. *Angewandte Chemie: International Edition*, **6**(5), 458–459.
7 Ashe, A. J. (1971) Phosphabenzene and arsabenzene. *Journal of the American Chemical Society*, **93**(13), 3293–3295.
8 Ashe, A. J. and Chan, W.-T. (1979) Preparation of 2-substituted arsabenzenes. *Journal of Organic Chemistry*, **44**(9), 1409–1413.
9 Wakita , K., Tokitoh, N., Okazaki, R. and Nagase, S. (2000) Synthesis and properties of an overcrowded silabenzene stable at ambient temperature. *Angewandte Chemie: International Edition*, **39**(3), 634–636.
10 Ashe, A. J., Diephouse, T. R. and El-Sheikh, M. Y. (1982) Stabilization of stibabenzene and bismabenzene by 4-alkyl substituents. *Journal of the American Chemical Society*, **104**(21), 5693–5699.

11 Nakata, N., Takeda, N. and Tokitoh, N. (2002) Synthesis and properties of the first stable germabenzene. *Journal of the American Chemical Society*, **124**(24), 6914–6920.

12 Mizuhata, Y., Noda, N. and Tokitoh, N. (2010) Generation of stannabenzenes and their properties. *Organometallics*, **29**(21), 4781–4784.

13 Thorn, D. L. and Hoffmann, R. (1979) Delocalization in metallocycles. *Nouveau Journal De Chimie*, **3**(1), 39–45.

14 Elliott, G. P., Roper, W. R. and Waters, J. M. (1982) Metallacyclohexatrienes or "metallabenzenes": Synthesis of osmabenzene derivatives and X-ray crystal structure of [$Os(CSCHCHCHCH)(CO)(PPh_3)_2$]. *Journal of the Chemical Society: Chemical Communications*, **14**, 811–813.

15 Fraser, P. J., Roper, W. R. and Stone, F. G. A. (1974) Carbene complexes of iridium, rhodium, manganese, chromium, and iron containing thiazolidinylidene and pyridinylidene ligands. *Journal of the Chemical Society: Dalton Transactions*, **7**, 760–764.

16 Clark, G. R., Roper, W. R. and Wright, A. H. (1982) Synthesis, structure, and reactions of $IrCl_3(CCl_2)(PPh_3)_2$: A compound with an iridium–carbon double bond ($L_nIr{=}CCl_2$). *Journal of Organometallic Chemistry*, **236**(1), C7–C10.

17 Clark, G. R., Marsden, K., Roper, W. R. and Wright, L. J. (1980) An osmium-carbyne complex. *Journal of the American Chemical Society*, **102**(21), 6570–6571.

18 Clark, G. R., Cochrane, C. M., Roper, W. R. and Wright, L. J. (1980) The interaction of an osmium–carbon triple bond with copper(I), silver(I) and gold(I) to give mixed dimetallocyclopropene species and the structures of $Os(AgCl)(CR)Cl(CO)(PPh_3)_2$. *Journal of Organometallic Chemistry*, **199**(2), C35–C38.

19 Collins, T. J. and Roper, W. R. (1977) Hydrido-thiocarbonyl complexes as precursors of low-valent thiocarbonyl complexes. *Journal of Organometallic Chemistry*, **139**(2), C56–C58.

20 Collins, T. J., Roper, W. R. and Town, K. G. (1976) Synthesis of low-valent thiocarbonyl complexes via 1, 2-elimination of methylthiol from cis-metal-hydrido-dithiomethylester complexes: $Os(CS)(CO)_2(PPh_3)_2$ and $IrCl(CS)(PPh_3)_2$. *Journal of Organometallic Chemistry*, **121**(2), C41–C44.

21 Clark, G. R., Marsden, K., Roper, W. R. and Wright, L. J. (1980) Carbonyl, thiocarbonyl, selenocarbonyl, and tellurocarbonyl complexes derived from a dichlorocarbene complex of osmium. *Journal of the American Chemical Society*, **102**(3), 1206–1207.

22 Frogley, B. J. and Wright, L. J. (2014) Fused-ring metallabenzenes. *Coordination Chemistry Reviews*, **270–271**, 151–166.

23 Elliott, G. P., McAuley, N. M., Roper, W. R. and Shapley, P. A. (1989) An osmium containing benzene analog, $Os(CSCHCHCHCH)(CO)(PPh_3)_2$, carbonyl(5-thioxo-1,3-pentadiene-1,5-diyl-C^1,C^5,S)bis(triphenylphosphine)osmium, and its precursors. In: *Inorganic Syntheses*, New York: John Wiley & Sons, Inc., 184–189.

24 Iron, M. A., Lucassen, A. C. B., Cohen, H., *et al.* (2004) A computational foray into the formation and reactivity of metallabenzenes. *Journal of the American Chemical Society*, **126**(37), 11699–11710.

25 Rickard, C. E. F., Roper, W. R., Woodgate, S. D. and Wright, L. J. (2001) Reaction between the thiocarbonyl complex, $Os(CS)(CO)(PPh_3)_3$, and propyne: Crystal structure of a new sulfur-substituted osmabenzene. *Journal of Organometallic Chemistry*, **623**(1–2), 109–115.

26 Orpen, A. G., Brammer, L., Allen, F. H., *et al.* (1989) Supplement: Tables of bond lengths determined by X-ray and neutron diffraction: Part 2: Organometallic

compounds and co-ordination complexes of the d-and f-block metals. *Journal of the Chemical Society: Dalton Transactions*, **12**, S1–S83.
27 Wright, L. J. (2006) Metallabenzenes and metallabenzenoids. *Dalton Transactions*, **15**, 1821–1827.
28 Chin, C. S., Park, Y., Kim, J. and Lee, B. (1995) Facile insertion of alkynes into Ir–P (phosphine) and Ir–As (arsine) bonds: Second and third alkyne addition to mononuclear iridium complexes. *Journal of the Chemical Society: Chemical Communications*, **15**, 1495–1496.
29 Vaska, L. and DiLuzio, J. W. (1961) Carbonyl and hydrido-carbonyl complexes of iridium by reactions with alchols: Hydrido complexes by reaction with acid. *Journal of the American Chemical Society*, **83**(12), 2784–2785.
30 Reed, C. A. and Roper, W. R. (1973) The iridium(I) cation, $[Ir(CO)(CH_3CN)(PPh_3)_2]^+$ and its substitution reactions. *Journal of the Chemical Society: Dalton Transactions*, **13**, 1365–1370.
31 Lu, G.-L., Roper, W. R., Wright, L. J. and Clark, G. R. (2005) A 2-iridathiophene from reaction between $IrCl(CS)(PPh_3)_2$ and $Hg(CHCHPh)_2$. *Journal of Organometallic Chemistry*, **690**(4), 972–981.
32 Dalebrook, A. F. and Wright, L. J. (2009) Annulation of an iridabenzene through formal cycloaddition reactions with organonitriles. *Organometallics*, **28**(18), 5536–5540.
33 Clark, G. R., Johns, P. M., Roper, W. R. and Wright, L. J. (2008) A stable iridabenzene formed from an iridacyclopentadiene where the additional ring-carbon atom is derived from a thiocarbonyl ligand. *Organometallics*, **27**(3), 451–454.
34 Dalebrook, A. F. and Wright, L. J. (2012) Chapter Three: Metallabenzenes and metallabenzenoids. In: *Advances in Organometallic Chemistry*, F. H. Anthony, J. F. Mark (eds), London: Academic Press, Elsevier, **60**, 93–177.
35 Zhu, J., Jia, G. and Lin, Z. (2007) Understanding nonplanarity in metallabenzene complexes. *Organometallics*, **26**(8), 1986–1995.
36 Elliott, G. P. and Roper, W. R. (1983) Metallacyclobutadiene complexes from ligand combination of cs and acetylene on osmium. *Journal of Organometallic Chemistry*, **250**(1), C5–C8.
37 Burrell, A. K., Elliott, G. P., Rickard, C. E. F. and Roper, W. R. (1990) The X-ray structure of $Os(CS)(PhCCPh)(PPh_3)_2$: A complex containing a four-electron donor acetylene ligand. *Applied Organometallic Chemistry*, **4**(5), 535–542.
38 Clark, G. R., Johns, P. M., Roper, W. R. and Wright, L. J. (2006) An osmabenzofuran from reaction between $Os(PhC{\equiv}CPh)(CS)(PPh_3)_2$ and methyl propiolate and the C-protonation of this compound to form a tethered osmabenzene. *Organometallics*, **25**(7), 1771–1777.
39 Chin, C. S., Won, G., Chong, D., *et al.* (2002) Carbon–carbon bond formation involving reactions of alkynes with group 9 metals (Ir, Rh, Co): Preparation of conjugated olefins. *Accounts of Chemical Research*, **35**(4), 218–225.
40 Chin, C. S. and Lee, H. (2004) New iridacyclohexadienes and iridabenzenes by [2+2+1] cyclotrimerization of alkynes and facile interconversion between iridacyclohexadienes and iridabenzenes. *Chemistry: A European Journal*, **10**(18), 4518–4522.
41 Chin, C. S., Lee, H. and Eum, M.-S. (2005) Iridabenzenes from iridacyclopentadienes: Unusual C–C bond formation between unsaturated hydrocarbyl ligands. *Organometallics*, **24**(20), 4849–4852.

42 Clark, G. R., O'Neale, T. R., Roper, W. R., *et al.* (2009) Stable cationic and neutral ruthenabenzenes. *Organometallics*, **28**(2), 567–572.

43 Xia, H., He, G., Zhang, H., *et al.* (2004) Osmabenzenes from the reactions of HC≡CCH(OH)C≡CH with $OsX_2(PPh_3)_3$ (X = Cl, Br). *Journal of the American Chemical Society*, **126**(22), 6862–6863.

44 Gong, L., Chen, Z., Lin, Y., *et al.* (2009) Osmabenzenes from osmacycles containing an η^2-coordinated olefin. *Chemistry: A European Journal*, **15**(25), 6258–6266.

45 Zhang, H., Wu, L., Lin, R., *et al.* (2009) Synthesis, characterization and electrochemical properties of stable osmabenzenes containing PPh_3 substituents. *Chemistry: A European Journal*, **15**(14), 3546–3559.

46 Yamazaki, H. and Aoki, K. (1976) A novel six-membered ruthenium metallocycle. *Journal of Organometallic Chemistry*, **122**(3), C54–C58.

47 Bruce, M. I., Hall, B. C., Skelton, B. W., *et al.* (2000) Some Pentamethylcyclopentadienyl-ruthenium derivatives of methyl propiolate. *Australian Journal of Chemistry*, **53**(2), 99–107.

48 Yang, J., Jones, W. M., Dixon, J. K. and Allison, N. T. (1995) Detection of a ruthenabenzene, ruthenaphenoxide, and ruthenaphenanthrene oxide: The first metalla aromatics of a second-row transition metal. *Journal of the American Chemical Society*, **117**(38), 9776–9777.

49 Zhang, H., Xia, H., He, G., *et al.* (2006) Synthesis and characterization of stable ruthenabenzenes. *Angewandte Chemie: International Edition*, **45**(18), 2920–2923.

50 Zhang, H., Feng, L., Gong, L., *et al.* (2007) Synthesis and characterization of stable ruthenabenzenes starting from HC≡CCH(OH)C≡CH. *Organometallics*, **26**(10), 2705–2713.

51 Lin, R., Zhang, H., Li, S., *et al.* (2011) New highly stable metallabenzenes via nucleophilic aromatic substitution reaction. *Chemistry: A European Journal*, **17**(15), 4223–4231.

52 Clark, G. R., Johns, P. M., Roper, W. R., *et al.* (2011) Regioselective mono-, di-, and trifunctionalization of iridabenzofurans through electrophilic substitution reactions. *Organometallics*, **30**(1), 129–138.

53 Bleeke, J. R., Xie, Y. F., Peng, W. J. and Chiang, M. (1989) Metallabenzene: Synthesis, structure, and spectroscopy of a 1-irida-3,5-dimethylbenzene complex. *Journal of the American Chemical Society*, **111**(11), 4118–4120.

54 Gilbertson, R. D., Weakley, T. J. R. and Haley, M. M. (1999) Direct synthesis of an iridabenzene from a nucleophilic 3-vinyl-1-cyclopropene. *Journal of the American Chemical Society*, **121**(11), 2597–2598.

55 Gilbertson, R. D., Lau, T. L. S., Lanza, S., *et al.* (2003) Synthesis, spectroscopy, and structure of a family of iridabenzenes generated by the reaction of Vaska-type complexes with a nucleophilic 3-vinyl-1-cyclopropene. *Organometallics*, **22**(16), 3279–3289.

56 Paneque, M., Posadas, C. M., Poveda, M. L., *et al.* (2003) Formation of unusual iridabenzene and metallanaphthalene containing electron-withdrawing substituents. *Journal of the American Chemical Society*, **125**(33), 9898–9899.

57 Álvarez, E., Paneque, M., Poveda, M. L. and Rendón, N. (2006) Formation of iridabenzenes by coupling of iridacyclopentadienes and alkenes. *Angewandte Chemie: International Edition*, **45**(3), 474–477.

58 Chin, C. S., Lee, H. and Oh, M. (1997) Reactions of iridium(III) compounds with alkynes in the presence of triethylamine: The first example of $M{-}CH{=}CH{-}^{+}NR_3$. *Organometallics*, **16**(4), 816–818.

59 Bleeke, J. R. (2001) Metallabenzenes. *Chemical Reviews*, **101**(5), 1205–1228.
60 Landorf, C. W. and Haley, M. M. (2006) Recent advances in metallabenzene chemistry. *Angewandte Chemie: International Edition*, **45**(24), 3914–3936.
61 Mandado, M., Otero, N. and Mosquera, R. A. (2006) Local aromaticity study of heterocycles using *n*-center delocalization indices: The role of aromaticity on the relative stability of position isomers. *Tetrahedron*, **62**(52), 12204–12210.
62 Clark, G. R., Lu, G.-L., Roper, W. R. and Wright, L. J. (2007) Stepwise reactions of acetylenes with iridium thiocarbonyl complexes to produce isolable iridacyclobutadienes and conversion of these to either cyclopentadienyliridium or tethered iridabenzene complexes. *Organometallics*, **26**(9), 2167–2177.
63 Rickard, C. E. F., Roper, W. R., Woodgate, S. D. and Wright, L. J. (2000) Electrophilic aromatic substitution reactions of a metallabenzene: Nitration and halogenation of the osmabenzene [Os{C(SMe)CHCHCHCH}I(CO)(PPh_3)$_2$]. *Angewandte Chemie: International Edition*, **39**(4), 750–752.
64 Clark, G. R., Ferguson, L. A., McIntosh, A. E., *et al.* (2010) Functionalization of metallabenzenes through nucleophilic aromatic substitution of hydrogen. *Journal of the American Chemical Society*, **132**(38), 13443–13452.
65 Menke, J. B. (1925) Brit. Pat. 235698, June 25.
66 Menke, J. B. (1925) Nitrieren mit Nitraten. *Recueil des Travaux Chimiques des Pays-Bas*, **44**(2), 141–149.
67 Menke, J. B. (1925) Nitrierung mit Nitraten: II. *Recueil des Travaux Chimiques des Pays-Bas*, **44**(3), 269–270.
68 Frogley, B. J., Dalebrook, A. F. and Wright, L. J. (2016) Regioselective nitration and/or halogenation of iridabenzofurans through electrophilic substitution. *Organometallics*, **35**(3), 400–409.
69 Einhorn, J., Demerseman, P. and Royer, R. (1983) Application en série benzofurannique d'un nouveau procédé de nitration par l'acide nitrique en présence de chlorure stannique. *Canadian Journal of Chemistry*, **61**(10), 2287–2290.
70 Dell, C. P. (2001) Science of synthesis. *Thieme Chemistry*, **10**, 11.
71 Rasala, D. and Gawinecki, R. (1992) ^{13}C NMR study of the substituent effects in ortho-substituted nitrobenzenes. *Magnetic Resonance in Chemistry*, **30**(8), 740–745.
72 Crivello, J. V. (1981) Nitrations and oxidations with inorganic nitrate salts in trifluoroacetic anhydride. *Journal of Organic Chemistry*, **46**(15), 3056–3060.
73 Gigante B., Prazeres A. O., Marcelo-Curto M. J., *et al.* (1995) Mild and selective nitration by claycop. *Journal of Organic Chemistry*, **60**(11), 3445–3447.
74 Cornélis, A., Delaude, L., Gerstmans, A. and Laszlo, P. (1988) A procedure for quantitative regioselective nitration of aromatic hydrocarbons in the laboratory. *Tetrahedron Letters*, **29**(44), 5657–5660.
75 Kurz, M. E., Zahora, E. P. and Layman, D. (1973) Benzoyl nitrate reduction with halide ions. *Journal of Organic Chemistry*, **38**(13), 2277–2281.
76 Iron, M. A., Martin, J. M. L. and van der Boom, M. E. (2003) Metallabenzene versus Cp complex formation: A DFT investigation. *Journal of the American Chemical Society*, **125**(43), 13020–13021.
77 Shi, C., Guo, T., Poon, K. C., *et al.* (2011) Theoretical study on the rearrangement of metallabenzenes to cyclopentadienyl complexes. *Dalton Transactions*, **40**(42), 11315–11320.

78 Schrock, R. R., Pedersen, S. F., Churchill, M. R. and Ziller, J. W. (1984) Formation of cyclopentadienyl complexes from tungstenacyclobutadiene complexes and the X-ray crystal structure of an η^3-cyclopropenyl complex, $W[C(CMe_3)C(Me)C(Me)](Me_2NCH_2CH_2NMe_2)Cl_3$. *Organometallics*, **3**(10), 1574–1583.

79 Jacob, V., Weakley, T. J. R. and Haley, M. M. (2002) Metallabenzenes and valence isomers: Synthesis and characterization of a platinabenzene. *Angewandte Chemie: International Edition*, **41**(18), 3470–3473.

80 Jacob, V., Landorf, C. W., Zakharov, L. N., *et al.* (2009) Platinabenzenes: Synthesis, properties, and reactivity studies of a rare class of metalla-aromatics. *Organometallics*, **28**(17), 5183–5190.

81 Jacob, V., Weakley, T. J. R. and Haley, M. M. (2002) Rearrangement of a σ-2-(Cycloprop-2-enyl)vinyl- to an η^3-Cyclopentadienylplatinum(II) Complex: Selective Protonolysis of the Platinum–Methyl Bond. *Organometallics*, **21**(24), 5394–5400.

82 Ferede, R. and Allison, N. T. (1983) Possible formation of ferrabenzene and its novel conversion to 1,3-diphenyl-2-methoxyferrocene. *Organometallics*, **2**(3), 463–465.

83 Ferede, R., Hinton, J. F., Korfmacher, W. A., *et al.* (1985) Possible formation of rhenabenzene: Lithium-halogen exchange reactions in η^1-4-bromo-1,4-diphenyl-1,3-butadienyl ligands to form rhenabenzene and ferrabenzene. *Organometallics*, **4**(3), 614–616.

84 Wu, H.-P., Lanza, S., Weakley, T. J. R. and Haley, M. M. (2002) Metallabenzenes and valence isomers: 3: Unexpected rearrangement of two regioisomeric iridabenzenes to an (η^5-Cyclopentadienyl)iridium(I) complex. *Organometallics*, **21**(14), 2824–2826.

85 Wu, H.-P., Weakley, T. J. R. and Haley, M. M. (2005) Regioselective formation of β-alkyl-α-phenyliridabenzenes via unsymmetrical 3-vinylcyclopropenes: Probing steric and electronic influences by varying the alkyl ring substituent. *Chemistry: A European Journal*, **11**(4), 1191–1200.

86 Wu, H.-P., Ess, D. H., Lanza, S., *et al.* (2007) Rearrangement of iridabenzvalenes to iridabenzenes and/or η^5-cyclopentadienyliridium(I) complexes: Experimental and computational analysis of the influence of silyl ring substituents and phosphine ligands. *Organometallics*, **26**(16), 3957–3968.

87 Johns, P. M., Roper, W. R., Woodgate, S. D. and Wright, L. J. (2009) Carbonylchlorido(1-methylsulfanylpenta-1,3-dien-1-yl-5-ylidene)bis(triphenylphosphane)osmium(II). *Acta Crystallographica Section E*, **E65**, m1319.

88 Johns, P. M., Roper, W. R., Woodgate, S. D. and Wright, L. J. (2010) Thermal rearrangement of osmabenzenes to osmium cyclopentadienyl complexes. *Organometallics*, **29**(21), 5358–5365.

89 Bleeke, J. R. (2007) Aromatic iridacycles. *Accounts of Chemical Research*, **40**(10), 1035–1047.

90 Cao, X.-Y., Zhao, Q., Lin, Z. and Xia, H. (2013) The chemistry of aromatic osmacycles. *Accounts of Chemical Research*, **47**(2), 341–354.

91 Chen, J. and Jia, G. (2013) Recent development in the chemistry of transition metal-containing metallabenzenes and metallabenzynes. *Coordination Chemistry Reviews*, **257**(17–18), 2491–2521.

92 Paneque, M., Poveda, M. L. and Rendón, N. (2011) Synthesis and reactivity of iridacycles containing the $TpMe_2Ir$ moiety. *European Journal of Inorganic Chemistry*, **1**, 19–33.

93 Paneque, M., Posadas, C. M., Poveda, M. L., *et al.* (2007) Metallacycloheptatrienes of iridium(III): Synthesis and reactivity. *Organometallics*, **26**(14), 3403–3415.

94 Wang, T., Li, S., Zhang, H., *et al.* (2009) Annulation of metallabenzenes: From osmabenzene to osmabenzothiazole to osmabenzoxazole. *Angewandte Chemie: International Edition*, **48**(35), 6453–6456.

95 Wang, T., Zhang, H., Han, F., *et al.* (2013) Cine-substitution reactions of metallabenzenes: An experimental and computational study. *Chemistry: A European Journal*, **19**(33), 10982–10991.

2

The First Iridabenzenes: Syntheses, Properties, and Reactions

John R. Bleeke

Department of Chemistry, Washington University, One Brookings Drive, St. Louis, Missouri, USA

2.1 Introduction

In 1865, Kekulé proposed the cyclic structure of benzene, and the concept of "aromaticity" was born [1, 2]. In 2017, as we celebrate the 152nd anniversary of Kekulé's remarkable insight, interest in aromaticity shows no signs of abating [3]. Aromatic chemistry continues to challenge both the synthesist and the theoretician, and provides one of the most fruitful interplays of theory and experiment in chemistry [4].

Although the term "aromaticity" remains somewhat nebulous, it is generally agreed that aromatic compounds are planar, cyclic, fully conjugated systems which possess delocalized π-electrons. This cyclically delocalized electronic structure confers special properties on aromatic compounds, including ring bond lengths which are intermediate between normal single and double bonds, diamagnetic ring currents (diatropicity) as evidenced by diamagnetic susceptibility exaltation and unusual ^1H NMR chemical shifts, and high thermodynamic stability [5–8]. Recently, a variety of computational studies have sought to address the question of aromaticity more quantitatively. Among the criteria that have been proposed in this context are absolute hardness [9, 10], nucleus-independent chemical shift (NICS) [11, 12], magnetic susceptibility anisotropy [13, 14], and aromatic stabilization energy (ASE) calculated using energy decomposition analysis [15, 16].

While benzene remains the archetypical aromatic compound, heterocyclic analogues of benzene in which a CH group is formally replaced by an isoelectronic heteroatom such as N, P, As, O^+ or S^+ have long been known to exhibit aromatic properties [17, 18]. In contrast, until fairly recently, little was known about metallacyclic benzenoid compounds, i.e. benzene analogues in which a CH group has been formally replaced by a transition metal and its associated ligands. In particular, it was unknown whether these so-called metallabenzenes would be stable and, if so, whether they would exhibit the hallmark features of aromaticity.

Metallabenzenes: An Expert View, First Edition. Edited by L. James Wright.

2.2 Basic Theory

Metallabenzenes differ from conventional aromatic compounds in that the π-bonding requires the involvement of metal *d*π-orbitals rather than the *p*π-orbitals used by main group elements. In a seminal 1979 paper, Thorn and Hoffmann were the first to address theoretically the issues of delocalized π-bonding and aromaticity in transition metal metallacycles [19]. On the basis of extended Hückel calculations, three hypothetical classes of metallabenzenes (Figure 2.1) were identified as good candidates for exhibiting the delocalized bonding characteristics of aromatic species. In each of these structures, the metal-ligand fragment is isolobal with CH [20, 21]. In their calculations, Thorn and Hoffmann treat the carbon portion of the metallacycle as a monoanionic ligand ($C_5H_5^-$). It serves as a $4e^-$ donor to the metal, and these electrons form the two M–C σ bonds. In addition, the carbon fragment provides four π-electrons which reside in relatively low-lying π-orbitals (1π and 2π; Figure 2.2) [22]. Two more π-electrons are contributed by the filled metal d_{xz} orbital (or hybrid of appropriate symmetry), which interacts with the empty 3π carbon fragment orbital (Figure 2.2). Hence, the metallabenzene ring possesses a total of six π-electrons and obeys the Hückel 4n + 2 rule.

On the basis of density functional theory (DFT) calculations, Schleyer has challenged this interpretation of the π-bonding in metallabenzenes [23]. He argues that *two* filled metal *d*-orbitals, d_{xz} and d_{yz} (or hybrids of appropriate symmetry), are strongly involved. Orbital d_{xz} interacts with empty 3π as described above, while orbital d_{yz} interacts with filled 2π to produce two new filled π-molecular orbitals (π-MOs). One of these new π-MOs is bonding between the metal and the α ring carbons, while the other is antibonding. The antibonding partner is somewhat stabilized by interaction with the empty carbon fragment orbital 4π* (Figure 2.2). Hence, Schleyer concludes that metallabenzenes possess eight π-electrons which are housed in 1π, d_{yz} + 2π, d_{yz} – 2π, and d_{xz} + 3π.[1] Interestingly, two of these π-orbitals (1π and d_{xz} + 3π) have Hückel character, while the remaining two have Möbius character (i.e. the interaction between the metal *d*-orbital, and the carbon fragment is a delta or face-to-face interaction). In situations like this involving both Hückel and Möbius interactions, the 4n + 2 rule no longer applies as a criterion for aromaticity [24]. Schleyer's interpretation is supported by a DFT study carried out by Zhu, Jia, and Lin [25]. They conclude that the filled d_{yz} – 2π orbital, which

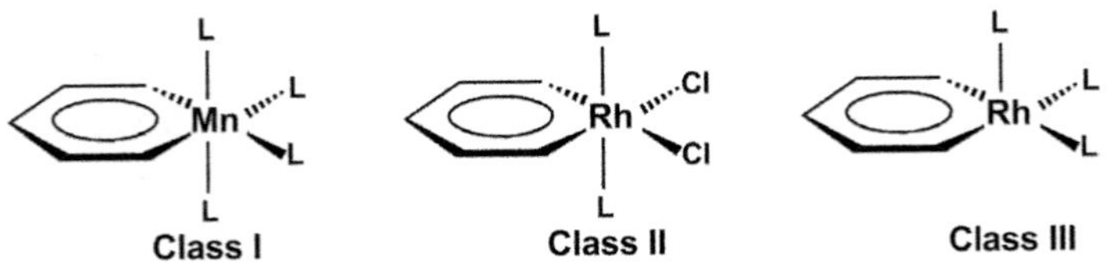

Figure 2.1 The three classes of stable metallabenzenes predicted by Thorn and Hoffmann [19]. L represents a neutral $2e^-$ ligand.

1 Note that Schleyer calculates the DFT MO's of the metallabenzenes directly rather than through fragment orbital interaction diagrams. However, it is straightforward to correlate the DFT MO's with those from a fragment orbital approach, and we have taken the liberty to do that here.

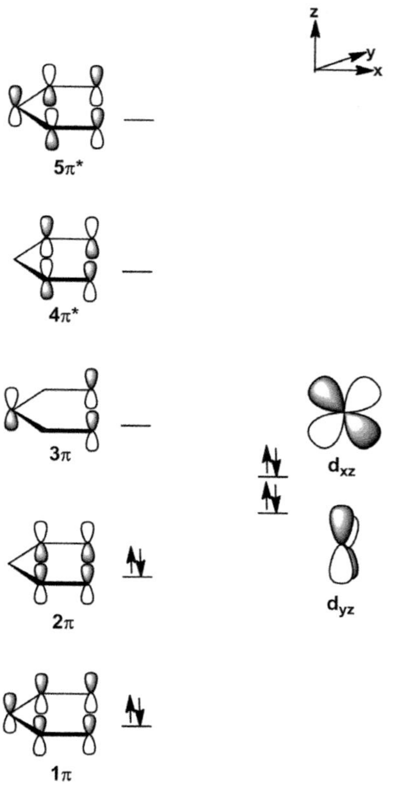

Figure 2.2 The five π-orbitals of the $C_5H_5^-$ carbon fragment (left) and the d_{xy} and d_{yz} metal orbitals (right) in a metallabenzene. Interactions between these orbitals give rise to a cyclically delocalized π electronic structure.

is antibonding between the metal and the α-ring carbons, is partially responsible for the non-planarity observed in some metallabenzenes (vide infra). By bending the metal out of the plane of the ring carbons, this antibonding interaction can be reduced.

Recently, Fernandez and Frenking investigated the electronic structure and bonding in $C_5H_5Rh(PH_3)_2(Cl)_2$ (Hoffmann's hypothetical Class II metallabenzene) using DFT [26]. They were surprised to find that there are actually *five* occupied π-orbitals which have contributions from both the $C_5H_5^-$ fragment and the metal center. In addition to the three bonding orbitals identified earlier (1π, d_{yz} + 2π, and d_{xz} + 3π), there are *two* orbitals that include an antibonding interaction between d_{yz} and 2π. In one of these, d_{yz} is bonding with respect to the chlorine *p*π-orbitals (d_{yz} + Cl(π) – 2π), while in the other, d_{yz} is antibonding with respect to the chlorine *p*π-orbitals (d_{yz} – Cl(π) – 2π). The antibonding nature of the d_{yz} – 2π interaction is mitigated somewhat by mixing with the vacant 4π* orbital of $C_5H_5^-$. Counting the electrons in all five of these occupied π-type orbitals, one can make the case that this is actually a 10π-electron system, which satisfies the 4n + 2 rule for aromatic compounds. However, it should be noted that Hoffmann's Class I and Class III compounds (Figure 2.1) do not include ancillary ligands with available π-electrons; these molecules will, in general, have a maximum of eight π-electrons.

2.3 Discovery of the First Stable Metallabenzenes

In 1982, the first example of a stable metallabenzene was reported by Roper [27]. Roper's "osmabenzene" (Figure 2.3) was prepared via a cyclization reaction involving a thiocarbonyl ligand and two ethyne molecules. The X-ray structure revealed a planar six-membered ring with no significant alternation in C–C bond lengths, supporting the idea of electron delocalization within the ring. The ^{1}H NMR spectrum, reported later [28], showed downfield chemical shifts for the ring protons at δ 13.95 (α H) and δ 7.28 (β, γ, and δ H's), consistent with an aromatic ring current.

Following Roper's initial communication, there was a hiatus in the field until the late 1980s, when we succeeded in synthesizing the first stable iridabenzene (see Figure 2.3), using a pentadienyl reagent as the source of ring carbon atoms and C–H bond activation as the key ring-forming step [29, 30]. The initially formed six-membered ring was an iridacyclohexadiene, which was "dehydrogenated" to the iridabenzene in two steps. This straightforward high-yield synthesis allowed us the unique opportunity to study in detail the physical and chemical properties of a stable metallabenzene. In this chapter, we summarize the results of that fascinating scientific journey.

2.4 Synthesis of Iridabenzene

In the fall of 1986, we discovered serendipitously that (pentadienyl)metal chemistry could provide a convenient synthetic route to unsaturated six-membered metallacycles [31, 32]. As shown in Scheme 2.1, we found that treatment of $(Cl)Ir(PEt_3)_3$ with potassium 2,4-dimethylpentadienide produces the *fac* isomer of the iridacyclohexa-1,3-diene complex **1a**, which slowly isomerizes to the thermodynamically more favorable *mer* isomer, **1b**, over the course of 24 h at 25°C. The probable mechanism of this metallacycle-forming reaction involves the intermediacy of 16e$^-$ (η^1-2,4-dimethylpentadienyl) $Ir(PEt_3)_3$ (**A**; Scheme 2.1), which undergoes intramolecular oxidative addition across a C–H bond on the terminus of the η^1-pentadienyl ligand. In a sense, this reaction is an organometallic actualization of Kekulé's famous dream in which "one of the snakes seizes hold of its own tail", and a ring is formed [33].

The X-ray crystal structure of **1b** (31) shows an essentially planar ring (none of the atoms deviates by more than 0.015 Å from the plane) and, as expected, there is a clear alternation in C–C bond lengths around the six-membered ring: C1–C2 = 1.33(1) Å, C2–C3 = 1.46(1) Å, C3–C4 = 1.33(1) Å, C4–C5 = 1.49(1) Å. The iridium–carbon bond distances are Ir–C1 = 2.085(6) Å and Ir–C5 = 2.189(6) Å, giving a ring circumference of

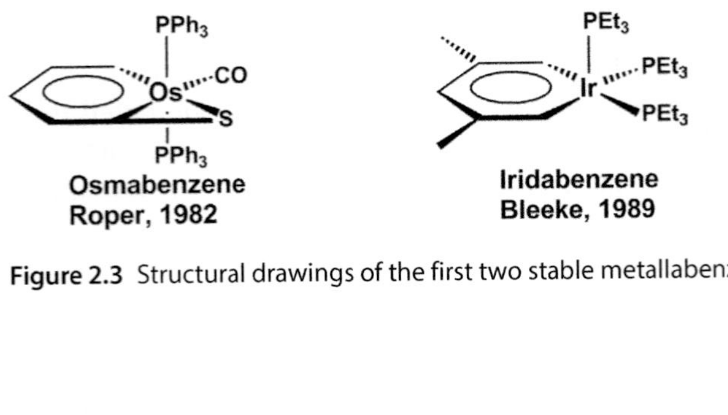

Figure 2.3 Structural drawings of the first two stable metallabenzenes.

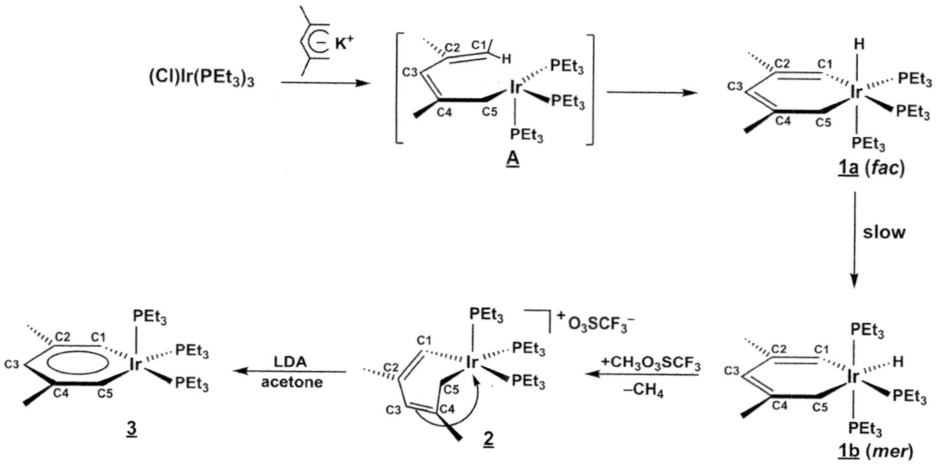

Scheme 2.1

9.88 Å. The angles around the ring are Ir–C1–C2 = 129.5(5)°, C1–C2–C3 = 125.2(7)°, C2–C3–C4 = 127.9(7)°, C3–C4–C5 = 126.7(6)°, C4–C5–Ir = 121.7(5)°, and C5–Ir–C1 = 89.0(3)°, summing to 720.0°, as expected for a planar hexagon. In the ^{1}H NMR, ring protons H1, H3, and the equivalent H5's resonate at δ 7.00, δ 5.93, and δ 2.68, respectively, while the metal-hydride appears at δ −13.10. In the ^{13}C{^{1}H} NMR, the ring carbons resonate at δ 138.5/129.0 (C2/C4), δ 128.0 (C3), δ 120.0 (C1), and δ −13.1 (C5). Carbon C1 is strongly coupled to the PEt_3 ligand that lies *trans* to it (J_{CP} = 76.5 Hz). The ^{31}P{^{1}H} NMR spectrum consists of a doublet (axial phosphines) and a triplet (equatorial phosphine), as expected for a *mer* coordination geometry.

Treatment of **1b** with methyl triflate ($CH_3O_3SCF_3$) results in abstraction of the metal-hydride ligand and coordination of the ring double bond C3=C4 to the iridium center, generating the cationic (1, 3, 4, 5-η)-pentadienediyl complex, **2** [31]. The crystal structure of **2** shows a rather weak, strained interaction between the ring double bond and the iridium atom. In solution, this species exhibits a dynamic process wherein the α-hydrogens (H1 and H5's) undergo chemical exchange. Hence, separate peaks due to H1 and the two H5's are well resolved in the ^{1}H NMR at −80°C, but when the temperature is raised, these peaks broaden and coalesce at about 10°C [34]. As shown in Scheme 2.2, this α-hydrogen exchange process is most likely metal-mediated, involving the 16e$^-$ "unhinged" iridacyclohexadiene **B** and the iridabenzene-hydride **C** [35, 36][2]. The "unhinging" (dissociation of C3=C4 from the iridium center) is presumably driven by steric crowding in **2**.

Returning to Scheme 2.1, treatment of compound **2** with acetone enolate (generated *in situ* from acetone and lithium diisopropylamide (LDA)) at −30°C results in α-proton abstraction and production of red **3**, "iridabenzene" [29, 34], the first example of a Hoffmann type III metallabenzene (Figure 2.1). Compound **3** is also produced upon

2 Hughes has also proposed an iridabenzene intermediate to explain phenyl group migration between the α-carbons of an iridacyclohexadiene ring.

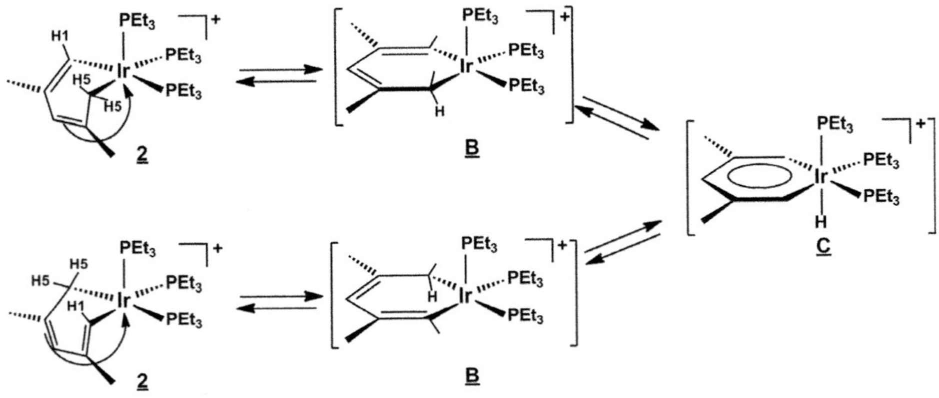

Scheme 2.2

treatment of **2** with sodium hydroxide in tetrahydrofuran (THF). While this deprotonation could occur directly from the α ring carbon (C5), it seems more likely that it occurs through one of the intermediates involved in the fluxional process (i.e. intermediates **B** or **C**; Scheme 2.2). Evidence for the involvement of these "unhinged" species in the deprotonation reaction comes from the analogous *tris*(PMe_3) system, discussed in Section 2.6.

2.5 Valence Bond Structures and Electron Counting for Iridabenzene

We use the inscribed circle notation for iridabenzene **3** as a shorthand for the two equivalent resonance structures drawn in Figure 2.4. In these valence bond structures, we generally count the carbene moiety of the $C_5H_3Me_2^-$ fragment as a neutral $2e^-$ σ-donor (or Fischer-type carbene) [37], while the vinyl moiety is an anionic $2e^-$ σ-donor. The metal is then an Ir(I)≡d^8 center, and the metal–carbon π interaction is a backbond from a filled metal *d*-orbital to an empty (acceptor) carbon *p*-orbital. Alternatively, the carbene moiety can be counted as a dianionic $4e^-$ donor (or Schrock-type carbene) [38], and the metal an Ir(III)≡d^6 center. In this view, the metal–carbon π interaction involves a filled carbon *p*-orbital and an empty (acceptor) metal *d*-orbital. These two bonding pictures are complementary, and each is supported by aspects of compound **3**'s reaction chemistry (vide infra).

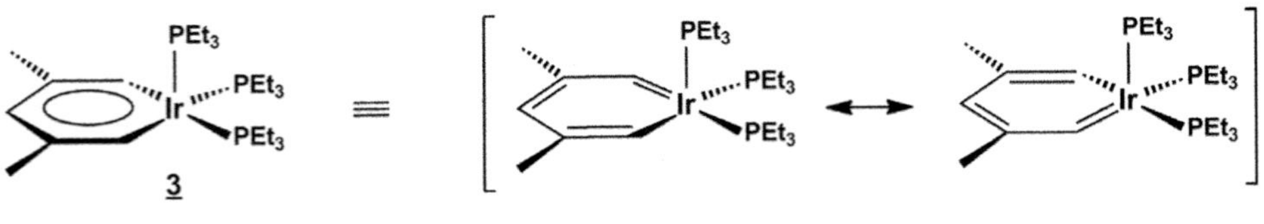

Figure 2.4 Equivalent resonance structures for iridabenzene, **3**.

2.6 The Tris(trimethylphosphine) Reaction System

Interestingly, the reactions described above in Section 2.4 are quite sensitive to phosphine ligand sterics; hence substitution of PMe_3 (cone angle = 118°) for PEt_3 (cone angle = 132°) [39] leads to significantly different (but instructive) chemistry. As shown in Scheme 2.3, treatment of $(Cl)Ir(PMe_3)_3$ with potassium 2,4-dimethylpentadienide produces an equilibrium mixture of ((1,2,5-η)-2,4-dimethylpentadienyl)$Ir(PMe_3)_3$ (**4**) and the *fac*-iridacyclohexa-1,3-diene complex, **5** [40]. Interconversion between **4** and **5** probably involves the intermediacy of $16e^-$ (η^1-2,4-dimethylpentadienyl)$Ir(PMe_3)_3$ (**D**; Scheme 2.3). The iridium center in this $16e^-$ species can either coordinate the terminal double bond of the pentadienyl (producing **4**) or oxidatively add across a C–H bond on the end of the η^1-pentadienyl (producing **5**). Unlike its PEt_3 analogue, **5** does not isomerize from *fac* to *mer*, apparently because of reduced steric strain in the *fac* isomer.

The position of the equilibrium shown in Scheme 2.3 can be pushed far to the right by refluxing the mixture in acetone; pure metallacycle **5** can be crystallized from the resulting isomerically enriched solution. Treatment of **5** with methyl triflate cleanly removes the hydride ligand, producing **6**, the *tris*(PMe_3) analogue of **2** (30). However, **6** does *not* undergo exchange of the α-hydrogens, presumably because this requires prior "unhinging" of the ring double bond from iridium (cf. Scheme 2.2). Steric crowding would promote this dissociation in the *tris*(PEt_3) complex but not in the *tris*(PMe_3)

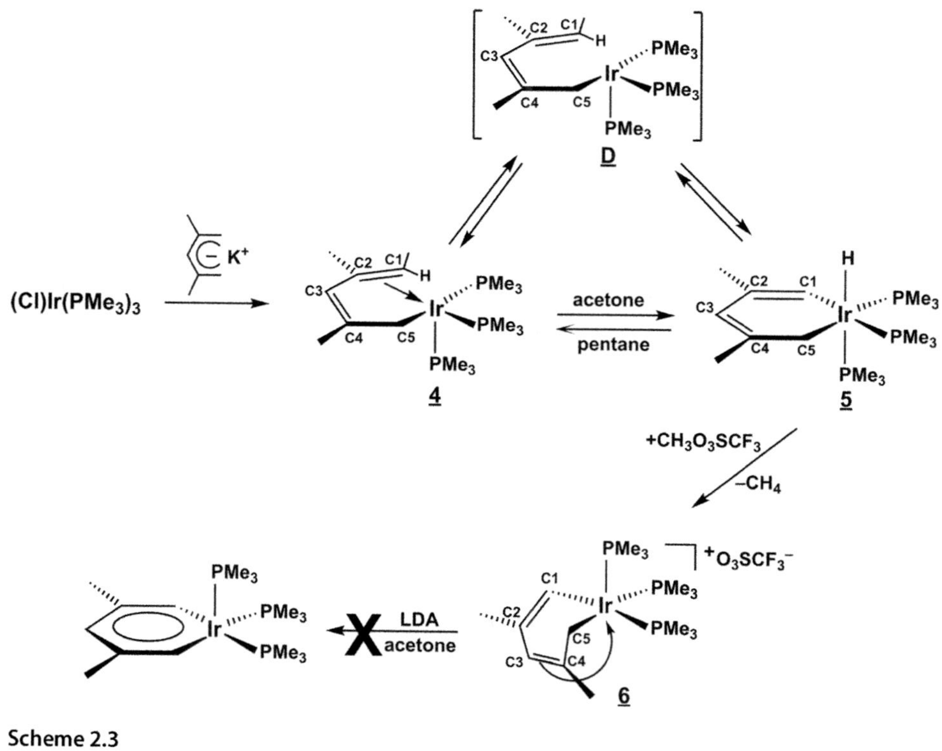

Scheme 2.3

system. In addition, **6** does *not* react with base to form a metallabenzene [41][3]. This observation lends credence to the idea that deprotonation proceeds through an "unhinged" 16e$^-$ iridacyclohexadiene intermediate, a transient iridabenzene-hydride, or even an agostic species on the pathway from one to the other.

2.7 Structure and Spectroscopy of Iridabenzene 3

Iridabenzene **3** crystallizes as a deep-red solid, and its X-ray structure is presented in Figure 2.5 [29]. The coordination geometry is square pyramidal, with C1, C5, P1, and P2 occupying the four basal sites and P3 residing in the axial site. The carbon portion of the metallacyclic ring (C1/C2/C3/C4/C5) is very nearly planar (mean deviation 0.014 Å), while the iridium center lies 0.24 Å out of this plane. The dihedral angle between planes C1/C2/C3/C4/C5 and Cl/Ir/C5 is 9.2°. The displacement of the iridium center from the ring plane probably results – at least in part – from steric interactions involving the basal phosphine ligands and the ring. However, an electronic effect may also contribute because iridabenzenes have a filled molecular orbital that is antibonding between Ir and the α-carbons (vide supra). A slight displacement of the iridium atom out of the ring plane would reduce that antibonding interaction [25]. Bonding within the ring is highly delocalized, and the bond lengths (C1–C2 = 1.37(1) Å, C2–C3 = 1.40(1) Å, C3–C4 = 1.37(1) Å, C4–C5 = 1.39(1) Å) are comparable to those found in benzene itself

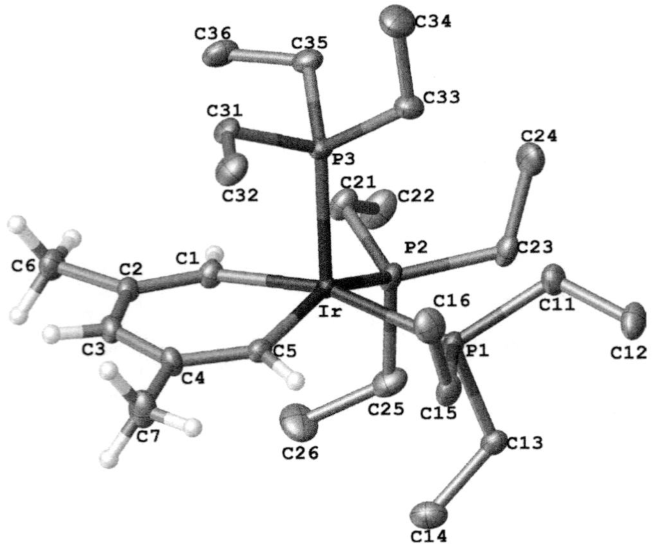

Figure 2.5 X-ray structure of iridabenzene, **3**, with phosphine H's omitted for clarity. The plot was created using the Olex2 software package [42].

3 Hughes has reported that a related rhodium compound, ((1,3,4,5,-η)-2,3-diphenylpentadienediyl) $Rh(PMe_3)_3^+$, is similarly unreactive toward deprotonation.

(1.398(9) Å) [43]. The iridium–carbon distances (Ir–Cl = 2.024(8) Å, Ir–C5 = 1.985(8) Å) are equivalent within experimental error and intermediate between a normal Ir–C_{vinyl} single bond (2.09 Å) (31), and a normal Ir–C double bond (1.87 Å) [44, 45]. The circumference of the iridabenzene ring is 9.54 Å, as compared to 9.88 Å in the non-aromatic precursor **1b**. The bond angles around the ring are Ir–C1–C2 = 131.2(6)°, C1–C2–C3 = 122.6(7)°, C2–C3–C4 = 125.3(7)°, C3–C4–C5 = 121.8(7)°, C4–C5–Ir = 132.9(6)°, and C1–Ir–C5 = 84.7(3)°, summing to 718.5°, close to the value of 720° required for a planar hexagon.

In the ^{1}H NMR spectrum of compound **3** in acetone-d_6, the signal for H1/H5 appears at δ 10.91, while H3 resonates at δ 7.18. By comparison, H1 and H3 of the iridacyclohexadiene precursor **1b** resonate at δ 7.00 and δ 5.93, respectively. The downfield shift of the H1/H5 protons in **3** reflects the carbene nature of the C1/C5 carbons (these carbons have 50% carbene character, based on simple resonance structures; Figure 2.4) and can be attributed primarily to the magnetic anisotropic influences of the large metal atom. However, such effects are strongly dependent on internuclear separation [46] and decrease rapidly for protons remote from the metal center. For example, in the ruthenium-vinylcarbene complex $(Cl)_2(PCy_3)_2Ru{=}CHCH{=}CH_2$, the chemical shifts of the protons along the vinylcarbene chain decrease from δ 19.06 for H_α to δ 8.11 for H_β to δ 6.25/6.01 for H_γ's [47]. Hence, we believe that the significant downfield chemical shift of H3 (δ 7.18) results mainly from the influence of an aromatic ring current rather than a metal-based anisotropy. The $^{13}C\{^{1}H\}$ NMR spectrum of **3** follows the trend seen in the ^{1}H NMR spectrum. Equivalent carbons C1 and C5 resonate quite far downfield at δ 167.6, while C2/C4 and C3 appear in the normal aromatic region at δ 132.0 and δ 129.9, respectively.

The $^{31}P\{^{1}H\}$ NMR signal for **3** is a sharp singlet, which shows no broadening even when the sample is cooled to −80°C. This indicates that **3** is stereochemically nonrigid and that the axial and basal phosphines are exchanging rapidly in solution, probably via a Berry-type process [48, 49]. In the Berry-type process (Scheme 2.4) the square-pyramidal complex isomerizes to a trigonal-bipyramidal intermediate in which one ring carbon atom and one basal phosphine ligand (P_B) assume *trans*-axial positions. This intermediate then reisomerizes back to a square-pyramidal complex, causing P_A/B_B exchange. As a result of the phosphine exchange process, the H1/H5 signal in the ^{1}H NMR spectrum of **3** appears as a binominal quartet with J_{HP} = 7.2 Hz. Likewise, the C1/C5 signal in the $^{13}C\{^{1}H\}$ NMR is a binomial quartet with J_{CP} = 28.1 Hz.

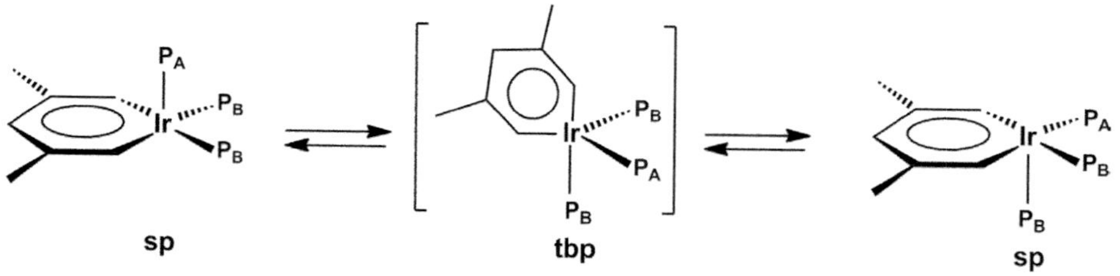

Scheme 2.4

2.8 Chemical Reactivity of Iridabenzene 3

Metallabenzenes have been implicated as intermediates in the formation of a variety of (cyclopentadienyl)metal complexes [50–56], and calculations by Martin and van der Boom [57, 58] suggest that most metallabenzenes are thermodynamically unstable with respect to this rearrangement. In fact, Haley [59–61] has shown that several isolated iridabenzenes will rearrange cleanly to (cyclopentadienyl)iridium complexes upon heating. The iridabenzenes most prone to undergo this reaction are those with very bulky groups (e.g. phenyl, *tert*-butyl and trimethylsilyl) on ring carbons C1 and C2. Interactions between these groups and with the metal-ligand fragment lead to carbene-vinyl migratory insertion and cyclopentadienyl ligand formation. In contrast, compound **3** (which has small substituents on C1 and C2) is stable indefinitely in the solid state or in solution at room temperature (under an N_2 atmosphere). Decomposition (to unidentifiable products) does occur slowly in refluxing benzene, but this process can be significantly retarded by the addition of excess PEt_3, suggesting that phosphine dissociation precedes decomposition. While **3** shows good thermal stability, it is also highly reactive toward a wide range of substrates. This reactivity is described in detail in the sections that follow.

2.8.1 Ligand Substitution

As shown in Scheme 2.5, iridabenzene **3** undergoes ligand replacement reactions with a variety of small $2e^-$ ligands L, including PMe_3, $P(OMe)_3$, and CO, to produce monosubstituted iridabenzenes **7a** (L = PMe_3), **7b** (L = $P(OMe)_3$), and **7c** (L = CO). At room temperature, only *one* PEt_3 ligand is replaced, even when excess L is employed. A kinetic study of the reaction of **3** with PMe_3 has shown clean first-order kinetics, consistent with a dissociative mechanism, where loss of PEt_3 from **3** is the rate-limiting step (34).

In solution, compounds **7a–c** exhibit *apparent* mirror-plane symmetry by NMR, even at −80°C. Hence, H1 and H5 appear equivalent, C1 and C5 appear equivalent, and so on. The $^{31}P\{^1H\}$ NMR spectra of **7a, b** consist of a doublet (PEt_3's) and triplet (PMe_3 or $P(OMe)_3$), while the $^{31}P\{^1H\}$ NMR spectrum of **7c** is a sharp singlet (PEt_3's). Several interpretations of these spectra are possible. First, the compounds could possess static symmetrical structures in which the unique ligand L resides in the unique axial coordination site of the square pyramid. Second, the compounds could be stereochemically nonrigid with the three ligands spending equal time in the three available coordination sites. Third, the compounds could be stereochemically nonrigid with the unique ligand L spending a disproportionate fraction of its time in the basal or axial coordination sites. As described below, a careful analysis of H–P coupling constants in these molecules strongly supports the last of these possibilities, with L residing preferentially in the basal plane.

By studying H–P coupling constants in (η^6-iridabenzene)$Mo(CO)_3$ complexes (in which intramolecular ligand exchange is completely arrested) [62], we have established that coupling between an α ring proton and the *cis* basal phosphine is large ($J_{HP} \approx 20–30$ Hz), while coupling of H_α to the *trans* basal phosphine or axial phosphine is small (J_{HP} < 5 Hz). Hence the magnitude of H_α–P coupling constants provides information about the phosphine coordination geometry. We find that in compound **7a** the H1/H5 ^1H NMR signal is a doublet (J_{HP} = 14.8 Hz) of triplets (J_{HP} = 4.1 Hz) centered at δ 10.85.

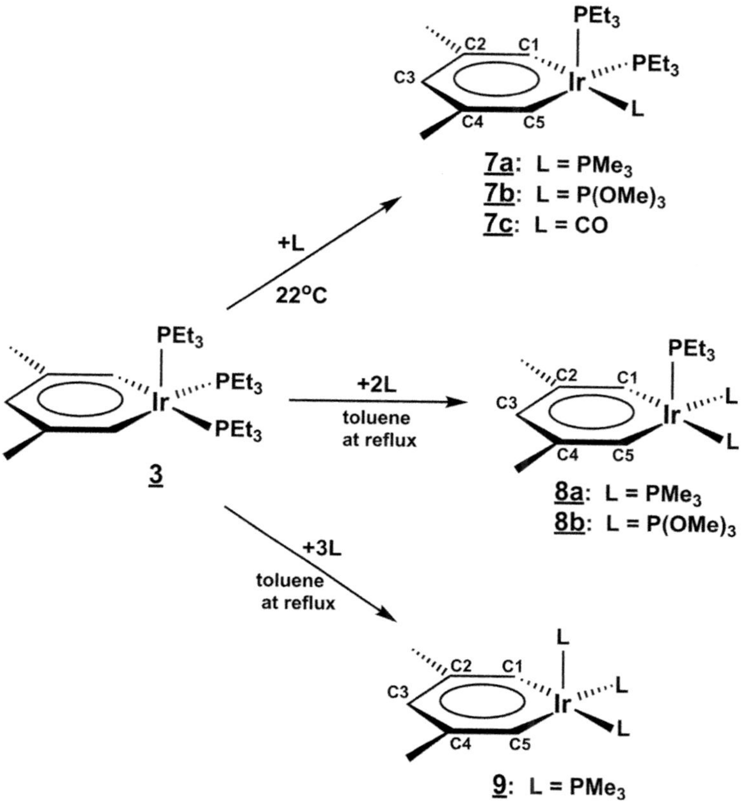

Scheme 2.5

Selective ^{31}P decoupling of the PMe_3 signal causes the doublet coupling to disappear, while selective ^{31}P decoupling of the PEt_3 signal removes the triplet coupling. Hence, PMe_3 is responsible for the 14.8 Hz coupling, while two rapidly exchanging PEt_3's are responsible for the 4.1 Hz coupling. From the magnitude of these numbers, it is clear the PMe_3 resides preferentially in the basal plane (and spends approximately half of its time *cis* to each α ring proton), while the PEt_3 ligands exchange between basal and axial sites (and spend approximately one-fourth of their time *cis* to each α ring proton). Similar analysis of H–P coupling constants in **7b**, **c** strongly suggests that they are isostructural with **7a**.

The ligand site preferences in these compounds are easily rationalized on steric grounds. In square-pyramidal coordination complexes, the axial site is less congested than the basal sites. For example, in the structure of **3**, the P_{axial}–Ir–P_{basal} angles are 105.0° and 103.0°, while the P_{basal}–Ir–P_{basal} angle is 93.0°. Hence, a bulky PEt_3 ligand resides preferentially in the roomier axial site.

While only one PEt_3 substitution is observed at room temperature, additional substitutions occur under more forcing conditions. Hence, when **3** is treated with two equivalents of PMe_3 or $P(OMe)_3$ in toluene at reflux, the *bis*(PMe_3) and *bis*($P(OMe)_3$) substitution products (**8a**, **b**, respectively) are formed (see Scheme 2.5). Like **7a–c**, these molecules are stereochemically nonrigid, and careful analysis of the NMR coupling constants confirms that the large PEt_3 ligand spends the majority of its time in the axial position. Treatment of **3** with excess PMe_3 in toluene at reflux generates the *tris*(PMe_3) substitution product (**9**; Scheme 2.5). In contrast, treatment of **3** with excess $P(OMe)_3$ under identical conditions yields only the *bis*($P(OMe)_3$) substitution product **8b**. Hence, there appears to be a strong electronic preference for retaining at least one good σ-donor ligand (i.e. one PR_3) in the iridabenzene coordination sphere.

The NMR spectra of compound **9** are virtually identical with those of **3**. The $^{31}P\{^1H\}$ NMR spectrum is a sharp singlet due to rapidly exchanging PMe_3 ligands. The H1/H5 signal is a binomial quartet (J_{HP} = 8.0 Hz) at δ 10.62, while H3 is a singlet at δ 7.10. Similarly, the C1/C5 signal is a binominal quartet (J_{CP} = 30.1 Hz) at δ 167.9, while C2/C4 and C3 resonate at δ 133.1 and δ 129.1, respectively. Compound **9** is extremely robust, showing little or no decomposition in refluxing toluene.

In the X-ray crystal structure of **9** [34], the square-pyramidal molecule resides on a crystallographically imposed mirror plane containing Ir, ring carbon C3, and the axial phosphorus atom. Bonding within the metallacyclic ring is fully delocalized; the Ir–C1, C1–C2, and C2–C3 distances are 2.008(7) Å, 1.390(10) Å, and 1.387(10) Å, respectively. Furthermore, the ring is even more planar than in compound **3**, probably as a result of reduced steric interactions with the basal phosphine ligands. The iridium center lies only 0.17 Å out of the plane of the ring carbon atoms (C1/C2/C3/C2a/C1a), while the dihedral angle between this plane and the C1/Ir/C1a plane is 6.7°. The internal angles around the ring sum to 719.2°.

When iridabenzene **3** is treated with excess carbon monoxide in toluene at reflux, a novel rearrangement occurs, generating the iridium-phenoxide compound **10** (see Scheme 2.6) (34). Although the detailed mechanism of this reaction has not been established, it probably involves CO insertion into an iridabenzene Ir–C_α bond to produce intermediate **E**, followed by ring closure (intermediate **F**), and finally metal migration to the oxygen center to produce **10**.

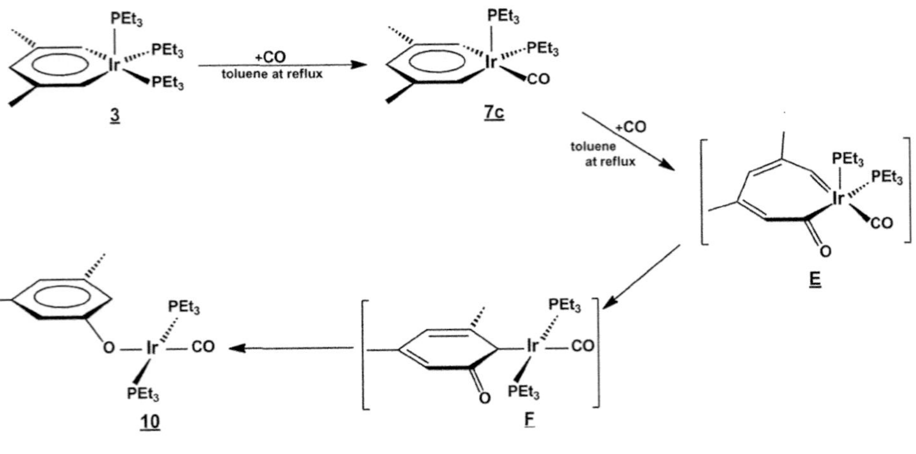

Scheme 2.6

2.8.2 Oxidative Addition/Oxidation

Iridabenzene **3** reacts slowly with excess H_2 gas at room temperature and 1 atm pressure to generate 2,4-dimethyl-1,3-pentadiene, 2,4-dimethyl-1-pentene, and iridium polyhydrides [34]. The principal hydride product is $(H)_5Ir(PEt_3)_2$, with smaller quantities of *mer-* and *fac-*$(H)_3Ir(PEt_3)_3$ being observed [63]. One reasonable mechanism for this reaction, outlined in Scheme 2.7, involves dissociation of PEt_3 (**G**), oxidative addition of H_2 to the iridium center (**H**), and hydride migration from iridium to an α-carbon of the ring (**I, J**).[4] Further hydrogen addition and migration steps would lead ultimately to the observed organic and inorganic products. The rate of hydrogenation reactions is drastically retarded by the addition of excess PEt_3 to the reaction mixture, suggesting that the phosphine dissociation step is rate-limiting. Furthermore, compounds **7a–c** do *not* react with H_2 at 25°C, presumably because $16e^-$ intermediates cannot be accessed in these cases.

Unlike benzene, iridabenzene does *not* undergo classical electrophilic aromatic substitution when treated with halogens. Instead, oxidative addition occurs at the iridium center.[5] For example, treatment of **3** with iodine (I_2) leads to rapid production of the olive-green oxidative addition product **11a** (see Scheme 2.8) [64]. Compound **11a** is the first representative of Hoffmann's predicted Class II metallabenzenes (Figure 2.1). If the carbene moieties in **3** and **11a** are both counted as a $2e^-$ neutral ligands (i.e. as Fischer carbenes), then the metal center has been oxidized from Ir(I)≡d^8 to Ir(III)≡d^6 in this reaction.

The solid-state structure of **11a** has been determined by X-ray crystallography and is presented in Figure 2.6 [34]. The complex adopts an octahedral coordination geometry in which the two PEt_3 ligands assume *trans*-diaxial positions, while the two iodide

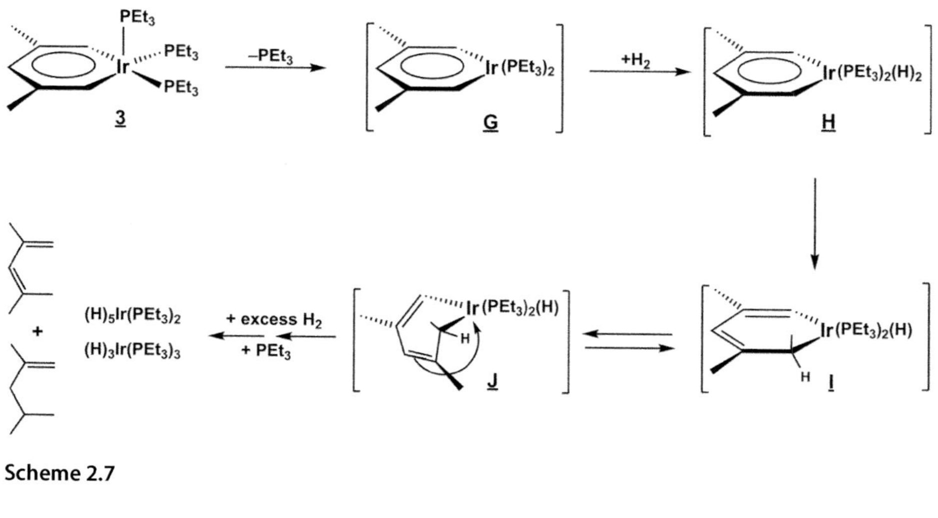

Scheme 2.7

4 Small quantities of **1b** are observed in the reaction solutions by NMR. This product is apparently generated by trapping of intermediate **I** (Scheme 2.7) with PEt_3. However, **1b** also reacts slowly with excess H_2 to produce iridium hydrides.

5 This occurs even when the reaction is run in the presence of a Friedel–Crafts catalyst. It should be noted, however, that several potential substitution sites (those *meta* to Ir) are blocked by methyl groups.

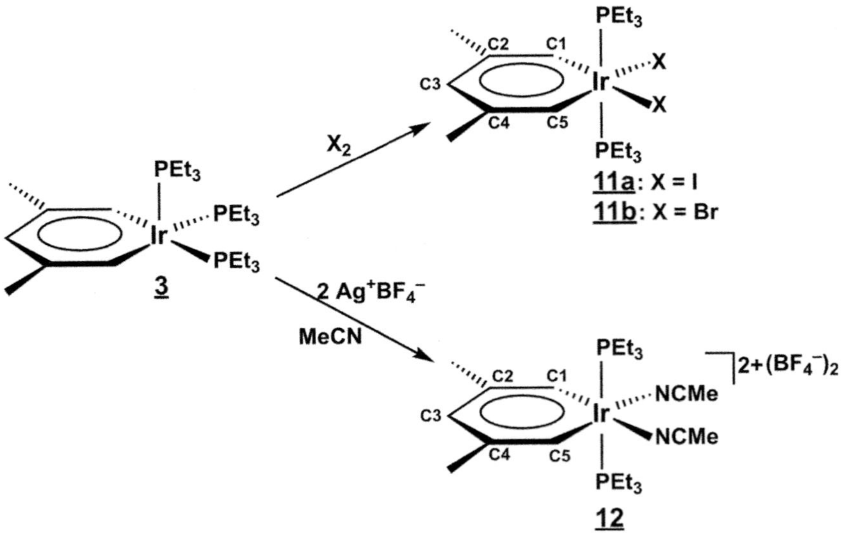

Scheme 2.8

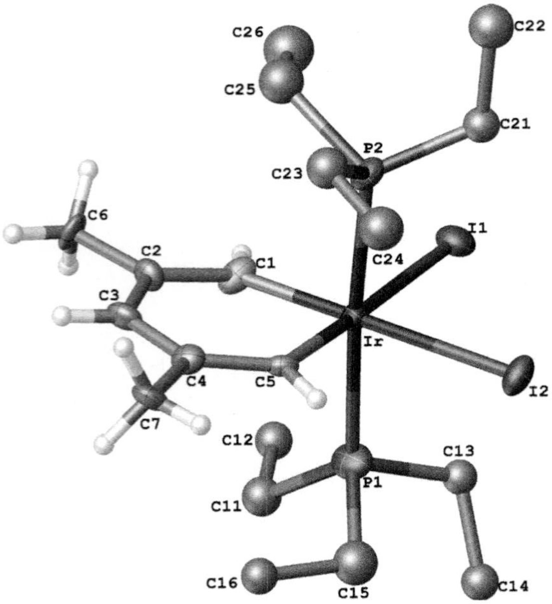

Figure 2.6 X-ray structure of the diiodide adduct of iridabenzene, **11a**, with phosphine H's omitted for clarity. The plot was created using the Olex2 software package [42].

ligands reside *trans* to C1 and C5 of the metallacycle. The aromatic ring is nearly planar, with the iridium atom lying just 0.13 Å out of the ring carbon plane (C1/C2/C3/C4/C5). The dihedral angle between the ring carbon plane and C1/Ir1/C5 is 5.3°. This enhanced planarity (vs. **3**) may result from steric pressure applied by the *trans*-diaxial phosphines.

Bonding within the ring is delocalized; the ring C–C bond distances average 1.39 Å, while the Ir–C bond distances average 1.95 Å.

The ^{1}H NMR signals for ring protons H1/H5 and H3 in **11a** are shifted even more dramatically downfield than those in **3**, appearing at δ 13.95 and δ 7.86, respectively, in acetone-d_6. In addition to the ring current and neighboring group anisotropic effects described earlier, these shifts probably reflect the inductive effect of an oxidized iridium center and two electronegative iodine atoms in the ring plane [65]. The $^{13}C\{^{1}H\}$ NMR spectrum of **11a** follows the same chemical shift pattern with C1/C5, C3 and C2/C4 appearing at δ 215.1, δ161.1 and δ 134.8, respectively. The H1/H5 and C1/C5 signals show essentially no coupling to the *trans*-diaxial phosphine ligands.

Treatment of iridabenzene **3** with bromine (Br_2) produces dark-blue oxidative addition product **11b** (see Scheme 2.8). The NMR spectra of **11b** are closely analogous to those of **11a**, strongly suggesting that the two compounds are isostructural. Iridabenzene **3** can also be oxidized with conventional oxidizing agents such as Ag^{+}. For example, treatment of **3** with two equivalents of $Ag^{+}BF_4^{-}$ in acetonitrile leads to clean production of the *bis*(acetonitrile) adduct, **12**. The ^{1}H NMR chemical shifts for H1/H5 and H3 in **12** are δ 13.02 and δ 8.16, respectively, while the ^{13}C chemical shifts for C1/C5, C3 and C2/C4 are δ 210.8, δ 172.3, and δ 138.3, respectively. The H1/H5 and C1/C5 signals show essentially no phosphorus coupling, indicating that the phosphines occupy *trans*-diaxial positions in the metal's octahedral coordination sphere.

2.8.3 Cycloaddition

Although cycloaddition is not a reaction type normally associated with aromatic compounds, iridabenzene **3** is ideally suited to participate in cycloaddition reactions because (a) it has a square-pyramidal coordination geometry with an open face that allows close approach by substrate molecules and (b) it possesses a reactive metalladiene moiety held rigidly in a cisoid geometry. Hence, **3** reacts with a wide variety of cycloaddition substrates, including olefins, heteroolefins, and heterocumulenes [66]. While some of these reactions almost certainly proceed via stepwise mechanisms, the majority are thought to occur in a concerted fashion (vide infra) [67].

As shown in Scheme 2.9, treatment of compound **3** with molecular oxygen, nitrosobenzene, maleic anhydride, or carbon disulfide leads to clean [4+2] cycloaddition reactions in which the substrate adds across iridium and C3 of the iridabenzene ring, producing compounds **13–16**, respectively. Similarly, **3** undergoes a cheleotropic cycloaddition with sulfur dioxide to produce adduct **17**. These reactions are all accompanied by a dramatic color change from red (iridabenzene) to light yellow (adducts) and generate the products in high yield.

The dioxygen reaction is reminiscent of the reactions of certain polycyclic aromatic compounds (e.g. anthracene) with O_2, which lead to internal peroxides [68]. However, unlike these organic reactions, which require singlet oxygen, compound **3** reacts with ground-state (triplet) oxygen. The reaction probably proceeds by initial transfer of an electron from **3** to O_2, producing the iridabenzene radical cation (**3**$^{+}$) and superoxide (O_2^{-}), which then recombine to generate the observed product.

Compounds **13**, **15**, and **17** have been characterized by X-ray crystallography [34, 64], and the structure of **13** is presented in Figure 2.7. In each structure, the cycloaddition substrate has added across the open face of square-pyramidal **3**, generating an

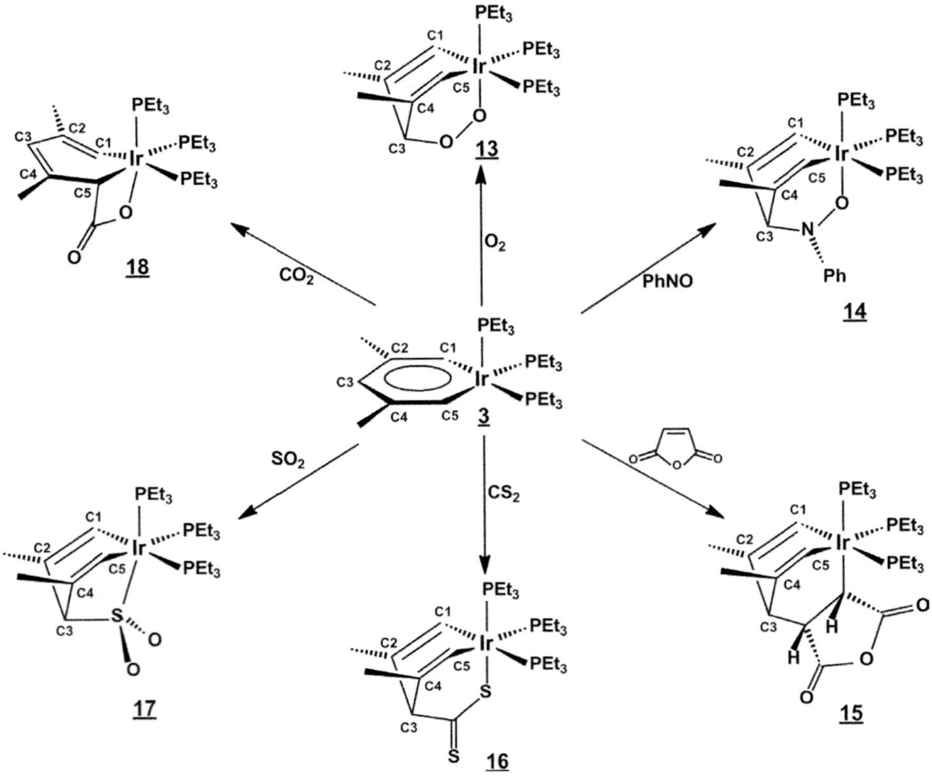

Scheme 2.9

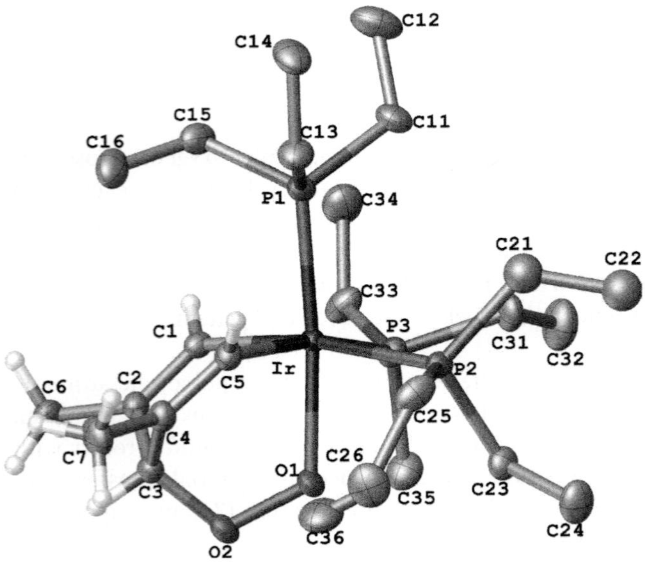

Figure 2.7 X-ray structure of the dioxygen adduct of iridabenzene, **13**, with phosphine H's omitted for clarity. The plot was created using the Olex2 software package [42]. (*See color plate section for the color representation of this figure.*)

octahedral adduct. The resulting iridacyclohexa-1,4-diene ring is boat-shaped, with Ir and C3 lying significantly out of the C1/C2/C4/C5 plane. In **13**, for example, Ir and C3 lie 0.92 Å and 0.65 Å, respectively, from the C1/C2/C4/C5 plane, while the dihedral angles made by planes C1/Ir/C5 and C2/C3/C4 with the C1/C2/C4/C5 plane are 37.6° and 48.9°, respectively. In **17**, the bending of the ends of the boat is even more severe, due to the demands of the five-membered rings. In this case, Ir and C3 lie 0.99 Å and 0.70 Å from the C1/C2/C2a/C1a plane (the molecule sits on a crystallographically imposed mirror plane), while the dihedral angles made by C1/Ir/C1a and C2/C3/C2a with Cl/C2/C2a/C1a are 40.0° and 55.8°, respectively. Bonding within the iridacyclohexa-1, 4-diene rings is localized as expected.

Compounds **13**, **14**, **16**, and **17** possess mirror-plane symmetry[6]; therefore, the $^{31}P\{^1H\}$ NMR spectra of these species consist of a doublet (two equivalent phosphines *trans* to the ring) and a triplet (unique phosphine *trans* to substrate). Adduct **15** possesses no symmetry, so that all three phosphines exhibit separate doublet-of-doublet resonances in the $^{31}P\{^1H\}$ NMR. In the $^{13}C\{^1H\}$ NMR spectra of **13–17**, the C1/C5 signals shift upfield to δ 122–136 from their position of δ 167.3 in **3**, reflecting the loss of carbene character. These signals are split by phosphorus coupling into doublet-of-triplet patterns with the large doublet coupling due to the *trans* PEt_3 and the small (apparent) triplet coupling due to the *cis* PEt_3's. The C3 signal in **13–17** shifts upfield to δ 58–101 from its position of δ 129.9 in **3**, reflecting the sp^3 hybridization of C3 in the adducts. Similarly, the H1/H5 signals in **13–17** shift upfield to δ 6.75–7.40 from their position of δ 10.91 in **3**, while the H3 signal moves upfield to δ 3.25–4.75 from δ 7.18 in **3**.

Unlike the cycloaddition reactions described above, iridabenzene **3** reacts with CO_2 to produce a [2+2] cycloaddition product (**18**; Scheme 2.9), probably because the short C=O bond cannot "stretch" across the Ir–C3 ring diagonal in the transition state of the [4+2] cycloadduct. (The C=O bond length in free CO_2 is 1.162 Å, while the C=S bond length in CS_2 is 1.554 Å.) [67]. The X-ray crystal structure of **18** [66] shows the iridacyclohexa-1,3-diene ring in a half-boat conformation with Ir lying 0.649 Å out of the C1/C2/C3/C4/C5 plane. The dihedral angle between plane C1/Ir/C5 and plane C1/C2/C3/C4/C5 is 24.4°. The $^{31}P\{^1H\}$ NMR spectrum of **18** consists of three separate doublet-of-doublet signals due to the three inequivalent PEt_3 ligands. The chemical shifts for C1 and H1 are similar to those seen in adducts **13–17**, but C3 and H3 are much more downfield (δ 126.0 and δ 5.5, respectively) because C3 retains its sp^2 hybridization in **18**. C5 and H5, in contrast, shift far upfield (to δ 23.8 and δ 3.2, respectively), reflecting C5's sp^3 hybridization. Both C1 and C5 exhibit the characteristic splitting patterns that results from phosphorus coupling.

Finally, treatment of **3** with nitrous oxide (N_2O) leads to ring contraction and formation of iridacyclopentadiene complex **19** (Scheme 2.10) [66]. The same product is obtained when **3** is reacted with amine *N*-oxides, including 4-methylmorpholine *N*-oxide and trimethylamine *N*-oxide. One reasonable pathway for this reaction involves initial formation of metallaepoxide **K**, which can also be viewed as an (η^2-aldehyde) metal complex. Rearrangement would lead to a metallacyclohexadienone hydride (**L**) which could retroinsert to produce intermediate **M**, and finally isomerize to iridacyclopentadiene **19**.

6 In the case of **14**, mirror-plane symmetry is apparently maintained by rapid inversion about the nitrogen center.

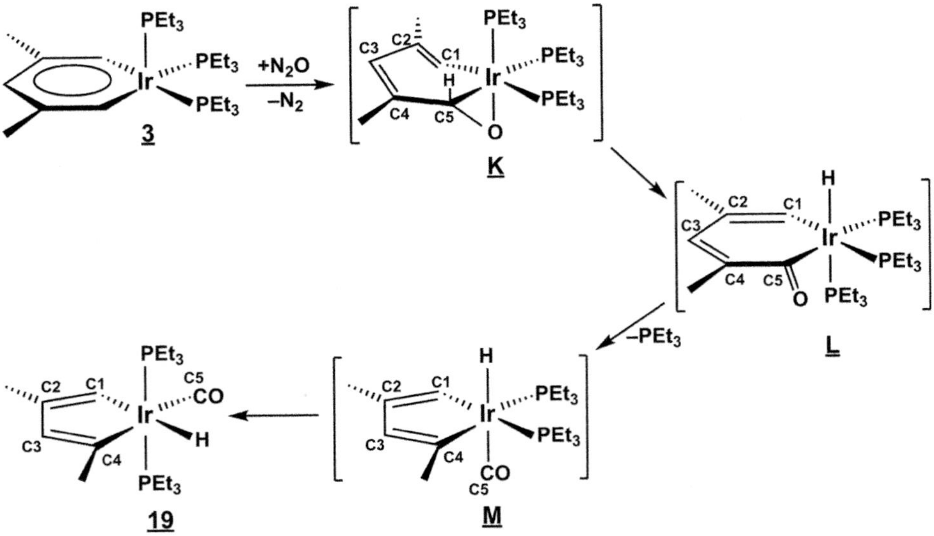

Scheme 2.10

2.8.4 Theoretical Study of Cycloaddition

Martin and van der Boom [67] have carried out a computational study on the cycloaddition reactions of an iridabenzene model compound, $(C_5H_5Ir)(PH_3)_3$, with carbon disulfide and carbon dioxide in order to obtain a better mechanistic understanding. They found that both of these substrates react in a concerted fashion. In the transition states, both bonds between the metallabenzene and the substrate are formed simultaneously. In the CS_2 cycloaddition, the substrate approaches the iridabenzene ring from the open face in a parallel orientation, and the iridabenzene HOMO and substrate LUMO have appropriate orbital symmetry for an allowed [4πs + 2πs] cycloaddition. A cartoon picture of that HOMO, which can be thought of as a bonding interaction between a metal d_{xz}/d_{z^2} hybrid orbital and the 3π orbital of the carbon fragment, is shown in Figure 2.8. Note that the wave function has opposite phases on Ir and C3, as required for matching the symmetry of CS_2's π* orbital (the LUMO).

The reaction of iridabenzene with CO_2, on the other hand, proceeds via a perpendicular approach of the CO_2 molecule to the Ir–C$_\alpha$ bond, as required for a [2πs + 2πa]

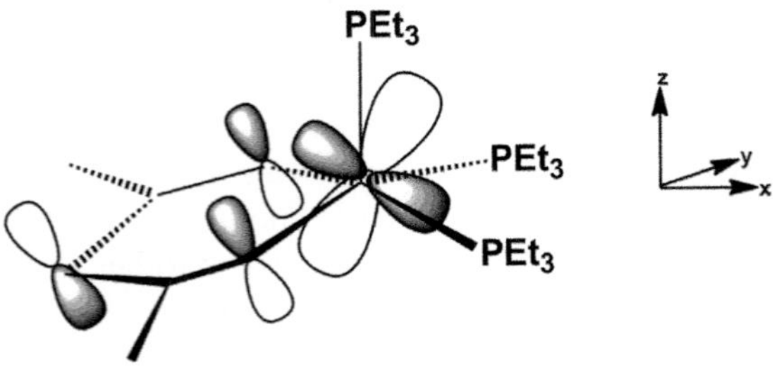

Figure 2.8 The highest occupied molecular orbital (HOMO) for iridabenzene **3**, which results from an interaction between a metal d_{xz}/d_{z^2} hybrid orbital and the 3π orbital of the carbon fragment.

cycloaddition [69]. The minimal steric demands of CO_2 and the presence of a second set of π orbitals may provide stabilization for this perpendicular orientation. As mentioned above, the main reason for the formation of the 1, 2 addition product in the CO_2 case is apparently the short C=O bond, which cannot easily span the ring in the transition state of the 1, 4 addition product. Although the observed [2+2] cycloadduct contains a strained four-membered ring, it is stabilized somewhat by the presence of conjugated C=C bonds within the six-membered ring.

2.8.5 Electrophilic Addition

Iridabenzene **3** reacts with electrophiles at the electron-rich α ring carbon atoms, C1/C5. Hence, treatment of **3** with one equivalent of triflic acid ($H^+O_3SCF_3^-$) results in clean protonation at C5[7] and production of the cationic (1, 3, 4, 5-η)-pentadienediyl complex, **2** (see Scheme 2.11) [34]. Similarly, treatment of **3** with acetylacetone ($pK_a \approx 9$) or with diethyl malonate ($pK_a \approx 13$) leads to protonation at C5. However, in these cases, the cationic pentadienediyl product is rapidly attacked by the acid anion, generating neutral *bis*(PEt_3) compounds **20** and **21**, respectively. Iridabenzene **3** does not react with acids having pK_a's above ~15, including water [70].

When **3** is treated with *two* equivalents of a strong acid such as $H^+O_3SCF_3^-$, protonation occurs at *both* C1 and C5, generating $[(\eta^5\text{-2,4-dimethylpentadienyl})Ir(PEt_3)_3]^{2+}$

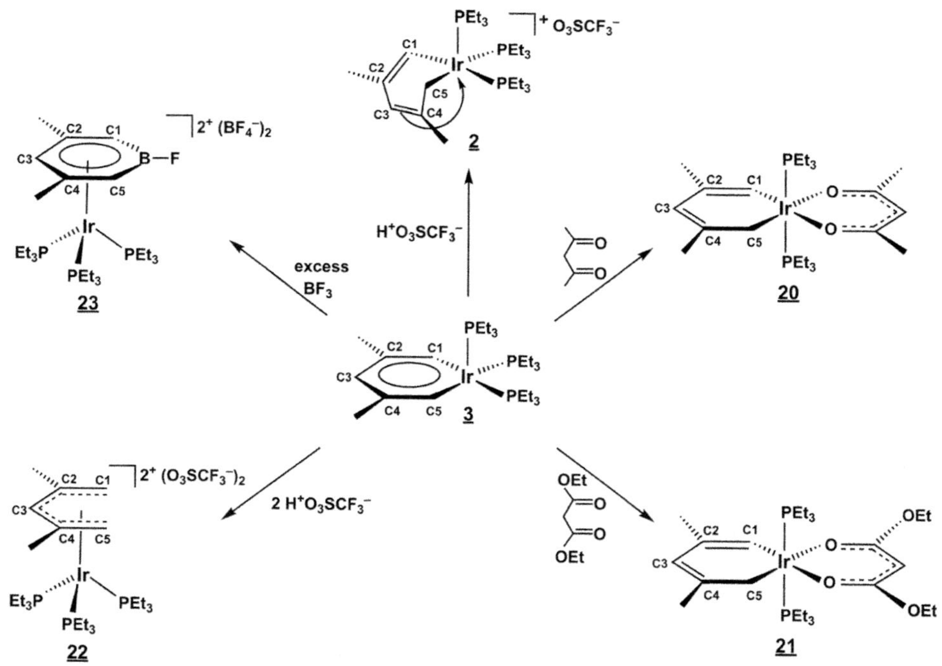

Scheme 2.11

7 We cannot rule out the possibility of initial protonation at the iridium center, followed by rapid migration to C5.

$(O_3SCF_3^-)_2$ (**22**; Scheme 2.11) [34]. In solution, compound **22** exhibits mirror-plane symmetry. Hence, the $^{31}P\{^1H\}$ NMR spectrum consists of a doublet (due to two equivalent phosphines located under the pentadienyl chain) and a triplet (due to the unique phosphine situated under the open pentadienyl mouth). These signals do not broaden significantly upon heating to 60°C, indicating that the η^5-pentadienyl ligand does not rotate with respect to the $Ir(PEt_3)_3$ framework on the NMR timescale.

Finally, treatment of iridabenzene **3** with excess boron trifluoride leads to the formation of a novel (η^6-borabenzene)metal complex, **23** (Scheme 2.11) [34]. Although the detailed mechanism of this reaction is not known, one reasonable pathway, shown in Scheme 2.12, involves an initial electrophilic attack of BF_3 at ring carbon C5, generating intermediate **N**. Migration of ring carbon C1 to boron with concomitant removal of fluoride (as BF_4^-) closes the boracyclohexadienyl ring, generating **O**. Finally, reaction of **O** with a third equivalent of BF_3 results in removal of another fluoride as BF_4^- and aromatization of the borabenzene ring.

In the X-ray crystal structure of **23** [34], the borabenzene ring exhibits a chair-like distortion from planarity, with boron lying 0.229 Å above the C1/C2/C4/C5 plane (away from the iridium center) and C3 lying 0.153 below the same plane (toward Ir). The dihedral angles made by planes C1/B/C5 and C2/C3/C4 with the C1/C2/C4/C5 plane are 15.9° and 12.8°, respectively. Among the six iridium-ring atom bonds, the Ir–C3 interaction is the shortest (2.267 Å) and the Ir–B bond is the longest (2.534 Å). Distances and angles within the ring are typical for metal-coordinated borabenzene [71]. The phosphine ligands on iridium are situated approximately under atoms B, C2, and C4 of the ring.

In acetone solution at room temperature, the η^6-borabenzene ligand rotates with respect to the $Ir(PEt_3)_3$ moiety. Hence, the $^{31}P\{^1H\}$ NMR spectrum of **23** at 25°C consist of a single resonance. However, as the sample is cooled, borabenzene rotation is slowed down and decoalescence of the ^{31}P NMR signal occurs. At −90°C, the stopped-exchange limiting spectrum – a doublet (intensity 2) and a triplet (intensity 1) – is observed.

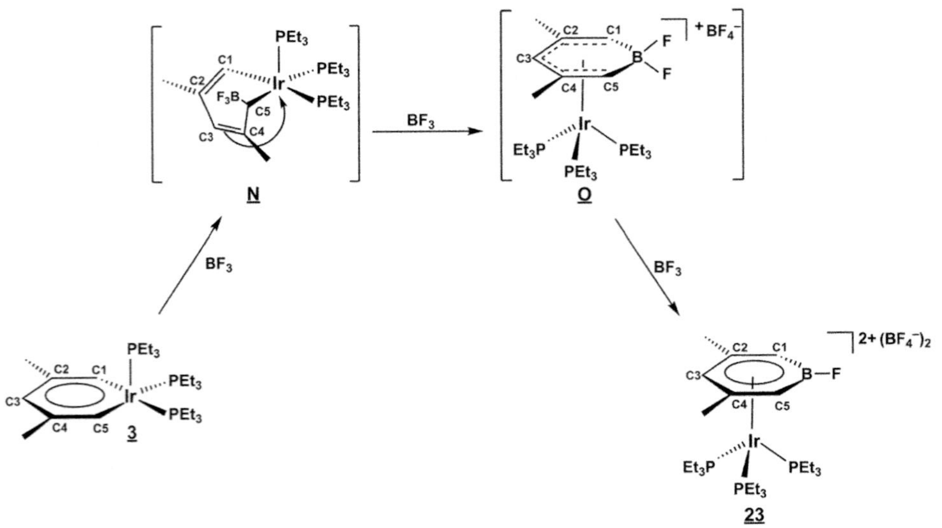

Scheme 2.12

2.8.6 Coordination to $Mo(CO)_3$

Iridabenzene **3** cleanly displaces *p*-xylene from (*p*-xylene)$Mo(CO)_3$ in THF solvent, producing the metal-coordinated metallabenzene complex (η^6-iridabenzene)$Mo(CO)_3$, **24** (see Scheme 2.13). The X-ray crystal structure of **24** (Figure 2.9 [62, 64]), shows that the square pyramidal coordination geometry of the parent iridabenzene is retained, although the $Mo(CO)_3$ moiety now occupies the formerly "open face" of the square pyramid. The bonding within the metallabenzene ring is still fully delocalized, but the average C–C and Ir–C bond distances (1.41 Å and 2.03 Å, respectively) are slightly longer than those in parent **3** [C–C (ave) = 1.38 Å, Ir–C (ave) = 2.00 Å], because π-electron density is being removed from the ring by the $Mo(CO)_3$ moiety. The circumference of the iridabenzene ring is 9.70 Å vs. 9.54 Å in **3**. The iridium atom in **24** resides 0.30 Å out of the plane of the five ring carbons (C1/C2/C3/C4/C5), and the dihedral angle between this plane and the plane C1/Ir/C5 is 11.7°. The corresponding numbers in **3** are 0.24 Å and 9.2°. The molybdenum atom is strongly π-complexed to all six atoms of the arene ring. The Mo–Ir distance is 2.978(1) Å, while the Mo–C_{ring} distances range from 2.318(10) Å to 2.404(9) Å; the shortest bond is with C3. The three carbonyl groups, when projected onto the arene plane, approximately eclipse C1, C3, and C5.

Unlike **3**, which undergoes a very low energy intramolecular phosphine exchange process in solution (vide supra), **24** is stereochemically rigid at iridium, even upon heating to 60°C. Hence, the $^{31}P\{^1H\}$ NMR spectrum of **24** consists of a doublet (equivalent basal phosphines) and a triplet (axial phosphine). The ^1H NMR signals for the ring protons in **24** shift upfield from their positions in **3**, as is normally observed when arenes coordinate to metal fragments [72]. Protons H1/H5 in **24** appear at δ 8.08 (vs. δ 10.91 in **3**), while H3 resonates at δ 6.25 (vs. δ 7.18 in **3**). The H1/H5 signal is a doublet (J_{HP} = 20.7 Hz) due to strong coupling to the *cis* basal phosphine. The ring

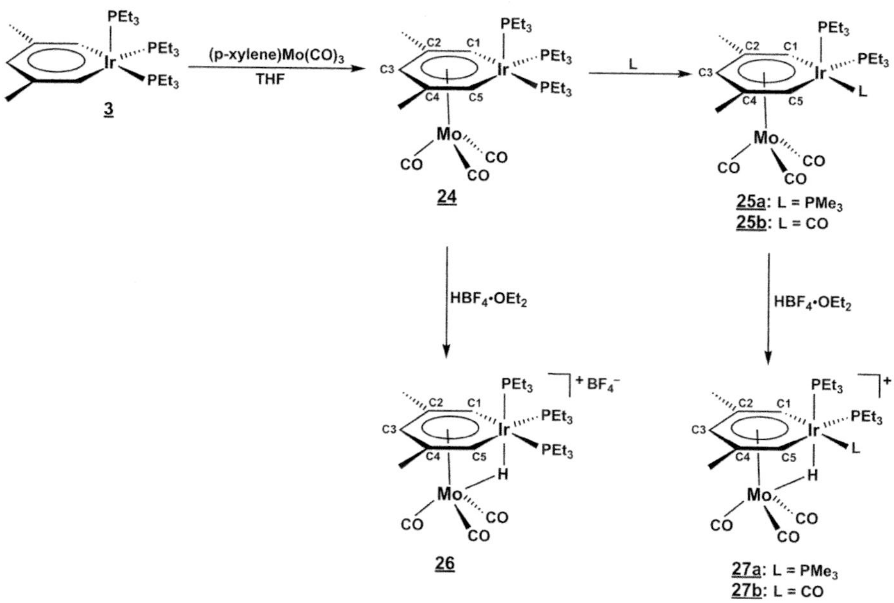

Scheme 2.13

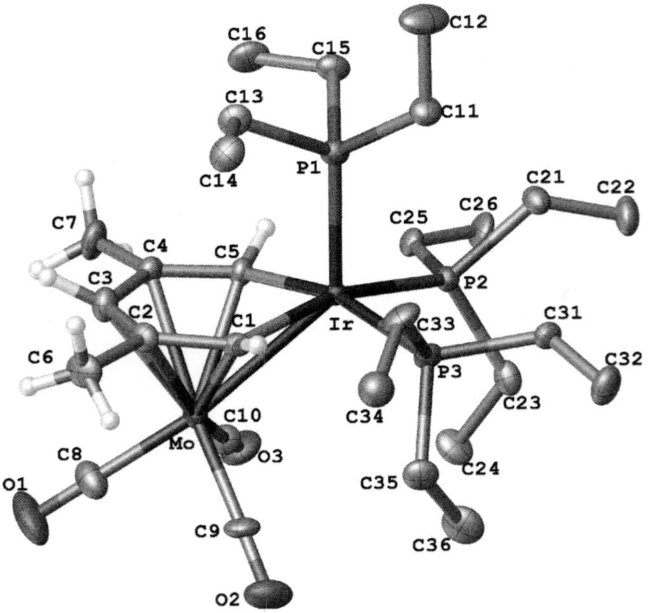

Figure 2.9 X-ray structure of (η^6-iridabenzene)Mo(CO)$_3$, **24**, with phosphine H's omitted for clarity. The plot was created using the Olex2 software package [42]. (*See color plate section for the color representation of this figure.*)

carbons also show an upfield ^{13}C NMR shift. Hence, C1/C5, C2/C4, and C3 resonate at δ 135.7, δ 107.9, and δ 99.5 in **24** vs δ 167.6, δ 132.0, and δ 129.9 in **3**, and the C1/C5 signal exhibits strong coupling to the *trans* basal phosphine (J_{CP} = 74.5 Hz).

As is typical for (η^6-arene)Mo(CO)$_3$ complexes [72], the metallabenzene ligand in **24** rotates freely with respect to the Mo(CO)$_3$ moiety. As a result, the carbonyl carbon atoms in **24** give rise to a sharp singlet in the ^{13}C NMR spectrum, even at −80°C, indicating a rotational barrier, $\Delta G^\ddagger$, of less than 8 kcal/mol. The infrared spectrum of **24** shows the presence of two intense ν(CO) bands (A_1 and E), as is characteristic of (η^6-arene) Mo(CO)$_3$ complexes. The very low energy of these bands (1918 and 1836 cm^{-1}) indicates substantial π-back-bonding from Mo to π* of the CO's, and reflects the extremely electron-rich nature of arene **3**. By comparison, the ν(CO) bands for (η^6-*p*-xylene) Mo(CO)$_3$ appear at 1975 and 1901 cm^{-1} [73]. Since the stability of (η^6-arene)Mo(CO)$_3$ complexes increases with increasing arene basicity [74], it is not surprising that **3** cleanly displaces organic arenes from (η^6-arene)Mo(CO)$_3$ complexes in THF solvent.

Like iridabenzene **3** itself, compound **24** undergoes ligand substitution reactions when treated with ligands that are less sterically demanding than PEt$_3$, although higher temperatures are required [62]. For example, treatment of **24** with PMe$_3$ or CO in refluxing THF or acetone results in clean replacement of one PEt$_3$ ligand, producing **25a, b** (see Scheme 2.13). These species can also be produced via arene exchange by treating preformed ligand-substituted iridabenzenes (**7a**, **c**) with (η^6-*p*-xylene)Mo(CO)$_3$ in THF. The iridabenzene ligands in these molecules are stereochemically rigid and, as expected, the unique ligand L resides in the more sterically crowded basal site. This ligand arrangement is evident from the ^{31}P{^{1}H} NMR spectra, which show separate resonances for the PEt_3 ligands in each case. In solution, the iridabenzene rings in **25a**, **b** undergo facile rotation with respect to the $Mo(CO)_3$ moiety, causing the CO ligands on molybdenum to appear to be equivalent by ^{13}C NMR, even at −80°C.

Coordination of iridabenzene **3** to $Mo(CO)_3$ reduces its reactivity by effectively blocking its open face. Thus, for example, complex **24** is stable in air for several hours, while **3** reacts with atmospheric oxygen in minutes. Interestingly, both **3** and **24** react with acids, but the site of addition differs. As described earlier, treatment of **3** with acid results in protonation at the α-carbon. Compound **24**, in contrast, protonates at the metal centers, producing a novel heterobimetallic μ-hydride complex, **26** (Scheme 2.13). The NMR spectra of **26** bear a close resemblance to those of **24**, except for the presence of a new hydride resonance at δ −13.30 in the ^1H NMR. This signal is split into a doublet of triplets (J_{HP} = 42.1 Hz, 11.3 Hz) with the doublet coupling due to the *trans*/axial phosphine and the triplet due to the basal phosphines. The relatively small magnitude of the *trans* coupling (42.1 Hz) is probably a consequence of the μ_2-bridging character of the hydride.

In protonated compound **26**, the metallabenzene ligand still rotates with respect to the $Mo(CO)_3$ moiety at room temperature. Hence, the Mo-bound carbonyls give rise to a single peak in the room temperature $^{13}C\{^1H\}$ NMR spectrum. However, the rotational barrier is substantially higher than that for **24**, and the motion can be arrested by cooling. At −30°C, the stopped exchange limit is achieved, and two sharp CO resonances in a 2:1 ratio are observed by $^{13}C\{^1H\}$ NMR. From the CO line shape changes, the free energy of activation ($\Delta G^{\ddagger}$) for arene rotation is calculated to be 13.7(4) kcal/mol.

The ligand-substituted iridabenzenes, **25a**, **b**, also protonate at the metal centers when treated with acid, producing compounds **27 a**, **b** (Scheme 2.13). In the ^1H NMR of **27a**, the hydride signal appears at δ −13.15 and is an apparent doublet of triplets (J_{HP} = 42.0 Hz, 11.4 Hz), while in **27b**, the hydride resonates at δ −12.91 and is a doublet of doublets (J_{HP} = 45.8 Hz, 10.0 Hz).

The X-ray crystal structure of protonation product **27a** [62] has been obtained (Figure 2.10), and can be compared to that of its neutral precursor, **25a** [62]. In **27a**, the Mo–Ir bond shortens to 2.854(2) Å from 2.950(1) Å in **25a**. Likewise, the Mo–C_{ring} bonds in **27a** shorten slightly, averaging 2.33 Å, as compared to 2.37 Å in **25a**. The metallabenzene ring in **27a** remains delocalized, but the Ir–C_α distances are lengthened slightly vs. those in **25a** (Ir–C (ave) = 2.06 Å in **27a** vs. 2.03 Å in **25a**), perhaps reflecting a weakening of the Ir–C_α π interaction upon protonation. The Ir atom in **27a** lies 0.20 Å out of the plane of the five ring carbons (C1/C2/C3/C4/C5) and the dihedral angle between this plane and the plane C1/Ir/C5 is 7.7°. The corresponding numbers for **25a** are 0.25 Å and 9.5°. The Ir–H and Mo–H bond lengths are 1.77(6) Å and 1.97(7) Å, respectively, while the Ir–H–Mo angle is 99(3)°. The hydride resides approximately *trans* to the axial phosphine on iridium but bends in slightly toward the molybdenum atom, generating a P1–Ir–H angle of 173(2)°. The influence of the *trans*-hydride ligand causes the Ir–P1 bond to lengthen to 2.323(2) Å in **27a** from 2.259 Å in **25a**.

2.9 Iridaphenol

The synthesis of the first example of an iridaphenol resulted directly from our reactivity studies of iridabenzenes with nitrous oxide [75]. As discussed earlier (Scheme 2.10), treatment of iridabenzene **3** with N_2O leads to the formation of an iridaepoxide (**K**),

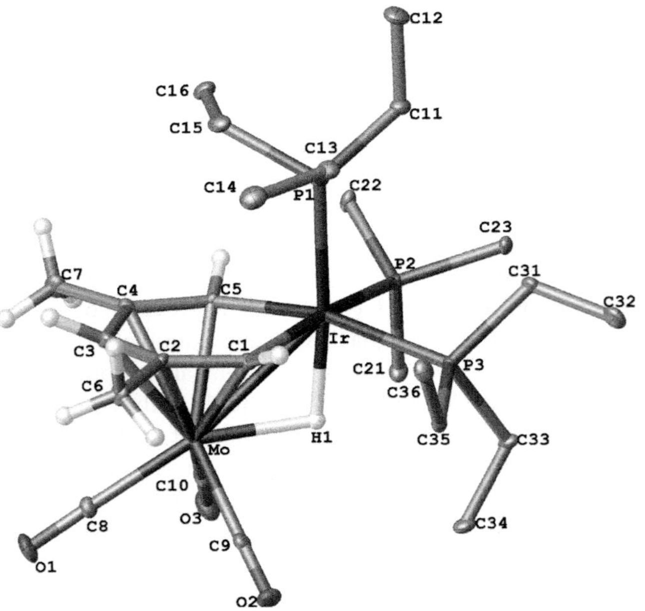

Figure 2.10 X-ray structure of the cation in protonated (η^6-iridabenzene)Mo(CO)$_3$, **27a**, with phosphine H's omitted for clarity. Note that the iridabenzene has one PMe_3 and two PEt_3 ligands. The plot was created using the Olex2 software package [42].

which can rearrange to an iridacyclohexadienone-hydride (**L**). This species can then rapidly lose a phosphine, retroinsert to form an iridacyclopentadiene-carbonyl-hydride (**M**), and isomerize to the final product, **19**. In the analogous reaction involving *tris*(PMe_3) iridabenzene **9** (see Scheme 2.14), the iridaepoxide **P** again leads to the

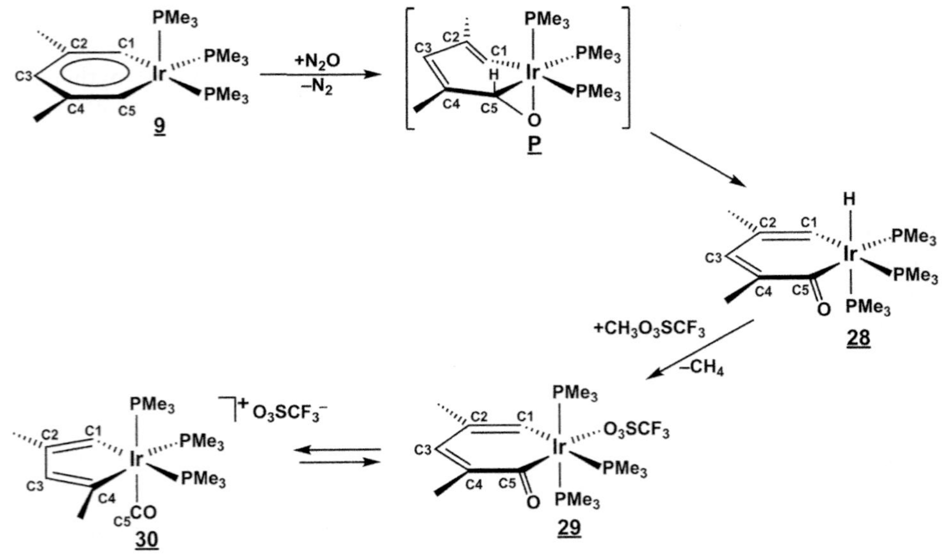

Scheme 2.14

iridacyclohexadienone-hydride species (**28**), but in this case it can be isolated, presumably because phosphine dissociation is much slower due to reduced steric interactions. Treatment of **28** with methyl triflate ($CH_3O_3SCF_3$) removes the hydride ligand, generating methane and iridacyclohexadienone **29**. When **29** is dissolved in acetone, it establishes an equilibrium with the iridacyclopentadiene-carbonyl compound **30**, a process which involves dissociation of the labile triflate ligand and reversible retroinsertion of the carbonyl group. At low temperature, the equilibrium favors **29**, and pure **29** can be crystallized from the mixture at −30°C as yellow blocks.

In the ^1H NMR spectrum of **29**, H1 resonates at δ 9.25 and is a singlet due to its *cis* relationship to the triflate ligand, while H3 appears as singlet at δ 6.66. In the $^{13}C\{^1H\}$ NMR, C1 resonates at δ 156.7 and is strongly coupled to the *trans* PMe_3 ligand (J_{CP} = 78.2 Hz). The remaining ring carbons are singlets at δ 188.7 (C5), δ 145.7 (C3), δ 127.9 (C2), and δ 127.1 (C4). In the X-ray crystal structure of **29** (75), bonding around the carbon portion of the ring is localized, as expected, with the following bond distances: C1–C2 = 1.332(26) Å, C2–C3 = 1.438(31) Å, C3–C4 = 1.340(31) Å, and C4–C5 = 1.523(27) Å. The iridium–carbon distances are Ir–C1 = 2.052(17) Å and Ir–C5 = 1.987(19) Å, while C5–O1 = 1.197(25) Å. The internal angles of the metallacycle sum to 719.7°, indicating a nearly planar ring.

Treatment of iridacyclohexadienone **29** with acids leads to protonation at the carbonyl oxygen and the production of stable iridaphenols. For example, when **29** is reacted with triflic acid ($H^+O_3SCF_3^-$) in THF, the solution immediately turns from yellow to dark-red-orange, signaling the formation of iridaphenol **31** (see Scheme 2.15) [75]. The aromatization of the ring in **31** is indicated by the downfield shifting of ring protons H1 and H3. H1 moves downfield to δ 10.54 (from δ 9.25 in **29**), while H3 shifts to δ 7.49 (from δ 6.66). Significant downfield shifts are also observed in the

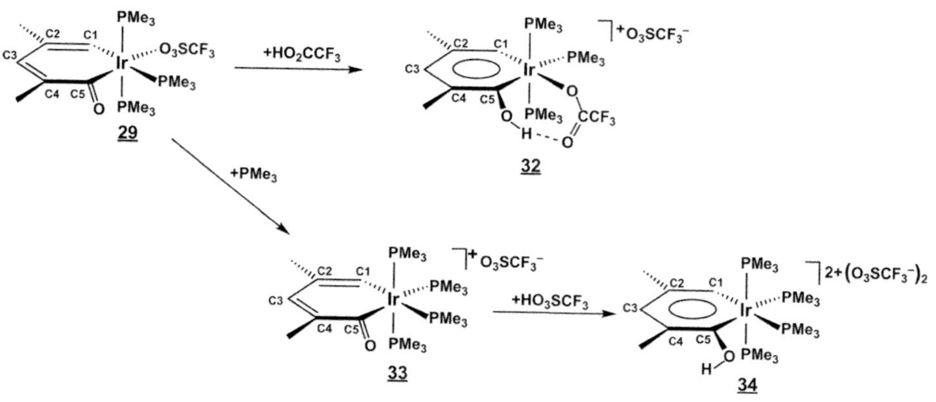

Scheme 2.15

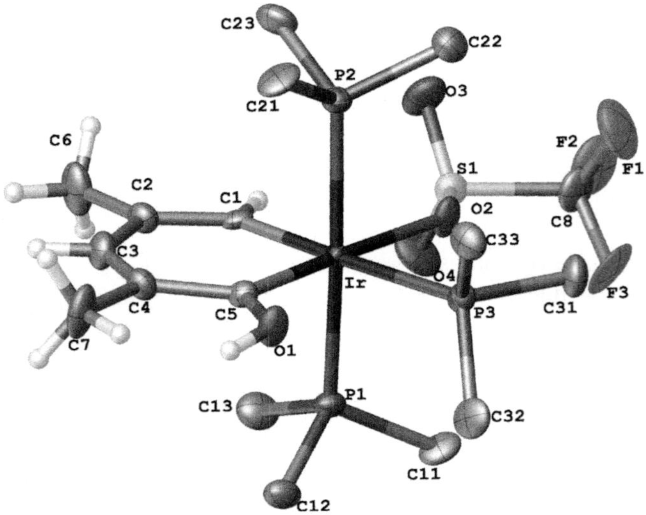

Figure 2.11 X-ray structure of iridaphenol, **31**, with phosphine H's omitted for clarity. The plot was created using the Olex2 software package [42]. *(See color plate section for the color representation of this figure.)*

$^{13}C\{^1H\}$ NMR for ring carbons C1, C3, and C5, which resonate at δ 179.0, δ 165.1, and δ 219.3, respectively. The C1 signal is a phosphorus-coupled doublet (J_{CP} = 76.9 Hz), indicating that it still lies *trans* to PMe_3, while C5 is a singlet due to its *trans* relationship to triflate. The phenol proton is observed as a broad resonance at δ 12–14 in the ^1H NMR.

The X-ray crystal structure of **31** (see Figure 2.11) shows that the ring carbon–carbon bonds have moved toward equalization. Bonds C1–C2, C2–C3, C3–C4, and C4–C5 exhibit distances of 1.352(25) Å, 1.406(26) Å, 1.355(26) Å, and 1.464(25) Å, respectively. The phenol carbon–oxygen bond (C5–O1) has lengthened to 1.331(20) Å (from 1.197(25) Å in precursor **29**). Furthermore, the iridium–carbon bonds, Ir–C1 and Ir–C5, have shortened to 2.031(16) Å and 1.916(16) Å, respectively, from their values of 2.052(17) Å and 1.987(19) Å in **29**, indicating significant metal participation in ring π-bonding. While the trend toward delocalization in **31** is clear, the small differences observed in bond distances within the ring appear to be real. This suggest that, of the two resonances shown in Figure 2.12 (top line), structure **I** contributes more strongly than **II** to the bonding in **31**, perhaps due to heteroatom (oxygen) stabilization of the metal carbene in **I**. In addition, the weak *trans* influence of the triflate ligand (*trans* to C5) reinforces the short Ir–C5 distance. The iridaphenol ring in **31** is nearly planar; the iridium atom resides less than 0.10 Å out of the best plane made by ring carbons C1/C2/C3/C4/C5, and the dihedral angle between this plane and the C1/Ir/C5 plane is a mere 3.8°. The sum of the six internal angles within the ring is 719.7°.

Treatment of iridacyclohexadienone **29** with trifluoroacetic acid ($H^+O_2CCF_3^-$) also leads to the production of an iridaphenol, **32** (see Scheme 2.15) [75]. This reaction is accompanied by exchange of the triflate ligand for a trifluoroacetate ligand, which coordinates *cis* to C5 (*trans* to C1), allowing intramolecular hydrogen bonding to occur between the phenol oxygen and the carbonyl oxygen of the trifluoroacetate group.

The position of the trifluoroacetate ligand in **32** is again evident from the $^{13}C\{^1H\}$ NMR spectrum. The C5 signal is now phosphorus-coupled (J_{CP} = 93.8 Hz), indicating its *trans* relationship to PMe_3, while C1 is a singlet (*trans* to trifluoroacetate). In the ^1H NMR of **32**, H1 and H3 resonate at δ 8.95 and δ 7.50, while the phenol proton appears at δ 17.50.

Bonding within the metallacycle in **32** is even more delocalized than in **31**. The Ir–C1 and Ir–C5 distances are nearly identical at 2.002(12) Å and 2.023(13) Å, respectively, while ring C–C bonds C1–C2, C2–C3, C3–C4, and C4–C5 exhibit distances of 1.371(19) Å, 1.427(21) Å, 1.398(22) Å, and 1.371(20) Å, respectively. Hence, it appears that resonance structures **III** and **IV** (Figure 2.12, bottom line) contribute almost equally to the bonding in compound **32**. Structure **III** may be stabilized by the presence of a heteroatom (oxygen) on the carbene carbon, while structure **IV** probably benefits from the weaker *trans* influence of the trifluoroacetate group. The metallacycle in **32** is nearly planar; the iridium atom lies only 0.10 Å out of the C1/C2/C3/C4/C5 plane, and the dihedral angle between this plane and the C1/Ir1/C5 plane is 4.0°. The sum of the six internal angles within the iridaphenol is 719.6°. As mentioned above, the placement of the trifluoroacetate ligand *cis* to C5 allows the phenol hydrogen to form an intramolecular hydrogen bond, and the observed distance of 2.576 Å between the phenol oxygen and the trifluoroacetate carbonyl oxygen is fully consistent with this type of interaction [76].

As shown in Scheme 2.15, treatment of iridacyclohexadienone **29** with PMe_3 leads to displacement of triflate by PMe_3 and production of the cationic *tetrakis*(PMe_3) iridacyclohexadienone (**33**). Treatment of **33** with triflic acid generates dicationic iridaphenol **34** [75]. Again, comparison of the ^1H NMR spectrum of **34** with that of precursor **33** shows significant downfield shifting of ring protons H1 and H3, consistent with aromatization of the ring. In particular, the H1 signal moves from δ 7.97 in **33** to δ 9.18 in **34**, while H3 shifts from δ 6.67 to δ 7.55. The phenol proton appears as a broad resonance in the δ 12–14 region. Trifluoroacetic acid, unlike triflic acid, does *not* react with iridacyclohexadienone **33**, a result that can be attributed to both the weaker acidity of trifluoracetic acid and the cationic charge of **33**.

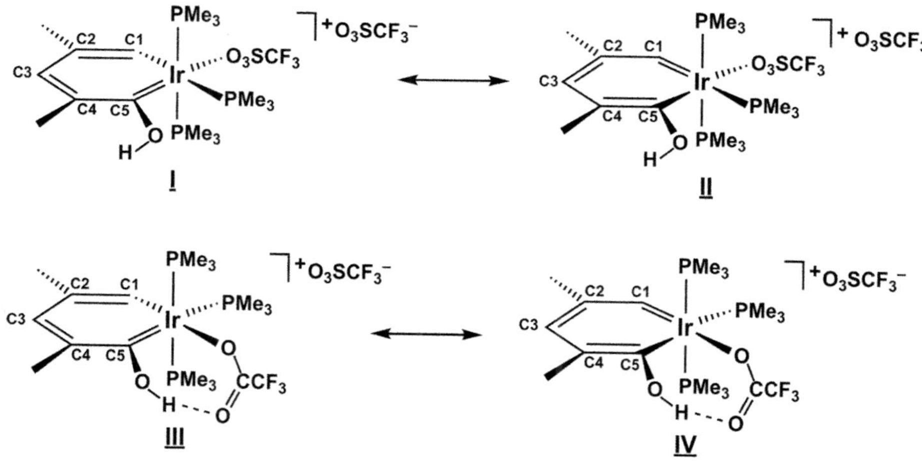

Figure 2.12 Resonance structures for iridaphenol **31** (top) and iridaphenol **32** (bottom).

2.10 Synthesis and Spectroscopy of Iridapyrylium

In 1991, having synthesized and begun to explore the reactivity of iridabenzene **3**, we thought it would be interesting to synthesize heteroatom-containing analogues of **3**, in order to compare the physical and chemical properties of these related species to those of **3**. To this end, we began to study the reactions of (Cl)Ir(PEt_3)$_3$ with potassium oxapentadienide reagents [77, 78] and, in 1993, published the synthesis of the dimethylirid apyran(dimethyliridaoxacyclohexa-1,3-diene) complex, **35** (see Scheme 2.16), from the reaction of (Cl)Ir(PEt_3)$_3$ with potassium 2,4-dimethyl-5-oxapentadienide [79]. Like the original synthesis of iridacyclohexadiene, **1**, this reaction involves C–H bond activation on the terminus of an η^1-2,4-dimethyl-5-oxapentadienyl ligand in intermediate **Q**. The $^{31}P\{^1H\}$ spectrum of **35** consists of a doublet and a triplet, indicating a *mer* arrangement of phosphines and planar metallacycle. In the ^1H NMR, H1, and H3 resonate at δ 6.04 and δ 4.70, respectively, while the metal-hydride appears far upfield at δ −24.46. In the $^{13}C\{^1H\}$ NMR, C1 resonates at δ 108.7 and is strongly coupled to the *trans* phosphine (J_{CP} = 75.3 Hz). The remaining carbons appear at δ 156.8 (C4), δ 127.4 (C2), and δ 97.4 (C3).

With compound **35** in hand, we attempted to convert it to the aromatic iridapyrylium, the oxygen-containing analogue of iridabenzene **3**. Recalling the synthesis of iridabenzene (Scheme 2.1), we reasoned that removal of the metal hydride would allow an oxygen lone pair to form a π-bond to the iridium center, aromatizing the ring system. Unfortunately, the reagent used to remove the hydride ligand in the iridabenzene synthesis, methyl triflate, did not react with iridapyran **35** in the same way. Instead, as shown in Scheme 2.17, it methylated the iridapyran ring at C3, generating an iridaoxacyclohexa-1,4-diene product, **36**. Eventually, we discovered that treatment of iridapyran **35** with oxidizing agents such as $Ag^+BF_4^-$ results in removal of the metal hydride and production of the desired iridapyrylium **37**, the first example of a stable metallapyrylium [80, 81]. As shown in Scheme 2.17, this oxidation reaction actually generates a clean 1:1 mixture of the iridapyrylium **37** and the C3-protonated ring compound **38**. In this reaction, one-half of the iridapyran molecules are doubly oxidized and transfer their metal "hydrides" as *protons* to the other half of the iridapyran molecules. The doubly oxidized proton donor molecules become iridapyryliums (**37**), while the proton acceptor molecules become the protonated rings (**38**). The detailed mechanism of this reaction has been worked out and is shown in Scheme 2.18 [81].

Note that the proton transfer occurs after the *first* oxidation from radical cation **R**. The resulting 17e$^-$ neutral species **S** is then oxidized again to iridapyrylium **37**. Although the yield of **37** is only 50%, the situation is not as dire as it seems, because **38** can be

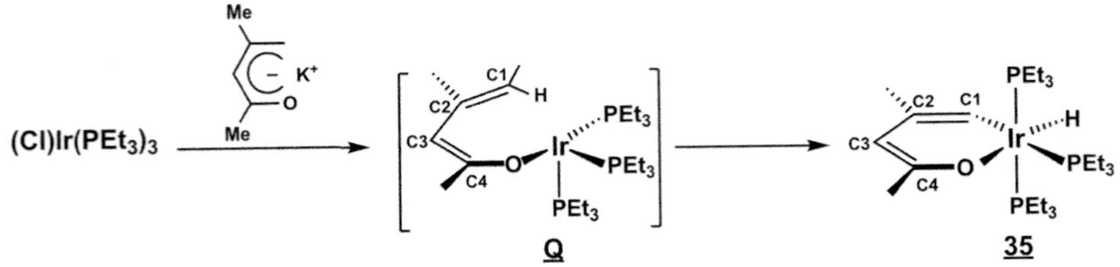

Scheme 2.16

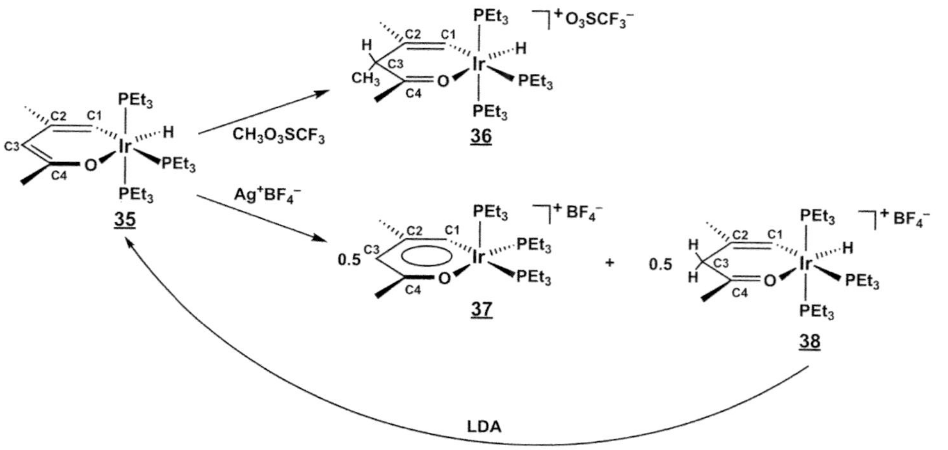

Scheme 2.17

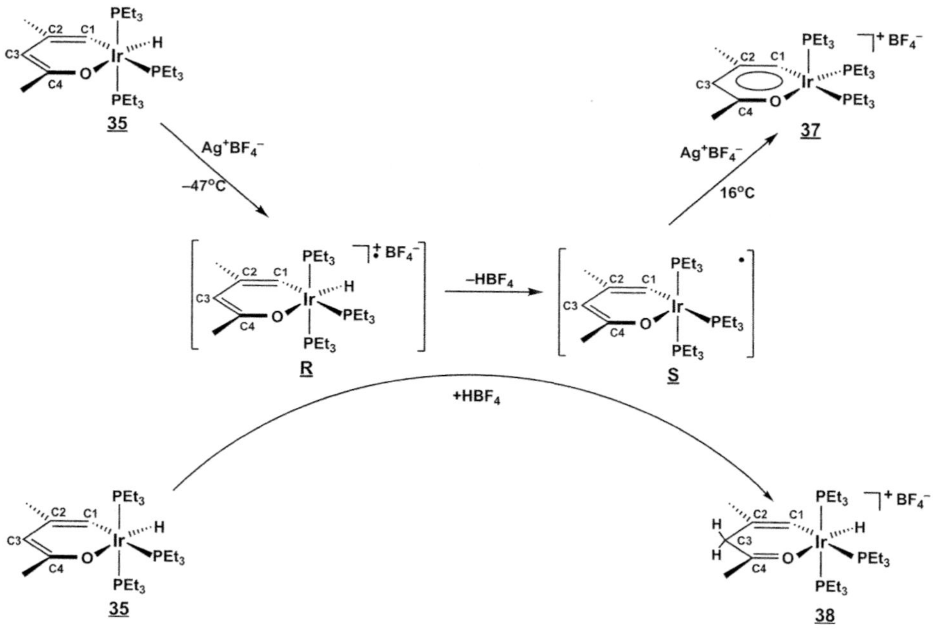

Scheme 2.18

cleanly deprotonated with LDA and, in this way, converted back to **35** for re-use. In practice, we treat the mixture of **37** and **38** with ½ equivalent of LDA in THF, which reacts selectively with **38** to regenerate ½ equivalent of **35**. This neutral species is then extracted with pentane, leaving behind pure **37** as a deep-purple solid.

The $^{31}P\{^{1}H\}$ NMR spectrum of iridapyrylium **37**, like that of iridabenzene **3**, consists of a sharp singlet, which does not broaden significantly even upon cooling to −90°C. This behavior is indicative of a low-energy fluxional process that exchanges the phosphines, probably a Berry-type process. In the ^1H NMR spectrum, ring protons H1 and H3 are shifted downfield to δ 9.35 and δ 6.45 from their positions of δ 6.04 and δ 4.70, respectively, in iridapyran precursor **35**. This downfield shifting indicates the participation of metal orbitals in ring π-bonding and the establishment of a ring current. The H1 signal is split into a characteristic binomial quartet (J_{HP} = 6.0 Hz) by the ^{31}P nuclei of the three rapidly exchanging phosphine ligands. In the ^{13}C NMR, the ring carbons also shift downfield from their positions in **35** to δ 170.4 (C4), δ 162.2 (C1), δ 147.3 (C2), and δ 112.2 (C3). The C1 signal is split into a phosphorus-coupled binomial quartet (J_{CP} = 22.7 Hz). Although the X-ray structure of **37** has not been obtained, the optimized geometry for the model compound, $[C_4H_4OIr(PH_3)_3]^+$, is square pyramidal, consistent with the X-ray structure of **3** [67].

2.11 Valence Bond Structures and Electron Counting for Iridapyrylium

One can write three reasonable resonance structures for iridapyrylium **37** (**I–III**; Figure 2.13). In structures **I** and **II**, the iridium center possesses 18 valence electrons, while in **III** it possesses 16 electrons. If the carbene moiety in **I** is counted as a neutral $2e^-$ donor (Fischer carbene) [37], the metal center is Ir(I)≡d^8. On the other hand, if the carbene in **I** is treated as a $4e^-$ dianion (Schrock carbene) [38], the metal center becomes formally Ir(III)≡d^6. (In each case, we count the carbonyl ligand as a neutral $2e^-$ donor.) In resonance structures **II** and **III**, the metal center is formally Ir(III)≡d^6, because both the alkoxy and vinyl ligands are counted as anions. However, these structures differ in the number of electrons donated by the alkoxy ligand. In **II**, the alkoxy ligand is a $4e^-$ donor (one oxygen lone pair is used to form the Ir–O π bond), while in **III**, it donates just $2e^-$.

2.12 Chemical Reactivity of Iridapyrylium 37

2.12.1 Ligand Addition

Because of the contribution from $16e^-$ resonance structure **III** (Figure 2.13), the Ir center in compound **37** is reactive toward a variety of $2e^-$ donor reagents, L, including hydride, methide, chloride, and trimethylphosphine [80, 81]. As shown in Scheme 2.19, the products of these ligand addition reactions are all octahedral iridapyran compounds with an $18e^-$ Ir(III) center. Of particular interest is the room temperature reaction of **37** with excess PMe_3, which leads to PMe_3 addition at Ir *and* complete

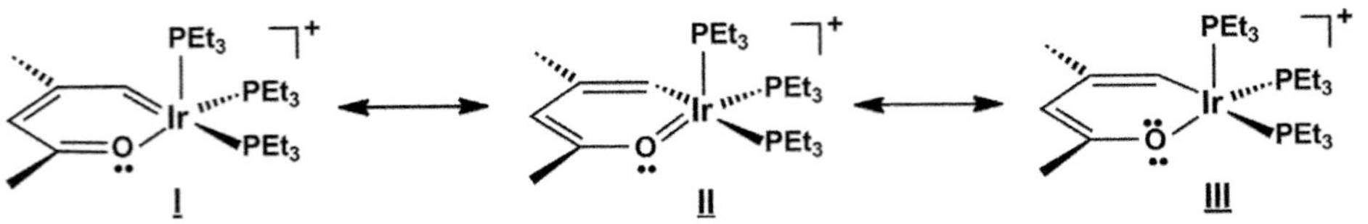

Figure 2.13 Resonance structures for iridapyrylium **37**. Structures **I** and **II** have $18e^-$ iridium centers while **III** has a $16e^-$ iridium center.

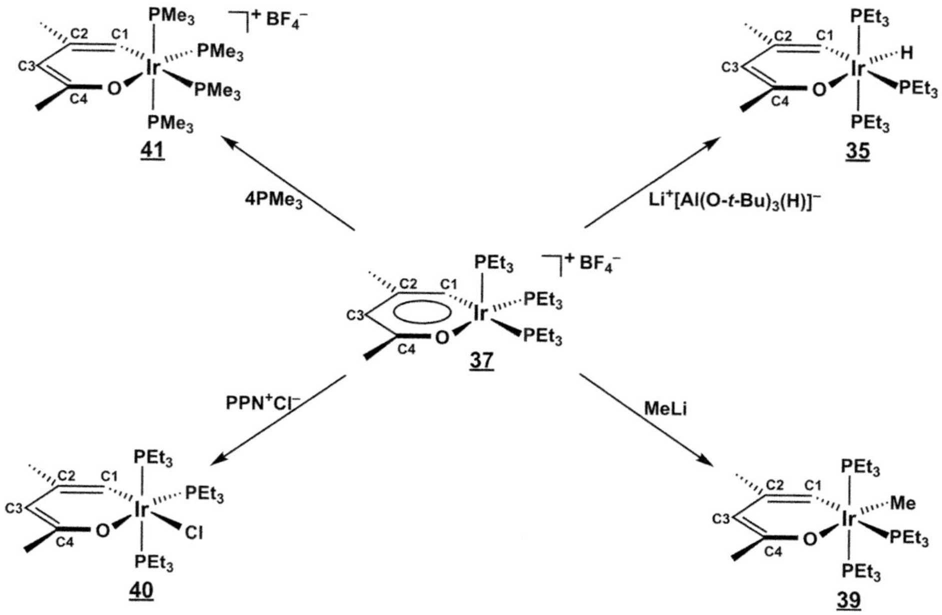

Scheme 2.19

replacement of the three bulky PEt_3 ligands with smaller PMe_3's, generating the *tetrakis*(PMe_3) product, **41**. This reaction contrasts sharply with the room temperature reaction of iridabenzene **3** with excess PMe_3, which results in the replacement of just one PEt_3 ligand with PMe_3 and retention of the aromatic ring system (cf. Scheme 2.5). The reason for this divergent behavior is that **3** reacts with PMe_3 by a *dissociative* mechanism (vide supra), while the reaction of **37** with PMe_3 involves a series of *associative* steps, made possible by the fact that the Ir–O π-electrons can be localized on the oxygen atom (resonance structure **III**), rendering the iridium atom a reactive $16e^-$ center.

2.12.2 Cycloaddition

As shown in Scheme 2.20, compound **37** undergoes [4+2] cycloaddition reactions similar to those described earlier for iridabenzene **3** [81]. For example, **37** reacts with the electron-poor olefin methyl acrylate and undergoes a cheleotropic cycloaddition with sulfur dioxide, producing adducts **42** and **43**, respectively. But, unlike iridabenzene **3**, **37** also reacts with a variety of terminal alkynes, including phenylacetylene, trimethylsilylacetylene, and methylpropiolate, to generate products exhibiting novel iridaoxabarrelene frameworks (compounds **44–46**). The $^{31}P\{^1H\}$ NMR spectra of alkyne adducts **44–46** each consist of three separate doublet-of-doublet patterns, as expected for a *fac* phosphine geometry. In the $^{13}C\{^1H\}$ NMR, the C1 signals shift upfield to δ 137.4–137.9 from their position of δ 162.2 in **37**, reflecting the loss of carbene character. These signals exhibit strong coupling to the *trans* phosphine. The C3 signals shift upfield to δ 67.5–72.0 from δ 112.2 in **37**, indicating the formation of an sp^3 center. Carbon C4, in contrast, shifts downfield to δ 217.6–219.7 from δ 170.4 in **37**, as expected for a

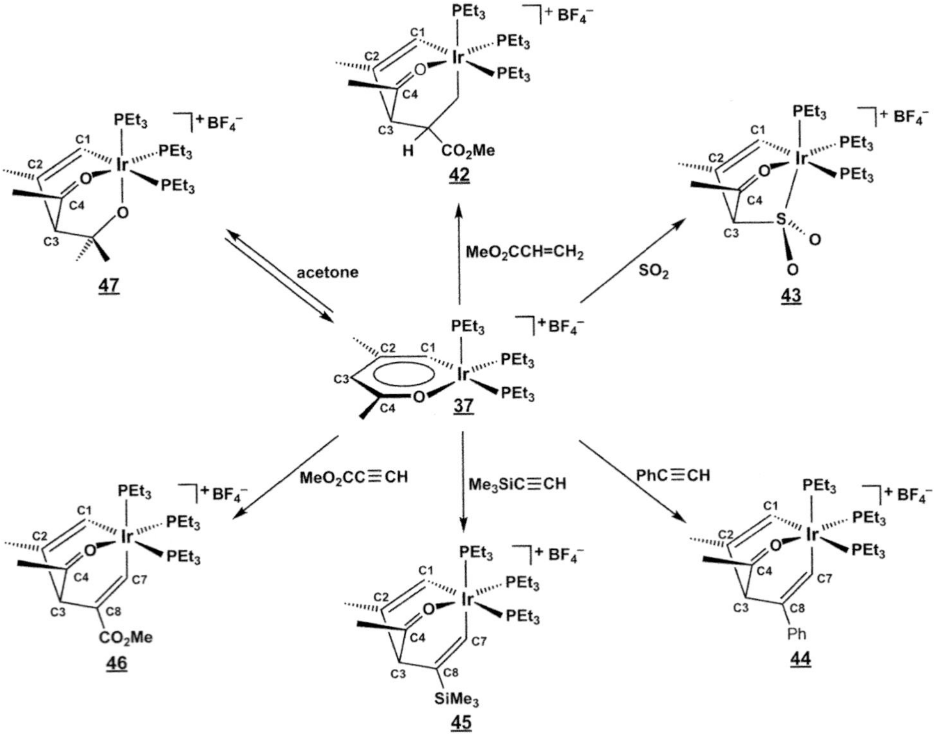

Scheme 2.20

carbonyl carbon. The signal for iridium-bound C7 (formerly an alkyne carbon) resonates in the range δ 144.6–169.4 and, like C1, shows strong phosphorus coupling. In the ^{1}H NMR, H1 shifts upfield to δ 7.10–7.19 from δ 9.35 in **37**, while H3 resonates at δ 5.12–5.50 vs. δ 6.45 in **37**. H7, the former alkyne hydrogen, resonates in the range δ 7.89–9.66. We had hoped that these iridaoxabarrelenes might be induced to release propyne, and produce new iridapyryliums, but, alas, they proved to be highly stable.

Another noteworthy cycloaddition is the reversible reaction of **37** with acetone, which we discovered accidentally while cooling a sample of **37** in d_6-acetone during a variable-temperature NMR experiment [80, 81]. At low temperature, the yellow acetone adduct (**47**) is stable, but when the mixture is warmed to room temperature, compound **37** is regenerated, as evidenced by the reappearance of its deep-violet color. The mechanism of this reaction has been explored by Martin and van der Boom by performing DFT calculations on an iridapyrylium model compound, $[C_4H_4OIr(PH_3)_3]^+$ [67, 82]. Unlike the concerted cycloaddition reactions of iridabenzene with CO_2 and CS_2, described earlier, the reaction of iridapyrylium with acetone occurs in a stepwise manner with prior coordination of an oxygen lone pair to the iridium center. This initial step is made possible by the contribution from 16e$^-$ resonance structure **III**. The cationic charge on iridapyrylium **37** also contributes to its reactivity toward the electronegative oxygen of acetone.

Perhaps the most surprising reaction of **37** results from treatment with nitrosobenzene (PhN=O). As shown in Scheme 2.21, the product, **48**, contains two metallacycles,

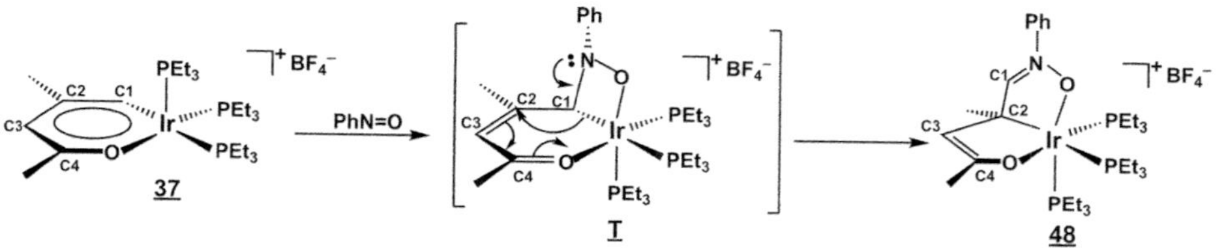

Scheme 2.21

which are fused along the Ir–C2 bond [80, 81]. Although the mechanism of this reaction is not known, the connectivity of the final product (i.e. nitrogen bonded to C1) suggests that the initial interaction may be a [2+2] cycloaddition involving the NO double bond of nitrosobenzene and the metal-carbene bond of **37**. Rearrangement of this strained cycloadduct, **T** (Scheme 2.21), could lead to the observed product, **48**.

2.13 Comparison of Iridabenzene 3 and Iridapyrylium 37

How, then, does iridapyrylium **37** compare to its iridabenzene analogue, **3**? Like **3**, it exhibits downfield ^{1}H NMR chemical shifts for its ring protons, consistent with the presence of an aromatic ring current. But unlike **3**, compound **37** reacts with 2e$^-$ donors at iridium in an *associative* manner, generating six-coordinate Ir(III) products with the iridapyran (iridaoxacyclohexa-1,3-diene) ring skeleton. This reactivity results primarily from the fact that the electrons in the Ir–O π bond of **37** can be localized on oxygen, rendering the iridium atom a reactive 16e$^-$ center. The positive charge on **37** also contributes to its enhanced reactivity toward anionic reagents such as hydride, methide, and chloride. Like **3**, compound **37** undergoes [4+2] cycloaddition reactions with a variety of unsaturated substrates. For example, both **3** and **37** react in a cheleotropic fashion with sulfur dioxide and both undergo cycloaddition with electron-poor olefins in Diels–Alder type processes. However, **37** also forms cycloadducts with a series of alkyne substrates that are completely unreactive toward **3**. The enhanced reactivity of **37** in this case probably results from the fact that a strong, localized carbon–oxygen double bond is formed in the product. It is also interesting to note that **37** undergoes *reversible* cycloaddition with ketones such as acetone. In this reaction, the formation of a carbon–oxygen double bond in the product is offset by the loss of a carbon–oxygen double bond in the ketone substrate. One final contrast involves the reactivity of **3** and **37** with nitrosobenzene. **3** reacts with this substrate in a standard [4+2] fashion, while **37** appears to initially form a [2+2] cycloadduct, which then rearranges to a novel product containing two fused five-membered rings. The reasons for this divergent behavior are not understood.

2.14 Synthesis and Spectroscopy of Iridathiabenzene

While studying the chemistry of oxygen-containing iridacycles, we began concurrently to investigate the analogous sulfur systems [83, 84] with the goal of synthesizing iridathiabenzene. At about the same time, several other research groups reported

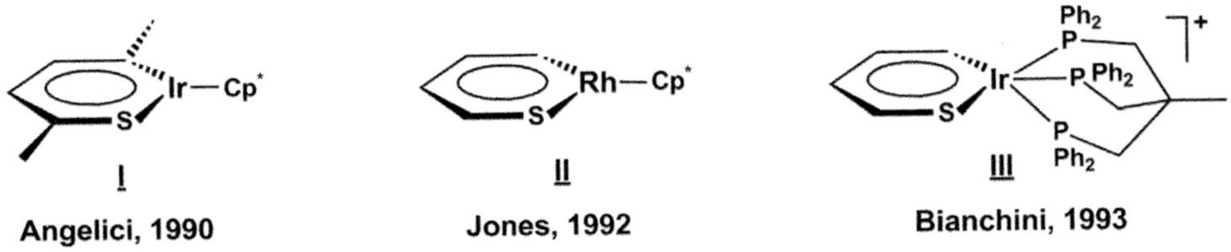

Figure 2.14 Previous examples of metallathiabenzenes. Compound **II** is unstable and undergoes further reactions. Cp* is the η^5-pentamethylcyclopentadienyl ligand.

syntheses of metallathiabenzenes via metal-mediated activation of a C–S bond in thiophene. Angelici [85, 86], for example, synthesized iridathiabenzene **I** (Figure 2.14), while Jones [87] reported the low-temperature detection of rhodium analogue **II**. Bianchini [88, 89], in turn, produced iridathiabenzene **III**, together with a close analogue containing a fused benzene ring.

Our synthetic approach to iridathiabenzene differed from these by employing a thiapentadienide reagent to supply the ring carbons and sulfur. We decided to use 2,3-dimethylthiapentadienide (rather than its 2,4-dimethylated analogue), because it could be easily made from the symmetrical starting material, 2,5-dihydro-3,4-dimethylthiophene [90] by base-initiated ring opening [91]. As shown in Scheme 2.22, treatment of $(Cl)Ir(PEt_3)_3$ with lithium 2,3-dimethyl-5-thiapentadienide produces dimethyliridathiacyclohexa-1,3-diene **49** via C–H bond activation on the terminus of an η^1-2,3-dimethyl-5-thiapentadienyl ligand in intermediate **U** [92, 93]. The X-ray crystal structure of **49** [93] shows an octahedral *mer* coordination geometry, an essentially planar ring (internal angles sum to 719.5°) and localized bonding within the carbon/sulfur portion of the ring (C1–C2 = 1.340(14) Å, C2–C3 = 1.456(18) Å, C3–C4 = 1.380(19) Å, C4–S = 1.731(12) Å). The Ir–C_{vinyl} and Ir–S bond distances of 2.093(10) Å and 2.433(3) Å, respectively, are typical of single bond lengths. The circumference of the ring is 10.43 Å. The ^{1}H NMR spectrum of **49** exhibits signals for ring protons H1 and H4 at δ 7.05 and δ 5.78, respectively, while the metal-hydride signal appears far upfield at δ −16.10. In the $^{13}C\{^1H\}$ NMR, the ring carbons resonate at δ 118.8 (C1), δ 127.3 (C2), δ 127.3 (C3), and δ 120.9 (C4), and the signal for C1

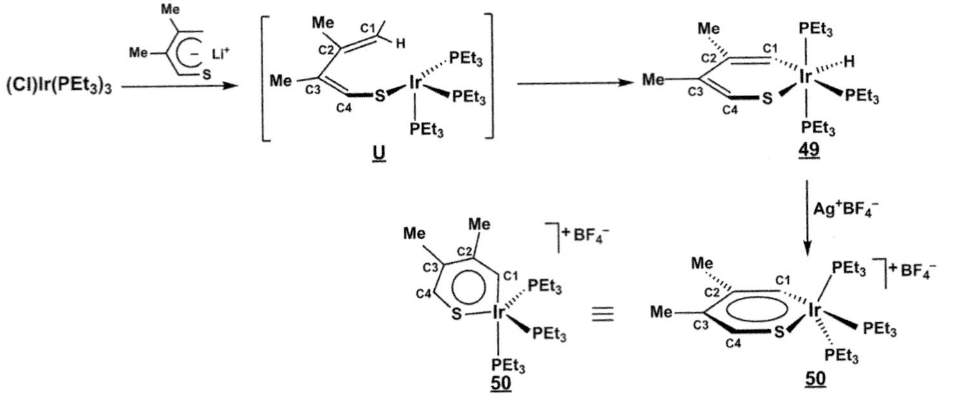

Scheme 2.22

shows characteristic doublet coupling (J_{CP} = 72.0 Hz) to the ^{31}P nucleus of the *trans* equatorial phosphine.

As in the synthesis of iridapyrylium (vide supra), treatment of yellow-orange **49** with silver tetrafluoroborate in THF leads to the immediate production of deep-red iridathiabenzene, **50** (see Scheme 2.22) [92, 93]. The ^{1}H NMR spectrum of **50** shows downfield shifting for ring protons H1 and H4, consistent with its formulation as an aromatic species. Proton H1 resonates at δ 10.36 and is a quartet (J_{HP} = 8.4 Hz) due to coupling to three equivalent ^{31}P nuclei (vide infra), while H4 resonates at δ 8.61 and is likewise a quartet (J_{HP} = 3.3 Hz). The ring carbons resonate at δ 165.1 (C1), δ 143.6 (C2), δ 134.9 (C3), and δ 130.0 (C4), and C1 appears as a phosphorus-coupled quartet (J_{CP} = 21.6 Hz). The ^{31}P{^{1}H} NMR signal for **50**, like that for iridabenzene **3** and iridapyrylium **37**, is a singlet at room temperature, indicating that the three phosphine ligands are exchanging rapidly in solution, probably via a Berry-type process. However, when the temperature is lowered to −90°C, the ^{31}P{^{1}H} signal broadens and ultimately resolves into two sharp resonances with an intensity ratio of 2:1. Simulation of the variable temperature ^{31}P{^{1}H} NMR spectra yields a $\Delta G^{\ddagger}$ value of 9.5(0.2) kcal/mol for the intramolecular phosphine exchange process.

2.15 Structure of Iridathiabenzene 50

The molecular structure of **50** has been determined by X-ray diffraction [92] and is presented in Figure 2.15. Consistent with the low-temperature ^{31}P{^{1}H} NMR spectrum, the molecule displays approximate (although not crystallographically imposed) mirror-plane symmetry.

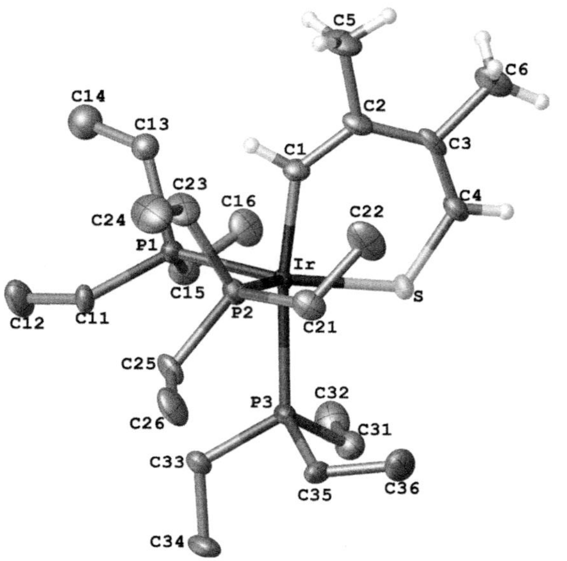

Figure 2.15 X-ray structure of the cation in iridathiabenzene, **50**, with phosphine H's omitted for clarity. The plot was created using the Olex2 software package [42]. (*See color plate section for the color representation of this figure.*)

The symmetry plane includes the metallacyclic ring and phosphorus atom P3, while bisecting the P1–Ir–P2 angle. In contrast to iridabenzene **3**, which adopts a square pyramidal coordination geometry in the solid state, the geometry of **50** is probably best described as a distorted trigonal bipyramid with C1 and P3 occupying the axial sites and S, P1, and P2 occupying the equatorial sites. However, the equatorial ligands are significantly distorted from idealized positions. Hence, the three equatorial L–Ir–L angles, which ideally should each be 120°, are actually 136.8(1)°, 127.2(1)°, and 95.9(1)°. Surprisingly, the smallest of these angles involves the two bulky PEt_3 ligands!

Bonding within the metallacycle in **50** is delocalized, consistent with aromatic character. The carbon–carbon bond distances have moved toward equalization (C1–C2 = 1.396(16) Å, C2–C3 = 1.415(15) Å, C3–C4 = 1.361(16) Å, C4–S = 1.713(12) Å), while bonds Ir–C1 and Ir–S (2.019(10) Å and 2.249(3) Å, respectively) have both shortened substantially from their values in **49** (2.093(10) Å and 2.433(3) Å, respectively), indicating significant metal participation in the ring π-bonding. The circumference of the ring has shrunk to 10.15 Å from its value of 10.43 Å in **49**. The nonmetal portion of the ring (C1/C2/C3/C4/S) is very nearly planar (mean deviation 0.015 Å), while the iridium center lies 0.185 Å out of this plane. The dihedral angle between planes C1/C2/C3/C4/S and C1/Ir/S is 7.0°. Although the solid-state structure of **50** is a distorted trigonal bipyramid, the square pyramidal coordination geometry is undoubtedly accessible in solution, and many of the reactions of **50** (vide infra) appear to proceed through that geometry.

2.16 Chemical Reactivity of Iridathiabenzene 50

2.16.1 Ligand Addition and Cycloaddition

Iridathiabenzene **50**, like iridapyrylium **37**, has three important resonance contributors (cf. Figure 2.13). Because of the contribution from the $16e^-$ resonance structure (**III**), **50** reacts with $2e^-$ donor ligands L at the iridium center to produce octahedral iridathiacyclohexa-1,3-diene products [92, 93]. For example, treatment of **50** with Cl^- produces **51** (Scheme 2.23) as an equilibrium mixture of *cis* and *trans* isomers (in a 40:60 ratio).

These species are thought to interconvert through a transient neutral iridathiabenzene, **V** (see Scheme 2.23). Similarly, treatment of **50** with excess PMe_3 yields the *tetrakis*(PMe_3) product **52** (Scheme 2.23), an analogue of the iridaoxacyclohexa-1,3-diene **41**, described earlier. Compound **50** is also capable of undergoing [4+2] cycloaddition reactions. For example, treatment with nitrosobenzene (PhN=O) leads to the formation of cycloadduct **53** (Scheme 2.23). The iridathiacyclohexa-1,4-diene ring in **53** is characterized by a relatively upfield chemical shift position for sp^3 carbon C3 (δ 50.8) and downfield shift positions for the thioaldehyde carbon C4 (δ 154.2) and hydrogen H4 (δ 8.31). Recall that nitrosobenzene gives a similar product (**14**) when reacted with iridabenzene **3** but the novel fused-ring product (**48**) when reacted with iridapyrylium **37**.

2.16.2 Coordination to Other Metals

Like iridabenzene **3**, iridathiabenzene **50** displaces *p*-xylene from (η^6-*p*-xylene)$Mo(CO)_3$ to produce [(η^6-iridathiabenzene)$Mo(CO)_3$]$^+$(BF_4^-), **54** (Scheme 2.24). The X-ray crystal

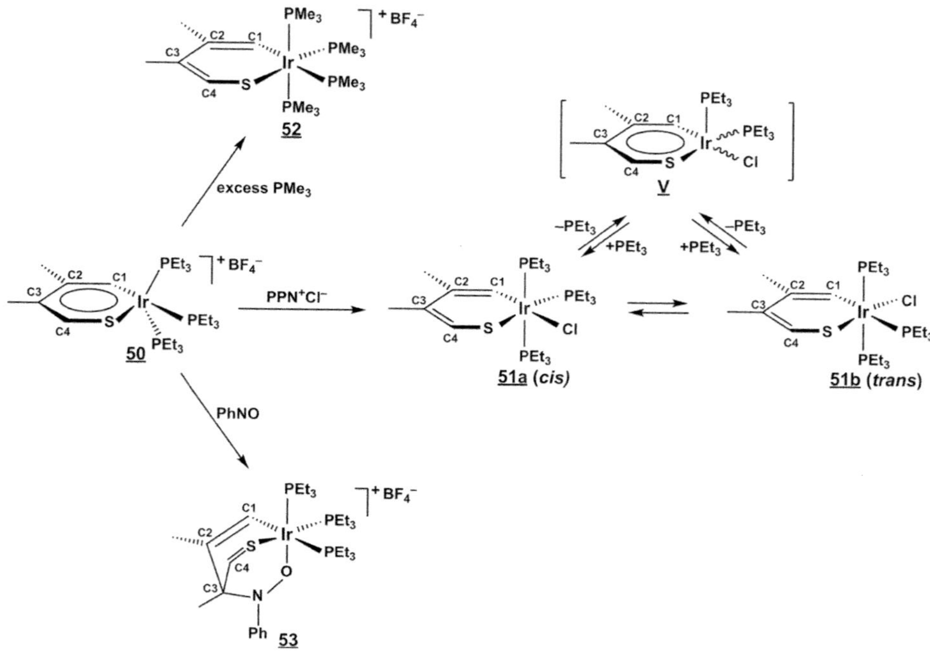

Scheme 2.23

structure of **54** [92] shows that the molybdenum atom is strongly π-complexed to the six atoms of the iridathiabenzene ring: the Mo–C_{ring} distances range from 2.382(17) Å to 2.459(16) Å, the Mo–S distance is 2.544(5) Å, and the Mo–Ir distance is 3.0180(16) Å. Bonding within the ring is still delocalized. However, the bond lengths in the complexed ring are slightly longer than those in the free iridathiabenzene, because π-electron density is being removed from the ring by the $Mo(CO)_3$ moiety. The circumference of the ring in **54** is 10.37 Å, as compared to 10.15 Å in **50**. The non-metal portion of the ring in **54** is close to planar, but the iridium atom is displaced 0.21 Å out of that plane and away from Mo. The dihedral angle between the planes C1/C2/C3/C4/S and C1/Ir/S is 7.7°. The corresponding numbers for **50** are 0.185 Å and 7.0°. The coordination geometry at the iridium center in **54** is best described as a square pyramid, where C1, S, and two PEt_3 P's define the basal ligand set while the third PEt_3 occupies the unique axial site. This geometry provides an accessible face for interaction with the $Mo(CO)_3$ moiety.

Unlike **50**, which undergoes intramolecular phosphine exchange in solution, **54** is stereochemically rigid. Hence, the $^{31}P\{^1H\}$ NMR spectrum at room temperature consists of three discrete signals for the three inequivalent phosphine ligands. In the ^1H NMR of **54**, the ring protons shift upfield from their positions in **50**, as is typical for coordinated arenes [72]. Protons H1 and H4 in **54** resonate at δ 8.27 and δ 6.59, respectively, as compared to δ 10.36 and δ 8.61 in **50**. H1 is strongly coupled to the *cis*-basal phosphine (J_{HP} = 15.6 Hz), while H4 also appears as a phosphorus-coupled doublet (J_{HP} = 9.6 Hz). Carbons C1, C2, C3, and C4 resonate at δ 129.0, δ 117.6, δ 109.5, and δ 89.0, respectively, and C1 exhibits strong coupling to the ^{31}P nucleus of the *trans*-basal phosphine (J_{CP} = 61.8 Hz). The three carbonyl ligands give rise to a single peak in the $^{13}C\{^1H\}$

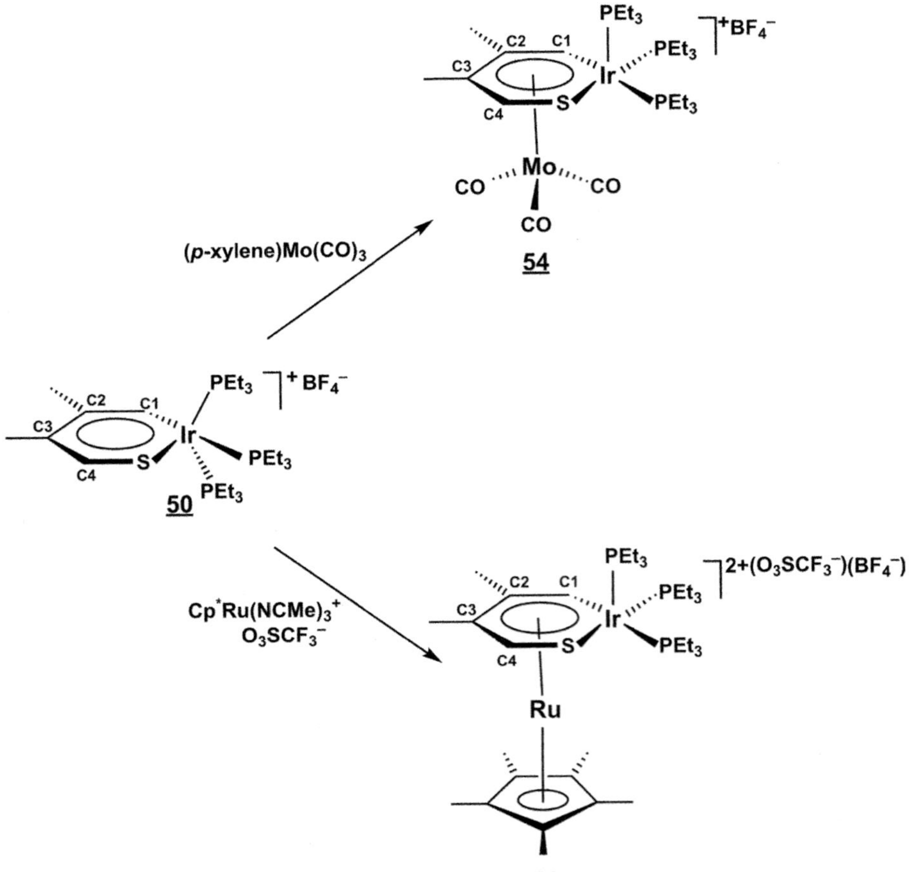

Scheme 2.24

NMR, indicating that the iridathiabenzene ring is rotating freely with respect to the $Mo(CO)_3$ moiety.

The infrared spectrum of **54** shows the presence of two intense γ(CO) bands A_1 and E at 1962 and 1895 cm^{-1}. By comparison, the γ(CO) bands for (η^6-iridabenzene) $Mo(CO)_3$ (**24**) appear at 1918 and 1836 cm^{-1}. The higher energy of the bands in **54** reflects the positive charge on **54** and, consequently, the reduced π-back-bonding into CO π* orbitals.

As shown in Scheme 2.24, iridathiabenzene **50** also cleanly displaces acetonitrile ligands from $[Cp^*Ru(NCMe)_3]^+(O_3SCF_3^-)$ to produce the "sandwich" compound, $[(\eta^6$-iridathiabenzene$)RuCp^*]^{2+}(O_3SCF_3^-)(BF_4^-)$, **55** [93]. The ^1H NMR spectrum of **55** is strikingly similar to that of **54** with ring protons H1 and H4 resonating at δ 8.83 and δ 6.33, respectively. H1 is strongly coupled to the *cis*-basal phosphine (J_{HP} = 15.6 Hz), while H4 also shows substantial phosphorus coupling (J_{HP} = 9.0 Hz). The ring carbon chemical shifts in **55** are also very similar to those in **54** with C1, C2, C3, and C4 resonating at δ 116.8, δ 110.6, δ 100.5, and δ 77.9, respectively, and C1 exhibiting strong coupling to the *trans*-basal PEt_3 ligand (J_{CP} = 62.7 Hz). In the $^{31}P\{^1H\}$ NMR spectrum of **55**, three separate signals are observed for the three inequivalent PEt_3 signals, again indicating stereochemical rigidity at iridium.

It should be noted that Angelici's iridathiabenzene, **I** (Figure 2.14), also forms π-complexes with other metal centers [94]. In particular, **I** cleanly displaces the acetonitrile ligands from $(MeCN)_3M(CO)_3$ (M = Cr, Mo, W) to produce (η^6-**I**)$M(CO)_3$. The molybdenum derivative can also be produced by displacement of toluene from (η^6-toluene)$Mo(CO)_3$. Under photolysis conditions, **I** displaces chlorobenzene from (η^6-chlorobenzene)$Fe(Cp)^+$ to produce the sandwich compound, (η^6-**I**)$Fe(Cp)^+$. These reactions, like those of **50**, demonstrate that iridathiabenzenes are often better ligands than conventional organic arenes for a variety of metal centers.

2.17 Comparison of Iridathiabenzene 50 and Iridapyrylium 37

Iridathiabenzene **50** and iridapyrylium **37** both exhibit the spectroscopic features consistent with the presence of an aromatic ring system. The X-ray crystal structure of **50** shows the expected delocalized bonding with Ir–C, Ir–S, C–S, and C–C bonds all intermediate in length between normal single and double bonds. Unfortunately, the solid-state structure of **37** has not been obtained, so a side-by-side structural comparison cannot be made. While the solid-state structure of **50** shows a distorted trigonal bipyramidal coordination geometry at Ir, the square pyramidal structure is easily accessible in solution (via a Berry-type process), and most of the reactions of both **50** and **37** appear to proceed through this geometry. Both **50** and **37** react associatively at Ir with small $2e^-$ donors such as PMe_3 and Cl^-, generating six-coordinate Ir(III) products with 1,3-cyclohexadiene ring skeletons. This reactivity results primarily from the fact that the electrons in the Ir–S or Ir–O π-bond can be localized on the heteroatom, rendering the iridium atom a reactive $16e^-$ center. But **50** also reacts with sterically bulky anionic reagents like *tert*-butylthiolate to displace a PEt_3 ligand and retain the aromatic ring structure in a five-coordinate environment (vide infra). Both **50** and **37** engage in [4+2] cycloaddition chemistry, although this type of reaction is less commonly observed for **50** than for **37**. The reason for this may be that the iridathiacyclohexa-1,4-diene cycloaddition product contains a C–S double bond, which is less stable than the C–O double bond produced in the analogous iridaoxacyclohexa-1,4-diene product. Iridathiabenzene **50** is able to displace conventional arenes from organometallic reagents, forming η^6-π complexes with $Mo(CO)_3$ and $(C_5Me_5)Ru^+$ metal-ligand fragments. Iridapyrylium **37**, in contrast, does not form π-complexes with organometallic reagents, perhaps reflecting a somewhat less robust aromaticity.

2.18 Synthesis and Structure of a Neutral Iridathiabenzene

While we were studying the reaction of **50** with chloride, we observed the transient existence of neutral iridathiabenzene **V** (Scheme 2.23) in polar solvents [93]. We reasoned that if the chloride were replaced with a bulkier anionic ligand, a stable neutral iridathiabenzene should result. When we treated **50** with sodium *tert*-butylthiolate,

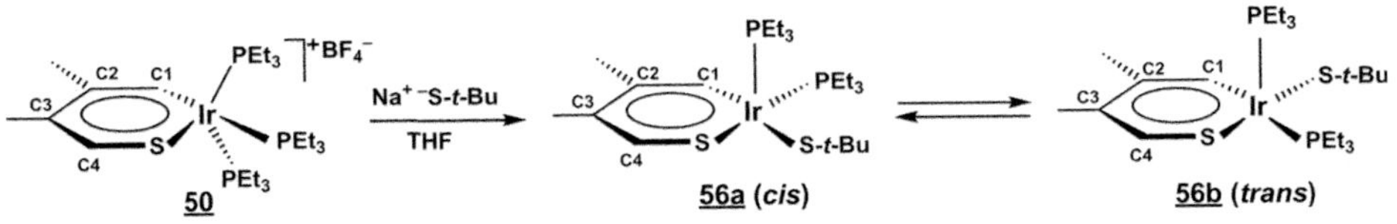

Scheme 2.25

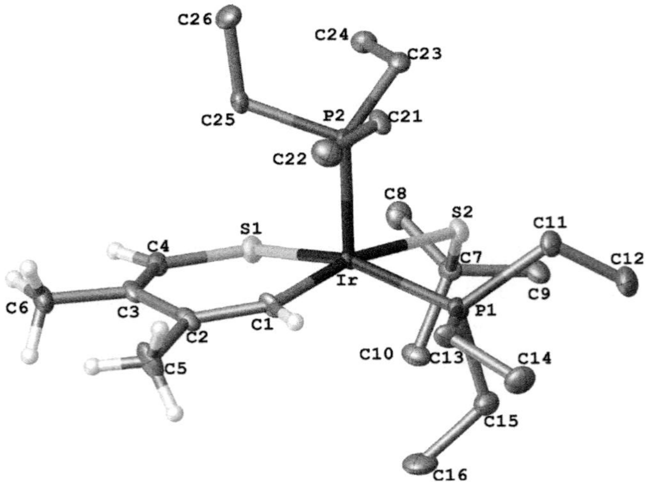

Figure 2.16 X-ray structure of neutral iridathiabenzene, **56a**, with phosphine and thiolate H's omitted for clarity. The plot was created using the Olex2 software package [42].

that is exactly what happened, and iridathiabenzene **56** (Scheme 2.25) was isolated as an equilibrium mixture of square pyramidal *cis* and *trans* isomers [95].

An X-ray crystal structure of the *cis* isomer, **56a** (Figure 2.16), has confirmed the square pyramidal coordination geometry, in which PEt_3 ligand P2 occupies the unique axial site, while ring sulfur S1, ring carbon C1, thiolate sulfur S2, and the remaining PEt_3 ligand (P1) occupy the four basal sites. The two sulfur atoms reside *cis* to one another (S1–Ir–S2 = 93.30(10)°). The nonmetal portion of the ring (C1/C2/C3/C4/S1) is nearly planar (mean deviation = 0.050 Å), while the Ir atom sits 0.27 Å out of this plane. The dihedral angle between the C1/C2/C3/C4/S1 plane and the S1/Ir/C1 plane is 9.9°. The Ir–C1 and Ir–S1(ring) bond lengths of 1.993(9) Å and 2.263(3) Å, respectively, are similar to those in **50** (2.019(10) Å and 2.249(3) Å), reflecting substantial π-bonding between these ring atoms. The bond lengths within the nonmetal portion of the ring are also similar to those in **50** (C1–C2 = 1.391(15) Å, C2–C3 = 1.455(16) Å, C3–C4 = 1.336(15) Å, C4–S1 = 1.711(11) Å), although C2–C3 is slightly longer and C3–C4 is slightly shorter than in **50**. The circumference of the ring is 10.15 Å, identical to that in **50**.

2.19 Spectroscopy of Neutral Iridathiabenzene 56

At room temperature in solution, isomers **56a** and **56b** rapidly interconvert, giving rise to averaged signals in the NMR. But upon cooling to −60°C, this interconversion is

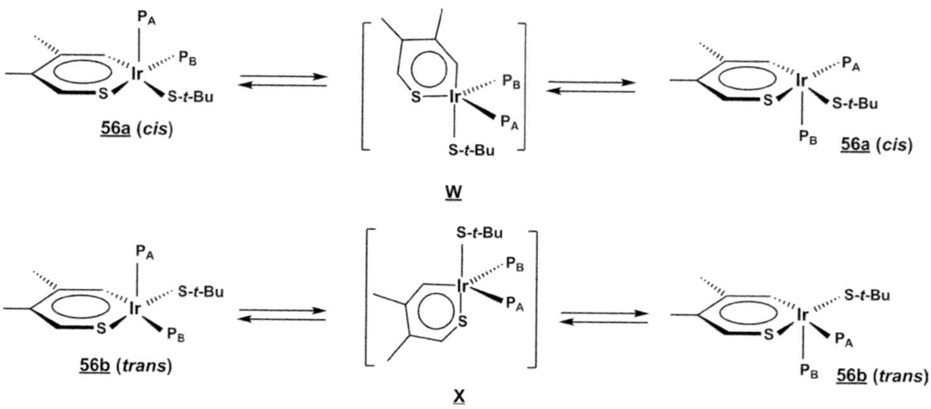

Scheme 2.26

halted and the two separate isomers can be observed by NMR. Curiously, the relative ratio of the isomers varies dramatically with solvent. In polar solvents such as methanol, **56a** is favored (**56a**:**56b** = 60:40), while in nonpolar solvents like methylcyclohexane, **56b** predominates (**56a**:**56b** = 10:90).

The *cis* isomer, **56a**, exhibits a single sharp peak in the $^{31}P\{^1H\}$ NMR at −60°C, indicating rapid exchange of the two PEt_3 ligands. In the ^1H NMR, H1 resonates at δ 9.79 and is a binomial triplet due to coupling to the two equivalent (exchanging) phosphines (J_{HP} = 7.7 Hz). Ring proton H4 resonates at δ 8.13 and also shows triplet coupling (J_{HP} = 5.2 Hz). In the $^{13}C\{^1H\}$ NMR at −60°C, the ring carbons resonate at δ 152.9 (C1), δ 139.1 (C2), δ 132.4 (C3), and δ 128.3 (C4). The *trans* isomer, **56b**, exhibits two equal intensity peaks in the $^{31}P\{^1H\}$ NMR at −60°C, indicating that it is not fluxional at this temperature. The ring protons, H1 and H4, appear at δ 11.02 and δ 7.88, respectively, while the ring carbons resonate at δ 171.6 (C1), δ 140.7 (C2), δ 132.2 (C3), and δ 126.0 (C4). C1 is strongly coupled (J_{CP} = 83.4 Hz) to the basal phosphine which is situated *trans* to it.

The difference in fluxional behavior exhibited by **56a** and **56b** reflects a difference in the energies of the intermediates that must be accessed in order for the exchange to occur via a Berry-type process. As shown in Scheme 2.26 (top line), the *cis* isomer **56a** exchanges through trigonal bipyramidal intermediate **W** in which ring carbon and thiolate sulfur occupy *trans*-diaxial sites. In contrast, the *trans* isomer **56b** must exchange through the trigonal bipyramidal intermediate **X** in which ring sulfur and thiolate sulfur are *trans*-diaxial (Scheme 2.26, bottom line). Why is **W** more accessible than **X**? We are not entirely sure, but it is interesting to note that the X-ray structure of **50** has the same geometry as **W** with the ring sulfur in an equatorial coordination site and the ring carbon in an axial site of a trigonal bipyramid.

2.20 Chemical Reactivity of Neutral Iridathiabenzene 56

Some of the reactions of neutral **56** are similar to those of its cationic relative, **50**. For example, as shown in Scheme 2.27, treatment of **56** with excess PMe_3 results in addition of PMe_3 to the iridium center *and* substitution of the remaining PEt_3's, producing

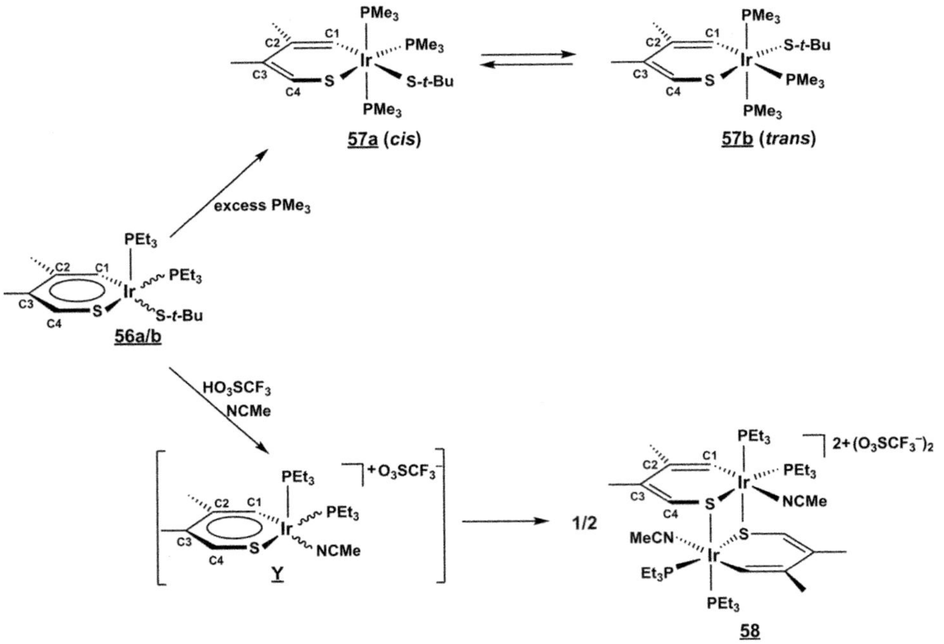

Scheme 2.27

iridathiacyclohexa-1,3-diene **57** as an equilibrium mixture of *cis* and *trans* isomers (**57a**:**57b** = 67:33). But other reactions diverge from those of **50**. For example, when **56** is treated with triflic acid in acetonitrile, the *tert*-butylthiolate sulfur is protonated, causing *tert*-butylthiol to dissociate from the iridium center. It is replaced by the much smaller acetonitrile ligand, and the resulting sterically destabilized iridathiabenzene (**Y**; Scheme 2.27) then dimerizes to **58** [95]. In contrast, cationic iridabenzene **50** is unreactive toward acidic reagents.

The reaction of **56** with $[Cp^*Ru(NCMe)_3]^+(O_3SCF_3^-)$ also differs from that of **50**. As shown in Scheme 2.28, a "sandwich" compound, **59**, is formed (as an 85:15 mixture of *cis* and *trans* isomers), but the X-ray crystal structure of **59a** [95] reveals that the ruthenium atom is not bonded to iridium! Instead, C1, C2, C3, C4, and S1 of the metallacycle are bonded to ruthenium in an η^5-fashion: the Ru–C_{ring} distances range from 2.154(5) Å

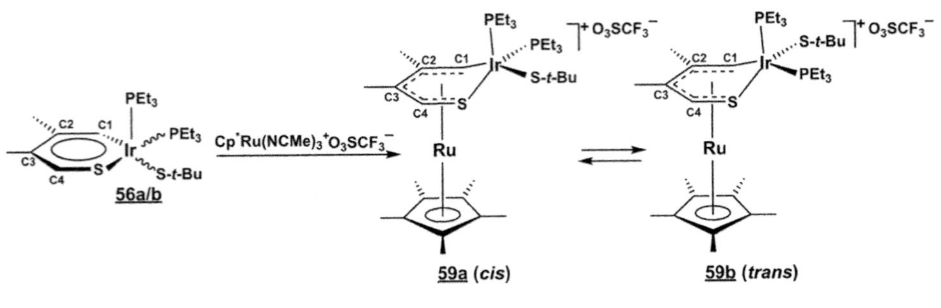

Scheme 2.28

to 2.254(6) Å, while Ru–S1 = 2.4316(15) Å. The Ir atom is pushed up and out of the plane by 0.71 Å[8], and the dihedral angle between C1/C2/C3/C4/S1 and C1/Ir/S1 is 26.1°. The large displacement of the iridium atom up and out of the ring plane is probably a response to unfavorable steric interactions between the η^5-C_5Me_5 ligand on Ru and the *tert*-butylthiolate and PEt_3 groups on Ir. Within the metallacycle, the Ir–C1 and Ir–S1 bonds have lengthened somewhat compared to their values in iridathiabenzene **56a**, reflecting a shift toward single bond character. In **59a**, these distances are 2.054(6) Å and 2.3079(13) Å, respectively, versus 1.993(9) Å and 2.263(3) Å in **56a**. The bonding within the remaining portion of the metallacycle is still delocalized with C1–C2 = 1.414(8) Å, C2–C3 = 1.445(8) Å, C3–C4 = 1.390(8) Å, and C4–S1 = 1.742(6) Å. Overall, the bonds within this portion of the ring are lengthened slightly over their values in **56a**, and the ring circumference has expanded to 10.35 Å from 10.15 Å in **56a**.

In the $^{31}P\{^1H\}$ NMR spectrum of **59a**, separate peaks are observed for the axial and basal phosphines, indicating that the fluxional process displayed by **56a** (vide supra) has been arrested upon coordination to ruthenium. In the ^1H NMR, H1 shifts upfield to δ 7.72 (from δ 9.79 in **56a**) and is a doublet (J_{HP} = 12.5 Hz) due to coupling to the *cis*-basal phosphine. Similarly, H4 shifts to δ 5.99 (from δ 8.13 in **56a**) and is also a doublet (J_{HP} = 8.5 Hz). The ring carbons likewise shift upfield in the $^{13}C\{^1H\}$ NMR to δ 101.6 (C1), δ 107.0 (C2), δ 99.4 (C3), and δ 75.7 (C4). In solution at room temperature, **59a** slowly establishes an equilibrium with its *trans* isomer, **59b**, but the rate of this process is much slower that that observed for the parent compound, **56a**. While isomers **56a, b** exhibit averaged NMR signals at room temperature (indicating rapid interconversion), the NMR signals for **59a, b** are sharp at room temperature.

The iridacycle in **59** can be viewed as a neutral 6e$^-$ donor, using two carbon–carbon double bonds and a sulfur lone pair to bond to ruthenium (see Figure 2.17). The Ru atom then bears the compound's formal positive charge and can be counted as a Ru(II)≡d^6 center. The iridium is formally 16 e$^-$ Ir(III) but may be stabilized by additional π-donation from the sulfur atom of the t-butylthiolate ligand. This view is supported by the observation that the Ir–S2 (thiolate) bond in **59a** has been shortened significantly

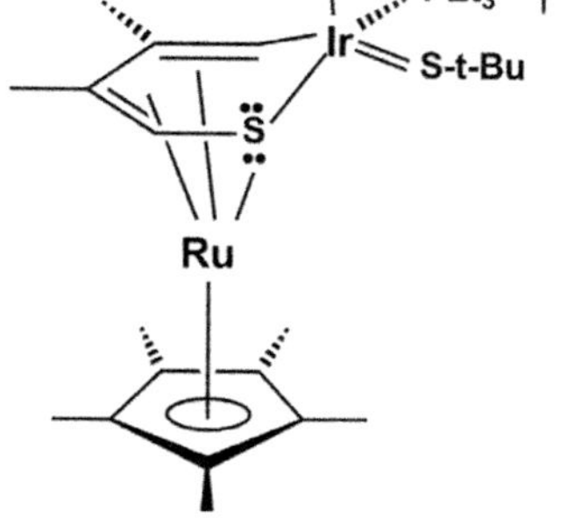

Figure 2.17 Dominant resonance structure for neutral iridathiabenzene coordinated to Cp*Ru$^+$ (compound **59a**).

8 The non-bonding Ru–Ir distance is 3.196 Å.

(by 0.074 Å) as compared to its length in **56a**. In fact, the ability of the thiolate sulfur (S2) to compensate for the loss of the Ir-ring sulfur (S1) π-interaction may be the key to the stability of the observed η^5-bonding mode in **59**.

2.21 Related Metal-Coordinated Metallabenzenes

As discussed earlier (Section 2.8.6), iridabenzene **3** can be coordinated to a $Mo(CO)_3$ moiety by arene exchange chemistry. However, a number of other metal-coordinated metallabenzenes have been synthesized directly, i.e. without prior generation of the *free* metallabenzene. In this section, we summarize the important work of other researchers who have synthesized metal-coordinated metallabenzenes directly from 2,4-dimethylpentadienyl-metal precursors.

In 1987, Ernst [96, 97] reported isolation of the metal-coordinated molybdenabenzene complex, **60a** (Scheme 2.29), from the reaction of two equivalents of (η^5-2,4-dimethylpentadienyl)$Mo(CO)_3^-$ with 1,2-diiodoethane. Although the detailed mechanism of this reaction is not known, one of the pentadienyl ligands loses a hydrogen atom from each of its terminal carbon atoms to become the carbon portion of the metallabenzene ring. This type of hydrogen loss from pentadienyl ligands has previously been observed in the mass spectra of "half-open" and "open" metallocenes [96]. Thermal conversions of pentadienyl ligands to cyclopentadienyl ligands with hydrogen elimination are also known [98–100]. In **60a**, the metallabenzene serves as a $6e^-$ neutral ligand, enabling the second molybdenum atom to achieve an $18e^-$ count.

The bonding within the carbon portion of the metallabenzene ring is delocalized, and the Mo–C_α bond distances (2.150(5) Å and 2.161(4) Å) suggest some metal participation in ring π-bonding. The metallabenzene Mo center resides 0.33 Å out of the plane made by C1/C2/C3/C4/C5, resulting in a dihedral angle of 11.6° between this plane and plane C1/Mo/C5. The Mo–Mo distance is 2.989(1) Å. In solution at room temperature, the molecule exhibits broad ^1H NMR line shapes, probably as a result of rotation (or oscillation) of the η^5-2,4-dimethylpentadienyl ligand. This motion can be arrested by cooling to −80°C, and the individual molybdenabenzene ring protons can be observed at δ 8.31 (H1), δ 5.74 (H3), and δ 6.91 (H5). It is not known whether the metallabenzene ligand can rotate with respect to the $Mo(CO)_3$ moiety.

Compound **60a** undergoes ligand substitution reactions when treated with trimethyl phosphite or *bis*(dimethylphosphino)ethane (Scheme 2.29). In each case, one of the CO ligands on the metallabenzene Mo atom is replaced with the phosphorus-based ligand,

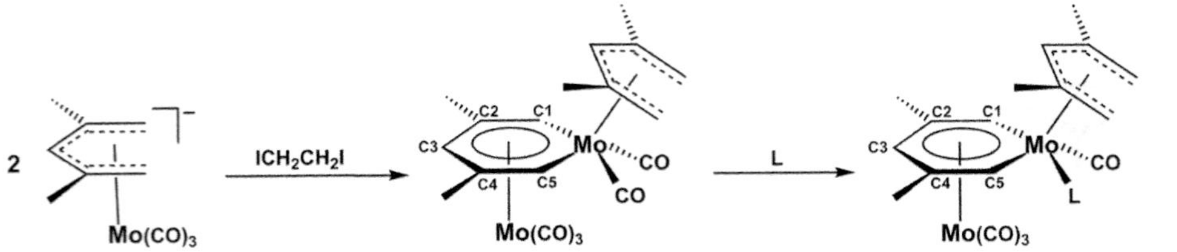

60a

60b: L = $P(OMe)_3$
60c: L = η^1-$Me_2PCH_2CH_2PMe_2$

Scheme 2.29

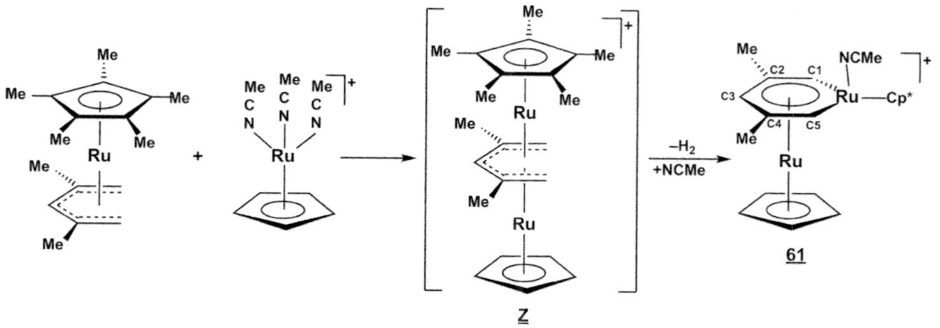

Scheme 2.30

generating **60b** and **60c**. The position of this substitution can be readily rationalized; the metallabenzene Mo atom has a higher oxidation state (Mo(II)) than the other Mo atom (Mo(0)) and hence is a less effective back-bonder to CO. As a result, the carbonyls on the metallabenzene Mo are less tightly bound and more easily substituted. Although X-ray structures of the ligand-substituted derivatives, **60b**, **c**, have not been obtained, NMR data are fully consistent with retention of the aromatic ring.

Salzer [101] produced metal-coordinated metallabenzene **61** (Scheme 2.30) by treating (η^5-2,4-dimethylpentadienyl)(η^5-C_5Me_5)Ru with the labile precursor [(η^5-C_5H_5)Ru($CH_3CN)_3]^+$ PF_6^-. The triple-decker complex **Z** may serve as an intermediate in this reaction. Loss of two hydrogen atoms from the terminal carbons of the central pentadienyl would lead to the coordinated ruthenabenzene product. The ruthenabenzene ligand serves as a $6e^-$ neutral donor, allowing the central ruthenium atom (which carries the compound's formal charge) to achieve an $18e^-$ count.

The ring metal atom in **61** is significantly displaced out of the plane of the ring carbons and away from the second Ru atom (0.63 Å), resulting in a dihedral angle of 27.0° between planes C1/C2/C3/C4/C5 and C1/Ru/C5. This large displacement is probably due primarily to strong steric interactions between the cyclopentadienyl and pentamethylcyclopentadienyl ligands (which lie in approximately perpendicular planes), but may also help to optimize the Ru–Ru bond length, which is 2.767(1) Å.

The metallabenzene ring in **61** sits on a crystallographically imposed mirror plane and the carbon–carbon bonds are fully delocalized (C1–C2 = 1.420(5) Å, C2–C3 = 1.425(5) Å). The Ru–C1 distance of 1.983 Å falls between normal Ru–C single and double bond lengths. In the ^{1}H NMR, the H1s resonate at δ 10.31, while H3 resonates at δ 5.47. The corresponding carbon shifts are δ 178.5 (C1) and δ 95.6 (C3).

A ruthenium-coordinated nickelabenzene complex, **62** (Scheme 2.31), has also been synthesized by Salzer, using a similar approach [102]. In this case (η^5-2,4-dimethylpentadienyl)(η^5-C_5Me_5)Ru is treated with [(η^5-C_5H_5)Ni(η^5-C_5H_5)Ni(η^5-C_5H_5)]$^+BF_4^-$, which serves as a labile source of the $14e^-$ fragment (η^5-C_5H_5)Ni^+. Again, a triple-decker species, **AA**, in which a 2,4-dimethylpentadienyl group serves as the central ligand, may be involved as an intermediate. The metallabenzene ligand in **62** again serves as a $6e^-$ neutral donor, enabling the central ruthenium center (formally a cation) to achieve an $18e^-$ count.

The metallabenzene ring in **62** is still somewhat non-planar, but the displacement of the nickel atom out of the ring and away from ruthenium (0.36 Å) is less pronounced

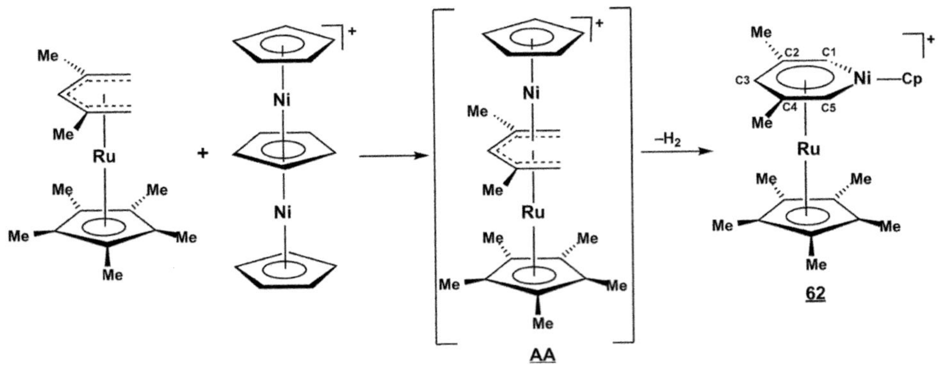

Scheme 2.31

than the displacement of Ru in **61** (0.63 Å). The dihedral angle between C1/C2/C3/C4/C5 and C1/Ni/C5 in **62** is 16.4° vs 27.0° for the analogous angle in **61**. Salzer [102] suggested that the reduced displacement results from the fact that the optimal Ru–Ni distance is shorter than the optimal Ru–Ru distance. However, Girolami [103] points out that, in general, shorter M–M bonds do *not* correlate with the smaller metal displacements and reduced dihedral angles in metal-coordinated metallabenzene rings. He concludes that the smaller displacement results from reduced steric interactions between the ligands on the two metal centers.

The C–C bonds within the ring in **62** are delocalized, and the Ni–C bond lengths of 1.867(9) Å and 1.838(9) Å are consistent with some partial double bond character. The Ru–Ni distance is 2.554(1) Å. In solution, the compound exhibits mirror-plane symmetry. The H1s and H3 resonate at δ 9.7 and δ 5.7, respectively, in the ^{1}H NMR, while the corresponding ^{13}C peaks appear at δ 167.7 and δ 91.9.

More recently, Salzer [104] succeeded in generating the first examples of *bis*(metallabenzene) sandwich compounds, **63a, b** (Scheme 2.32). This was accomplished by treating an open ruthenocene, (η^5-2,4-dimethylpentadienyl)$_2$Ru or (η^5-2,3,4-trimethylpentadienyl)$_2$Ru, with $Ru_3(CO)_{12}$ in a 2:1 ratio at 145°C. The *mono*(metallabenzene) compound (**64**; Scheme 2.32) could also be isolated in 1% yield when a 1:1 ratio of (η^5-2,4-dimethylpentadienyl)$_2$Ru to $Ru_3(CO)_{12}$ was used. This suggests that the metallabenzenes in **63** are produced in two consecutive steps, with **64** being an intermediate on the way to **63**. The electron counting in **63** and **64** is somewhat different than that in the previously discussed metal-coordinated metallabenzenes. If the metallabenzene ligands are counted as neutral 6e$^-$ donors, the central ruthenium is then formally 20e$^-$ in **63** and 19e$^-$ in **64**. Hence, a better approach in this case is to assign the metallabenzene ligands a negative charge and count them as anionic 6e$^-$ donors, much like borabenzene ($C_5H_5BH^-$). The central ruthenium ion is then Ru(II)≡d^6 and has an 18 valence electron count. The ruthenium atoms in the anionic metallabenzene ligands are formally Ru(0)≡d^8 and are likewise 18e$^-$ centers, if the carbene moieties are counted as 2e$^-$ neutral (Fischer-type) ligands (Figure 2.18, resonance structures **I** and **II**).

In the X-ray crystal structure of **63a**, the two metallabenzene rings are syn-eclipsed, and the ruthenium ends tilt toward each other by 18°. The ruthenium atoms, however,

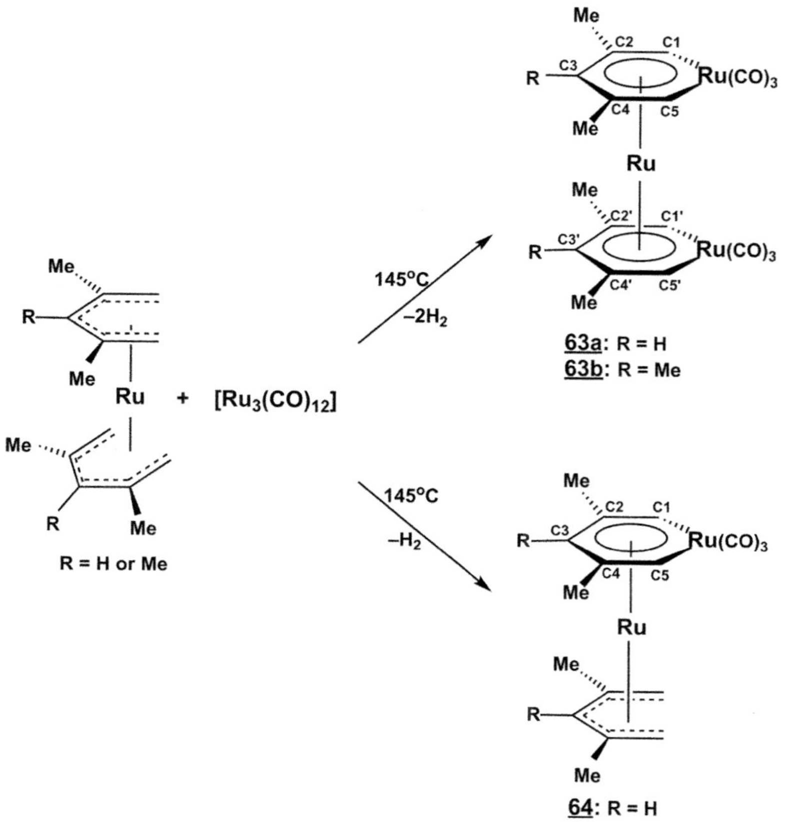

Scheme 2.32

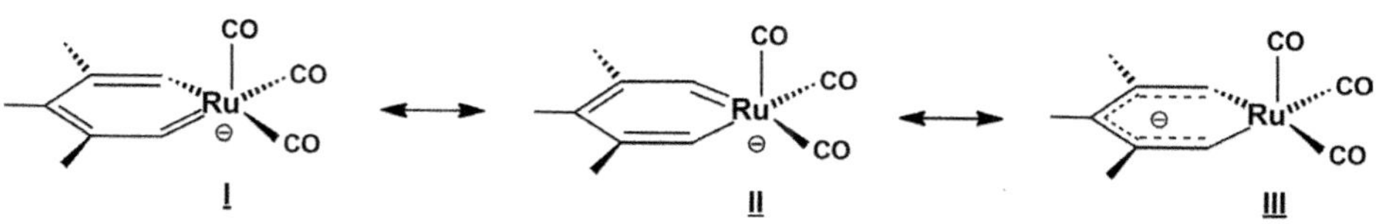

Figure 2.18 Resonance structures for the anionic metallabenzene ligand in Salzer's *bis*(metallabenzene) and *mono*(metallabenzene) sandwich compounds, **63** and **64**.

are displaced slightly out of their metallabenzene rings and away from each other; the dihedral angles between planes C1/C2/C3/C4/C5 and C1/Ru/C5 are 15.1° and 12.9° for the two rings. The bond distances from the central Ru to the two metallabenzene Ru's are 2.831(2) Å and 2.812(2) Å. The syn-eclipsed conformation allows for a possible interaction *between* the two metallabenzene ruthenium atoms. Their separation is 3.382(2) Å, longer than a covalent bond but shorter than the sum of the van der Waals radii. The carbon–carbon bonds within the metallabenzene rings in **63a** are delocalized, as expected, but the Ru–C bonds (2.076(3) Å and 2.062(3) Å) are significantly longer than those in **61** (1.983 Å).

The ^1H NMR spectrum of **63a** is unusual in that the α-H's of the ring are not as downfield shifted as is typically observed for a coordinated metallabenzene. The H1's appear at δ 5.62 and are, in fact, slightly upfield from H3 (δ 6.23). The corresponding carbon shifts are δ 128.3 (C1) and δ 94.4 (C3). These unusual shifts may be due to the negative charge on the metallabenzene ligands, and, in particular, the contribution from an η^5-pentadienyl-like resonance structure (**III** in Figure 2.18), in which that negative charge resides on the carbon portion of the ring. This resonance structure would also explain the longer Ru–C bond distances observed in **63a** (vide supra).

Using a different synthetic approach, Girolami [103, 105] succeeded in generating the first example of a metallabenzene complex with no substituents (only H's) on its carbon framework. As shown in Scheme 2.33, this species (**65**) was isolated from the reaction of a methylene-bridged diruthenium precursor with two equivalents of ethyne. Cyclization of the ruthenium-methylene moiety with the acetylenes, together with loss of HCl, leads to the production of **65**. When the reaction is carried out with ethyne-d_2, the tetradeuterated metallabenzene product is obtained. One of the two hydrogen atoms of the bridging methylene group is retained, and it resides exclusively in an *ortho* position of the ruthenabenzene ring. This strongly suggests that the elimination of HCl takes place *after* the insertion of both acetylene molecules into the ruthenium-methylene unit and rules out the intermediacy of a symmetrical metallacyclobutadiene intermediate. As in **63** and **64**, the metallabenzene ligand in **65** is probably best regarded as an anion. In this view, the central ruthenium ion is then Ru(II)≡d^6 and has an $18e^-$ count, while the metallabenzene ruthenium center is likewise an $18e^-$ Ru(II)≡d^6 ion.

In the X-ray crystal structure of **65**, the ring ruthenium atom sits 0.32 Å out of the ring carbon plane and away from the central ruthenium, leading to a dihedral angle of 13° between planes C1/C2/C3/C4/C5 and C1/Ru/C5. The Ru–Ru bond distance is 2.8983(3) Å. The C–C bonding within the ring is fully delocalized (C1–C2 = 1.406(4) Å, C2–C3 = 1.417(3) Å), while the Ru–C1 distance is 1.970(3) Å, consistent with some partial double bond character.[9]

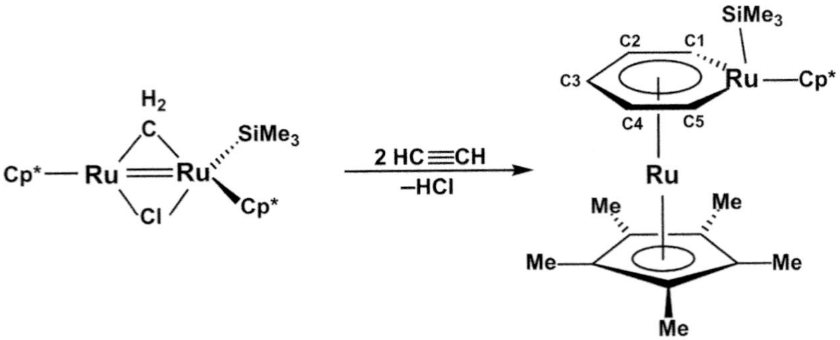

65

Scheme 2.33

9 The molecule sits on a crystallographically imposed mirror plane, which includes Ru and C3 of the metallabenzene ligand.

The ring protons in **65** resonate downfield in the ^{1}H NMR spectrum, with H1, H2, and H3 appearing at δ 9.75, δ 4.94, and δ 4.32, respectively. The corresponding ^{13}C resonances are at δ 177.0, δ 86.8, and δ 83.4. Even though the coordinated metallabenzene in **65** is anionic (like those in **63** and **64**), its structural and spectroscopic features are much more similar to those of **61**. This, in turn, suggests that the η^5-pentadienyl-like resonance structure (**III**; Figure 2.18) is less of a contributor for **65**.

2.22 Aromaticity

We return now to the question of whether iridabenzene (and, by extension, iridapyrylium and iridathiabenzene) are, in fact, aromatic. This is a difficult question to answer definitively because aromaticity can be measured or calculated in a wide variety of ways, and an unambiguous picture does not always emerge. The criteria that are most commonly used to evaluate aromaticity include electronic structure, structural (geometric) features, magnetic properties, stability, and chemical reactivity. Each of these is briefly discussed in the sections that follow.

2.22.1 Electronic Structure

The electronic structure of iridabenzene **3** [58] closely resembles that of benzene itself. The three metallabenzene bonding orbitals, 1π, d_{yz} + 2π, d_{xz} + 3π (cf. Figure 2.2)[10], are similar to the familiar benzene doughnut-shaped orbital and the degenerate HOMO's of benzene. However, the situation is a bit more complicated because another orbital of π-symmetry (d_{yz} – 2π) can contribute an additional two electrons, bringing the total π-electron count to eight rather than the normal six. But because two of the π-interactions (those involving d_{yz}) are of Möbius symmetry, the familiar Hückel 4n + 2 rule no longer applies as a criterion for aromaticity. Furthermore, when halide ligands are added to the metal center, as in compound **11**, the halide *p*π-orbitals can contribute to the ring π-bonding, increasing the π-electron count to 10. Further discussion of the electronic structure of metallabenzenes can be found in the final chapter of this book.

2.22.2 Structural (Geometric) Features

Iridabenzenes exhibit the delocalized bonding that is a requirement for aromaticity. The carbon–carbon bond lengths within the rings are intermediate between single and double bonds and similar to those found in benzene. Significantly, the iridium–carbon bond lengths are also intermediate between single and double bonds, indicating that the metal center is participating fully in ring π-bonding through its *d*-orbitals. While aromatic rings are generally planar, the iridium atoms in iridabenzenes sit slightly out of the plane of the five ring carbon atoms. In iridabenzene **3**, for example, the displacement is 0.24 Å. In **9**, where three PMe_3 ligands have replaced the three PEt_3 ligands, the displacement is 0.17 Å. In octahedral **11**, where large PEt_3 ligands reside above and below the plane, the displacement is a mere 0.13 Å. The reason for the non-planarity of iridabenzenes is not known. There may be a steric effect involving the ring and ligands on the metal center. But an electronic effect may also contribute, because iridabenzenes

10 In iridabenzene **3**, a hybrid metal orbital involving d_{xz} and d_{z^2} interacts with 3π.

possess a filled orbital that is antibonding between the metal and the ring α-carbons ($d_{yz} - 2\pi$). A slight displacement of the metal from the ring plane can reduce this interaction [25]. It is also possible that in square pyramidal molecules like **3** and **9** the participation of the d_{z^2} orbital in ring π-bonding may cause the displacement [58]. In any event, these small deviations from planarity do not appear to compromise the π-conjugation within the metallacyclic ring [25].

2.22.3 Magnetic Properties

One of the hallmarks of aromaticity is the diamagnetic ring current (diatropicity) that results from the cyclic, delocalized π-electrons. This effect is observed experimentally by NMR; protons outside the aromatic ring are deshielded and shifted downfield by the induced magnetic field, while protons inside the ring are shifted upfield. Our iridabenzenes show this effect. H1 and H5, which sit directly adjacent to the metal, are shifted far downfield as a result of a metal-based anisotropy and ring current. H3 is shifted downfield as well and resides far enough away from the metal that its deshielding can be attributed solely to the ring current effect. It is instructive to compare the H3 chemical shift in iridabenzene **3** (δ 7.18) to that in iridacyclohexa-1,3-diene **1b**, its non-aromatic precursor (δ 5.93). The aromatization of the ring has a clear effect on the chemical shift.

Schleyer has introduced a computational method, nucleus-independent chemical shift (NICS), to evaluate aromaticity [11]. This method involves placing a dummy atom above the ring and calculating its NMR chemical shift. A negative value indicates an aromatic system, while a positive value is obtained for an anti-aromatic system. Using $C_5H_5Ir(PH_3)_3$ as a model compound for iridabenzene **3**, van der Boom and Martin [58] calculated NICS values for a dummy atom suspended at 1.0 Å above the center of the ring or in the center of the ring. The values obtained were –8.8 ppm and –3.7 ppm, consistent with aromaticity. (The corresponding values for benzene were –10.0 ppm and –7.7 ppm.) Another NICS study [106] involving model compounds for four of our iridabenzenes, $C_5H_5Ir(PH_3)_3$ (model for **3**), $C_5H_5Ir(PH_3)_2(CO)$ (model for **7c**), $C_5H_5Ir(PH_3)_2(Cl)_2$ (model for **11**), and $[C_5H_5Ir(PH_3)_2(NCMe)_2]^{2+}$ (model for **12**) also reached the conclusion that these molecules exhibit diatropic ring currents and magnetic shielding properties consistent with aromaticity.

Another magnetic criterion for aromaticity is diamagnetic susceptibility exaltation, Λ, which is the difference between the calculated diamagnetic susceptibility of a cyclic, conjugated system and that of a corresponding localized system [107]. A negative diamagnetic susceptibility exaltation indicates aromaticity. The iridabenzene model compound, $C_5H_5Ir(PH_3)_3$, gave a Λ value of –10.2 cgs-ppm, consistent with aromaticity [108]. (The corresponding value for benzene is –10.9 cgs-ppm.)

A final magnetic criterion for aromaticity, the magnetic susceptibility anisotropy, ΔX, arises from the fact that diamagnetic susceptibility for aromatic molecules is much greater in the direction normal to the ring plane than in the directions parallel to it [13, 14]. Large negative numbers are expected for aromatic compounds, but the model compound for **3** gave a ΔX of 93.9 cgs-ppm [58] vs. –61.7 cgs-ppm for benzene.

2.22.4 Stability

Another key criterion of aromaticity is the stability of an aromatic compound with respect to an acyclic or cyclic non-conjugated reference compound, the aromatic stabilization energy (ASE). One approach is to look at homodesmotic reactions, where there are the same number and type of bonds (including hybridization) in reactants and products, but the products include a cyclic, conjugated (potentially) aromatic species. The energies of products and reactants are compared, with the difference being the homodesmotic reaction aromatic stabilization energy (HASE). A negative value for HASE indicates an aromatic product. Yang [108] calculated the HASE for the iridabenzene model compound $C_5H_5Ir(PH_3)_3$ and obtained a value of −61.7 kJ/mol (vs −163.0 kJ/mol for benzene), indicating aromaticity.

A second approach is to use energy decomposition analysis (EDA) within the density functional theory (DFT) framework [15]. Using this method, calculated $\Delta E\pi$ values for cyclic, conjugated compounds are compared to those of appropriate acyclic reference systems. The difference in $\Delta E\pi$'s defines the ASE. Model compounds for four of our iridabenzenes were evaluated: $C_5H_5Ir(PH_3)_3$ (model for **3**), $C_5H_5Ir(PH_3)_2(CO)$ (model for **7c**), $C_5H_5Ir(PH_3)_2(Cl)_2$ (model for **11**), and $[C_5H_5Ir(PH_3)_2(NCMe)_2]^{2+}$ (model for **12**). The calculated ASA values were 22.4 kcal/mol, 23.2 kcal/mol, 33.5 kcal/mol, and 8.7 kcal/mol, respectively, as compared to 42.5 kcal/mol for benzene [16, 26].

Another measure of stability is a compound's absolute hardness, η, which is defined as half of the HOMO/LUMO gap [9, 10]. An early study [109], based on Extended Hückel calculations, concluded that iridabenzene **3** had an absolute hardness of 0.60 eV (compared to 2.27 eV for benzene) and was not aromatic. However, a later study used Hartree–Fock calculations on structures optimized with the DFT method and came to the opposite conclusion. The model compound for **3**, $C_5H_5Ir(PH_3)_3$, had a η value of 4.22 eV (vs. 6.47 eV for benzene), consistent with aromaticity [108].

2.22.5 Chemical Reactivity

Some of the reactions of our iridabenzenes are consistent with expectations for aromatic species and others are not. In its reaction with (η^6-*p*-xylene)$Mo(CO)_3$, for example, iridabenzene **3** behaves as expected for an electron-rich arene, displacing *p*-xylene to form (η^6-iridabenzene)$Mo(CO)_3$. On the other hand, **3** reacts with halogen reagents *at the metal center*, rather than undergoing conventional electrophilic aromatic substitution on the carbon portion of the ring. But the most pervasive reaction that compound **3** undergoes is cycloaddition, primarily [4+2] cycloaddition. While cycloaddition may at first seem unusual for an aromatic species, it should be remembered that other aromatics, including anthracene, readily undergo Diels–Alder reactions with substrates like maleic anhydride [110]. The driving force for the anthracene reactions is the strengthening of aromaticity in the two outside rings. The driving force for the iridabenzene reactions is the formation of stable octahedral Ir(III) complexes.

2.22.6 Conclusion

While aromaticity remains a complex and somewhat ill-defined phenomenon, the preponderance of evidence suggests that the iridabenzenes described in this chapter are aromatic. While it is difficult to compare directly the aromaticity of metal-containing rings with organic rings, recent calculations suggest that aromatic stabilization energies for our iridabenzenes range from 79% of the ASE for benzene (compound **11**) to 20% of the ASE for benzene (compound **12**) with the original iridabenzene (compound **3**) falling in between at 53% [26]. Undoubtedly, additional attempts to quantify the aromaticity of iridabenzenes will follow, and we await those results with interest.

2.23 Final Word

The scientific journey described in this chapter has been both exciting and joyful. I am especially indebted to all of the students who have contributed to this work. Without them, none of this would have happened. An academic scientist leaves two important legacies, the science that he does and the students that he trains. Of these, the latter usually has the more lasting impact.

It has been gratifying to see the field of metallabenzenes grow and flourish over the years since the initial discoveries of the 1980s. Important contributions are now being made by groups from around the globe. I am reminded that science has the ability to bring the world together and, as scientists, we should seize that opportunity. This book is a step in that direction.

References

1 Kekulé, A. (1865) Sur la constitution des substances aromatiques. *Bull. Soc. Chim.*, **3**: 98–111.
2 Kekulé, A. (1865) Note sur quelques produits de substitution de la benzine. *Bull. Acad. R. Belg.*, **19**: 551–563.
3 Hoffmann, R. (2015) Perspective: The many guises of aromaticity. *Am Sci.*, **103**(1): 18–22.
4 Schleyer, P. R. (2001) Introduction: Aromaticity. *Chem. Rev.*, **101**(5): 1115–1118.
5 Badger, G. M. (1969) *Aromatic Character and Aromaticity*. London: Cambridge University Press.
6 Bergman, E. D. and Pullman, B. (eds) (1971) *Aromaticity, Pseudo-Aromaticity, Anti-Aromaticity*. New York: Academic Press.
7 Garratt, P. J. (1986) *Aromaticity*. New York: John Wiley & Sons, Inc.
8 Lloyd, D. (1989) *The Chemistry of Conjugated Cyclic Compounds*. Chichester: John Wiley & Sons, Ltd.
9 Zhou, Z. X., Parr, R. G., and Garst, J. F. (1988) Absolute hardness as a measure of aromaticity. *Tetrahedron Lett.*, **29**(38): 4843–4846.
10 Zhou, Z. X. and Parr, R. G. (1989) New measures of aromaticity: Absolute hardness and relative hardness. *J. Am. Chem. Soc.*, **111**(19): 7371–7379.
11 Schleyer, P. V., Maerker, C., Dransfeld, A., *et al.* (1996) Nucleus-independent chemical shifts: A simple and efficient aromaticity probe. *J. Am. Chem. Soc.*, **118**(26): 6317–6318.
12 Schleyer, P. V., Jiao, H. J., Hommes, N. J. R. V., *et al.* (1997) An evaluation of the aromaticity of inorganic rings: Refined evidence from magnetic properties. *J. Am. Chem. Soc.*, **119**(51): 12669–12670.
13 Schmalz, T. G., Norris, C. L., and Flygare, W. H. (1973) Localized magnetic susceptibility anisotropies. *J. Am. Chem. Soc.*, **95**(24): 7961–7967.
14 Fleischer, U., Kutzelnigg, W., Lazzeretti, P., and Muhlenkamp, V. (1994) Iglo study of benzene and some of its isomers and related molecules: Search for evidence of the ring current model. *J. Am. Chem. Soc.*, **116**(12): 5298–5306.

15 Cappel, D., Tullmann, S., Krapp, A., and Frenking, G. (2005) Direct estimate of the conjugative and hyperconjugative stabilization in diynes, dienes, and related compounds. *Angewandte Chemie: International Edition*, **44**(23): 3617–3620.

16 Fernandez, I. and Frenking, G. (2011) Direct estimate of aromaticity with energy decomposition analysis. *Open Organic Chemistry Journal*, **5**: 79–86.

17 Ashe, A. J. (1978) Group-5 heterobenzenes. *Accounts Chem. Res.*, **11**(4): 153–157.

18 Jutzi, P. (1975) New element carbon (p-p)π bonds. *Angewandte Chemie: International Edition in English*, **14**(4): 232–245.

19 Thorn, D. L. and Hoffmann, R. (1979) Delocalization in metallocycles. *Nouveau Journal De Chimie: New Journal of Chemistry*, **3**(1): 39–45.

20 Hoffmann, R. (1981) Theoretical organometallic chemistry. *Science*, **211**(4486): 995–1002.

21 Hoffmann, R. (1982) Building bridges between inorganic and organic chemistry (Nobel lecture). *Angewandte Chemie: International Edition*, **21**(10): 711–724.

22 Jorgensen, W. L. and Salem, L. (1973) *The Organic Chemist's Book of Orbitals*. New York: Academic Press.

23 Schleyer, P. V. and Wang, Z. X. Personal communication.

24 Heilbronner, E. (1964) Hückel molecular orbitals of Möbius-type conformations of annulenes. *Tetrahedron Lett.*, **29**: 1923–1928.

25 Zhu, J., Jia, G. C., and Lin, Z. Y. (2007) Understanding nonplanarity in metallabenzene complexes. *Organometallics*, **26**(8): 1986–1995.

26 Fernandez, I. and Frenking, G. (2007) Aromaticity in metallabenzenes. *Chemistry: A European Journal*, **13**(20): 5873–5884.

27 Elliott, G. P., Roper, W. R., and Waters, J. M. (1982) Metallacyclohexatrienes or metallabenzenes: Synthesis of osmabenzene derivatives and X-ray crystal-structure of $Os(CSCHCHCHCH)(CO)(PPh_3)_2$. *Journal of the Chemical Society: Chemical Communications*, **14**: 811–813.

28 Elliott, G. P., McAuley, N. M., Roper, W. R., and Shapley, P. A. (1989) An osmium containing benzene analog, $Os(CSCHCHCHCH)(CO)(PPh_3)_2$, carbonyl(5-thioxo-1, 3-pentadiene-1,5-diyl-C^1,C^5,S)-bis(triphenylphosphine)osmium, and its precursors. *Inorganic Syntheses*, **26**: 184–189.

29 Bleeke, J. R., Xie, Y. F., Peng, W. J., and Chiang, M. (1989) Metallabenzene: Synthesis, structure, and spectroscopy of a 1-irida-3,5-dimethylbenzene complex. *J. Am. Chem. Soc.*, **111**(11): 4118–4120.

30 Bleeke, J. R. (1991) Metallabenzene chemistry. *Accounts Chem. Res.*, **24**(9): 271–277.

31 Bleeke, J. R. and Peng, W. J. (1987) Pentadienyl-metal-phosphine chemistry: 12: Synthesis, structure, and spectroscopy of iridacyclohexadiene complexes. *Organometallics*, **6**(7): 1576–1578.

32 Bleeke, J. R., Peng, W. J., Xie, Y. F., and Chiang, M. Y. (1990) Metallacyclohexadiene and metallabenzene chem: 3: Synthesis, characterization, and reaction chemistry of a family of 1-iridacyclohexa-2,4-diene complexes. *Organometallics*, **9**(4): 1113–1119.

33 Kekulé, A. (1890) Benzolfest: Rede. *Berichte der Deutschen Chemischen Gesellschaft*, **23**: 1302–1311.

34 Bleeke, J. R., Behm, R., Xie, Y. F., *et al.* (1997) Synthesis, structure, spectroscopy, and reactivity of a metallabenzene. *Organometallics*, **16**(4): 606–623.

35 Hughes, R. P., Trujillo, H. A., and Rheingold, A. L. (1993) α-C-H and α′-C-C bond-cleavage in an iridacyclohexadiene: Interchange of α-hydrogen and α′-phenyl

substituents without accompanying skeletal rearrangement. *J. Am. Chem. Soc.*, **115**(4): 1583–1585.

36 Hughes, R. P., Trujillo, H. A., Egan, J. W., and Rheingold, A. L. (2000) Iridium-promoted reactions of carbon–carbon bonds: Skeletal rearrangement of a vinylcyclopropene during iridacyclohexadiene formation and subsequent isomerization of iridacyclohexadienes via α, α′-substituent migrations. *J. Am. Chem. Soc.*, **122**(10): 2261–2271.

37 Fischer, E. O. and Maasbol, A. (1964) On the existence of a tungsten carbonyl carbene complex. *Angewandte Chemie: International Edition*, **3**: 580–582.

38 Schrock, R. R. (1974) Alkylcarbene complex of tantalum by intramolecular α-hydrogen abstraction. *J. Am. Chem. Soc.*, **96**(21): 6796–6797.

39 Tolman, C. A. (1970) Phosphorus ligand exchange equilibria on zerovalent nickel: A dominant role for steric effects. *J. Am. Chem. Soc.*, **92**(10): 2956–2965.

40 Bleeke, J. R., Boorsma, D., Chiang, M. Y., *et al.* (1991) Pentadienyl-metal-phosphine chemistry: 23: Pentadienyl-iridium-phosphine chemistry: Survey of the reactions of pentadienide reagents with $ClIrL_3$ complexes. *Organometallics*, **10**(7): 2391–2398.

41 Hughes, R. P., Trujillo, H. A., Egan, J. W., and Rheingold, A. L. (1999) Skeletal rearrangement during rhodium-promoted ring opening of 1,2-diphenyl-3-vinyl-1-cyclopropene: Preparation and characterization of 1,2- and 2,3-diphenyl-3,4-pentadienediyl rhodium complexes and their ring closure to a 1,2-diphenylcyclopentadienyl complex. *Organometallics*, **18**(15): 2766–2772.

42 Dolomanov, O. V., Bourhis, L. J., Gildea, R. J., *et al.* (2009) Olex2: A complete structure solution, refinement and analysis package. *J. Appl. Cryst.*, **42**: 339–341.

43 Bacon, G. E., Curry, N. A., and Wilson, S. A. (1964) A crystallographic study of solid benzene by neutron diffraction. *Proc. R. Soc. London: Ser. A.*, **279**: 98–110.

44 Clark, G. R., Roper, W. R., and Wright, A. H. (1982) Synthesis, structure, and reactions of $IrCl_3(CCl_2)(PPh_3)_2$: A compound with an iridium carbon double-bond ($L_nIr{=}CCl_2$). *J. Organomet. Chem.*, **236**(1): C7–C10.

45 Fryzuk, M. D., Macneil, P. A., and Rettig, S. J. (1985) Photoinduced α-hydrogen elimination of an iridium(III) dialkyl: Formation of an isolable iridium methylidene. *J. Am. Chem. Soc.*, **107**(23): 6708–6710.

46 McConnell, H. (1957) Theory of nuclear magnetic shielding in molecules: I: Long-range dipolar shielding of protons. *Journal of Chemical Physics*, **27**: 226–229.

47 Schwab, P., Grubbs, R. H., and Ziller, J. W. (1996) Synthesis and applications of $RuCl_2(=CHR')(PR_3)_2$: The influence of the alkylidene moiety on metathesis activity. *J. Am. Chem. Soc.*, **118**(1): 100–110.

48 Berry, R. S. (1960) Correlation of rates of intramolecular tunneling processes, with applications to some group V compounds. *Journal of Chemical Physics*, **32**: 933–938.

49 Cotton, F. A., Wilkinson, G., Murillo, C. A., and Bochmann, M. (1999) *Advanced Inorganic Chemistry*, 6th edn. New York: Wiley-Interscience.

50 Ferede, R. and Allison, N. T. (1983) Possible formation of ferrabenzene and its novel conversion to 1, 3-diphenyl-2-methoxyferrocene. *Organometallics*, **2**(3): 463–465.

51 Schrock, R. R., Pedersen, S. F., Churchill, M. R., and Ziller, J. W. (1984) Formation of cyclopentadienyl complexes from tungstenacyclobutadiene complexes and the X-ray crystal-structure of an η^3-cyclopropenyl complex, $W[(C(CMe_3)C(Me)C(Me)]$-$(Me_2NCH_2CH_2NMe_2)Cl_3$. *Organometallics*, **3**(10): 1574–1583.

52 Ferede, R., Hinton, J. F., Korfmacher, W. A., *et al.* (1985) Possible formation of rhenabenzene. Lithium-halogen exchange reactions in η^1-4-bromo-1,4-diphenyl-1,

3-butadienyl ligands to form rhenabenzene and ferrabenzene. *Organometallics*, **4**(3): 614–616.

53 Mike, C. A., Ferede, R., and Allison, N. T. (1988) Evidence for rhenaphenanthrene formation and its conversion to fluorenone. *Organometallics*, **7**(7): 1457–1459.

54 Yang, J., Yin, J. G., Abboud, K. A., and Jones, W. M. (1994) Metallaindenes of molybdenum, tungsten, and ruthenium. *Organometallics*, **13**(3): 971–978.

55 Yang, J., Jones, W. M., Dixon, J. K., and Allison, N. T. (1995) Detection of a ruthenabenzene, ruthenaphenoxide, and ruthenaphenanthrene oxide: The first metalla aromatics of a 2nd-row transition-metal. *J. Am. Chem. Soc.*, **117**(38): 9776–9777.

56 Hughes, R. P. and Kowalski, A. S. (1998) Flash vacuum thermolysis of η^5-oxocyclohexadienyl complexes of ruthenium to give η^5-cyclopentadienyl ligands. *Organometallics*, **17**(2): 270–273.

57 Iron, M. A., Martin, J. M. L., and van der Boom, M. E. (2003) Metallabenzene versus Cp complex formation: A DFT investigation. *J. Am. Chem. Soc.*, **125**(43): 13020–13021.

58 Iron, M. A., Lucassen, A. C. B., Cohen, H., *et al.* (2004) A computational foray into the formation and reactivity of metallabenzenes. *J. Am. Chem. Soc.*, **126**(37): 11699–11710.

59 Wu, H. P., Lanza, S., Weakley, T. J. R., and Haley, M. M. (2002) Metallabenzenes and valence isomers: 3: Unexpected rearrangement of two regioisomeric iridabenzenes to an (η^5-cyclopentadienyl)iridium(I) complex. *Organometallics*, **21**(14): 2824–2826.

60 Wu, H. P., Weakley, T. J. R., and Haley, M. M. (2005) Regioselective formation of β-alkyl-α-phenyliridabenzenes via unsymmetrical 3-vinylcyclopropenes: Probing steric and electronic influences by varying the alkyl ring substituent. *Chemistry: A European Journal*, **11**(4): 1191–1200.

61 Wu, H. P., Ess, D. H., Lanza, S., *et al.* (2007) Rearrangement of iridabenzvalenes to iridabenzenes and/or η^5-cyclopentadienyliridium(I) complexes: Experimental and computational analysis of the influence of silyl ring substituents and phosphine ligands. *Organometallics*, **26**(16): 3957–3968.

62 Bleeke, J. R., Bass, L. A., Xie, Y. F., and Chiang, M. Y. (1992) Metallacyclohexadiene and metallabenzene chemistry: 6: Chemistry of an (η^6-metallabenzene)metal complex, [(η^6-IrCHC(Me)CHC(Me)CH)(PEt_3)$_3$]$Mo(CO)_3$. *J. Am. Chem. Soc.*, **114**(11): 4213–4219.

63 Muetterties, E. L (ed.) (1971) *Transition Metal Hydrides*. New York: Marcel Dekker.

64 Bleeke, J. R., Xie, Y. F., Bass, L., and Chiang, M. Y. (1991) Chemical-reactivity of a metallabenzene. *J. Am. Chem. Soc.*, **113**(12): 4703–4704.

65 Becker, E. D. (2000) *High Resolution NMR*, 3rd edn. San Diego: Academic Press.

66 Bleeke, J. R., Behm, R., Xie, Y. F., *et al.* (1994) Metallacyclohexadiene and metallabenzene chemistry: 10: Cycloaddition reactions of a metallabenzene. *J. Am. Chem. Soc.*, **116**(9): 4093–4094.

67 Iron, M. A., Martin, J. M. L., and van der Boom, M. E. (2003) Cycloaddition reactions of metalloaromatic complexes of iridium and rhodium: A mechanistic DFT investigation. *J. Am. Chem. Soc.*, **125**(38): 11702–11709.

68 Rigaudy, J. (1968) Photooxydation des derives aromatiques. *Pure Appl. Chem.*, **16**: 169–186.

69 Woodward, R. B. and Hoffmann, R. (1970) *The Conservation of Orbital Symmetry*. New York: Academic Press.

70 House, H. O. (1972) *Modern Synthetic Reactions*, 2nd edn. Menlo Park, CA: W. A. Benjamin.

71 Herberich, G. E. and Ohst, H. (1986) Borabenzene metal-complexes. *Advances in Organometallic Chemistry*, **25**: 199–236.

72 Wilkinson G., Stone F. G. A., and Abel E. W. (eds) (1982) *Comprehensive Organometallic Chemistry*. Oxford: Pergamon Press.

73 Barbeau, C. and Turcotte, J. (1976) Theoretical dipole-moments of $Mo(CO)_6$ derivatives. *Canadian Journal of Chemistry: Revue Canadienne De Chimie*, **54**(10): 1612–1616.

74 Muetterties, E. L., Bleeke, J. R., and Sievert, A. C. (1979) Arene transition-metal chemistry: 3: Arene exchange phenomena. *J. Organomet. Chem.*, **178**(1): 197–216.

75 Bleeke, J. R. and Behm, R. (1997) Synthesis, structure, and reactivity of iridacyclohexadienone and iridaphenol complexes. *J. Am. Chem. Soc.*, **119**(36): 8503–8511.

76 Stout, G. H. and Jensen, L. H. (1989) *X-Ray Structure Determination: A practical guide*, 2nd edn. New York: John Wiley & Sons, Ltd.

77 Heiszwolf, G. J. and Kloosterziel, H. (1967) The distribution of charge in ambient anions. *Recl. Trav. Chim. Pays-Bas.*, **86**: 807–808.

78 Bleeke, J. R., Haile, T., and Chiang, M. Y. (1991) Synthesis of iridaoxacyclohexadiene and iridacyclopentenone complexes via C–H bond activation. *Organometallics*, **10**(1): 19–21.

79 Bleeke, J. R., Haile, T., New, P. R., and Chiang, M. Y. (1993) Oxapentadienyl iridium phosphine chemistry: Synthesis of oxygen-containing iridacycles via C-H bond activation. *Organometallics*, **12**(2): 517–528.

80 Bleeke, J. R. and Blanchard, J. M. B. (1997) Synthesis and reactivity of a metallapyrylium. *J. Am. Chem. Soc.*, **119**(23): 5443–5444.

81 Bleeke, J. R., Blanchard, J. M. B., and Donnay, E. (2001) Synthesis, spectroscopy, and reactivity of a metallapyrylium. *Organometallics*, **20**(2): 324–336.

82 Iron, M. A., Martin, J. M. L., and van der Boom, M. E. (2003) Mechanistic aspects of acetone addition to metalloaromatic complexes of iridium: A DFT investigation. *Chem. Commun.*, **1**: 132–133.

83 Bleeke, J. R. and Ortwerth, M. F., and Chiang, M. Y. (1992) Thiapentadienyl iridium phosphine chemistry: Synthesis of sulfur-containing iridacycles via C-H bond activation. *Organometallics*, **11**(8): 2740–2743.

84 Bleeke, J. R., Ortwerth, M. F., and Rohde, A. M. (1995) Pentadienyl-metal-phosphine chemistry: 30: Thiapentadienyl-iridium-phosphine chemistry. *Organometallics*, **14**(6): 2813–2826.

85 Chen, J. B., Daniels, L. M., and Angelici, R. J. (1990) New modes of thiophene coordination and reactivity: Structures of Cp*Ir(η^2-thiophene), an iridathiabenzene, and Cp*Ir(η^4-thiophene-BH_3). *J. Am. Chem. Soc.*, **112**(1): 199–204.

86 Chen, J. B., Daniels, L. M., and Angelici, R. J. (1990) Reactions of the η^2-isomers and η^4-isomers of Cp*Ir(2,5-dimethylthiophene) with phosphines, CO and H_2. *Polyhedron*, **9**(15–16): 1883–1891.

87 Jones, W. D. and Dong, L. Z. (1991) Insertion of rhodium into the carbon sulfur bond of thiophene: Mechanism of a model for the hydrodesulfurization reaction. *J. Am. Chem. Soc.*, **113**(2): 559–564.

88 Bianchini, C., Meli, A., Peruzzini, M., *et al.* (1993) Opening, desulfurization, and hydrogenation of thiophene at iridium: An experimental study in a homogeneous phase. *J. Am. Chem. Soc.*, **115**(7): 2731–2742.

89 Bianchini, C., Meli, A., Peruzzini, M., *et al.* (1994) HDS model systems: Coordination, opening, and hydrogenation of benzo[*b*]thiophene at iridium. *J. Am. Chem. Soc.*, **116**(10): 4370–4381.

90 Pettett, M. G. and Holmes, A. B. (1985) Functionalization of alkenes by a cycloaddition-cycloreversion sequence: 2: Anionic cycloreversion reactions of 2, 5-dihydrothiophene derivatives. *Journal of the Chemical Society: Perkin Transactions*, **1**(6): 1161–1165.

91 Kloosterziel, H., Van Drunen, J. A. A., and Galama, P. (1969) Base-induced electrocyclic opening of some heterocyclic compounds. *J. Chem. Soc. D.*, **15**: 885.

92 Bleeke, J. R., Hinkle, P. V., and Rath, N. P. (1999) Synthesis, structure, spectroscopy, and reactivity of a metallathiabenzene. *J. Am. Chem. Soc.*, **121**(3): 595–596.

93 Bleeke, J. R., Hinkle, P. V., and Rath, N. P. (2001) Metallacyclohexadiene and metallabenzene chemistry: Part 17: Synthesis, structure, spectroscopy, and reactivity of a metallathiabenzene. *Organometallics*, **20**(10): 1939–1951.

94 Chen, J. B., Young, V. G., and Angelici, R. J. (1995) Iridathiabenzene: A novel η^6 heteroaromatic π-ligand. *J. Am. Chem. Soc.*, **117**(23): 6362–6363.

95 Bleeke, J. R., Hinkle, P. V., Shokeen, M., and Rath, N. P. (2004) Synthesis, structure, spectroscopy, and reactivity of a neutral iridathiabenzene. *Organometallics*, **23**(17): 4139–4149.

96 Kralik, M. S., Rheingold, A. L., and Ernst, R. D. (1987) (Pentadienyl)molybdenum carbonyl chemistry: Conversion of a pentadienyl ligand to a coordinated metallabenzene complex. *Organometallics*, **6**(12): 2612–2614.

97 Kralik, M. S., Rheingold, A. L., Hutchinson, J. P., *et al.* (1996) Synthesis, characterization, and reaction chemistry of (pentadienyl)molybdenum carbonyl complexes. *Organometallics*, **15**(2): 551–561.

98 Mann, B. E., Manning, P. W., and Spencer, C. M. (1986) The preparation of (η-C_5H_5)-$RuCl(PPh_3)_2$ from $RuHCl(PPh_3)_3$ and penta-1,4-diene: Carbon carbon bond formation by dehydrogenation. *J. Organomet. Chem.*, **312**(3): C64–C66.

99 Kirss, R. U. (1992) Electrocyclic reactions of open metallocenes: Carbon carbon bond formation during thermolysis of bis(2,4-dimethyl-1,3-pentadienyl) ruthenium and bis(2,4-dimethyl-1,3-pentadienyl)osmium. *Organometallics*, **11**(2): 497–499.

100 Kirss, R. U., Quazi, A., Lake, C. H., and Churchill, M. R. (1993) Synthesis of unsymmetrical ruthenocenes by gas-phase electrocyclic reactions of pentadienylruthenium complexes. *Organometallics*, **12**(10): 4145–4150.

101 Bosch, H. W., Hund, H. U., Nietlispach, D., and Salzer, A. (1992) General-route to the half-open metallocenes C_5Me_5Ru(pentadienyl) and C_5Me_5Ru(diene)Cl: X-ray structures of an optically-active half-open metallocene and of a dimetallic ruthenabenzene complex. *Organometallics*, **11**(6): 2087–2098.

102 Bertling, U., Englert, U., and Salzer, A. (1994) Triple-decker to metallabenzene: A new-generation of sandwich complexes. *Angewandte Chemie: International Edition*, **33**(9): 1003–1004.

103 Lin, W. B., Wilson, S. R., and Girolami, G. S. (1997) Carbon-carbon bond formation promoted by organoruthenium complexes: The first unsubstituted π-metallabenzene complex, $Cp^*{}_2Ru_2(\eta^2{:}\eta^5\text{-}C_5H_5)(SiMe_3)$, and synthesis of the tetramethyleneethane complex $Cp^*{}_2Ru_2$ ($\eta^3{:}\eta^3\text{-}C_6H_8)Cl_4$. *Organometallics*, **16**(11): 2356–2361.

104 Englert, U., Podewils, F., Schiffers, I., and Salzer, A. (1998) The first homoleptic metallabenzene sandwich complex. *Angewandte Chemie: International Edition*, **37**(15): 2134–2136.

105 Lin, W. B., Wilson, S. R., and Girolami, G. S. (1993) The first unsubstituted metallabenzene complex $[Ru_2(\eta^5\text{-}C_5Me_5)_2(\eta^2,\eta^5\text{-}C_5H_5)(SiMe_3)]$. *Journal of the Chemical Society: Chemical Communications*, **3**: 284–285.

106 Periyasamy, G., Burton, N. A., Hillier, I. H., and Thomas, J. M. H. (2008) Electron delocalization in the metallabenzenes: A computational analysis of ring currents. *J. Phys. Chem. A.*, **112**(26): 5960–5972.

107 Dauben, H. J. J., Wilson, J. D., and Laity, J. L. (1968) Diamagnetic susceptibility exaltation as a criterion of aromaticity. *J. Am. Chem. Soc.*, **90**: 811–813.

108 Huang, Y. Z., Yang, S. Y., and Li, X. Y. (2004) An investigation of the aromaticity of transition metal heterocyclic complexes by conventional criteria and indices of aromaticity. *J. Organomet. Chem.*, **689**(6): 1050–1056.

109 Chamizo, J. A., Morgado, J., and Sosa, P. (1993) Organometallic aromaticity. *Organometallics*, **12**(12): 5005–5007.

110 Alder, O. D. K. (1928) Synthesen in der hydroaromatischen reihe. *Liebigs Annalen der Chemie*, **460**: 98–122.

3

Metallabenzenes and Valence Isomers via the Nucleophilic 3-Vinylcyclopropene Route

Michael M. Haley

Department of Chemistry & Biochemistry and Materials Science Institute, University of Oregon, Eugene, Oregon, USA

3.1 Project Origin and Inspiration (A Nod to Binger, Bleeke, Grubbs, Hughes, and Roper)

During the late summer of 1992, I was beginning to hone the research proposals for my job applications to become an assistant professor. I felt comfortable with the two "solid" ideas I had—dehydrobenzo[*n*]annulenes as organic materials (worked great) [1] and enediyne macrocycles with multiple "warheads" (failed miserably)—but I still felt I needed a third topic. My PhD research had been on cyclopropene and cycloproparene chemistry. I really liked working with strained and/or aromatic molecules, but clearly I needed to push in a new direction to distinguish myself from my PhD mentor's work. My postdoctoral stay at Berkeley had exposed me to the rapidly developing area of organometallic chemistry. "OK," I thought. "What about removing the nitrogen atom in pyridine and instead incorporating a transition metal complex as part of an aromatic ring?"

The basic idea sounded simple enough, but how to accomplish this was another matter as I approached the problem as only an organic chemist would—design a fancy organic ligand! If one bond fixes the aromatic ring (**1**) and extracts the organometallic fragment, the resultant C_5 fragment (**2**) is a *cis*-vinyl anion on one end and a vinylcarbene on the other end (Scheme 3.1). The former could be generated by lithium-halogen exchange and the latter by ring opening of a cyclopropene [2]; therefore, a suitably substituted *Z*-3-(2-halovinyl)cyclopropene (e.g. **3**) should be a perfect ligand for metallabenzene synthesis.

One thing I quickly learned upon further literature research was that organometallic complexes react readily with cyclopropenes to afford products via η^2-coordination to the double bond or σ-insertion into the strained single bond to give metallabicyclobutanes or metallacyclobutenes, respectively [3]. The Binger [4, 5] and Grubbs [6–8] groups had observed these very species, along with metallabutadienes, depending upon the choice of metal complex (Scheme 3.2). It did not take a great deal of imagination to link the ends of the metallabutadiene with a –CH=CH– bridge and thus generate a metallabenzene. I should also note that it was this early exploratory work

Metallabenzenes: An Expert View, First Edition. Edited by L. James Wright.

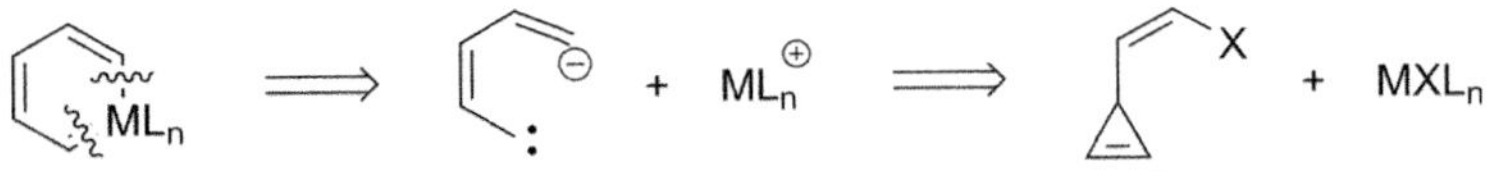

Scheme 3.1

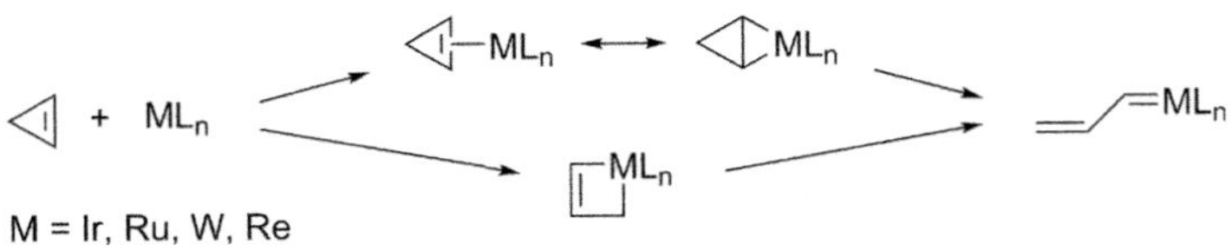

Scheme 3.2

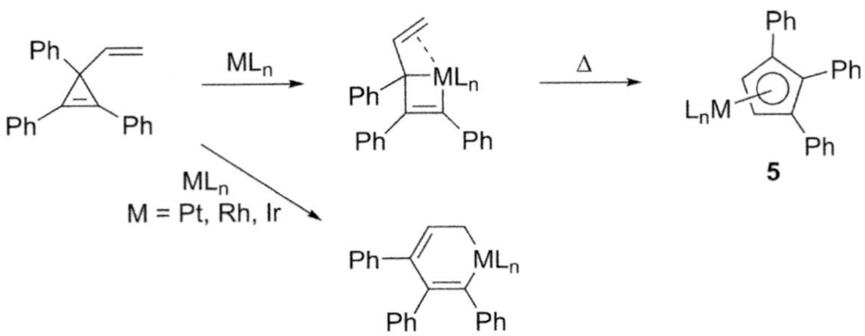

Scheme 3.3

with 3,3-diphenylcyclopropene that ultimately led to the development in the Grubbs group of efficient ring-opening and ring-closing metathesis catalysts [9].

Of course, while digging through the literature to bolster my nascent idea, I encountered the related papers of Hughes. His group at Dartmouth had done some beautiful work in the 1980s reacting 3-vinylcyclopropenes with a number of different metal complexes [10, 11]. Like the Binger/Grubbs studies, differing metal complexes afforded several isomeric organometallic structures (Scheme 3.3). The most interesting and relevant of these was cyclohexadiene **4** [12–14]. It was clear the Dartmouth team came close to preparing a metallabenzene but was never able to oxidize **4** to the fully aromatic species. More often than not, cyclopentadienyl complexes such as **5** were isolated as the predominant product.

There were exactly *two* known metallabenzenes in the literature in 1992: osmabenzene **6** and iridabenzene 7 (Figure 3.1). Complex **6**, reported by Roper in 1982 [15], was the first unambiguous isolation of a metallabenzene and appeared just three years after Thorn and Hoffman postulated such aromatics to be stable species in their seminal 1979 theory paper [16]. Arene **6** was isolated from the formal [2+2+2] cycloaddition of two ethyne molecules to $[Os(CO)(CS)(PPh_3)_3]$ [17]. Aside from the unusual proton chemical shift of the hydrogen atom *ortho* to the osmium (13.95 ppm), the remaining three peaks at 7.28 ppm resonated in the typical range for aromatic protons. Importantly, there was minimal bond length alternation of the carbon backbone as shown by X-ray

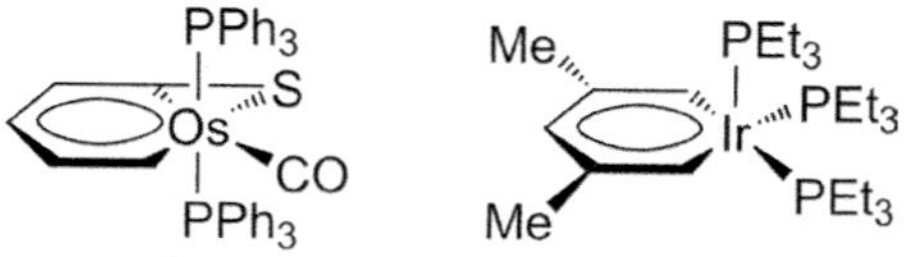

6 7

Figure 3.1 Chemical structures of the only two known metallabenzenes before 1992.

crystallography. The reader is referred to the Wright/Roper review in Chapter 1 for much, much more on their work.

In 1989, Bleeke reported the synthesis of the first stable iridabenzene (7) utilizing 2,4-dimethylpentadienide as the source of the carbon backbone [18]. Using the resultant iridabenzene as a starting point, Bleeke examined the chemistry of these complexes and found that iridabenzenes in general do not behave as typical aromatics: (1) electrophiles react preferentially with the electron rich metal center rather than electrophilic aromatic substitution pathways and (2) dienophiles undergo cycloaddition reactions with iridabenzenes [19]. Interestingly, if one trades out the phosphine in the starting iridium complex $[IrCl(PEt_3)_3]$ with PMe_3, deprotonation of the intermediate iridacyclohexadiene fails; thus, iridabenzene formation and all subsequent transformations must come from the tris-PEt_3 complex [20]. Nonetheless, the reader should consult the Bleeke review in Chapter 2 to learn more on these pioneering studies.

Considering the aforementioned studies as a whole, I was guardedly optimistic that the *Z*-3-(2-halovinyl)cyclopropene route to metallabenzenes stood a good chance of working as posited (Scheme 3.4). Lithium-halogen exchange on **3** followed by introduction of the metal fragment should lead to σ-vinyl complex **8**. Once tethered, the metal could react quickly with the strained ring to directly afford the metallabenzene (**1**). If we were lucky or operated at low temperatures, we might be able to isolate unknown metallabenzene valence isomers such as a metallabenzvalene (**9**) or a metalla-Dewar benzene (**10**). Upon heating, these should convert to **1** as the putative thermodynamic sink. In the worst-case scenario, **1** would not be stable and further react to give a cyclopentadienyl complex such as **11**. Before this chapter is finished, I will present evidence for all five possible products.

3.2 Ligand Synthesis (An Exercise in Over-Engineering)

Fortunately for me, my job search during the fall of 1992/winter of 1993 resulted in several offers and I decided to move up Interstate 5 to the University of Oregon in Eugene. Graduate student Rob Gilbertson joined my group in winter 1994 and I convinced him that a general route for metallabenzene synthesis was a great idea for a PhD thesis project. I was certain the chemistry in Scheme 3.4 would work. All Rob needed to do was prepare a suitably substituted vinylcyclopropene. I had done cyclopropene chemistry as a grad student, so how hard could this task be, right?

Wrong! As the tongue-in-cheek subtitle implies, coming up with an efficient, reliable, and scalable synthesis of the vinylcyclopropene ligand was a protracted, arduous four-year process. Between Rob, me, and several undergraduate co-workers, we prepared well over 100 different cyclopropenes (most unpublished and not shown here) before

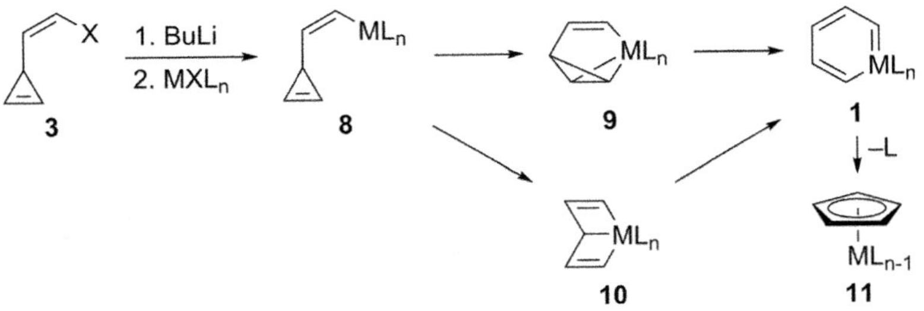

Scheme 3.4

we found the best route. We were quickly able to make **3** (X = Br) using the vacuum gas-solid reaction apparatus I had used as a graduate student; however, the *Z:E* selectivity was poor and we could never prepare more that 20–30 mg of material at a time [21]. That certainly was not going to work if this was to be a general route to metallabenzenes. We needed grams of ligand, not milligrams.

We next investigated use of cyclopropenylium salts as the cations could be prepared in multigram quantities and their reactions with electrophiles were in general high yielding. As shown in Scheme 3.5, we tried multiple variations of the theme. At the time it was reported that one could generate the Grignard or lithiate of a *cis*-dianion equivalent, but in our hands we obtained either low or no yield, or more often observed isomerization to the deleterious trans-substituted product [22]. One could easily make a wide variety of 3-ethynyl-substituted molecules [23]; however, all of the stereoselective reductions known in the mid-1990s (e.g. Schwarz's reagent, Lindlar's catalyst) completely failed to react with the triple bond and more often would reduce the strained cyclopropene double bond instead [24]. Attempts to use a variety of 1,2-benzenedianion equivalents gave very low yields of products as degradation pathways involving benzyne predominated [22]. In the end, we elected to introduce the vinyl group one carbon at a time, starting by addition of nucleophilic cyanide anion. This worked very well, affording the tetrasubstituted cyclopropenes in 85–90% yield.

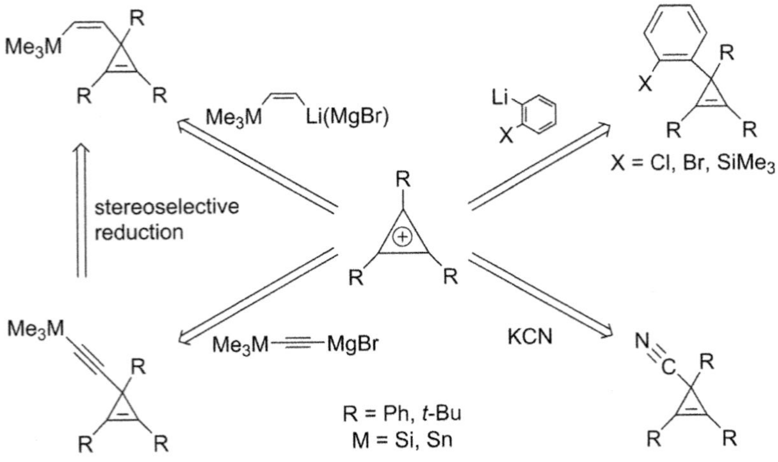

Scheme 3.5

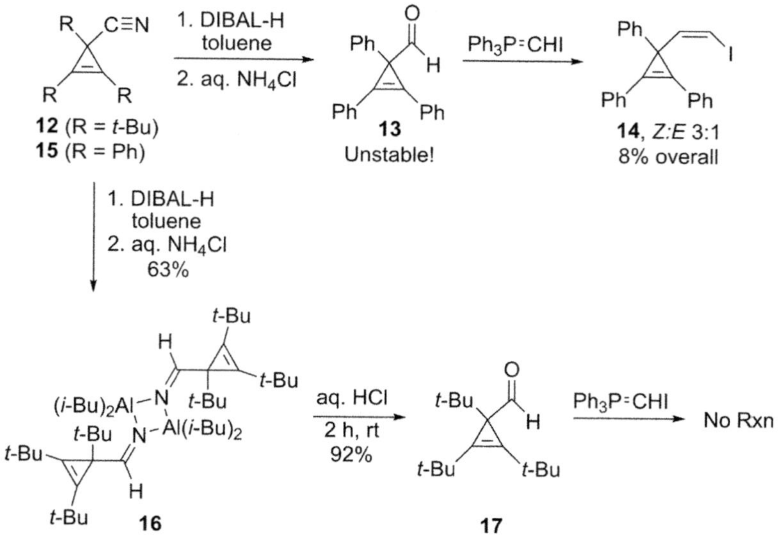

Scheme 3.6

We envisioned the last two steps of ligand synthesis to be reduction of the nitrile to the aldehyde followed by a Wittig reaction. Once again, this was easier said than done. DIBAL-H reduction of nitrile **12** furnished aldehyde **13** as an unstable solid (Scheme 3.6) [25]. We quickly took **13** forward and, after the Wittig reaction, obtained a dismal 8% yield of iodide **14** that was 3:1 *Z*:*E* [22]. DIBAL-H reduction of nitrile **15** afforded a white crystalline solid. When Rob showed me the proton NMR data, I accused him of doing a poor job of purifying the material as the spectrum clearly showed residual isobutyl peaks. After rechromatographing and recrystallizing the material, Rob obtained the same spectrum. Fortunately, the crystals were large enough for X-ray diffraction. Imagine our surprise when we obtained the structure of aldimine **16** [26]!

In hindsight, it is not too surprising that the brief hydrolysis step to give **13** failed in the case of **16**. With four *iso*-butyl and six *tert*-butyl groups, the aldimine is severely crowded sterically. Only after stirring with 10% HCl solution for 2 h did Rob isolate aldehyde **17**. As you would know it, the steric bulk that impeded aldimine hydrolysis shut down the Wittig reaction altogether, affording recovered **17** [22].

What finally worked was omitting the extra substituent on C3. This required yet another new approach to the cyclopropene skeleton. The reaction of ethyl diazoacetate with diphenylacetylene had been examined in the early 1960s and the subsequent chemistry of the carboxylic acid functionality was well developed. Starting from Weinreb amide **18**, first prepared by the Hughes group (Scheme 3.7), reduction with $LiAlH_4$ afforded aldehyde **19**, which fortunately was much more stable than triphenyl analogue **13**. Although the Wittig reaction to install the *cis*-bromo group in **20** again gave poor *Z*:*E* selectivity, [22] the corresponding reaction to generate a *cis*-iodide furnished **21** in excellent yield (81%) with outstanding selectivity (>20:1 *Z*:*E*). We at last had our desired ligand, and we could make it in 1–2 g quantities at a time. Would the chemistry as proposed in Scheme 3.4 work? It was time to find out.

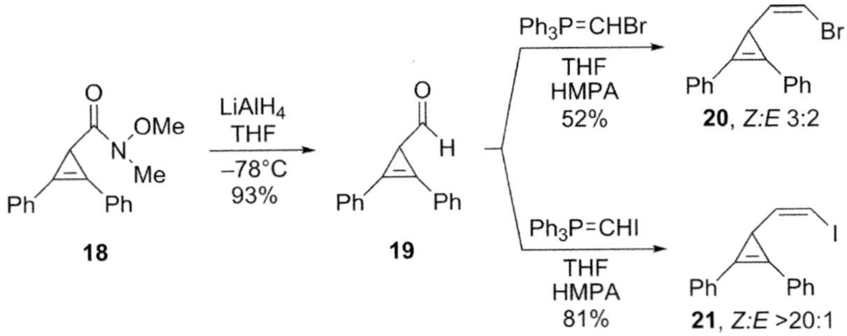

Scheme 3.7

3.3 Iridabenzenes and Valence Isomers (Success after Six Long Years)

From the time the idea was first envisioned and followed by the protracted synthesis of the C_5 ligand, nearly six years had elapsed. Many, many more metallabenzenes had been reported in the literature over that period, but most included a heteroatom (or two) or were coordinated to another organometallic fragment [27]. Most were serendipitous discoveries, and nearly all were produced as singular examples. The only additional, confirmed example of a "true" metallabenzene—a metallacycle with one metal and five carbons without additional stabilizing elements—was reported by Jones and Allison in 1995, but the resultant ruthenabenzene decomposed above −50°C to an η^3-cyclopentadienyl complex [28]. With only three "true" metallabenzenes in the literature, I was confident we could still make a splash.

3.3.1 Iridabenzene/Iridabenzvalene Synthesis

One thing we learned from all our synthetic adventures attempting to prepare the ligand was that vinyl anions are incredibly sensitive to even the slightest hint of residual water; thus, we had to flame dry EVERYTHING—flasks, syringes, cannula needles—to minimize the protonated vinylcyclopropene side product. The second thing we learned in my group was that organometallic chemistry can be a very tricky business. Coming from an organic background, it made good sense to keep the organometallic starting reagents we chose to study simple, and commercially available if at all possible. With these caveats in mind, and given the success Bleeke had with iridium, we opted to first examine one of the simplest and best-studied sources of iridium, namely Vaska's complex.

Lithium–halogen exchange of vinyl iodide **21** at −78°C was followed by addition of Vaska's complex (Scheme 3.8). At first nothing appeared to happen, but upon warming to room temperature the reaction changed color from yellow to orange to deep red (which also happened to be the color of **7**). Rob worked up the reaction and obtained the proton NMR spectrum of the crude material. In addition to the presence of residual protonated vinylcyclopropene, the telltale peak at 10.4 ppm strongly implicated formation of iridabenzene **22a** (R = Ph), a result corroborated by the ^{13}C and ^{31}P NMR data. Son of a gun—my idea had worked [29]!

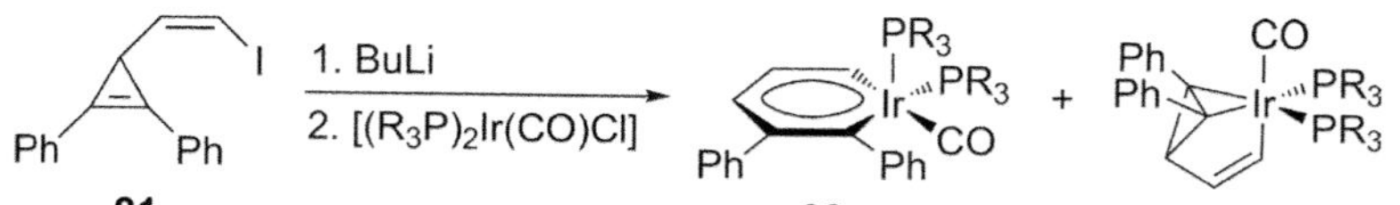

Scheme 3.8

Concurrent with these initial studies, Rob had prepared a series of Vaska-type complexes with varying phosphines, as they are easily accessible compounds. He next tried the transformation in Scheme 3.8 with the less bulky PMe_3 analogue of Vaska's complex, but was initially disappointed as the reaction mixture remained yellow-orange; however, his excitement returned once he obtained the NMR data for the yellow crystals he isolated. Although there were no signals indicative of an iridabenzene, the data suggested formation of a *symmetrical* structure. Rob had isolated the σ-vinyl/η^2-cyclopropene complex **23b** (R = Me), the first confirmed example of an "irida-benzvalene," a valence isomer of **22**. He subsequently showed that this molecule could rearrange to the corresponding iridabenzene **22b** in nearly quantitative yield either by heating in solution or by treatment with Ag(I) salts [30].

Varying the phosphine ligand provided key insights into the factors determining which isomer could be isolated (Table 3.1) [31]. Assisted by masters student Tom Lau and undergraduate Seren Lanza, Rob found that phosphines with smaller cone angles [32] afforded more of iridabenzvalene **23**. Comparison of PMe_3 (entry **b**) and P*i*-Bu_3 (entry **f**) shows that the smaller PMe_3 ligand furnished 100% iridabenzvalene, while the larger P*i*-Bu_3 yielded only 65%. Additionally, ligands possessing electron-rich groups resulted in an initial product ratio favoring the iridabenzvalene. For example, while PPh_3 led to exclusive formation of iridabenzene **22a**, P*i*-Bu_3 gave an initial product ratio of *ca.* 2:1 benzvalene **23f** to benzene **22f**. The cone angles on these two ligands are 145 and 143°, respectively, so it is likely that electronic effects are primarily responsible for the differing ratios. Steric effects apparently are secondary in this case, as these

Table 3.1 Product ratios for the synthesis of iridabenzene **22** and iridabenzvalene **23** with varying phosphines.

Entry	Phosphine	Cone angle[a]	Initial pdt ratio 22:23[b]	Yield (%) 22	Yield (%) 23	Yield (%) 23→22
a	PPh_3	145	100:0	66	—	—
b	PMe_3	118	0:100	—	54	97
c	PMe_2Ph	122	2:98	58	—	—
d	PEt_3	132	0:100	—	51	95
e	$PMePh_2$	136	30:70	56	—	—
f	P*i*-Bu_3	143	35:65	49	—	—
g	P(*p*-MeOPh)$_3$	145	100:0	61	—	—

a) Reference [32].
b) Determined by ^{1}H NMR spectroscopy.

eventually factor in as the C_6D_6 solution of the P*i*-Bu_3 product mixture converts completely overnight at room temperature to **22f**, while a C_6D_6 solution of the PMe_3 iridabenzvalene **23b** is indefinitely stable at room temperature. It is worth noting that use of Vaska-style complexes possessing even bulkier phosphines such as P*i*-Pr_3 (cone angle 160°) and PCy_3 (cone angle 170°) failed to generate metallacyclic species.

In addition to isolating iridabenzenes with different phosphine ligands, Heping Wu, a talented postdoctoral associate who was Rob's successor on the project, successfully synthesized a number of iridabenzenes with different ring substituents via unsymmetrically substituted cyclopropene precursors **24a–e** (Figure 3.2). Similar to our initial studies, iridabenzvalenes—such as **25**, **29**, and **33**—could be isolated from the reaction mixtures in addition to iridabenzenes depending not on the phosphine but rather on the nature of the R group. In certain cases, isomerization to the iridabenzene was highly regioselective, forming the α-phenyl isomer **26** over the β-isomer **27**. In addition to the iridacycles, Heping's studies marked the first time we isolated cyclopentadienyl complexes—such as **28**, **31**, and **35**—which we believed originated from migratory carbene insertion of the corresponding iridacycle (vide infra).

Our studies based on 1-alkyl-2-phenylcyclopropene ligands **24a–d**, prepared in a manner analogous to original ligand **21**, were the most interesting and complex of the various systems we explored [33]. Fortunately, ligand **24d** was "well behaved" and permitted a rather thorough examination of the various structures that could be produced within the iridabenzene manifold. Reaction of lithiated **24d** with Vaska's complex

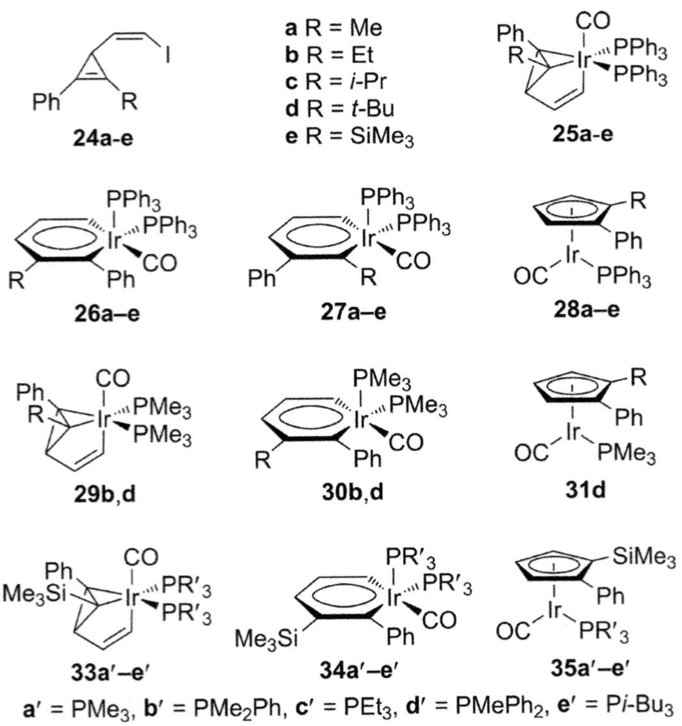

Figure 3.2 Series of iridabenzene, iridabenzvalene, and cyclopentadienyl complexes synthesized from Vaska-type complexes and unsymmetrically-substituted 3-vinylcyclopropenes **24**.

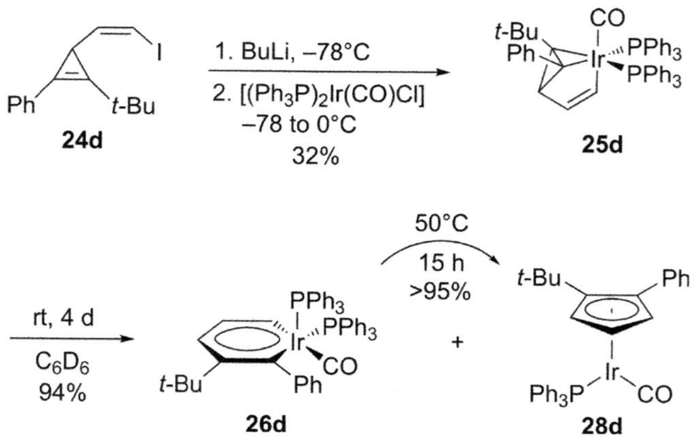

Scheme 3.9

furnished benzvalene **25d** (Scheme 3.9). Although stable in the solid state, a C_6D_6 solution of **25d** at room temperature began to isomerize immediately, and after four days furnished iridabenzene **26d** and cyclopentadienyl complex **28d** as a 3:1 mixture in 94% combined yield. The regiochemistry of **26d** was confirmed by X-ray analysis to be the β-*t*-Bu isomer, with none of the corresponding α-*t*-Bu regioisomer **27d** detected. Unlike **22a–g**, **26d** was surprisingly labile as heating a C_6D_6 solution at 50°C quantitatively converted the arene to cyclopentadienyl **28d** via a process that exhibited first-order kinetics.

We initially thought that undetected α-*t*-Bu regioisomer **27d** might be the source of **28d** via very facile rearrangement; however, we disproved this hypothesis by examination of the kinetics of the isomerization of **25d** and of **26d** to cyclopentadienyl **28d**. Heping found that at 20°C the isomerization rate of **26d** to **28d** was faster than the corresponding rate of **25d** to **28d**, meaning that the initially formed cyclopentadienyl complex **28d** originated from the rearrangement of the β-*t*-Bu isomer **26d** and not from α-*t*-Bu isomer **27d**. It therefore appears that the valence isomerization of unsymmetrical benzvalene **25d** to benzene **26d** is highly regioselective.

Reactions of cyclopropenes **24a–c** with Vaska's complex revealed the influence of the alkyl substituent on both formation of **25** and its subsequent isomerization to **26** [33]. Proton NMR data showed that each of the crude mixtures consisted of both iridabenzvalene and iridabenzene. Unfortunately, iridabenzvalenes **25a–c** could not be isolated because of their relatively rapid conversion to iridabenzenes **26a–c**, which, unlike **26d**, were stable for over 48 h at 75°C. Interestingly, very minor amounts of the α-alkyl regioisomers **27a–c** were detected in the crude products by NMR spectroscopy, but they could not be isolated in pure form. While these and the subsequent results (vide infra) seem to suggest that regioisomers **26** and **27** could be formed by different mechanisms, the exact origin(s) of the preferential formation of **26** is uncertain.

In contrast to Vaska's complex, reaction of cyclopropenes **24b, d** with less bulky and more electron rich $[Ir(CO)Cl(PMe_3)_2]$ furnished benzvalenes **29b, d** as the only products, which were stable at room temperature [33]. Isomerization at 75°C converted **29b** completely to β-Et regioisomer **30b** after 4 h. Under the same conditions, however, *t*-Bu analogue **29d** gave a mixture of unreacted benzvalene, benzene **30d**, and cyclopentadienyl complex **31d** in a 5:10:1 ratio along with partial decomposition; prolonged heating led to complete decomposition to unidentified materials. Interestingly, we never saw evidence for formation of the α-alkyl regioisomer of **30**; thus, the smaller size of PMe_3 than PPh_3 apparently does not alter the regioselectivity of the isomerization. The influence of the Et and *t*-Bu substituents on the stability and isomerization rate of **29** is similar to that observed for the corresponding PPh_3 analogues **25**. Additionally, the higher stability of β-*t*-Bu substituted **30d** compared to analogue **26d** suggests that the PMe_3 ligand does stabilize the iridabenzene. Considering the whole "alkyl" series, the trend of the isomerization rate of alkyl-substituted benzvalenes to benzenes is **25a** = **25b** > **25c** > **25d** > **29b** > **29d**, which is in agreement with increase of both electronic donation and sterics of the alkyl group as well as electron donation and/or reduced sterics of the phosphine ligand. Conversely, the order of iridabenzene stability is **30b** > **26a** = **26b** = **26c** > **30d** > **26d**, indicating that a decrease of sterics of the alkyl group, electron donation, and/or sterics of the phosphine ligand enhance the stability of the iridabenzene.

Heping next explored the reactivity of vinylcyclopropenes containing one or two trimethylsilyl ($SiMe_3$) moieties with Vaska-type complexes [34]. Treatment of the vinyllithiate of **24e** with Vaska's complex gave a 47% yield of **25e**, **27e**, and **28e** in a 10:2:3 ratio (Scheme 3.10). Benzvalene **25e** and cyclopentadienyl **28e** were isolated cleanly either by addition of MeI to the mixture or by heat, respectively; once again, the pure α-regioisomer (i.e. **27e**) could not be isolated. In contrast to **25a–d**, silyl benzvalene **25e** is stable at room temperature; however, similar to **25a–d**, heating a solution of **25e** afforded only β-silyl regioisomer **26e** and not α-silyl regioisomer **27e**. Benzene **26e** is more stable than **27e**, but it also undergoes carbene migratory insertion to give **28e**. While it seems reasonable that the α-silyl isomer is also an intermediate in the conversion of **25e** to **28e**, **27e** was not detected during the isomerization.

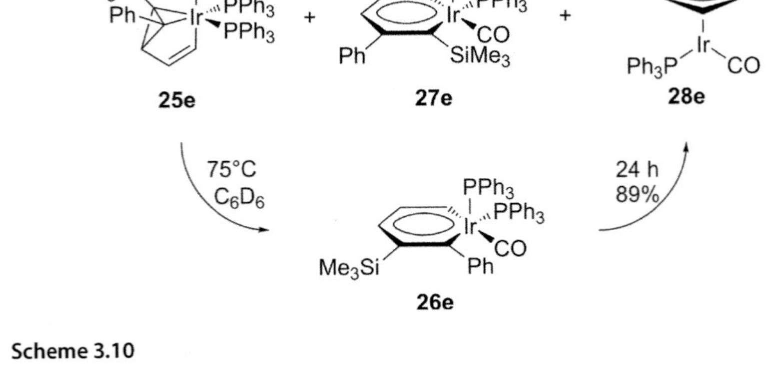

Scheme 3.10

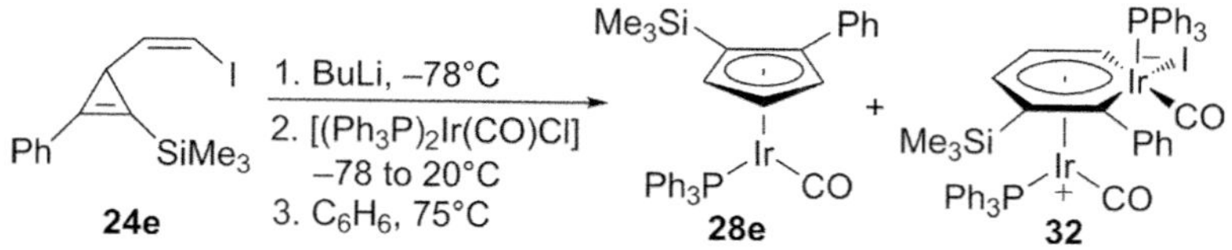

Scheme 3.11

During the course of this study, we tried to simplify the preparation of **28e** by performing the thermolysis reaction without isolation of **25e** [35]. As shown in Scheme 3.11, this did indeed work in roughly 40% yield; however, we isolated a new, bright-red complex in low yield (*ca.* 5%). The presence of two CO resonances in the ^{13}C NMR spectrum, two CO stretches in the IR spectrum, two phosphorus peaks in the ^{31}P NMR spectrum, and significantly upfield-shifted iridabenzene CH resonances all suggested an η^6-dinuclear complex. X-ray crystallography showed that complex **32** is indeed an η^6-diiridium species that can be formulated as the zwitterionic structure shown in Scheme 3.11, where both Ir centers possess 18 electrons. Admittedly, we were quite surprised by this structure as **32** was the only dinuclear species we ever observed up to this point. Although it is not clear how the two iridium fragments are formed, it is likely that an adventitious Ir(I) fragment intercepts either **25e** or **26e**/**27e**, which after replacement of PPh_3 with I^- affords **32**.

In the case of the Vaska-type complexes, reaction with lithiated ligand **24e** furnished only the corresponding benzvalenes **33a′-e′** (Scheme 3.12) as relatively stable, pale-yellow oils [35]. Interestingly, their rearrangement/decomposition at 75°C appears to be related simply to the size of the phosphine cone angle. In the case of **33a′–c′** (cone angle ≤ 132°), there was no change in the ^1H NMR spectra over the 24 h period; however, like **25e**, benzvalene **33e′** (both with cone angles ≥143°) converted to the analogous Cp complex **35e′**. Benzvalene **33d′** (cone angle 136°) seemed to be the "sweet spot" as after 24 h at 75°C the NMR spectrum showed a *ca.* 3:1:1 mixture of **33d′**:**34d′**:**35d′**; continued heating showed that the mixture could be converted to only **35d′** after *ca.* 90 h. It is worth noting the difference in reactivity between P*i*-Bu_3 and PPh_3 with ligand **24e**. Although both possess similar cone angles, reaction with

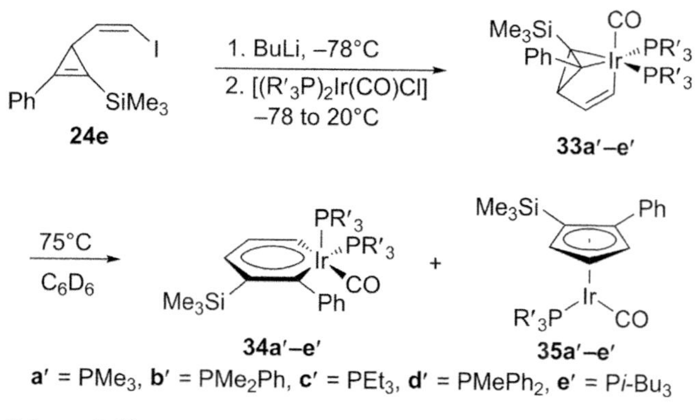

Scheme 3.12

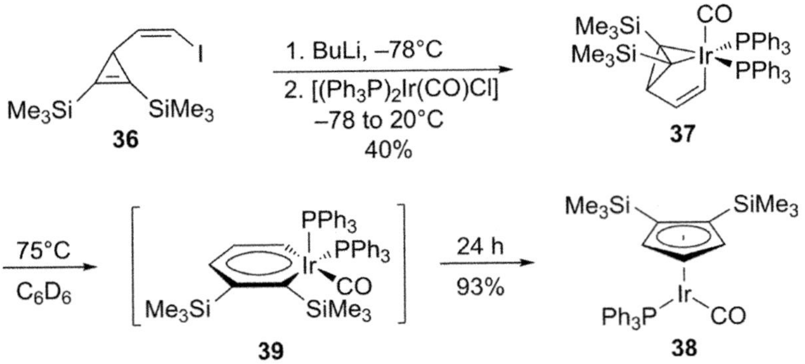

Scheme 3.13

[$(Pi\text{-}Bu_3)_2Ir(CO)Cl$] generated only **33e′** at room temperature, in contrast to the three products produced in the reaction with Vaska's complex (**25e**, **27e**, **28e**) under the same conditions. Once again, this difference highlights the significant role that the electron-donating ability of the phosphine ligands plays in benzvalene stability.

Taking silyl-substitution one further, Heping examined the reaction of bis(silyl) cyclopropene **36** (Scheme 3.13) [35]. Following our regular synthetic pathway, **37** could be isolated in 40% yield. This is the most stable PPh_3-containing benzvalene isolated to date, as it required heating at 75°C for 36 h to isomerize/rearrange into cyclopentadienyl complex **38** in essentially quantitative yield. While benzene **39** was never detected in this process, its intermediacy is presumed based on our related studies (vide supra). Incorporation of successive $SiMe_3$ groups seems to have two effects: the σ-donating ability of the Si atoms appears to stabilize benzvalene formation, yet the π-accepting nature of Si along with steric bulk of the $SiMe_3$ group(s) destabilizes the corresponding benzenes.

3.3.2 Mechanisms of Iridabenzvalene Isomerization and Iridabenzene Rearrangement

One aspect that intrigued us for quite some time was the mechanism (or mechanisms) by which the iridabenzvalenes rearrange into iridabenzenes, as one can envision two potential pathways (Scheme 3.14). One route begins by dissociation of the cyclopropene from the metal center followed by oxidative addition within σ-vinyl complex **40** to generate the Dewar-type metallabenzene **41**, which quickly rearranges into **42**. Evidence for this pathway lies in the ability of donor solvents (acetone, MeCN) to rapidly increase the rate of isomerization [22]. Hughes reported similar types of reactivity in the reaction of perfluorinated cyclopropenes with transition metal complexes, isolating intermediate metallacyclobutenes [36, 37]. Additionally, Grubbs *et al.* postulated a metal-assisted mechanism for the opening of 3,3-diphenylcyclopropenes based on the isolation of an Ir-coordinated iridacyclobutene [6]. Our isolation of Ir-coordinated iridabenzene **32** hints at this latter possibility.

Alternatively, direct rearrangement of benzvalene **43** into benzene **42** could be possible, similar to the valence isomerization of the purely C_6H_6 hydrocarbon analogue. DFT calculations by van der Boom and Martin suggest that this latter pathway,

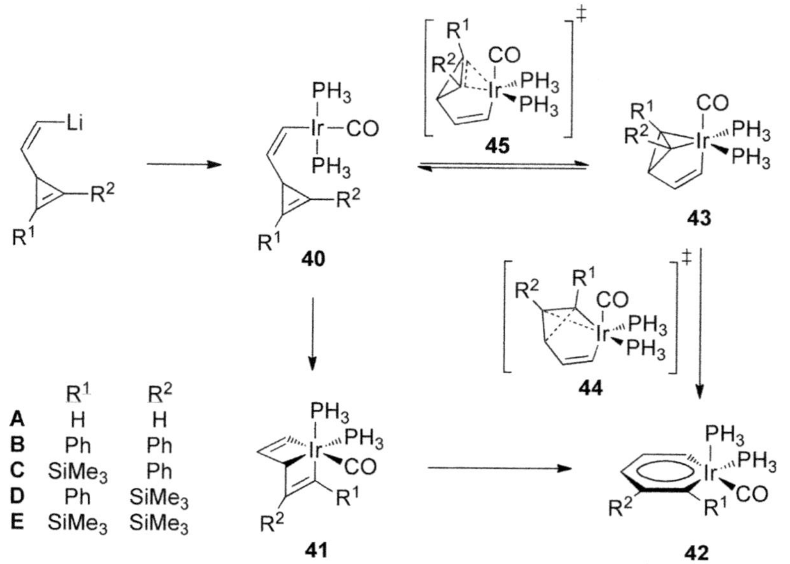

Scheme 3.14

predicted to occur by twisting of the iridacyclopropane ring into the planar metallacycle via transition state **44** (Scheme 3.14), is the more feasible mechanism [38]. Starting from computationally simplified Vaska's complex and vinylcyclopropene ligand where $R^1 = R^2 = H$, two isomers of **40A** that vary by orientation of the cyclopropene ring were identified: the *syn* isomer, where the cyclopropene ring points toward Ir center, is calculated to be 1.8 kcal mol^{-1} higher energy than the *anti* isomer. With this lower energy σ-complex defined as zero, they found that the transition state (**45A**) to generate iridabenzvalene **43A** is only 3.5 kcal mol^{-1} higher in energy. Benzvalene **43A** lies in an energy well at −23.1 kcal mol^{-1}. From the iridabenzvalene, it is a 39.5 kcal mol^{-1} barrier to form iridabenzene **42A** directly via transition state **44A**. While no transition state was found for the formation of Dewar benzene **41A**, its energy was calculated to be 51.3 kcal mol^{-1} above **43A**, suggesting that the Dewar benzene mode of isomerization would be considerably higher in energy.

In light of the differing reactivity of the trimethylsilyl-substituted metallacycles versus the analogous Ph-substituted systems described above, we computationally reexamined the energy manifold [35]. Unsurprisingly, iridabenzvalenes **43B-E** are 8–10 kcal mol^{-1} less stable than **43A** due to steric interactions of the metal center with the substituents on C1 and C2 of the ligand. The calculations clearly reveal the influence of the C1/C2 substituents on the isomerization process: whereas the transition state barrier in **44B** is lowered by 6.3 kcal mol^{-1} compared to **44A** because of the weaker η^2-interaction in **43B**, the barrier involving transition state **44E** is 5.1 kcal mol^{-1} higher compared to **43A**. The electronic effects of two $SiMe_3$ groups help stabilize the benzvalene through σ donation. Unfavorable steric effects also come into play in **44E** as the Si-Me/Si-Me interactions are reduced to 2.29 Å compared to 2.45 Å in **42E**. The σ donation of the $SiMe_3$ group also explains the regioselective formation of **26e**. Isomerization of **43** involves breaking the Ir–C2 and C1–C3 bonds along with twisting the C2 atom of the iridacyclopropane ring to afford planar iridabenzene **42**. In the case of **44C** vs. **44D**, the transition state barriers are comparable to that of **44A**, with **44D** favored by only 0.6 kcal mol^{-1}; however, the transition state geometries reveal a striking difference, where the Ir–C2 bond length in **44D** is 2.46 Å versus a considerably longer 2.59 Å value in **44C**. In addition, the Mulliken charges for C1 and C2 in **44D** are +0.25*e* and −0.29*e*, respectively, the magnitudes of which are considerably reduced compared to the analogous charges of −0.56*e* and +0.78*e* in **44C**; thus, strong preference is given to formation of the β-silyl regioisomer as the $SiMe_3$ group stabilizes the developing positive charge at C2.

Many of the above arguments for stabilization of benzvalenes **43** led to destabilization of iridabenzenes **42** and thus their rearrangement to η^5-cyclopentadienyliridium complexes **46** [35, 38, 39]. This deleterious reaction involves carbene migratory insertion through transition state **47**, giving η^1-Cp complexes **48**, followed by phosphine loss to yield **46** (Scheme 3.15). Based on our calculations, and as depicted in TS **47**, C1 (with substituent R^1) is best described as a metal–carbon double bond whereas C5 is more appropriately classified as a vinyl anion. The large dihedral angles between R^1 and R^2 torque C1 such that its p-orbital is aligned with the vinyl anion orbital at C5, permitting facile C–C bond coupling. This result also suggests that electronic effects at C1 will influence bond formation to a much greater extent than would substituents at C2–C5. The difference between the reactivity of an α-silyl (**42C**) versus an α-phenyl (**42D**) is representative: TS structure **47C** is 5 kcal mol^{-1} lower in energy than that in **47D**, thus reflecting the direct electronic effects on C1. In the case of **42E**/**47E**, both the α-silyl electronic effect as well as steric effects of the adjacent α,β-bis(silyl) come into play. Benzene **42E** is non-planar with a *ca.* 37° dihedral between the silyl groups. Transition structure **47E**, which has a very low barrier of 15.6 kcal mol^{-1}, minimizes this repulsive interaction and ultimately affords Cp-complex **46E** where the dihedral angle decreases to *ca.* 4°. Given these considerations, it is not surprising then that benzene **39** is never observed. The reader is referred to the latest computational assessments as described by Frenking and Fernandez-Lopez in Chapter 7.

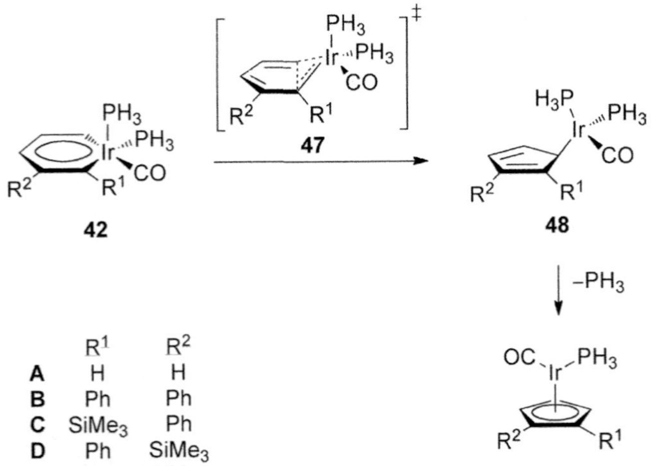

	R^1	R^2
A	H	H
B	Ph	Ph
C	$SiMe_3$	Ph
D	Ph	$SiMe_3$
E	$SiMe_3$	$SiMe_3$

Scheme 3.15

3.3.3 Iridabenzene/Iridabenzvalene Spectroscopic Properties

The synthesis of an extended family of structurally related iridabenzenes allows for detailed comparison of their spectroscopic properties. All iridabenzenes are easily identified by the NMR resonance of the proton on the carbon *ortho* to the metal (H5), a number that varies between 10.41 ppm for **22a** and 11.30 ppm for **22d** (Table 3.2). The downfield shift of these characteristic resonances has long been attributed to the anisotropy of the neighboring metal center [27]. This effect quickly dissipates for the protons on the carbons *meta* (H4) and *para* (H3) to the metal. Similar to substituted arenes, the *para* proton is next most influenced by the metal fragment, resonating between 7.69–8.78 ppm. The proton *meta* to the Ir center is least affected and resonates the furthest upfield of the metallacycle protons at 7.38–7.91 ppm. These latter two

Table 3.2 Selected NMR data for iridabenzenes **22**, **26**, **30** and **63**[a].

compd	Phosphine/ substuent	^{1}H NMR (δ) H3	H4	H5	^{13}C NMR (δ) C1	C5	C6	^{31}P NMR (δ) P
22a[b]	PPh_3	7.99	7.69	10.41	187.60	187.43	201.09	16.44
22a[c]	PPh_3	8.44	7.79	10.79	187.74	187.49	203.26	17.88
22b[b]	PMe_3	7.75	7.49	11.10	189.25	176.58	189.44	−40.54
22b[c]	PMe_3	8.25	7.91	11.00	189.95	176.04	189.44	−40.27
22c[b]	PMe_2Ph	7.82	7.53	11.05	188.56	179.68	190.77	−26.90
22d[b]	PEt_3	7.69	7.40	11.30	189.87	176.45	192.36	−2.08
22e[b]	$PMePh_2$	7.98	7.66	10.93	187.36	181.93	197.14	−5.20
22f[b]	P*i*Bu_3	7.71	7.51	10.89	190.54	177.28	194.09	−0.29
22g[b]	P(*p*-MeOPh)$_3$	7.96	7.38	10.43	188.51	187.11	198.18	13.01
26a[c]	Me	8.30	7.75	10.62	187.46	183.92	204.97	18.03
26b[c]	Et	8.29	7.79	10.61	187.31	184.95	204.17	18.65
26c[c]	*i*Pr	8.41	7.86	10.60	187.65	185.52	203.90	18.79
26d[c]	*t*Bu	8.69	7.83	10.54	186.68	186.55	208.32	18.24
26e[c]	$SiMe_3$	8.78	—[d]	10.85	188.00	188.93	—[e]	17.13
30b[c]	Et	8.13	7.90	10.92	188.83	174.20	181.02	−38.56
30d[c]	*t*Bu	8.46	7.88	10.87	—[e]	—[e]	—[e]	−38.79
63a[c]	MeO	8.32	7.91	11.09	189.53	176.05	191.76	−38.32
63b[c]	Me	8.31	7.89	11.07	189.61	176.27	191.85	−38.22
63c[c]	F	8.16	7.86	11.03	189.43	176.41	189.32	−38.02
63d[c]	Cl	8.09	7.82	11.00	189.30	176.50	188.60	−37.85

a) Atom labeling as shown in Figure 3.3.
b) Data acquired in CD_2Cl_2.
c) Data acquired in C_6D_6.
d) Resonance obscured by other signals.
e) Resonance not assigned.

resonances are more in line with those found in electron-withdrawn systems such as pyridine and benzonitrile. Interestingly, in C_6D_6 the *meta* and *para* protons are shifted 0.4–0.5 ppm downfield compared to the same protons in CD_2Cl_2, but the analogous comparison of the *ortho* proton shows that it generally resonates 0.1–0.2 ppm upfield.

The ^{13}C NMR spectra demonstrate comparable anisotropy trends (Table 3.2). Carbons attached to the Ir center resonate between 174–190 ppm with the Ph-substituted carbon C1 always more downfield shifted from the unsubstituted carbon C5, with only one minor exception (**26e**). The resonances of the remaining carbon atoms in the iridacycle (C2–C4, not shown) appear in the "normal" aromatic region of 128–145 ppm. While the chemical shift of the carbons *meta* to the metal remains relatively unaffected by changes to the ligand sphere of **22** ($\Delta\delta < 1.4$ ppm), carbons C5 (*ortho*) and C3 (*para*) to the metal vary much more. For example, replacement of an alkyl group with a phenyl group on the phosphine results in a deshielding of C3 ($\Delta\delta = 4.5$ ppm) and C5 carbon atoms ($\Delta\delta = 11$ ppm). The carbonyl carbon exhibits comparable variability ($\Delta\delta = 11.5$ ppm). The substituted carbon C1 does not show this dependence on the ligand sphere ($\Delta\delta = 1.4$ ppm) and in fact shifts in the opposite direction. For the "alkyl" series **26**/**30**, replacement of PPh_3 with PMe_3 on the Ir center leads to an upfield shift of the C3 ($\Delta\delta = 5.5$ ppm), C5 ($\Delta\delta = 10.8$ ppm), and C6 ($\Delta\delta = 23.2$ ppm) atom resonances, respectively, which can be rationalized in terms of electronic influences. The stronger donating ability of PMe_3 compared to PPh_3 results in a more electron-rich metal center, which in turn increases the electron density of the iridacycle and thus the carbons shift upfield. On the other hand, small shifts for C1 ($\Delta\delta = 1.5$ ppm), C2 ($\Delta\delta = 1.4$ ppm), and C4 ($\Delta\delta = 0.2$ ppm) are observed, indicating the lack of sensitivity of these carbon atoms toward the phosphines.

The phosphorus resonances appear as sharp singlets in the ^{31}P NMR spectra, even when cooled to −80°C, demonstrating that the axial and basal phosphines are exchanging rapidly in solution via the well-known Berry pseudorotation process. Interestingly, substituted carbon C1 exhibits ^{31}P–^{13}C coupling, whereas unsubstituted C5 does not. We believe this indicates that the smaller CO ligand resides a disproportionate amount of the time near substituted C1, probably due to steric crowding, even though the ligand sphere is dynamic in solution. Bleeke demonstrated an analogous preference through careful correlation of ^{31}P–^{1}H coupling constants in a structurally related iridabenzene [19].

It is rather fortunate (i.e. we got lucky!) that nearly all of the possible structures that can be generated by our synthetic methodology possess distinct ^{1}H NMR signals that are easily diagnostic as to what compounds a crude reaction mixture likely contains. For the iridabenzvalenes, the bridgehead cyclopropyl proton (H3) appears as a characteristic broad singlet (more like an ill-defined multiplet) in the range of 3.2–3.8 ppm. In addition, two complex multiplets in the region of 6.0–6.8 ppm and 6.7–7.2 ppm (the latter sometimes obscured by overlapping phenyl peaks) belong to the alkene resonances H4 and H5, respectively. If the cyclopropene double bond does not coordinate and instead affords just a σ-vinyl metal complex (vide supra), then H3 resolves into a doublet. For the η^5-cyclopentadienyl complexes, sharp, well-resolved signals for the three ring hydrogens appear around 4.2–5.4 ppm. Residual vinyllithiate that is protonated during work-up of the reaction has a characteristic signal that appears as a ddd at around 6.0–6.2 ppm.

3.3.4 Iridabenzene/Iridabenzvalene Solid-State Structures

X-ray diffraction studies of five of our iridabenzenes reveal some key structural features of metallabenzenes. The molecular structure of **26a** is shown in Figure 3.3; comparison of the bond lengths and bond angles of **22a, d** and **26a, b, d** is given in Table 3.3. The metallacycles possess a distorted coordination geometry in-between trigonal bipyramidal and square pyramidal. The C–C bond distances in the central ring (1.334–1.427 Å) are essentially equal and thus can be regarded as evidence of delocalization of the π-electrons. The mean C–C bond lengths are very close, ranging from 1.383 to 1.392 Å. The Ir–C1 and Ir–C5 bond lengths (avg. 2.02 Å) are intermediate between Ir–C single and double bonds. The metallacycles are essentially planar (sum of bond angles 719.1–720.0°) with the Ir center tilted out of the five carbon backbone, with dihedral angles from 1.2° for **26a** up to 6.4° for **22d**. The X-ray analyses verify that the alkyl groups on **26a, b, d** are at the β-position to the Ir center, in agreement with the spectroscopic assignment. Contrary to our expectations based on reactivity (vide supra), the bulky *t*-Bu group does not induce greater torsion strain in the iridabenzene ring of **26d**, at least not in the solid state.

The solid-state structures of the previously unknown benzvalene isomers are also of substantial interest. The molecular structure of **37** is shown in Figure 3.4, and selected bond lengths and bond angles of **23b, d, 25d, 37**, and **62a–d** are given in Table 3.4. Around the Ir center the benzvalenes possesses a trigonal-bipyramidal coordination composed of carbonyl, η^2-cyclopropene double bond, two phosphine, and σ-vinyl ligands. The Ir–C1 and Ir–C2 bond distances (2.134–2.220 Å) are typical of Ir–C single bonds. The C1–C2 bond length (1.405–1.453 Å) is characteristic of transition metal-olefin complexes that have significant back-bonding contribution from the metal center (i.e. metallacyclopropane). That said, the longer Ir–C bond lengths, the shorter C1–C2 bond length, and the smaller C1–Ir–C2 bond angle and Ir–C1–C2–C3

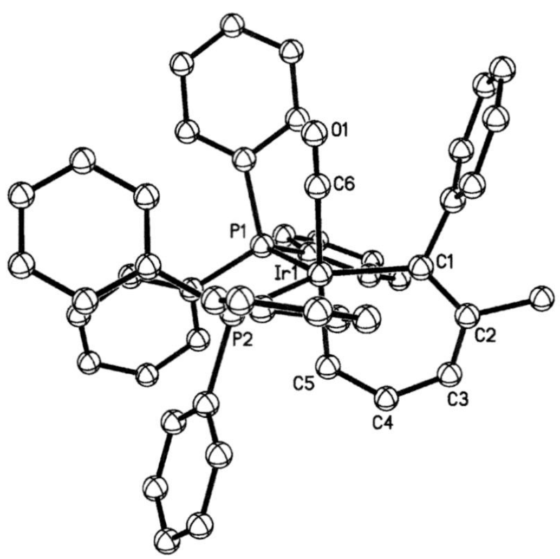

Figure 3.3 Molecular structure of iridabenzene **26a**; thermal ellipsoids are drawn at the 25% probability level.

Table 3.3 Selected bond lengths [Å] and bond angles [°] for iridabenzenes.

	22a	**22d**	**26a**	**26b**	**26d**
Ir–C1	2.021	2.047	2.020	2.029	2.054
Ir–C5	2.025	2.004	2.012	2.023	2.000
C1–C2	1.409	1.427	1.423	1.413	1.407
C2–C3	1.410	1.372	1.335	1.390	1.382
C3–C4	1.377	1.386	1.382	1.393	1.427
C4–C5	1.334	1.381	1.408	1.360	1.343
C=C mean	1.383	1.392	1.387	1.389	1.390
Ir–C1–C2	128.8	127.1	129.1	130.0	130.8
C1–C2–C3	123.6	123.1	123.5	122.1	119.5
C2–C3–C4	124.2	127.9	125.9	126.2	128.2
C3–C4–C5	125.5	122.5	126.6	124.3	123.9
C4–C5–Ir	130.7	130.4	126.4	130.3	130.2
C1–Ir–C5	86.9	88.1	88.5	87.0	87.4
sum of angles	719.7	719.1	720.0	719.9	720.0
C5–Ir–C6	165.0	167.8	172.5	171.8	172.1
P1–Ir–P2	99.2	104.8	102.5	102.4	105.6
P1–Ir–C1	112.6	101.9	119.8	121.0	137.5
P2–Ir–C1	148.2	153.0	137.7	136.3	116.5
dihedral angle[a]	3.7	6.4	1.7	1.2	1.4

a) The angle between plane 1 (C1, Ir, C5) and plane 2 (C1, C2, C3, C4, C5).

dihedral angle in **25d** all suggest that the η^2-interaction of the cyclopropene π-bond with the Ir center is weaker than in the analogous complexes listed in Table 3.4. This hypothesis is confirmed by the fact that **25d** isomerizes at room temperature to **26d**, whereas the other benzvalenes require elevated temperatures to isomerize (vide supra).

3.4 Platinabenzenes (How You Get Your Chemistry on a Beer Coaster)

In the summer of 2001, Volker Jacob joined the lab as an AvH Foundation-funded post-doctoral researcher. With Heping already working on iridium, Volker decided to explore other metals to see how general, or not, our route to metallabenzenes might be. He therefore moved one column to the right and went to work on platinum.

3.4.1 "Irrational" Platinabenzene Synthesis

Volker first explored "asymmetric" $L_2Pt(X)Y$ complexes, with the idea that only ligand X would be substituted by the vinylcyclopropene unit. In the next step Y would be removed from the σ-vinyl compound, allowing for facile coordination of the

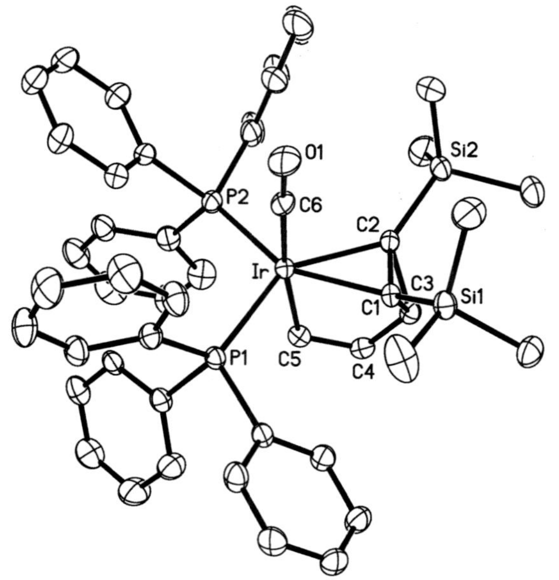

Figure 3.4 Molecular structure of iridabenzvalene **37**; thermal ellipsoids are drawn at the 25% probability level.

Table 3.4 Selected bond lengths [Å] and bond angles [°] for iridabenzvalenes.

	23b	**23d**	**25d**	**37**	**62a**	**62b**	**62c**	**62d**
Ir–C1	2.146	2.172	2.189	2.173	2.134	2.146	2.145	2.146
Ir–C2	2.143	2.159	2.220	2.189	2.161	2.139	2.156	2.149
Ir–C5	2.095	2.092	2.184	2.074	2.098	2.086	2.112	2.092
C1–C2	1.447	1.440	1.405	1.446	1.453	1.450	1.445	1.446
C1–C3	1.529	1.519	1.555	1.539	1.542	1.533	1.535	1.529
C2–C3	1.526	1.533	1.566	1.543	1.535	1.524	1.534	1.528
C3–C4	1.466	1.481	1.401	1.457	1.481	1.471	1.478	1.465
C4–C5	1.322	1.328	1.375	1.330	1.327	1.339	1.310	1.322
C1–Ir–C2	39.4	38.8	37.1	38.7	39.5	39.6	39.3	39.3
C5–Ir–C6	178.9	177.2	174.0	172.6	179.0	177.2	178.7	178.7
P1–Ir–P2	104.1	109.4	110.4	106.5	104.8	107.9	104.8	103.8
P1–Ir–C1	112.6	101.9	119.8	121.0	110.1	108.4	104.9	108.5
P2–Ir–C1	148.2	153.0	137.7	136.3	143.1	147.2	148.4	145.9
dihedral angle[a]	116.3	116.2	109.9	113.4	104.9	108.6	115.8	116.1

a) The angle between plane 1 (C1, C2, C3) and plane 2 (C7, C1, C2, C13).

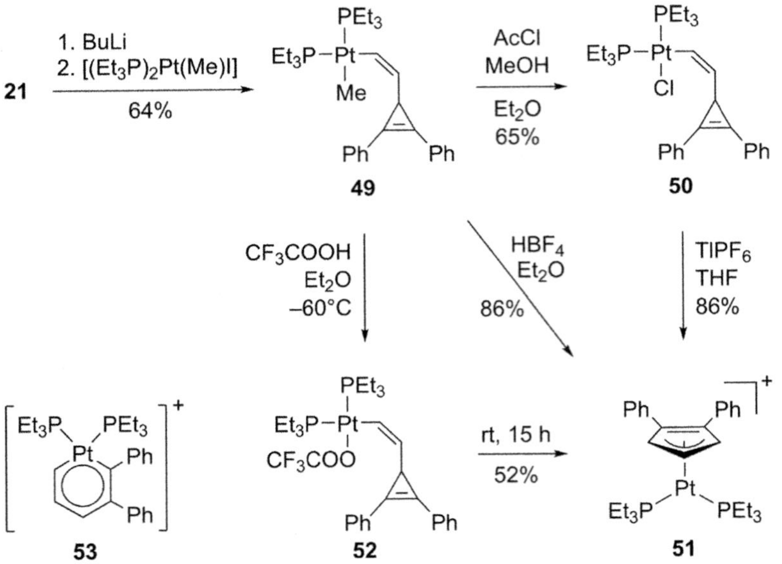

Scheme 3.16

cyclopropene double bond. Based on our experience with iridium, we assumed either a cationic platinabenzvalene or isomerized products like a cationic platinabenzene would be isolated (in theory). In practice, the platinum complexes had other plans. Addition of the lithiate of **21** to $[(PEt_3)_2Pt(Me)I]$ resulted in the displacement of the primary leaving group I and thus formation of σ-vinyl complex **49** (Scheme 3.16), as confirmed by X-ray crystallography [40]. Although it is well known that 16-electron Pt(II) prefers a square planar geometry, I was hoping that the strained cyclopropene moiety would coordinate to afford a platinabenzvalene, but nope, the Pt center was not interested. To make the species even more reactive, protonolysis of **49** with HBF_4 opened a coordination site by removal of the secondary leaving group Me. Likewise, substitution with Cl to give **50** followed by chloride abstraction with $TlPF_6$ generated the same 14-electron Pt(II) species. In both reactions, only complex **51** was isolated, an interesting molecule itself as the solid state showed that the molecule possessed a slipped Cp ring and thus preferred η^3-coordination. Similar to the reactivity Heping was observing, we suspected a platinabenzene intermediate; thus, Volker carried out the protonolysis with trifluoroacetic acid at −60°C to afford **52**, which was stable only at low temperatures. Upon warming to room temperature the reaction furnished a 3:2 mixture of **51**$[O_2CCF_3]$ and the trans isomer of **52**. Even though careful monitoring of the reaction by ^{1}H NMR spectroscopy provided no direct evidence for a platinabenzene, it seemed reasonable to assume the intermediacy of an unstable cationic species such as **53**, which readily undergoes carbene migratory insertion to give a Cp complex like **51**.

Discouraged by the above results, we took a step back and next utilized a simpler source of platinum, namely $[PtCl_2(cod)]$. This did the trick, as Volker was able to prepare and characterize successfully metallacycle **54**, the first example of a platinabenzene (Scheme 3.17) [41]. Although isolated in a dismal 7% yield, the NMR and

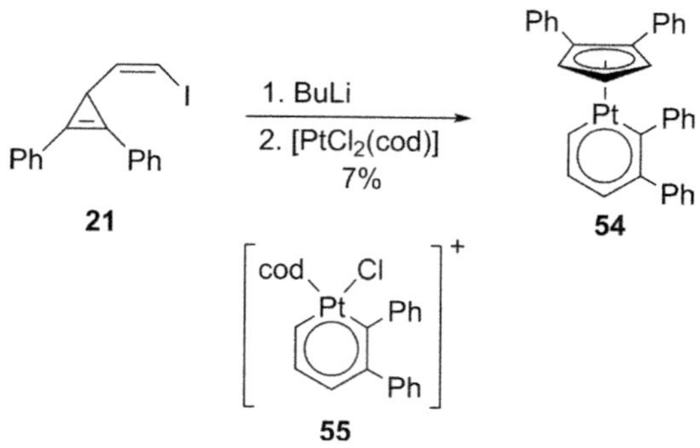

Scheme 3.17

X-ray data clearly confirmed its structure. A particularly intriguing aspect of this reaction is that both the metallacycle and cyclopentadienyl moiety must be derived from the starting 3-vinylcyclopropene. Based on related systems (vide supra), it is likely that formation of the cyclopentadienyl ligand occurs first by decomposition of an intermediate platinabenzene (e.g. **55**) before addition of the second equivalent of lithiated **21**, though admittedly a wide range of mechanistic pathways can be envisaged for the formation of **54**.

Van der Boom and Martin also examined computationally the stability of several platinabenzene complexes [39]. Their results clearly show that the presence of a cyclopentadienyl ligand stabilizes the platinabenzene by increasing the energy barrier of (and thus inhibiting) carbene migratory insertion. In the case of $[C_5H_5Pt(PH_3)_2{}^+]$, the computational analog of **53**, the transition state barrier to forming $[(\eta^5\text{-Cp})Pt(PH_3)_2{}^+]$ was only 24.0 kcal mol^{-1}, which corroborates the experimental results shown in Scheme 3.16. On the other hand, with a Cp unit as part of the metal complex, the transitional state barrier of $[C_5H_5PtCp]$, the computational analog of **54**, to form $[(\eta^3\text{-Cp})_2Pt]$ was considerably higher at 45.9 kcal mol^{-1}.

3.4.2 "Rational" Platinabenzene Synthesis

The isolated products of our first two studies along with the computations all suggested that the next set of experiments should begin with a Cp unit already a part of the starting Pt species. Volker initially examined known [Cp*Pt(CO)Cl] [42]; however, its reaction with lithiated **21** afforded unreacted Pt complex and the protonated ligand, with only trace amounts of platinabenzene. Exchanging Cl^- with more labile I^- as leaving group furnished previously unknown [Cp*Pt(CO)I], synthesized from the platinum dimer $[Cp^*Pt(CO)]_2$. Volker reacted lithiated **21** with [Cp*Pt(CO)I] to give platinabenzene **56a** along with its immediate precursor, σ-complex **57a** (Scheme 3.18) [43]. Both structures were confirmed by X-ray crystallography. Room temperature C_6D_6 solutions of pure **57a** rearranged cleanly and quantitatively to **56a** over a three-day period with no evidence of any intermediate Pt-containing species; the only observable Cp* proton resonances corresponded to **56a** and **57a**. A reasonable mechanism for this transformation is slow loss of CO from the 18-electron species **57a**, followed by coordination of the cyclopropene π-bond and then rapid rearrangement of the strained three-membered ring to afford the aromatic platinacycle.

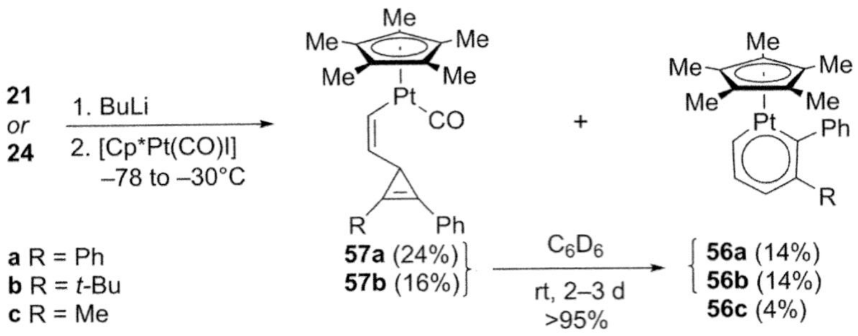

Scheme 3.18

Given the ready availability of unsymmetrical cyclopropene ligands **24a–e**, Christopher Landorf, a new graduate student and Volker's successor, next examined the potential for regioselective platinabenzene formation, much like we had seen for the iridabenzene series. Starting with bulky R = *t*-Bu, this worked as we hoped, yielding a 1:1 mixture of **56b** and **57b** (Scheme 3.18) [43]. As before, σ-complex **57b** cleanly and quantitatively rearranged to **56b**. Unfortunately, our attempts to extend our "rational" synthetic approach to the other alkyl analogues of **24** went all wrong. While we could isolate minute amounts of **56c**, the corresponding Et- and *i*-Pr-substituted platinacycles remained elusive [44]. NMR spectroscopy of the reaction mixtures revealed formation of numerous side products, two of which were Boag's dimer, the immediate precursor to [Cp*Pt(CO)I], and the regenerated starting vinyl iodides **21**/**24**. Control experiments showed that these latter compounds were indeed reformed during the reaction and did not originate from incomplete lithium–iodine exchange. The major difference between **56a** and **56b** versus **56c** (and the other elusive analogues) was the presence of hydrogen atoms on the carbon atoms directly attached to the platinacycle; thus, one could envisage a number of potential deleterious decomposition mechanisms. Alas, we never found a good "work around," so I had Christopher move on to exploring platinabenzene reactivity.

As alluded to earlier and discussed more in the computational chapter of Frenking, the Cp-substituted platinabenzenes are quite stable molecules; solutions of **54** and **56a** exhibited no decomposition upon heating at 100°C for several days [44]. Platinacycle **56a** is unreactive with protic solvents such as H_2O or MeOH. Slow decomposition occurs when solutions are exposed to O_2 over the course of a week. Exposure to elemental Br_2 led to more rapid decomposition, but in both cases, no specific decomposition products were isolated. The latter result contrasts with those reported for the Ir(I) iridabenzenes, where oxidative addition to the metal is usually observed (vide infra). Ir(I) iridabenzenes also react with dienophiles, and that proved to be the case here. Whereas there was no reaction with DMAD, reaction of **56a** with maleic anhydride afforded cyclo-adduct **58** (Scheme 3.19). Platinacycle **56a** also reacted with [Mo(tol)$(CO)_3$] to give **59**, only our second example of an η^6-dinuclear complex (along with **32**).

3.4.3 Platinabenzene Spectroscopic Properties and Solid-State Structures

The NMR spectroscopic data for **54** and **56** exhibit the characteristic chemical shifts similar to other metallabenzenes. The proton *ortho* to the Pt center (H5) produces a

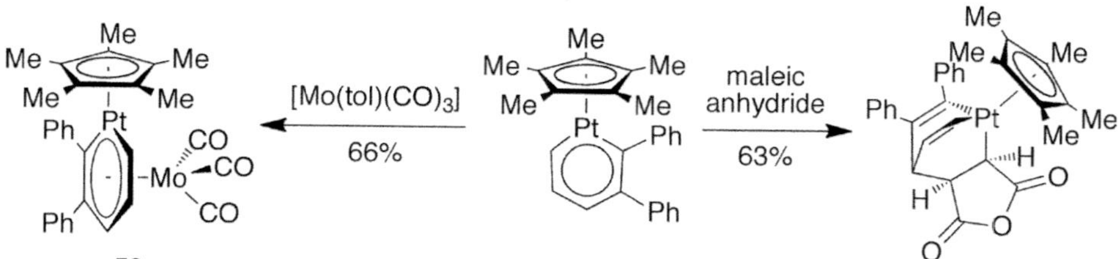

Scheme 3.19

[195]Pt-coupled peak at 12.76 ppm for **54** and in the range of 11.83–12.09 ppm for **56a–c**. The upfield shift for the latter platinacycles is attributable to the increased electron density on the metal center due to the donating Cp* unit. The *meta* and *para* protons afford peaks in the region of 7.4–7.6 and 8.2–8.5 ppm, respectively. Two peaks in the range of 200–204 and 188–195 ppm in the ^{13}C NMR data are due to the Ph-substituted (C1) and unsubstituted *ortho* (C5) carbons, respectively.

We secured solid-state structures of all four bright-red platinabenzenes. The molecular structure of **56b** is shown in Figure 3.5, and selected bond lengths and bond angles for **54** and **56a–c** are given in Table 3.5. Unlike many other metallabenzenes, the platinabenzenes are highly planar structures (deviation from mean plane ≤ 0.02 Å, sum of angles ~ 719.8–720.0°). The Pt–C bonds are 1.926–1.937 and 1.951–1.975 Å in length, which compare well with Pt–C bonds in other known Pt(II) carbene complexes. The C–C bonds in the metallacycle have an average length of 1.382–1.396 Å with no appreciable bond alternation. The Cp ring in each system is η^5-coordinated to Pt, with Pt–C_{Cp} distances in the range of 2.257–2.342 Å.

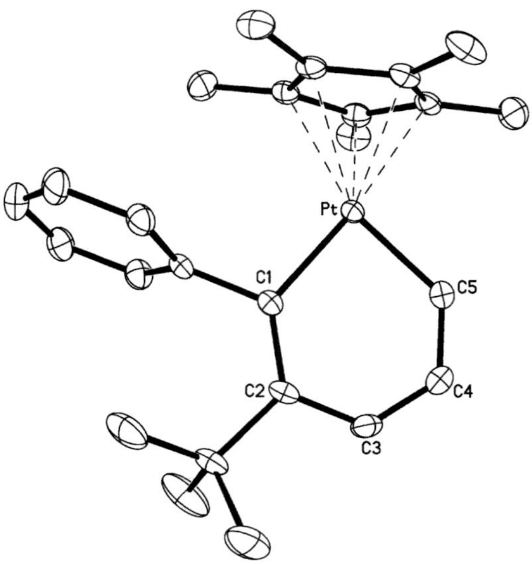

Figure 3.5 Molecular structure of platinabenzene **56b**; thermal ellipsoids are drawn at the 25% probability level.

Table 3.5 Selected bond lengths [Å] and bond angles [°] for platinabenzenes.

	54	**56a**	**56b**	**56c**
Pt–C1	1.959	1.951	1.975	1.951
Pt–C5	1.929	1.937	1.926	1.932
C1–C2	1.387	1.401	1.406	1.424
C2–C3	1.392	1.395	1.389	1.405
C3–C4	1.381	1.363	1.387	1.374
C4–C5	1.364	1.387	1.350	1.379
C=C mean	1.382	1.387	1.383	1.396
Pt–Cp[a]	1.965	1.959	1.957	1.948
Pt–C1–C2	129.2	129.2	129.8	129.7
C1–C2–C3	122.6	122.3	124.1	120.7
C2–C3–C4	124.8	125.2	126.9	125.5
C3–C4–C5	124.1	124.5	119.8	125.2
C4–C5–Pt	130.0	128.9	129.8	127.8
C1–Pt–C5	89.3	89.7	89.4	91.0
sum of angles	720.0	719.8	719.8	719.9

a) Distance from the Pt center to the centroid of the Cp ring.

By the time Christopher graduated in June 2007, it was obvious that we had hit an impasse with the platinabenzenes, as for reasons still unknown, the chemistry just stopped working. I was grateful for the attention the work had received, especially garnering the cover on *Angewandte Chemie* in the late summer of 2002 for the initial synthesis of **54**. Admittedly, that cover was dreamt up one beer-filled evening at the Canadian Society of Chemistry meeting in Vancouver in June 2002... Fast-forward a decade and I had just opened a package from Wiley-VCH. Imagine my surprise and delight to find out that that very cover had been chosen as one of six to be depicted on beer coasters given away by *Angewandte Chemie* as part of its 125th birthday celebration (Figure 3.6). Professors often like to show their journal covers during their presentations to impress the audience with how important their work is (or more likely that they can afford the exorbitant prices journals charge for covers), but very, very few people can claim their work is so important to make it onto a beer coaster!

3.5 Odds and Sods (Ones that Got Away)

When a research project reaches its inevitable conclusion, there are almost always some results that unfortunately never see the light of day. Reasons vary—a student leaves before completing the draft manuscript, the draft manuscript is not very good and requires significant re-writes from an over-committed advisor, difficulty in consistently reproducing a result, waiting for that all-important crystal structure, collaborators never coming through with their promised "results," etc. The following three short

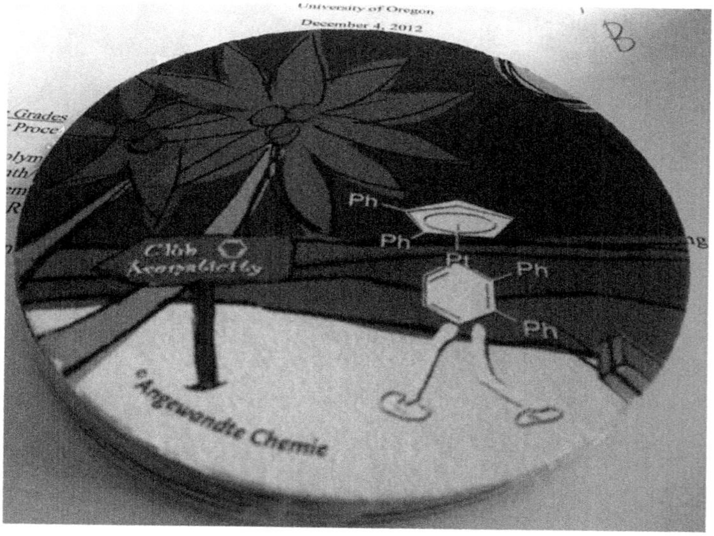

Figure 3.6 The Haley group's platinabenzene beer mat/coaster. (*See color plate section for the color representation of this figure.*)

stories suffer from one or more of these pitfalls. Given the purpose of this chapter—to give the reader a personal account of our metallabenzene research—I thought it appropriate to divulge these last few skeletons in the closet.

3.5.1 16-Electron, High Oxidation State Iridabenzenes

The computational studies by van der Boom *et al.* suggested that metallabenzenes with higher metal oxidation states would be more resistant to formation of Cp derivatives via the reductive elimination pathway [39]. We were intrigued by this prediction as prior results in our group [22, 30] showed that iridabenzenes can be oxidized with Ag(I) salts to create high oxidation state iridabenzenes. Similarly, Bleeke and coworkers demonstrated that treatment of 7 with halogens also afforded octahedral Ir(III) species [19]. Surprisingly, we had not tried this particular transformation on our systems, so I tasked Dan Chase, Christopher's successor, to examine the reactivity of **22a** with bromine and iodine.

Addition of one equivalent of either halogen to a THF solution of **22a** at −78°C produced dark-brown solutions from which bluish-brown crystals of **60a–b** (Scheme 3.20) could be isolated—sometimes. For reasons that are still not clear, the reaction proved to be wildly variable—sometimes we would get nice crystals, more often brown sludge. Solutions of the reaction mixtures would decompose rapidly upon exposure to air and water, and any attempt to clean up the mixtures also led to decomposition. When we did obtain crystals, yields varied wildly, and even then, their NMR spectra contained more signals that they should. That said, we were convinced of some degree of success as the ^{1}H NMR spectra exhibited the typical downfield resonances of iridabenzene ring protons (H5: 13.42 and 12.97 ppm and H3: 8.48 and 8.41 ppm for **60a** and **60b**, respectively). Unlike Bleeke's case, where both PEt_3 groups were retained, integration of the PPh_3 protons indicated only one phosphine unit was attached to the iridium core.

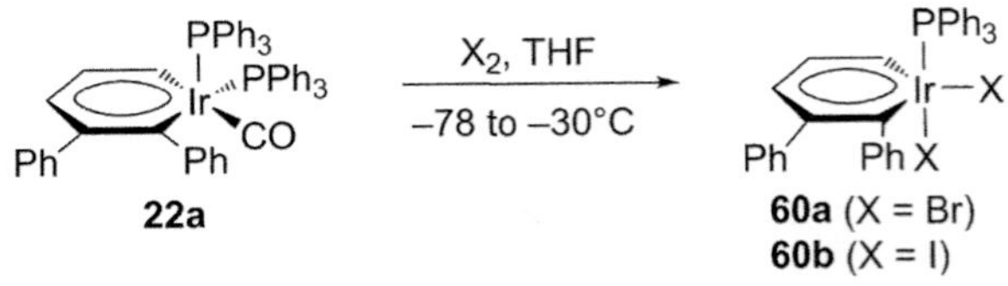

Scheme 3.20

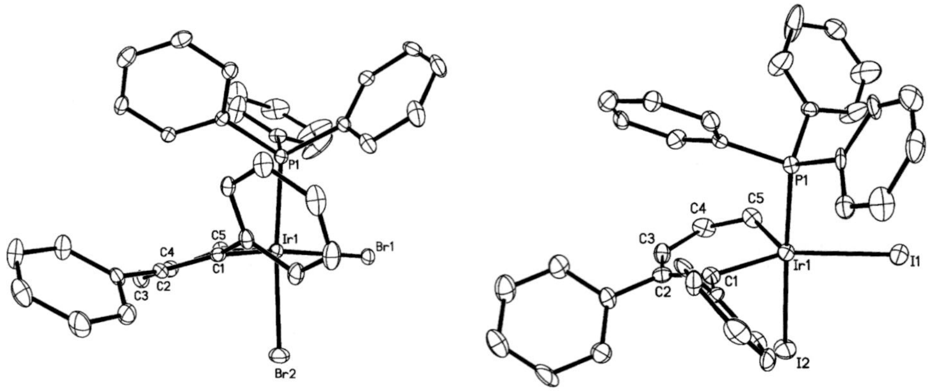

Figure 3.7 Different views of the molecular structure of isotypic iridabenzenes **60a** (left) and **60b** (right); thermal ellipsoids are drawn at the 25% probability level.

We know the reactions did work, as we obtained the X-ray crystal structures of both halogen-containing iridacycles (Figure 3.7) [45]. Both structures confirmed that the molecules were five-coordinated Ir complexes with only one PPh_3 group bound to the iridium center, unambiguously proving that **60a, b** were coordinatively unsaturated, 16-electron, high oxidation state iridabenzenes. Atoms P1, Br1/I1, Br2/I2, and C1 form the basal plane of the distorted square pyramid geometry with atom C5 in the apical position. The C–C bond lengths in the iridacycles vary between 1.360 and 1.402 Å, with averages of 1.387 and 1.382 Å in **60a** and **60b**, respectively. Both Ir–C bond lengths (1.958/1.903 and 1.963/1.913 Å for Ir–C1/Ir–C5, respectively, in **60a** and **60b**) are shorter than those in typical Ir(I) iridabenzenes (2.00–2.05 Å in Table 3.3), illustrating again the effects of the higher Ir(III) oxidation state. The steric bulk of the two halogen atoms, the PPh_3 group, and the Ph unit located on C1 all likely contribute to the presence of two unusual structural features: (1) as can be seen in the structure of **60a**, the iridacycle in both molecules deviates significantly from planarity. Whereas the dihedral angles of the Ir(I) iridabenzenes in Table 3.3 vary from 1.2 to 6.4°, these increase to 14.9° in **60b** and 17.2° in **60a**. (2) The loss of CO plus one PPh_3 group means there is an open coordination site in **60a, b** located equatorially to the Ir center. Normally occupied by CO (e.g. Figure 3.3), its absence translates into extremely large Br1/I1–Ir–C1 bond angles of 158.5 and 156.0°. For comparison, the Br1/I1–Ir–C5 bond angles are substantially smaller at 110.5 and 113.6°).

Sadly, the manuscript for this work was completely written but never submitted since we could not obtain reliable yields and clean NMR spectra. Fortunately, when asked to write this chapter, I made certain at least to get the X-ray structures out into the public domain in *Acta Cryst. E* in October 2015 [45]. While the 16-electron iridabenzene project did not pan out in the end, to Dan's credit, his work on our nascent (at the time) indenofluorene project paid off in spades, but that's a different story [46].

3.5.2 Hammett Plot of Iridabenzvalene Isomerization

Despite overwhelming computational evidence to the contrary, I still longed (obsessed?) to prove that a Dewar metallabenzene might still be a possible, even isolable, intermediate. I have already shown over the previous sections that we could obtain stable analogues of **1**, **8**, **9**, and **11** in Scheme 3.4. Where was experimental evidence for **10**? Although all computations strongly implicated the direct isomerization mechanism, our initial isomerization studies of **23b** showed that the process went faster in donating solvents. Since the first step involving a Dewar iridabenzene would be dissociation of the η^2-bound cyclopropene, we thought this observation might be consistent with the Dewar iridabenzene route.

We performed simple Mulliken charge calculations of the two possible transition states, which revealed that the charge densities at the cyclopropene double bond varied depending upon which mechanism was employed. For the direct isomerization pathway, the charge on the cyclopropenyl carbon that becomes C1 is more cationic in the transition state, while, for the Dewar iridabenzene mechanism, this carbon becomes more anionic. As both of these are formally benzylic carbon atoms, we envisaged that a linear-free energy analysis might provide some experimental insights. It seemed plausible that, if the benzylic position were becoming more cationic in the transition state, a Hammett plot should produce a negative ρ value, indicating that a direct isomerization was occurring. Alternatively, a positive ρ value would indicate that the benzylic position was becoming more anionic and that the Dewar iridabenzene mechanism was occurring. Christopher therefore set out to prepare a series of structurally related iridabenzvalenes and determine their rates of isomerization.

The syntheses of ligands **61a–d** (Scheme 3.21) were strictly analogous to the preparation of **21** in Scheme 3.7—carbene addition to the diarylacetylene, ester saponification, Weinreb amide formation via the acid chloride, $LiAlH_4$ reduction, and Wittig reaction [47]. Reaction as before with BuLi followed by the PMe_3 analogue of Vaska's complex gave the *para*-substituted phenyl benzvalenes **62a–d** in 24–32% isolated yield. Christopher obtained X-ray structures for all four compounds, two of which are shown in Figure 3.8; however, as can be seen in Table 3.4, as well as comparison with the data for **23b** (i.e. R = C_6H_5 in **62**), no meaningful trends in bond lengths and bond angles between the molecules could be discerned with the error limits of the data.

Thermolysis of iridabenzvalenes **62a–d** at 58°C in toluene-d_8 and monitoring the isomerization to benzenes **63a–d** by proton NMR spectroscopy, however, did afford

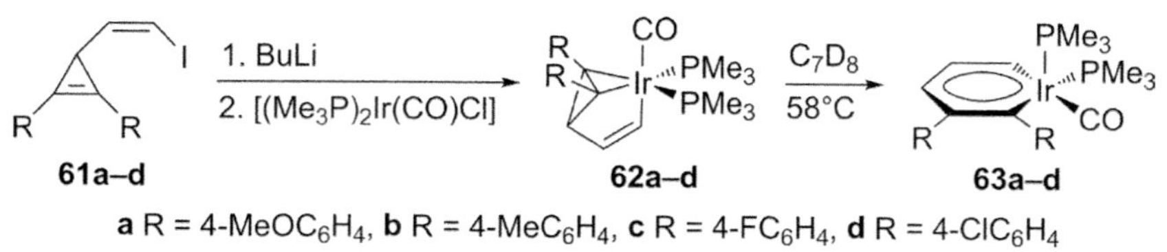

Scheme 3.21

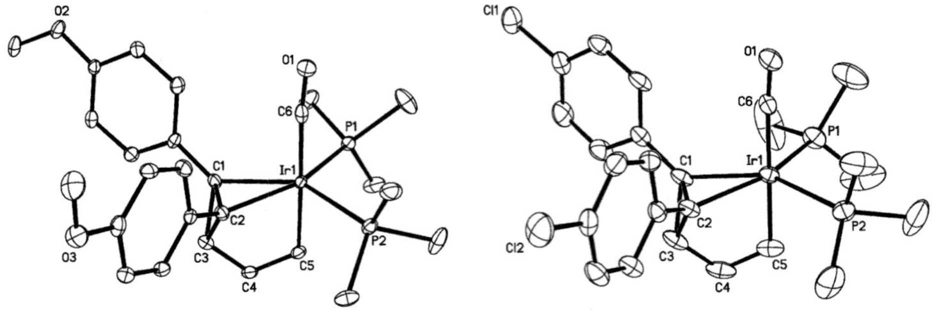

Figure 3.8 Molecular structures of *para*-substituted iridabenzvalenes **62a** (left) and **62d** (right); thermal ellipsoids are drawn at the 25% probability level.

meaningful data. The Hammett plot was made versus σ^+ values (Figure 3.9) due to the benzylic nature of the position in question (a plot versus σ gave a poor correlation), and in general, the experimental trends agreed with calculated trends. Particularly interesting is the divergence from linearity for the isomerization of **62d**. Including the data for **62d** (dotted line) gives an R^2 correlation of 0.934, whereas leaving these data out (solid line) improves R^2 to 0.977 with an intercept at essentially 0. Admittedly, both plots give strongly negative ρ values (−0.669 and −0.753), demonstrating unequivocally that electron-donating substituents enhance the rate of isomerization. This evidence, in combination with the computational predictions, suggests that the direct isomerization route must indeed be the correct mechanism.

Still, I was puzzled by the data for **62d**. Could substituents with positive σ^+ values afford a positive ρ value, validating a second possible mechanism? Sadly, we could

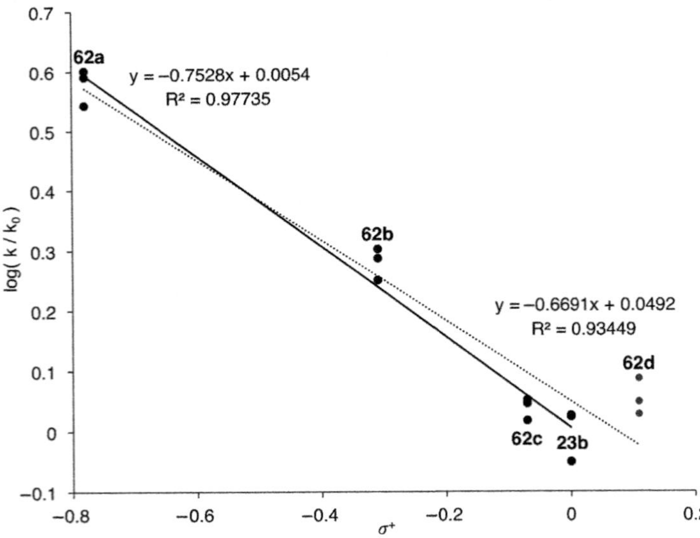

Figure 3.9 Hammett plot for the isomerization of iridabenzvalenes **23b** and **62a–d**. The dotted line includes the data for **62d**, whereas the solid line (better fit) does not.

never make derivatives of **61** with strong electron-withdrawing substituents in spite of many attempts [47]. We would have had to completely redesign ligand synthesis as many such electron-withdrawing functionalities (e.g. ester, nitro) could not withstand the “harsh” reagents we used (e.g. BuLi, $LiAlH_4$). Our best hope was trifluoromethyl groups on the outer phenyls. Unfortunately, their inclusion completely shut down carbene addition to the diarylacetylene. Alas, we will never know if we would have had a disjointed Hammett plot.

3.5.3 Rhodabenzvalene and Putative Dewar Rhodabenzene

Aside from iridium and platinum, the only other metal we seriously investigated was rhodium. Forays with gold and ruthenium gave σ-vinyl (unreactive) and η^5-Cp (too reactive) complexes, respectively, so we decided that moving up from third-row iridium to second-row rhodium was worth extended efforts. Also, there were no metallabenzenes incorporating 4d transition metals at that time [27].

Rob tried early on to react the lithiate of ligand **21** with the Rh analog of Vaska's complex. One could see in the proton NMR spectrum the telltale signs of a metallacycle, but all attempts to isolate the material furnished either uncharacterizable decomposition products or extremely low yields of Cp complexes similar to **28**; thus, we quickly set the Rh work aside [22]. Heping resumed in the fall of 2001 using $[RhCl(CO)(PMe_3)_2]$, which to our delight afforded rhodabenzvalene **64a** (Scheme 3.22), the Rh analogue of **23b** [48]. Although moderately stable in the solid state, **64a** is unstable in solution. Whereas a C_6D_6 solution of **64a** did not exhibit any noticeable change over two weeks at −30°C, the sample decomposed completely to an unidentified mixture overnight at room temperature. Utilizing silyl cyclopropenes **24e** and **36** under the same reaction condition furnished trimethylsilyl analogues **64b** and **64c**, respectively, as pale-yellow oils [49]. Complex **64b** was stable in C_6D_6 solution for a few hours at room temperature, but also decomposed to an unidentified mixture. Fortunately, bis(trimethylsilyl) analogue **64c** was stable in solution for over 48 h at 20°C, and could be converted to Cp complex **65c** in 15 h at 50°C.

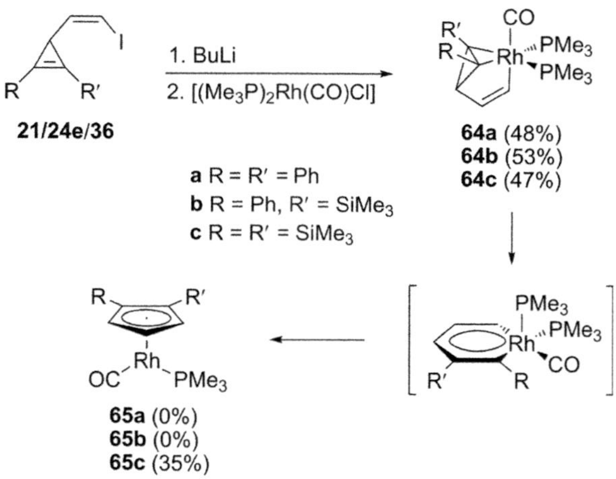

Scheme 3.22

X-ray diffraction of a crystal of **64a** revealed that most bond lengths and bonds angles were similar to **23b** and related iridabenzvalenes [48]. The main difference between the two structures was a shorter C1–C2 bond length—1.414 in **64a** vs. 1.447 Å in **23b**—which is similar to that in **25d** (1.405 Å). Whereas the shorter bond in **25d** can be attributed to steric interactions, for **64a** the shorter bond is a consequence of differing binding affinities for the cyclopropene π-bond, as second-row metals (Rh) are known to back-bond less efficiently than third-row metals (Ir). The weaker coordination also translated into a much larger P1–M–P2 angle, which was nearly 7° bigger in **64a** (111.0°) than **23b** (104.1°). While formation of the η^2-cyclopropene-/σ-vinyl-bound Rh(I) center in **64a–c** was quite rare, the molecules were simply too unstable. Unlike with the analogous Ir work, we never observed any indications of PMe_3-containing rhodabenzenes, and only ligand **36** furnished an isolable CpRh-complex (**65c**) [49].

Undeterred, Heping switched to [RhCl(cod)(P*i*-Pr_3)] as the starting Rh source. We reasoned that the steric bulk of *i*-Pr_3P could help stabilize the Rh(I) complexes, similar to the σ-vinyl Rh(I) species prepared by Werner [50–52]. Reaction with ligand **24d** gave σ-vinyl Rh complex **66** (Scheme 3.23), as observed by ^{1}H and ^{31}P NMR spectroscopy [49]. Similar to the Pt analogues, the key identifying proton signals were a triplet at 6.01 ppm due to alkene H4 and a doublet at 3.20 ppm due to the cyclopropyl H3. Molecule **66** was not stable/isolable and instead transformed into Rh complex **67a** overnight at room temperature. The structure of **67a** was confirmed by X-ray crystallography, in which the Rh(I) center was coordinated by both η^5-cyclopentadienyl and η^4-1, 5-cyclooctadiene ligands.

Things became much more interesting once Heping switched to original ligand **21** [49]. Under the same reaction conditions he isolated a new Rh complex as a brown solid in 37% yield. Although quite air-sensitive, the molecule exhibited moderate thermal stability. It could be converted cleanly to η^5-Cp Rh-complex **67b** at 75°C over 18 h in 75% yield (Scheme 3.23). We were puzzled as the ^{1}H NMR spectrum of the crude reaction mixture contained no evidence of any of the other metallated species previously isolated and described in this chapter (metallabenzene, metallabenzvalene, σ-vinyl

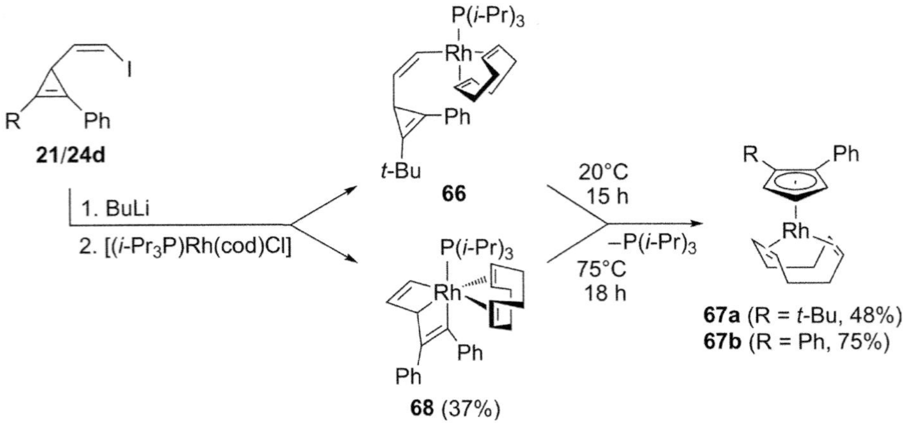

Scheme 3.23

complex, Cp-complex). Once purified, the new rhodacycle showed proton resonance signals at δ 8.70, 5.44, and 5.13 ppm due to the three CH groups. In my mind, it could be only one thing: Dewar rhodabenzene **68**! The complex multiplet at 8.70 ppm was the sp^2-CH fragment σ-bound to the Rh atom. A pseudo-quartet at 5.13 ppm was the sp^3-CH fragment because both the coupling pattern and low-field chemical shift suggested this group should also be bound to the Rh atom. The doublet of doublets signal at 5.44 ppm was the other sp^2-CH group not attached directly to Rh. The four sp^2-CH protons in (cod) appeared as a broad signal at 4.71 ppm. All other proton signals due to the cod methylene and two Ph group were easily assigned. In the ^{13}C NMR spectrum, two doublet of doublets signals at quite low-field 156.70 (J = 34.2, 7.7 Hz) and 133.42 (J = 44.3, 7.1 Hz) ppm were assigned to the two sp^2-C atoms directly attached to the Rh center due to coupling with both the Rh and P NMR-active nuclei. Again, these spectroscopic features matched nothing we had observed before. This fact, plus the isolation of Cp-complex **67b** upon thermolysis, indirectly implicated Dewar rhodabenzene **68** as the brown, crystalline intermediate.

So why didn't we publish this study? Simple—no X-ray crystal structure of **68**. Heping tried and tried and tried to obtain suitable crystals, all to no avail. While I was absolutely convinced (and still am to this day) that the mystery intermediate was a Dewar rhodabenzene, convincing much more skeptical reviewers was going to be a very difficult process. I was simply unwilling to attempt to put this out into the literature without an X-ray crystal structure to back us up. Given the very unusual metallacyclic structures the Xia group has isolated over the last decade (see Chapter 6), a Dewar metallabenzene really does not seem that far-fetched. Perhaps some computational chemists can analyze the reaction pathway in Scheme 3.23, but for now, I truly consider this result as the one that sadly got away.

3.6 Conclusion (So Long, and Thanks for All the Fish)

By the summer of 2008, it was painfully clear that my group's efforts in metallabenzene chemistry needed to come to an end. Progress had ground to a halt and the students needed positive results for their theses. Even though I did not want to, I made the decision it was time to move on and thus I killed the project. While that left a few research questions hanging, it was absolutely the correct decision.

Writing this chapter provided a really nice chance to look back at what we were able to accomplish, and especially to revisit some science that did not quite get out the door. I stated early on that I would present evidence for all five possible products in Scheme 3.4, and, with the exception of putative Dewar rhodabenzene **68**, we obtained definitive NMR data and multiple crystal structures of the other four species—metallabenzene, metallabenzvalene, σ-vinyl metal-complex, and Cp metal-complex. We introduced new metals (Rh, Pt) and were the first (I think) to show the complexity of the reaction manifold that makes up metallabenzenes and their structural isomers/analogues. Specifically, the chemistry in Scheme 3.9 demonstrated it was possible to observe the isomerization of a metallabenzvalene to the corresponding metallabenzene, followed by its decomposition to an η^5-Cp metal-complex. While our synthetic approach was not as "general" as I had hoped for, we did succeed in opening a new route to these unusual metallacycles, and we had some fun along the way too!

3.7 Acknowledgements

I am deeply indebted to the many University of Oregon postdoctoral, graduate, and undergraduate students who worked on this project, most of whom were named throughout this story. Without their hard work and tireless efforts, none of this chemistry would have been possible. Dr. Timothy Weakley and Dr. Lev Zakharov were instrumental in obtaining all of the X-ray structural data presented herein. I also thank my metallabenzene "colleagues," all of whom have contributed to this book. Each one of us brought a unique angle to the field, and very importantly, we were supportive of one another's efforts. Finally, I must thank the National Science Foundation (CHE-0075246 and -0647252) for supporting our studies on metallabenzenes and their valence isomers.

References

1 Spitler, E. L., Johnson, II, C. A., and Haley, M. M. (2006) Renaissance of Annulene Chemistry. *Chemical Reviews*, **106**, 5344–5386.
2 Bishop, K. C. (1976) Transition metal catalyzed rearrangements of small ring organic molecules. *Chemical Reviews*, **76**, 461–486.
3 For a general review of cyclopropene-transition metal complexes, see: Goldschmidt, Z. (1995) Organometallic Derivatives of Cyclopropanes and Their Reactions, in Z. Rappoport (ed.), *The Chemistry of the Cyclopropyl Group: Volume 2*. New York: John Wiley & Sons, Ltd: 495–695.
4 Binger, P. Müller, P., Benn, R., and Mynott, R. (1989) Vinylcarbene Complexes of Titanocene. *Angewandte Chemie: International Edition in English*, **28**, 610–611.
5 Binger, P., Müller, P., Hermann, A. T. *et al.*, (1991) Metallacyclobutene aus η^2-Cyclopropen-Komplexen des Titanocens und Zirkonocens. *Chemische Berichte*, **124**, 2165–2170.
6 Li, R. T., Nguyen, S. T., Grubbs, R. H., and Ziller, J. W. (1994) Reactions of 3,3-Diphenylcyclopropene with Iridium(I) Complexes: Probing the Mechanism of Cyclopropene Rearrangements at Transition Metal Centers. *Journal of the American Chemical Society*, **116**, 10032–10040.
7 Nguyen, S. T., Johnson, L. K., and Grubbs, R. H. (1992) Ring-opening metathesis polymerization (ROMP) of norbornene by a Group VIII carbene complex in protic media. *Journal of the American Chemical Society*, **114**, 3974–3975.
8 Gagne, M. R., Grubbs, R. H., Feldman, J., and Ziller, J. W. (1992) Catalytic activity of a well-defined binuclear ruthenium alkylidene complex. *Organometallics*, **11**, 3933–3935.
9 Grubbs, R. H., Miller, S. J., and Fu, G. C. (1995) Ring-Closing Metathesis and Related Processes in Organic Synthesis. *Accounts of Chemical Research*, **28**, 446–452.
10 Hughes R. P. and Robinson, D. J. (1989) Transition-metal-promoted activation of carbon-carbon bonds. A new synthetic route to substituted ruthenocene derivatives via ring expansion reactions of 3-vinyl-1-cyclopropenes. *Organometallics*, **8**, 1015–1019.
11 Hughes, R. P., Trujillo, H. A., Egan Jr., J. A., and Rheingold, A. L. (2000) Iridium-Promoted Reactions of Carbon–Carbon Bonds. Skeletal Rearrangement of a Vinylcyclopropene during Iridacyclohexadiene Formation and Subsequent Isomerization of Iridacyclohexadienes via α,α'-Substituent Migrations. *Journal of the American Chemical Society*, **122**, 2261–2271.

12 Grabowski, N. A., Hughes, R. P., Jaynes, B. S., and Rheingold, A. L. (1986) Stepwise transition metal promoted ring expansion reactions of vinylcyclopropenes to give cyclopentadienes and cyclohexa-2,4-dienones. The first example of a 1-metallacyclohexa-2,4-diene complex, {[Pt-CH_2-CH=C(Ph)-C(Ph)=C(Ph)](PPh_3)$_2$}. *Journal of the Chemical Society: Chemical Communications*, 1694–1695.

13 Egan Jr., J. W., Hughes, R. P., and Rheingold, A. L. (1987) Transition-metal-promoted ring-opening reactions of vinylcyclopropenes. 1,2,3,5-η-Penta-2,4-dienediyl- and 1, 5-η-penta-2,4-dienediyl(1-metallacyclohexa-2,4-diene) complexes of rhodium(III) and iridium(III) and their conversion to (η^5-cyclopentadienyl)hydridometal compounds. *Organometallics*, **6**, 1578–1581.

14 Hughes, R. P., Trujillo, H. A., Egan Jr., J. W., and Rheingold, A. L. (1999) Skeletal Rearrangement during Rhodium-Promoted Ring Opening of 1,2-Diphenyl-3-vinyl-1-cyclopropene. Preparation and Characterization of 1,2- and 2,3-Diphenyl-3,4-pentadienediyl Rhodium Complexes and Their Ring Closure to a 1,2-Diphenylcyclopentadienyl Complex. *Organometallics*, **18**, 2766–2772.

15 Elliott, G. P., Roper, W. R., and Waters, J. M. (1982) Metallacyclohexatrienes or "metallabenzenes." Synthesis of osmabenzene derivatives and X-ray crystal structure of [Os(CSCHCHCHCH)(CO)(PPh_3)$_2$]. *Journal of the Chemical Society: Chemical Communications*, 811–813.

16 Thorn, D. L. and Hoffmann, R (1979) Delocalization in metallocycles. *Nouveau Journal de Chemie*, **3**, 39–45.

17 Elliot, G. P., McAuley, N. M., and Roper, W. R. (1989) An Osmium Containing Benzene Analog, Os(CSCHCHCHCH)(CO)(PPH_3)$_2$, Carbonyl(5-Thioxo-1,3-Pentadiene-1,5-Diyl-C^1,C^5,*S*)-Bis(Triphenylphosphine)Osmium, and its Precursors. *Inorganic Syntheses*, **26**, 184–189.

18 Bleeke, J. R., Xie, Y.-F., Peng, W.-J., and Chiang, M. (1989) Metallabenzene: synthesis, structure, and spectroscopy of a 1-irida-3,5-dimethylbenzene complex. *Journal of the American Chemical Society*, **111**, 4118–4120.

19 Bleeke, J. R., Behm, R., Xie, Y.-F., *et al.*, (1997) Synthesis, Structure, Spectroscopy, and Reactivity of a Metallabenzene.*Organometallics*, **16**, 606–623.

20 Bleeke, J. R. (1991) Metallabenzene chemistry. *Accounts of Chemical Research*, **24**, 271–277.

21 Haley, M. M., Biggs, B., Looney, W. A., and Gilbertson, R. D. (1995) Synthesis of alkenyl- and alkynylcyclopropenes. *Tetrahedron Letters*, **36**, 3457–3460.

22 Gilbertson, R. D. and Haley, M. M., unpublished results.

23 Gilbertson, R. D., Weakley, T. J. R., and Haley, M. M. (2000) Preparation, X-ray Crystal Structures, and Reactivity of Alkynylcyclopropenylium Salts. *Journal of Organic Chemistry*, **65**, 1422–1430.

24 Gilbertson, R. D., Wu, H.-P., Gorman-Lewis, D., *et al.* (2004) An unusually robust triple bond: synthesis, structure and reactivity of 3-alkynylcyclopropenes. *Tetrahedron*, **60**, 1215–1223.

25 Padwa, A., Akiba, M., Chou, C. S., and Cohen, L. (1982) Photochemistry of cyclopropene derivatives. Synthesis and photorearrangement of a 3-acyl-substituted cyclopropene. *Journal of Organic Chemistry*, **47**, 183–191.

26 Gilbertson, R. D., Haley, M. M., Weakley, T. J. R., *et al.* (1998) Synthesis and Characterization of an Unusually Stable Dialkylaluminum Aldimine Complex: X-ray Crystal Structure of trans-[(1,2,3-(*t*-Bu)$_3$-*cyclo*-C_3)CHNAl(*i*-Bu)$_2$]$_2$. *Organometallics*, **17**, 3105–3107.

27 Bleeke, J. R. (2001) Metallabenzenes. *Chemical Reviews*, **101**, 1205–1227.
28 Yang, J., Jones, W. M., Dixon, J. K., and Allison, N. T. (1995) Detection of a Ruthenabenzene, Ruthenaphenoxide, and Ruthenaphenanthrene Oxide: The First Metalla Aromatics of a Second-Row Transition Metal. *Journal of the American Chemical Society*, **117**, 9776–9777.
29 Gilbertson, R. D., Weakley, T. J. R., and Haley, M. M. (1999) Direct Synthesis of an Iridabenzene from a Nucleophilic 3-Vinyl-1-cyclopropene. *Journal of the American Chemical Society*, **121**, 2597–2598.
30 Gilbertson, R. D., Weakley, T. J. R., and Haley, M. M. (2000) Synthesis, Characterization, and Isomerization of an Iridabenzvalene. *Chemistry: A European Journal*, **6**, 437–441.
31 Gilbertson, R. D., Lau, T. L. S., Lanza, S., *et al.* (2003) Synthesis, Spectroscopy, and Structure of a Family of Iridabenzenes Generated by the Reaction of Vaska-Type Complexes with a Nucleophilic 3-Vinyl-1-cyclopropene. *Organometallics*, **22**, 3279–3289.
32 Tolman, C. A. (1977) Steric effects of phosphorus ligands in organometallic chemistry and homogeneous catalysis. *Chemical Reviews*, **77**, 313–348.
33 Wu, H.-P., Weakley, T. J. R., and Haley, M. M. (2005) Regioselective Formation of β-Alkyl-α-phenyliridabenzenes via Unsymmetrical 3-Vinylcyclopropenes: Probing Steric and Electronic Influences by Varying the Alkyl Ring Substituent. *Chemistry: A European Journal*, **11**, 1191–1200.
34 Wu, H.-P., Lanza, S., Weakley, T. J. R., and Haley, M. M. (2002) Metallabenzenes and Valence Isomers. 3. Unexpected Rearrangement of Two Regioisomeric Iridabenzenes to an (η^5-Cyclopentadienyl)iridium(I) Complex. *Organometallics*, **21**, 2824–2826.
35 Wu, H.-P., Ess, D. H., Lanza, S., *et al.* (2007) Rearrangement of Iridabenzvalenes to Iridabenzenes and/or η^5-Cyclopentadienyliridium(I) Complexes: Experimental and Computational Analysis of the Influence of Silyl Ring Substituents and Phosphine Ligands. *Organometallics*, **26**, 3957–3968.
36 Hughes, R. P., King, M. E., Robinson, D. J., and Spotts, J. M. (1989) Stereoselective oxidative additions of a carbon-carbon σ-bond in tetrafluorocyclopropene to iridium(I) complexes. *Journal of the American Chemical Society*, **111**, 8919–8920.
37 Hemond, R. C., Hughes, R. P., Robinson, D. J., and Rheingold, A. L. (1988) Activation of a fluorinated carbon-carbon bond by oxidative addition of tetrafluorocyclopropene to platinum(0). The first example of a perfluorometallacyclobutene. *Organometallics*, **7**, 2239–2241.
38 Iron, M. A., Lucassen, A. C. B., Cohen, H., *et al.* (2004) A Computational Foray into the Formation and Reactivity of Metallabenzenes. *Journal of the American Chemical Society*, **126**, 11699–11710.
39 Iron, M. A., Martin, J. M. L., and van der Boom, M. E. (2003) Metallabenzene versus Cp Complex Formation: A DFT Investigation. *Journal of the American Chemical Society*, **125**, 13020–13021.
40 Jacob, V., Weakley, T. J. R., and Haley, M. M. (2002) Rearrangement of a σ-2-(Cycloprop-2-enyl)vinyl- to an η^3-cyclopentadienylplatinum(II) Complex. Selective Protonolysis of the Platinum–Methyl Bond. *Organometallics*, **21**, 5394–5400.
41 Jacob, V., Weakley, T. J. R., and Haley, M. M. (2002) Metallabenzenes and Valence Isomers. Synthesis and Characterization of a Platinabenzene. *Angewandte Chemie: International Edition*, **41**, 3470–3473.

42 Boag, N. M., Quyoum, R., and Rao, K. M. (1992) Nucleophilic substitution of $[Pt(\eta^5\text{-}C_5Me_5)(CO)X]$(X = Cl, Br); isolation of a ring-slipped intermediate. *Journal of the Chemical Society: Chemical Communications*, 114–115.

43 Landorf, C. W., Jacob, V., Weakley, T. J. R., and Haley, M. M. (2004) Rational Synthesis of Platinabenzenes. *Organometallics*, **23**, 1174–1176.

44 Jacob, V., Landorf, C. W., Zakharov, L. N., *et al.* (2009) Platinabenzenes: Synthesis, Properties, and Reactivity Studies of a Rare Class of Metalla-aromatics. *Organometallics*, **28**, 5183–5190.

45 Chase, D. T., Zakharov, L. N., and Haley, M. M. (2015) Crystal structures of two unusual, high oxidation state, 16-electron iridabenzenes. *Acta Crystallographica*, **E71**, 1315–1318.

46 Frederickson, C. K., Rose, B. D., and Haley, M. M. (2017) Explorations of the Indenofluorenes and Expanded Quinoidal Analogues. *Accounts of Chemical Research* **50**, 977–987.

47 Landorf, C. W., Zakharov, L. N., and Haley, M. M., unpublished results.

48 Wu, H.-P., Weakley, T. J. R., and Haley, M. M. (2002) Metallabenzenes and Valence Isomers. 5. Synthesis and Structural Characterization of a Rhodabenzvalene: A Rare η^2-Cyclopropene/σ-Vinylrhodium(I) Complex. *Organometallics*, **21**, 4320–4322.

49 Wu, H.-P., Weakley, T. J. R., and Haley, M. M., unpublished results.

50 Wiedemann, R., Steinert, P., Schäfer, M., and Werner, H. (1993) General method for stereoselective coupling of an alkyl, aryl, or vinyl group with a vinylidene unit at a transition-metal center. *Journal of the American Chemical Society*, **115**, 9864–9865.

51 Werner, H., Wiedemann, R., Stainert, P., and Wolf, J. (1997) Rhodium(I)-Assisted Stereoselective Coupling of an Alkyl, Aryl or Vinyl Group with a Vinylidene Ligand: A Novel Synthetic Route to π-Allyl and π-Butadienyl Rhodium Complexes. *Chemistry: A European Journal*, **3**, 127–137.

52 Wiedemann, R., Wolf, J., and Werner, H. (1996) Stable Carbonyl- and Thiocarbonylrhodium(I) Complexes Containing σ-Bonded Phenyl and Vinyl Groups as Ligands. *Chemische Berichte*, **129**, 29–31.

4

Iridabenzenes and Iridanaphthalenes with Supporting Tris(pyrazolyl)borate Ligands

Margarita Paneque and Nuria Rendón

Instituto de Investigaciones Químicas (IIQ), Departamento de Química Inorgánica and Centro de Innovación en Química Avanzada (ORFEO-CINQA), Consejo Superior de Investigaciones Científicas (CSIC) and Universidad de Sevilla, Sevilla, Spain

4.1 Introduction

Iridium compounds stabilized by polydentate ligands of the type tris(pyrazolyl)borate [1], exhibit a rich reactivity toward a variety of organic substrates. In particular, the Me-substituted tris(3,5-dimethylpyrazolyl)borate, Tp^{Me2}, ligand has proven to be a good agent for the stabilization of the reaction products and also of many reaction intermediates. The geometry of the Tp^{Me2} ligand and the hard nature of the N donors promote the easy transformation of simple Ir(I) precursors of type $Tp^{Me2}Ir(L)_2$ (L = for instance an olefin) into more stable Ir(III) derivatives of composition $Tp^{Me2}Ir(X)_2L'$, normally through an activation process. Provided L′ is a labile molecule, these Ir(III) products generate (normally upon warming) unsaturated reactive intermediates of composition $[Tp^{Me2}Ir(X)_2]$, which are active species in many activation processes. In particular, unsaturated ligands like olefins [2], alkynes [3], aldehydes [4], nitriles [5], etc. have provided entrance to new types of species or reactions involving C–C (or C–X) bond formation. The *fac* disposition of the three co-ligands facilitates interactions between them and transformation into new moieties that are still situated in the same three positions. This has resulted in the formation of a large variety of metallacyclic or metallabicyclic species [6, 7]. Amongst them, we have isolated several iridabenzenes and other iridaaromatic species. The first examples were obtained in serendipitous ways, but later we were able to design reliable routes for the preparation of iridaaromatics. Nevertheless, we still often encountered unexpected results during these reactions following small changes to the nature of the reagents, and this vividly illustrates the rich chemistry these systems are able to experience.

An important issue concerning the chemistry of metallabenzenes is to determine whether these complexes deserve to be described as "aromatic" derivatives. Aromaticity is not a directly observable or strictly measurable quantity, but rather it is the set of structural, spectroscopic, chemical, and energetic properties that, all together, are indicative of a characteristic of a (broad) type of compounds called "aromatic" [8].

Usually, structural issues, such as bond equalization within the ring; spectroscopic characteristics, such as deshielded proton resonances; and chemical behavior, like

Metallabenzenes: An Expert View, First Edition. Edited by L. James Wright.

participation in electrophilic aromatic substitutions, are usually invoked for the classification of this type of metallacyclic derivatives into the extended family of aromatics. Some of these questions will be discussed in this chapter.

This chapter describes the studies we have carried out that focus on metallaaromatic derivatives with supporting tris(pyrazolyl)borate ligands. Section 4.2 collects the different procedures used as synthetic methods, while Sections 4.3, 4.4, and 4.5 contain, respectively, the reactivity studies carried out and the structural and spectroscopic data characteristic of the species discussed in Section 4.2. In the schemes, the delocalized π-electrons of benzene rings are represented in general with Kekulé structures, but when fused to a metallabenzene, forming a metallanaphthalene, they are represented by circles, as are the metallaaromatics. Proposed intermediate compounds in the schemes are shown between brackets and are labelled with a capital letter, starting from "A" in each scheme, followed by the scheme number. When the same intermediate appears in two different schemes, it is always designated with the name corresponding to the first time it is shown. The yields indicated in the schemes for the different compounds correspond to the yield of product formed (normally measured spectroscopically).

4.2 Synthetic Routes to Iridaaromatic Derivatives with Supporting Tris(pyrazolyl)borate Ligands

4.2.1 Oxidatively Induced Ring Contraction

The first examples of iridaaromatics obtained in the $Tp^{Me2}Ir$ system were the result of a totally unexpected reaction, observed during the study of the coupling of alkynes (particularly dimethyl acetylenedicarboxylate, DMAD) with Ir(I) and Ir(III) precursors. Starting with the butadiene complex $Tp^{Me2}Ir(\eta^4\text{-}CH_2{=}C(Me)C(Me){=}CH_2)$ (**1**) or the phenyl derivative $Tp^{Me2}Ir(Ph)_2N_2$ (**2**), we had prepared the aquo-iridacycloheptatriene compounds **3** and **4** (Scheme 4.1). These species are quite thermally stable, both in the solid state and in solution, but the presence of the labile H_2O ligand (coming from adventitious water present in the reaction mixtures) makes them reactive toward other Lewis bases. Thus, water can be substituted with ligands, such as PMe_3 or CO, but oxygen is also capable of displacing the H_2O ligand and mild warming of solutions of **3** or **4** in the presence of oxygen promotes the formation of ketone derivatives **5** or **6**, respectively [9, 10].

Different oxidizing agents were tested for these reactions, and using the stronger ^{t}BuOOH we realized that, in spite of the high thermal stability of the ketones **5** and **6** (which, once isolated, are stable up to 150°C), the oxidation process does not stop at the ketone derivatives but continues forward to break the bicyclic ligand and generate the iridabenzene **7** and the iridanaphthalene **8**, respectively. A methyl oxalate ligand is formed in each case, and the reactions are proposed to proceed through a Baeyer–Villiger oxidation of the C(=O)–C(R) bond upon formation of unstable ester intermediates (**A2** and **B2**), which rearrange to the observed aromatic product by C–O rupture and bond reorganization (Scheme 4.2).

When compounds **7** and **8** were isolated, the family of metallaaromatics was still small, but they had already captured the attention of the chemistry community. A few

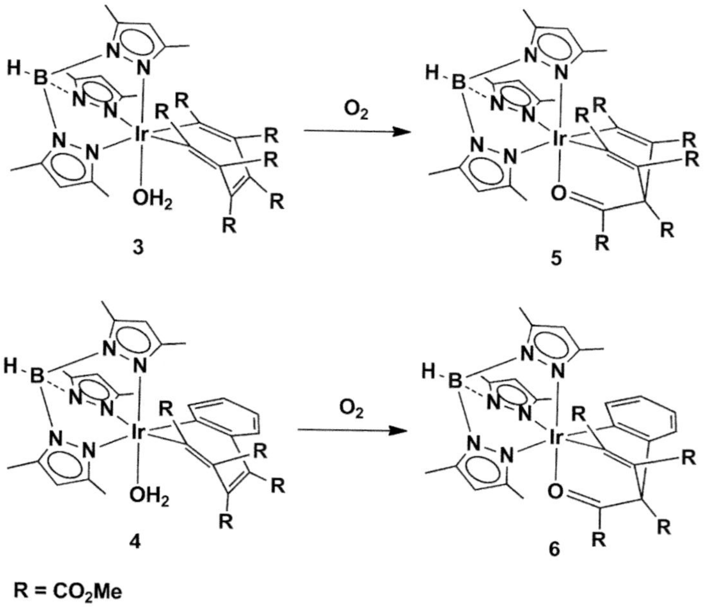

Scheme 4.1

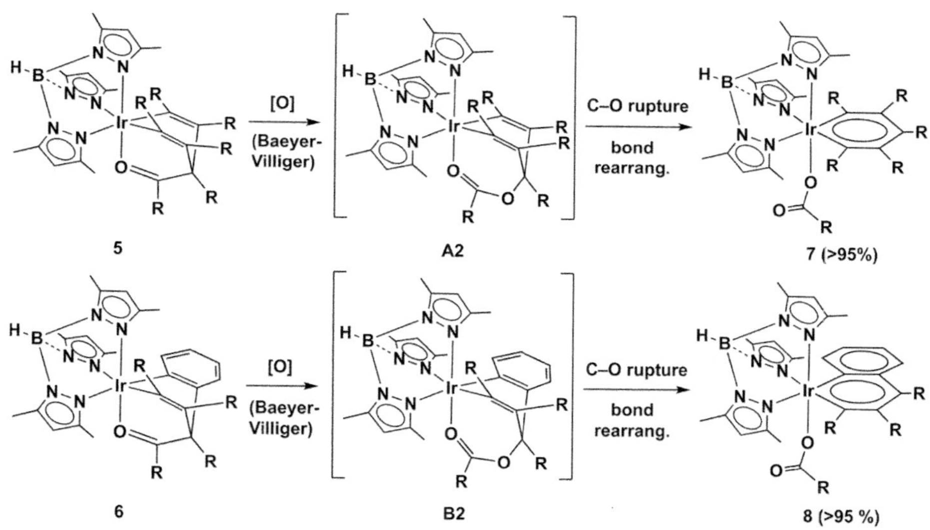

Scheme 4.2

types of compounds were already known for different metals, comprising the first example reported, an osmabenzene [11], for whose formation a thiocarbonyl ligand is involved; a highly symmetrical iridabenzene with stabilizing trialkyl phosphine ligands [12]; Pt [13] and Ir [14] metallabenzenes formed using as the precursor for the C_5

moiety of the ring an Li derivative of 3-vinyl-1-cyclopropene, and some other related derivatives.

The isolation of complexes **7** and **8** was a hint that we still knew relatively little about this growing and interesting group of products. First, compounds **7** and **8** were quite unexpected in view of the presence of a considerable number of electron-withdrawing substituents on the metallaaromatic ring, when it had been predicted that π-donors, especially those *ortho* and *para* to the metal, would stabilize aromatic metallacycles [15]. Nevertheless, not only did we prepare later other metallaaromatic species substituted with electron-accepting groups (compounds **25–28**, **32–33**, **35–39**, and **49** below), but also other authors published species bearing electron-withdrawing groups, although at the less influential *meta* positions, such as phosphonium substituted osmabenzenes and ruthenabenzenes [16, 17]. This indicated that metallaaromatics could be stabilized by a much broader range of substituents on the ring.

Second, iridanaphthalene **8** was the first example of a higher metallaaromatic derivative. Still, only a few examples of this benzene-fused metallaaromatic structure are known, but the results gathered in this chapter suggest that they could be conveniently obtained following routes analogous to those employed for metallabenzenes, provided a precursor already containing the required benzo fused ring is available.

In addition to compound **8** and the β-iridanaphthalenes discussed below (Sections 4.2.2.2.2 and 4.2.2.3.1), two other types of metallanaphthalenes are known: a family of Cp*-iridium derivatives **9** [18] and the dimer cationic osmium complex **10** [19] of Figure 4.1.

In both cases, these compounds are formed by *ortho*-metallation of a pendant Ph group on an appropriate carbon chain previously coordinated to the metal center.

Finally, with respect to the relevance of compounds **7** and **8**, it can be mentioned that, despite being heavily substituted, they have been shown to experience reactions that support their aromatic nature (see Section 4.3).

4.2.2 Synthesis of Iridaaromatics by Ring Expansion Reactions

Although the first derivatives of metallaaromatic nature we prepared with the $Tp^{Me2}Ir$ moiety were derived from the rather unusual oxidatively induced ring contraction described in Section 4.2.1, in accord with the more widely observed methods used in the literature, the majority of the Tp^{Me2}-iridaaromatic compounds we have prepared

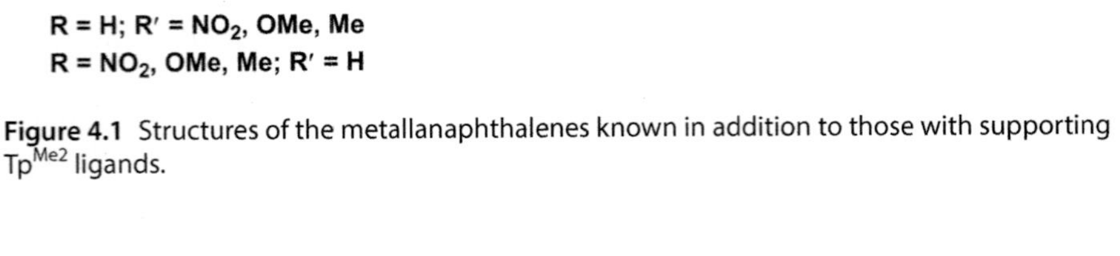

9

10

R = H; R′ = NO_2, OMe, Me
R = NO_2, OMe, Me; R′ = H

Figure 4.1 Structures of the metallanaphthalenes known in addition to those with supporting Tp^{Me2} ligands.

involved ring-expansion procedures. Within this class of methods, the most common transformation is the insertion of a C atom of a donor ligand into the M–C bond of an adjacent metallacyclopentadiene, which in turn is most often formed by the coupling of two alkynes with cyclization at a metal center. As the last step, aromatization then leads to the metallaaromatic ring.

The majority of $Tp^{Me2}Ir$ derivatives with metallaaromatic character were prepared through the insertion of an (unobserved) carbene ligand into an iridacyclopentadiene, generated through different ways, as shown in the subsequent sections. An exception to this is the case described in Section 4.2.2.1, where the addition of acid to a complex with vinylidene and iridacyclopentene ligands results in the formation of an iridabenzene.

4.2.2.1 Protonation of a Vinylidene-Iridacyclopentene

A unique type of ring-expansion reaction for the synthesis of Tp^{Me2}-iridaaromatics is the case of the reaction described in this section, where the substrate is an iridacyclopentene rather than an iridacyclopentadiene. Iridacyclopentenes are much less common than iridacyclopentadienes [20] but an easy entrance to one of these derivatives in the $Tp^{Me2}Ir$ system is via compound **1**. This species—an $18e^-$ Ir(I) derivative with the Me-substituted Tp^{Me2} ligand and η^4-coordinated 2,3-dimethylbutadiene—has been shown to generate an Ir(III) moiety by formal oxidative addition of the diene, which transforms it into a 2,3-dimethyl-2-butene-1,4-diyl ligand. Lewis bases like CO or PMe_3 have permitted the isolation and full characterization of species of composition $Tp^{Me2}Ir(CH_2{-}CH{=}CH{-}CH_2)(L)$ (L = CO, PMe_3) [21]. Alkynes, like diphenyl acetylene, upon reaction with **1** promote a complex transformation that is initiated by the insertion of the alkyne into one of the Ir–C bonds of this intermediate species [22]. In contrast, DMAD reacts with **1** displacing the butadiene, two molecules of DMAD forming an iridacyclopentadiene moiety [23], which is the precursor of the iridacycloheptatriene product **3** described above. In contrast, bis(trimethylsilyl)acetylene (BTMSA) neither inserts into the M–C bond nor displaces the butadiene, but reacts differently [24]. The ability of the $SiMe_3$ group to migrate allows the formation of a vinylidene ligand. Indeed, a widely used approach to transition metal vinylidene complexes uses alkyne derivatives as precursors. In the very favorable cases of $RC{\equiv}CSiMe_3$ and $Me_3SiC{\equiv}CSiMe_3$ ligands of type $M{=}C{=}C(R)(SiMe_3)$ and $M{=}C{=}C(SiMe_3)_2$, respectively, are often formed. In our context, reaction between the precursor $Tp^{Me2}Ir(\eta^4\text{-}CH_2{=}C(Me)C(Me){=}CH_2)$ (**1**) and $Me_3SiC{\equiv}CSiMe_3$ yields an Ir(III) compound with Tp^{Me2}, 2,3-dimethyl-2-butene-1,4-diyl and vinylidene ligands (complex **11** in Scheme 4.3).

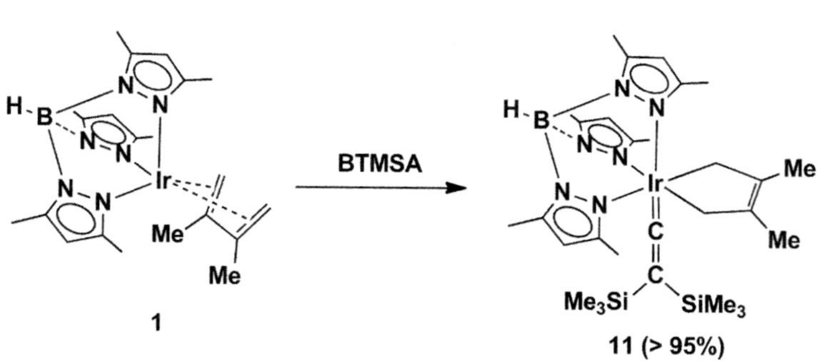

Scheme 4.3

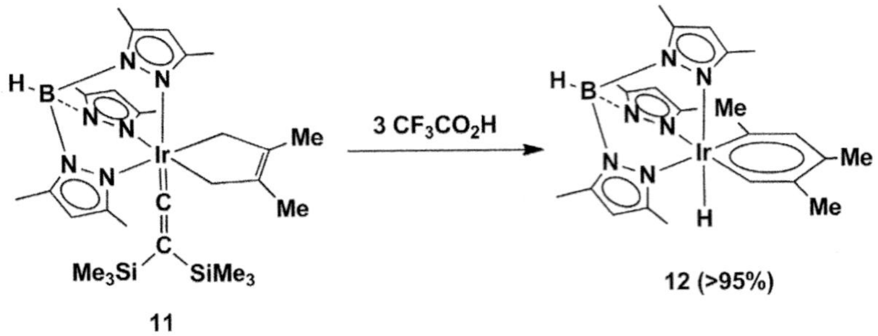

Scheme 4.4

Formation of the iridium vinylidene **11** requires a 1,2-silyl shift on the coordinated alkyne and the above-mentioned electronic reorganization within the diene to transform it into a butenediyl ligand with accompanying oxidation of the iridium center from +1 to +3. Complex **11** is quite thermally stable in solution and can be heated (note that it is formed at 80°C) without decomposition or transformation into other derivatives (NMR evidence). However, as expected, it does exhibit reactivity typical of vinylidenes. The C_α atom in transition metal vinylidene complexes is electron poor, whereas the M=C_α double bond and the C_β atom are electron rich. The latter is capable of being attacked by electrophiles and therefore protonation of this iridium vinylidene was studied. Unexpectedly, it was found that reaction with trifluoroacetic acid yields the iridabenzene derivative **12** directly (Scheme 4.4). This product is formed by the coupling of the C_2 fragment from the vinylidene with the iridacycle, promoted by means of the acid added.

This reaction was found to be highly dependent on the proton source used as well as on the reaction solvent. When HCl, HBF_4, $HBAr^F_4$ ($[H(OEt_2)_2][B(3,5\text{-}C_6H_3(CF_3)_2)_4]$) or triflic acid were employed, complex mixtures of products were obtained. On the other hand, formation of the iridabenzene using trifluoroacetic acid is favored in THF over Et_2O, *ca.* 80% isolated yield in the former and mixed with many unidentified byproducts in the latter solvent. The reaction requires 3 equiv. of acid to proceed.

A reasonable reaction pathway to iridabenzene **12** is depicted in Scheme 4.5. The first step is the protonation of the C_β of the vinylidene linkage and hydrolysis of the $SiMe_3$ groups to form an undetected cationic ethylidyne, which then rearranges, by migration of one of the methylene termini of the butenediyl to the carbyne carbon atom, to give the second unobserved intermediate **B5**. Lastly, the dehydrogenation of the non-conjugated iridacyclohexadiene to the iridabenzene structure is proposed to take place by an α-H elimination and loss of a proton (in either order).

The pathway proposed in Scheme 4.5 for the formation of compound **12** is a reasonable mechanistic proposal for the reaction, but it could also be considered that the moiety that inserts into the metallacycle is a hydrolyzed vinylidene (=C=CH_2) instead of a carbyne (≡C^+–CH_3), by comparison with the examples mentioned below. Nevertheless, several issues play against this. First, as has been mentioned, vinylidene compound **11** is very stable even at high temperatures without noticeable insertion of the cumulene ligand into the Ir–C bonds of the metallacycle. Second, a vinylidene ligand would hardly persist as such in the presence of an excess of acid in the reaction

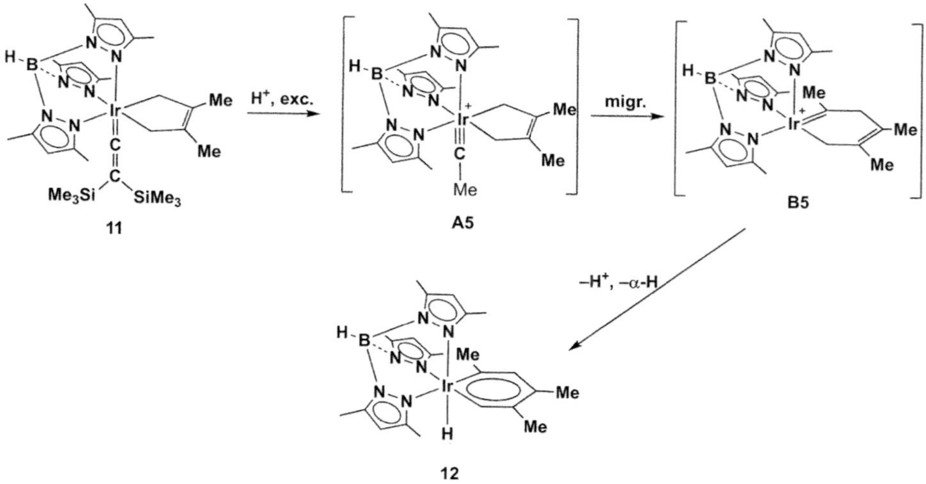

Scheme 4.5

mixture, so it is very likely the vinylidene is protonated to a carbyne ligand before insertion into the iridacycle. Finally, the reaction requires an excess of acid to proceed, as also happens with the related derivative $Tp^{Me2}Ir(Ph)_2[=C=C(SiMe_3)_2]$ (**13**) which is prepared by the reaction of the precursor $Tp^{Me2}Ir(Ph)_2N_2$ (**2**) and $Me_3SiC{\equiv}CSiMe_3$ (Scheme 4.6). In this case the product obtained is $Tp^{Me2}Ir(Ph)(X)(=C(Me)(Ph))$ (Scheme 4.6) where X stands for the anion of the acid if coordinative (e.g. Cl or CF_3CO_2) or a molecule of solvent if an acid such as HBF_4 is used and the corresponding cationic

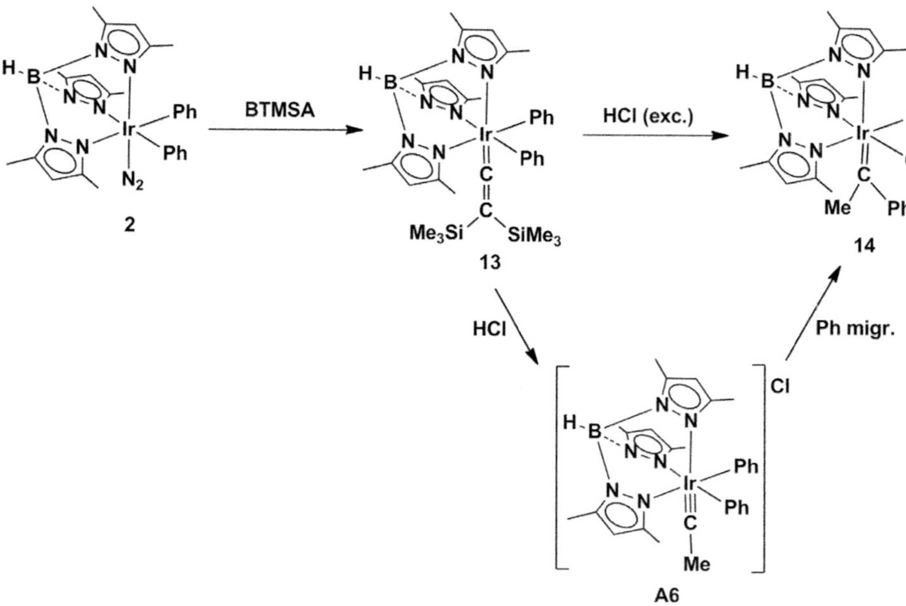

Scheme 4.6

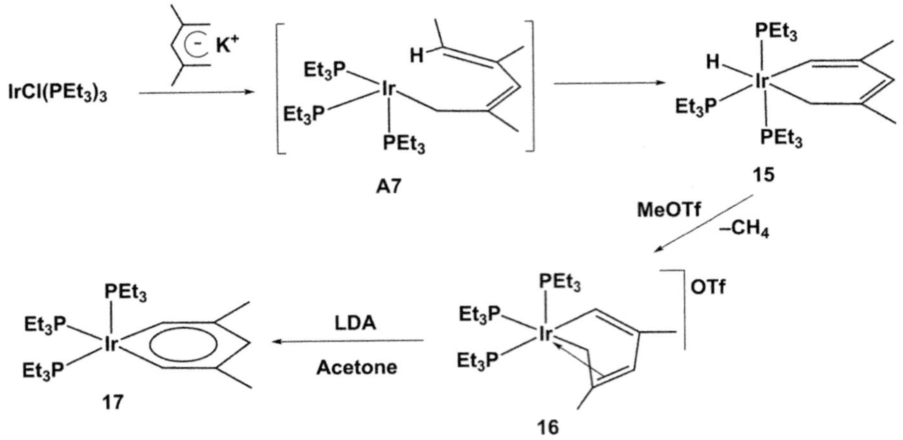

Scheme 4.7

product is formed. The carbene ligand present in the final compound, Ir=C(Me)(Ph), suggests the formation of the carbyne intermediate Ir≡C–Me, which is generated upon reaction of the vinylidene ligand with acid, and subsequent migration of one of the Ir-bonded Ph moieties to the carbyne carbon. The isolation of **14** strongly supports the mechanism proposed in Scheme 4.5.

There are other examples of dehydrogenation of metallacyclohexadienes to form metallabenzenes. For instance, compound **15** formed by the reaction of $IrCl(PEt_3)_3$ with the 2,4-dimethylpentadienide anion [12] is induced to aromatization by initial addition of MeOTf, resulting in the removal of the hydride ligand, and subsequent addition of LDA to promote the deprotonation of the ring. This last reagent is required to form iridabenzene **17** (Scheme 4.7). In comparison with this system, it is proposed that the purported intermediate **B5** spontaneously deprotonates without the need of an added base.

A related reaction for the formation of iridabenzenes is via protonation of $Ir(C_4H_4)(C_2Ph)(CO)(PPh_3)_2$ (**18**), which is a complex with an iridacyclopentadiene moiety and a σ-phenylethynyl ligand (Scheme 4.8) [25]. It was proposed that, upon reaction with HBF_4, protonation of the alkynyl ligand occurs to produce a coordinated vinylidene which readily inserts into one of the Ir–C bonds of the metallacycle to give the iridacyclohexadiene product, **19**. Aromatization of this species requires further addition of acid, resulting in protonation of the *exo* =C(H)(Ph) unit and generation of the required

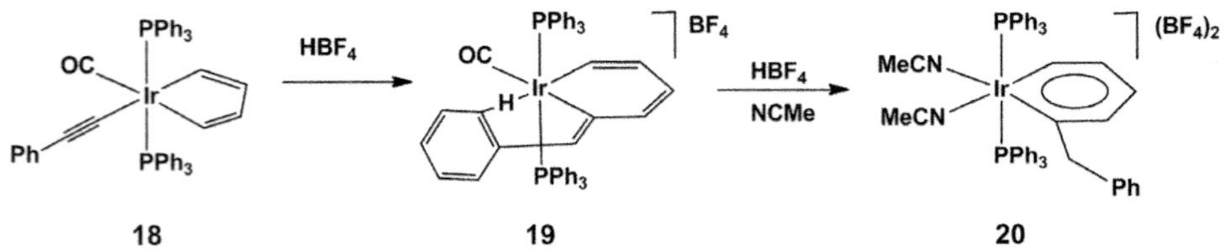

Scheme 4.8

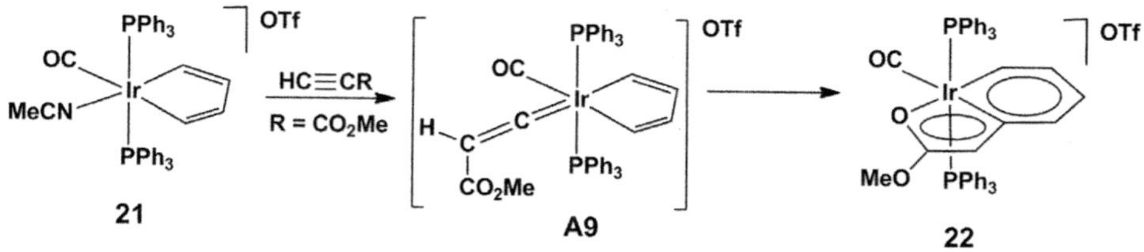

Scheme 4.9

additional unsaturation in the ring. For the case of compound **12**, owing to the presence of only one double bond in the original iridacyclopentene ring (instead of the two contained in the iridacyclopentadienes), aromatization requires in the final steps loss of two H atoms from the ring: one through migration to iridium and the other through loss as a proton.

Other procedures for the syntheses of metallabenzenes have also been proposed to occur by vinylidene insertion into metallacyclopentadienes. The use of methyl propiolate as the alkyne precursor of the vinylidene leads to the formation of a variety of fused-ring metallabenzenes, known as "metallabenzofurans." For instance, derivatives of Ir [26] and Os [27] have been prepared following reaction pathways related to the one exemplified for iridium in Scheme 4.9. None of these reactions requires the presence of acid to proceed.

Isolation of compound **12** shed light on some unresolved questions in our previous examples due to the nature of the metallaaromatics thus far prepared. In particular, the presence of an H atom on the C_α of the iridacycle served as a diagnostic tool to further evaluate the carbene character (and hence the electron delocalization along the ring) of this carbon (see Section 4.5). Also, this was our first example where there are only electron-donating ligands on the carbon atoms of the metallabenzene, unlike the previously prepared derivatives. This fact suggested that the $Tp^{Me2}Ir$ moiety not only could support a broad variety of iridaaromatic structures but also could drive a variety of reactions toward such structures.

4.2.2.2 Preparation of Tp^{Me2}-Iridaaromatics by Reaction of Olefins with Iridacyclopentadienes

4.2.2.2.1 Iridabenzenes

Besides the two somewhat serendipitous methods already described for the synthesis of Tp^{Me2}-iridaaromatic compounds, some more iridabenzenes have been prepared by the reaction of iridacyclopentadienes with olefins.

Two different aquo-Tp^{Me2}-iridacyclopentadiene structures have been chosen for this study, one fully substituted with CO_2Me groups (**23**) and a second with a hydrogen atom in the β-position (**24**) (Schemes 4.10 and 4.12). They result respectively from the coupling of two molecules of DMAD or one DMAD and one methyl propiolate using as precursor the Ir(I) butadiene derivative **1**. While ethylene reacts with each of them leading to products derived from the 1,2-insertion of the olefin into one of the two Ir–C bonds (regioselectively into the one adjacent to the CH moiety for the case of compound **24**), larger olefins give rise to the formation of iridabenzenes, although the composition of the final compounds varies from one precursor to another.

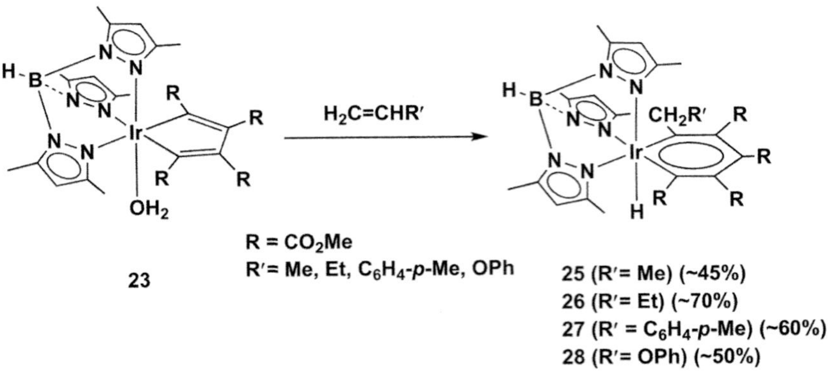

Scheme 4.10

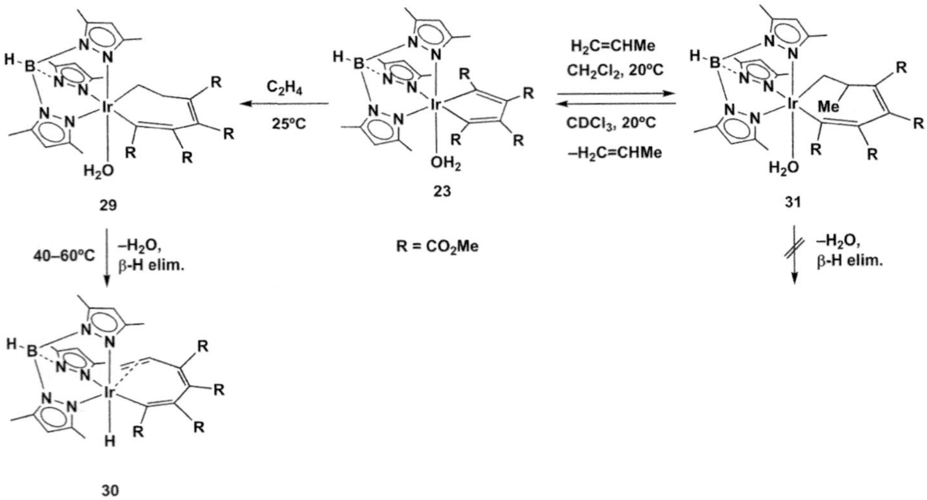

Scheme 4.11

Treatment of the fully substituted iridacyclopentadiene **23** with an excess of propene, 1-butene, *p*-methylstyrene, or phenylvinylether in CH_2Cl_2 at 60–80°C for 12 h gives the iridabenzenes **25**, **26**, **27**, and **28** in 45–70% spectroscopic yields (Scheme 4.10) [28, 29].

A pertinent detail is the reaction, at room temperature, of **23** and propene, which occurs with initial formation of the iridacycloheptadiene **31** (Scheme 4.11), which has been fully characterized by spectroscopy and single-crystal X-ray structure determination. The reaction is highly regio- and stereoselective, and stereochemical issues are of great relevance in the reactions under discussion. The analogous ethylene-insertion derivative **29** evolves by β-H elimination, but compound **31** has the opposite stereochemistry than required to undergo this elimination, so it eventually reverts to **23** and converts into the final product **25** upon heating the reaction mixture.

The moderate spectroscopic yield of the corresponding iridabenzene **25** is explained by the formation of other byproducts which derive from the opposite propene

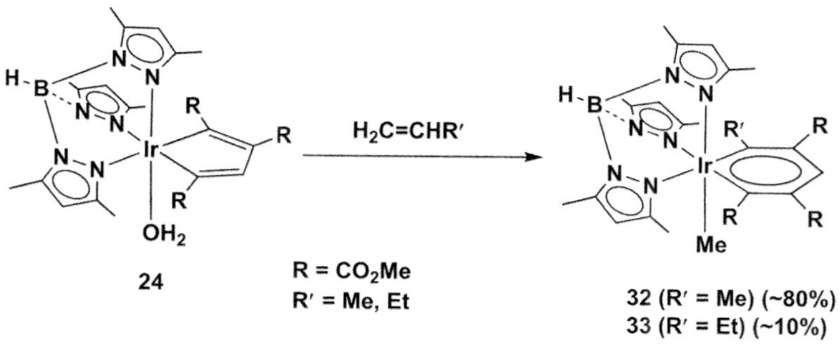

Scheme 4.12

1,2-insertion (with the Me in the Ir-C_α atom), and further evolution of the corresponding iridacycloheptadiene [30]. This is a general trend for the reactions of iridacyclopentadienes **23** and **24** with larger olefins (propene, butene, phenylvinyl ether, and *p*-methylstyrene, the two latter only studied for **23**), which normally yield variable amounts of products derived from a 1,2-olefin insertion, formed in admixtures with the corresponding iridabenzenes.

The less-substituted iridacyclopentadiene **24** reacts with an excess of propene, with formation of a new metallabenzene structure in 80% spectroscopic yield (Scheme 4.12). It contains an unexpected methyl ligand bound to the metal center, indicating a fragmentation of the olefin has taken place during the reaction. Accordingly, the substituent on the newly generated Ir–C_α bond of the iridacycle is a Me group (instead of a C_2 unit as could have been expected). Only one isomer is observed, with the CH moiety of the iridacycle located in the *para* position, showing the high regioselectivity of the process. The reaction with 1-butene takes place with similar results, with a C–Et unit in the α position of the iridabenzene, although in this case the reaction is not so clean and only 10% of this product is formed.

Generation of this series of metallabenzenes requires a more complex reaction pathway than a simple olefin insertion process, which has been seen to yield other types of products, and we proposed, as a key step, the isomerization of the corresponding olefin into a carbene ligand in the coordination sphere of iridium.

Scheme 4.13 shows the mechanisms proposed for the case of propene for both metallacycles, which include four different steps: substitution of the H_2O ligand with the corresponding olefin, alkene isomerization into an alkylidene unit (propylidene for **23** or dimethylcarbene for **24**), stereo and regioselective migratory carbene insertion into an Ir–C bond of the metallacycle, and an α-hydride or an α-Me elimination process.

The first of the steps proposed, the rearrangement of the alkene in the coordination sphere of a metal, is an unusual process, but there are quite a few examples in the literature [31–35]. In the case under study, it could be possible that the steric pressure exerted by the Tp^{Me2} ligand and the iridacycle moiety is the reason for this isomerization. The following steps proposed, namely migratory insertion of the carbene into the M–C bonds and α-H elimination, are well-known elementary steps. The α-Me elimination required for the formation of **32** is an example of the less-common α-elimination of aryl or alkyl groups, but once more there are literature precedents for this type of reaction [36–38].

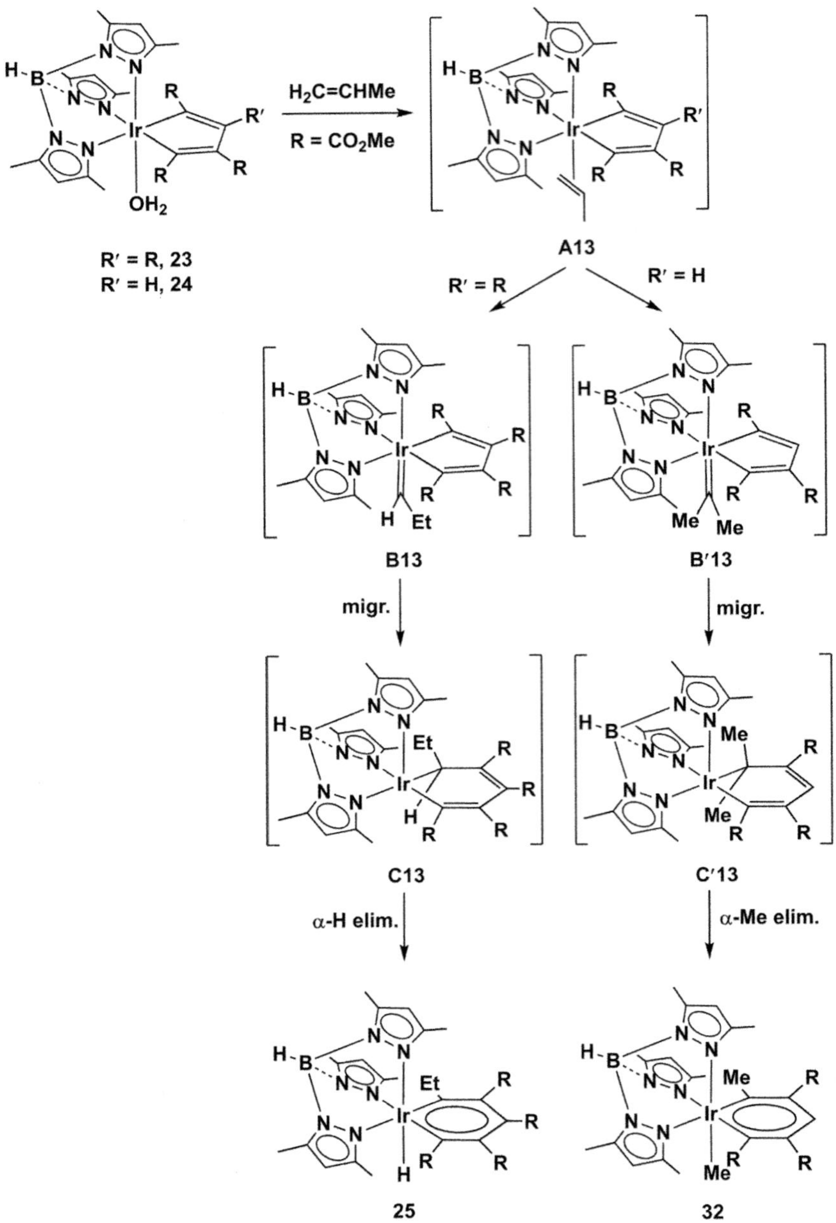

Scheme 4.13

It is not clear why the $=CMe_2$ unit appears to be favored over the linear propylidene, =CH(Et), in the reaction involving the less-substituted iridacycle, leading to compound **32**. In this case, the migration of the alkenyl carbon atom onto the alkylidene is regioselective as the only one involved is that adjacent to the CH group.

4.2.2.2.2 Iridanaphthalenes

This (isomerized) olefin insertion into the Ir–C bond of iridacyclopentadienes as the strategy to prepare iridabenzenes has also been applied with success for the preparation of iridanaphthalenes. We were inspired by the formation of iridanaphthalene **8** in a way analogous to the formation iridabenzene **7** upon the use of a precursor with a fused benzene ring. We had in hand the aquo-α-iridaindene **34**, a quite reactive product resulting from the reaction of $Tp^{Me2}Ir(Ph)_2N_2$ (**2**) and 1 equiv. of DMAD. Addition of a large amount of water (≥10 equiv.) to this reaction, under a N_2 atmosphere, is the key for the isolation of this intermediate since the formation of the seven-membered metallacycle **4** (which forms along with a symmetric isomer with the benzo-fused ring located in positions 3 and 4 of the seven-membered iridacycle) is very favored even in the presence of only one equivalent of DMAD, with unreacted starting material being observed in the crude mixture in that case.

This aquo-metallacycle **34** reacts with an excess of olefins $CH_2{=}CHR'$ (R′= H, Me, C_6H_4-*p*-Me, OPh) yielding β-iridanaphthalenes **35–38** in good yields (60–80%) (Scheme 4.14) [28].

The starting metallacycle **34** possesses two different types of M–C bonds, and the results of the reactions once more show that the insertion is chemoselective as only the products from the insertion of the incoming ligand into the Ir–C(aromatic) bond are observed. This selectivity has allowed for the isolation of the first examples of β-metallanaphthalenes (see Figure 4.2).

These reactions probably proceed in a manner similar to that proposed for the synthesis of iridabenzenes **25–28** (exemplified for **25** in Scheme 4.13). First, the olefin-adduct

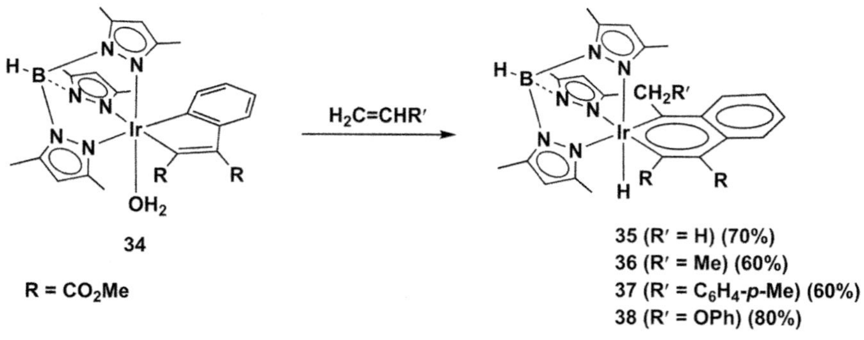

Scheme 4.14

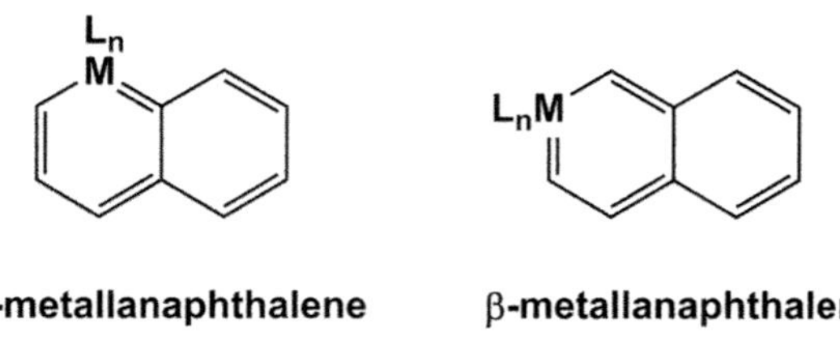

Figure 4.2 General structure for α-metallanaphthalenes and β-metallanaphthalenes.

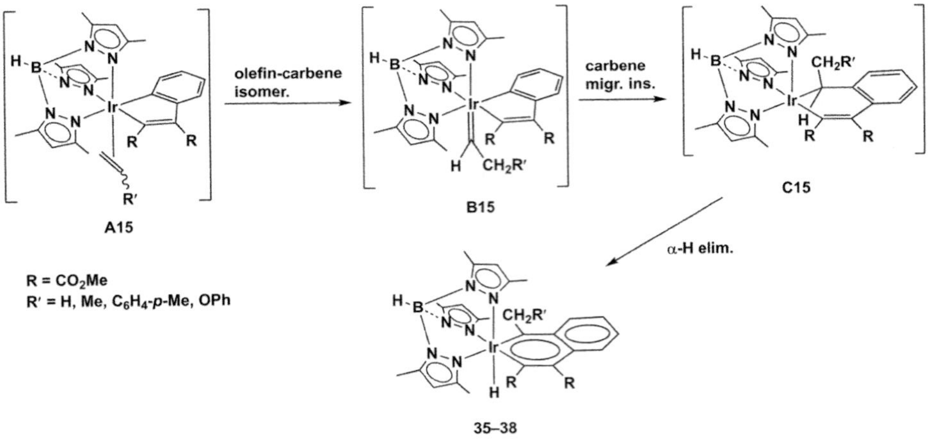

Scheme 4.15

is formed upon displacement of the highly labile H_2O ligand. This coordinated olefin transforms into an alkylidene ligand and then inserts selectively into the Ir–C(phenylic) bond; α-H elimination allows the formation of unprecedented β-metallanaphthalenes (Scheme 4.15).

It is worth noting that insertion into the Ir–alkenyl bond would have provided the α-metallanaphthalenes which were not observed. As shown in Scheme 4.2 and Figure 4.1, all the metallanaphthalenes prepared before the publication of compounds **35–38** have the metal atom located in the α-position of the naphthalene structure, at variance with derivatives **35–38**.

Another distinguishing issue in the reactions of Scheme 4.14 in comparison to those encountered in the case of Schemes 4.10 and 4.12 (iridacyclopentadienes with no fused benzene ring) is the fact that ethylene also forms an iridanaphthalene. This indicates, in accord with our proposal, that, in spite of being much less prone to isomerization, ethylene converts into a carbene ligand prior to insertion into the metallacycle.

In accord with the mechanism proposed all the β-iridanaphthalenes **35–38** have a hydride ligand, as a consequence of the isomerization of the corresponding olefin $H_2C{=}CHR$ to the carbene $=C(H)(CH_2R)$. At variance with this, the reaction with $H_2C{=}CHOEt$ is much more complex and affords three main species (Scheme 4.16). One of them (compound **39**, formed in *ca.* 21% spectroscopically determined yield) contains a Me ligand bound to iridium instead of H, a composition reminiscent of compounds **32** and **33** of Scheme 4.12. The same proposal is hence invoked to explain the formation of this product: the isomerization of the coordinated olefin is of type $H_2C{=}CHOEt \rightarrow =C(Me)(OEt)$. This may be due to the higher stabilization of this carbene in comparison with the alternative, $=C(H)(CH_2OEt)$, due to the OEt group directly bonded to the carbene carbon. However, it is not regiospecific as the latter isomerization also takes place (see formation of **41** below). Following analogous steps to those of Scheme 4.15, stereoselective insertion of the carbene into the Ir–C(phenylic) bond and subsequent α-Me elimination would eventually yield the observed product **39**.

It is also worthwhile to comment on the nature of the other two products isolated in the reaction depicted in Scheme 4.16. One of them, **40**, is the straightforward result of

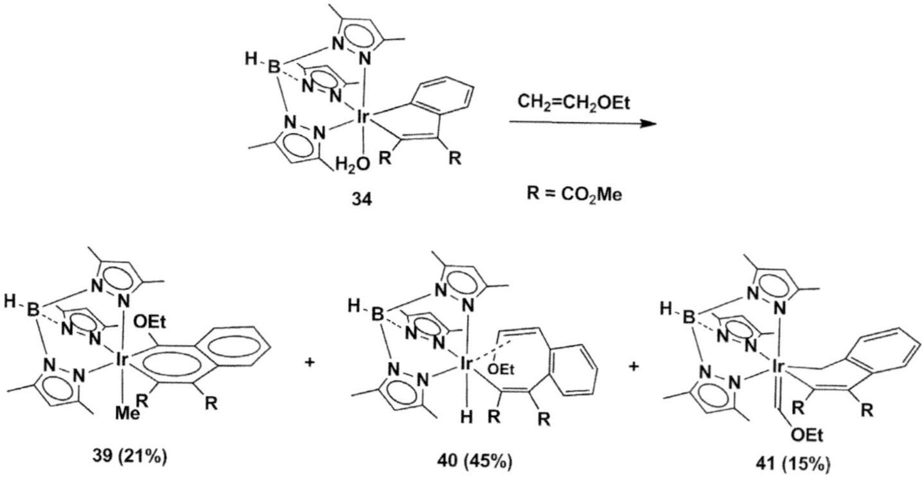

Scheme 4.16

the 1,2-olefin insertion into the Ir–C(phenylic) bond, followed by a β-H elimination reaction. This finds precedent in the reactions of the iridacycles **23** and **24** with ethylene (see, for instance, Scheme 4.11), and in other reactions with related compounds previously published [23]. As mentioned above, compounds related to **40** were also observed in very low yield in the reactions of **23** and **24** with larger olefins. In the case of the reaction with **34**, it becomes the more abundant one.

The other species, compound **41**, is clearly the result of another olefin-to-carbene isomerization/insertion/elimination pathway, for which we propose the sequence of events shown in Scheme 4.17.

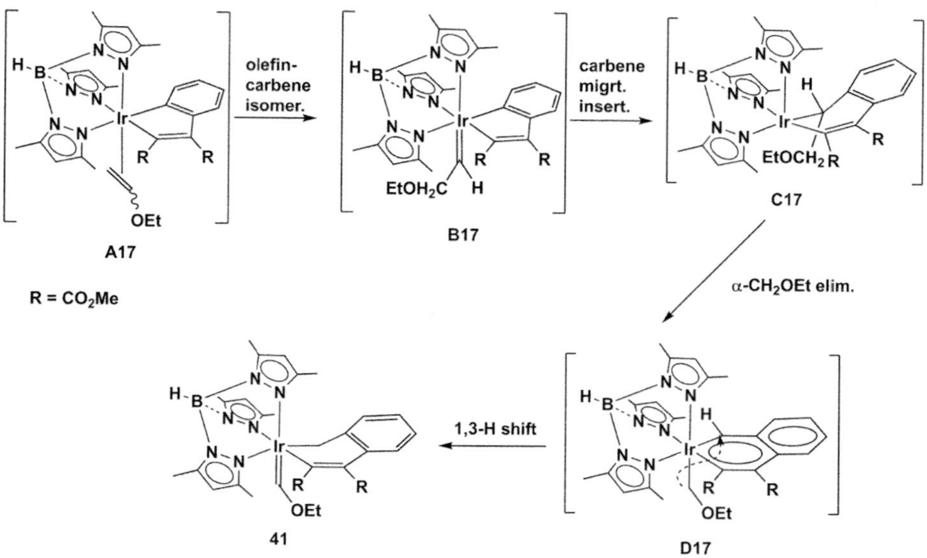

Scheme 4.17

This mechanism would imply an isomerization of the type $H_2C{=}CHOEt \rightarrow {=}C(H)(CH_2OEt)$, as for the former olefins, but in this case the insertion into the Ir–C(phenylic) bond would have the opposite stereochemistry. In this way, the group on the α-carbon of the metallacycle situated in the appropriate position to migrate would be the $-CH_2OEt$ moiety, instead of the H atom as in the previous case. The result of this elimination should be the purported iridanaphthalene **D17** of Scheme 4.17, which is not observed as a product. We invoke the higher stability of Fischer-type carbenes in these $Tp^{Me2}Ir(III)$ systems in comparison to Schrock carbenes to be the reason for the 1,3-H shift that would yield the isolated product, **41**, even overcoming the partial loss of aromaticity associated to the **D17** → **41** transformation. The same reasoning can explain why we do not observe an analogous 1,3-H shift from the Me ligand to the Ir=C(OEt) carbon atom in the β-iridanaphthalene **39** (Scheme 4.18).

It is pertinent to make a final comment on these reactions that involve the formation of iridabenzenes and iridanaphthalenes through the reaction of an olefin and an iridacyclopentadiene, and which proceed via a type of ring-expansion reaction. The key step proposed for these processes is the transformation of the olefin into a carbene ligand within the coordination sphere of the iridium atom. In all these reactions the isolation of the corresponding intermediates with a coordinated carbene ligand has been elusive, probably due to the high reactivity associated with such functionalities. Nevertheless, we have recently isolated a stable product featuring this isomerization. The reactions of the iridaindene **34** with butadiene and 2,3-dimethylbutadiene result in the formation of Diels–Alder products [39]. These reactions were proposed to occur by a [4+3] cycloaddition process (after water dissociation) involving the diene, the Ir atom and the β-(alkenyl)-carbon of the iridacycle. Cyclopentadiene, the most typical diene for Diels–Alder reactions, fails to lead to an analogous result, and instead generates a carbene complex,

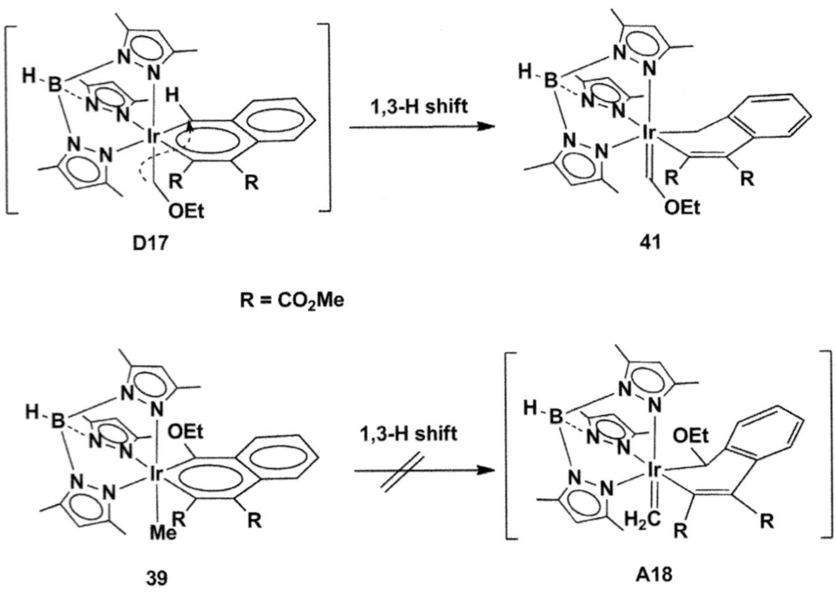

Scheme 4.18

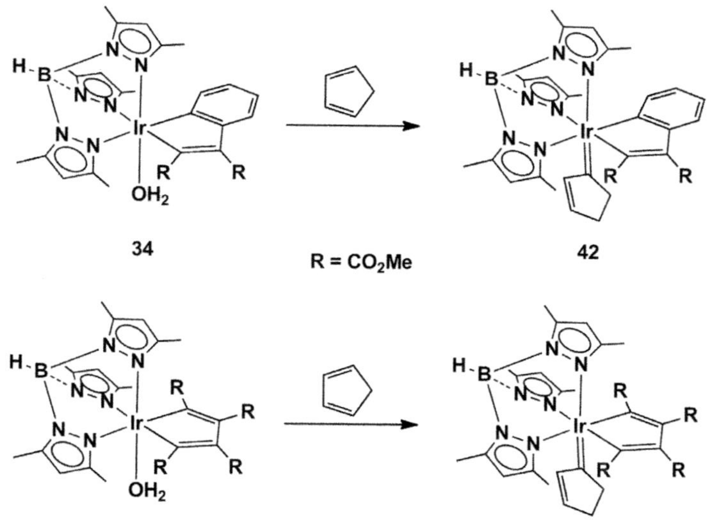

34 42 23 43

Scheme 4.19

42 (Scheme 4.19), stable up to at least 120°C. DFT calculations suggested steric hindrance to be the reason for the lack of reactivity of this compound. An analogous derivative **43** was also obtained from the related iridacycle **23**, both compounds representing most notable examples of the rarely observed olefin-to-alkylidene rearrangement, thus demonstrating the feasibility of this isomerization in the coordination sphere of these $Tp^{Me2}Ir$ derivatives.

4.2.2.3 Other Procedures for Ring Expansion Reactions

4.2.2.3.1 Reaction of a Tp^{Me2}-Iridaindene with Bis(trimethylsilyl)acetylene

One of the routes reported in the literature for the synthesis of iridabenzenes consists of the reaction of an iridacyclopentadiene with a terminal alkyne and subsequent addition of acid (Scheme 4.20) [40].

In contrast to the related process of Scheme 4.9, in which the starting material possesses a labile NCMe ligand, this reaction proceeds through protonation of the iridacycle by the alkyne, with formation of an alkynyl ligand, alkynyl protonation to vinylidene by the acid and eventual coupling of the two carbon chains and aromatization by 1,3-H shift in either way.

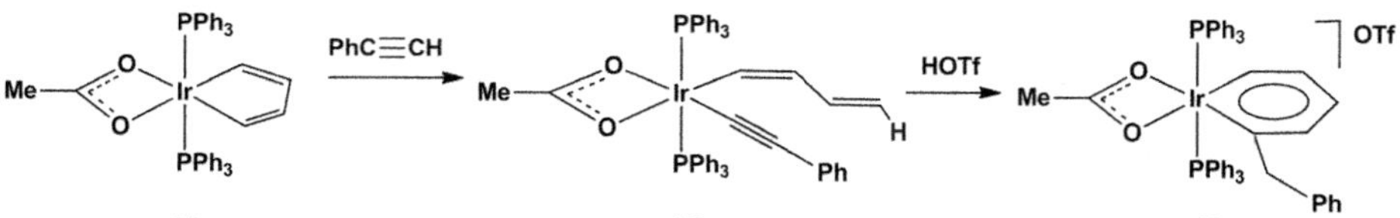

44 45 46

Scheme 4.20

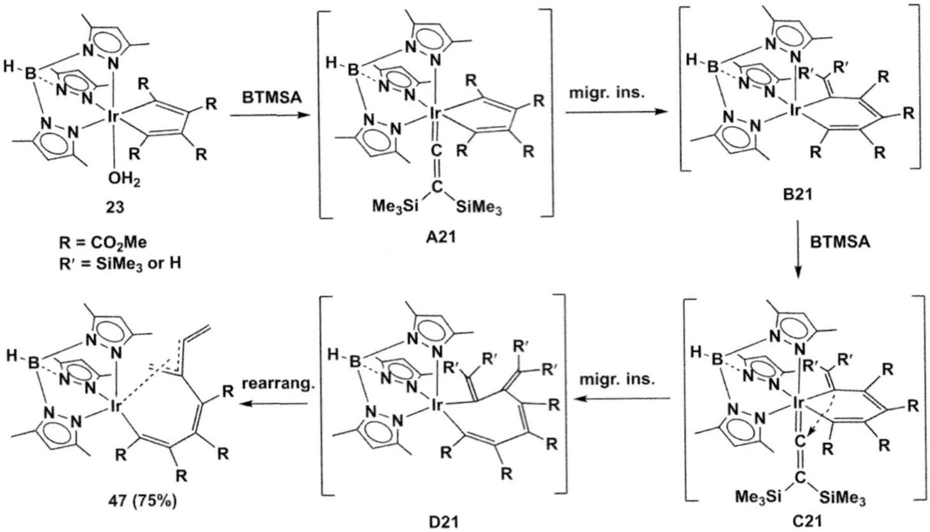

Scheme 4.21

In view of these reactions in Schemes 4.9 and 4.20, it could be interesting to see what happens with different metallacyclopentadienes and terminal alkynes, and if the design of a general procedure for the synthesis of metallabenzenes could be proposed. Our studies have shown that for the $Tp^{Me2}Ir$ system the reactions give quite different results.

The reactions of iridacyclopentadiene **23** with internal alkynes proceed with insertion of the unsaturated fragment into one of the Ir–C bonds, leading to either iridacycloheptatrienes [9, 10] or unusual bicycles of type metallobicyclo[3.2.0]-heptatriene [6], depending on the nature of the alkyne used. In contrast, when BTMSA is used as the reagent, a new iridacycle with four more C atoms but no silyl groups is formed (**47**). This is the result of the incorporation of two molecules of BTMSA and concomitant hydrolysis of the $SiMe_3$ groups [41]. The mechanism proposed (Scheme 4.21) includes the stepwise insertion of two vinylidene ligands formed from BTMSA (like in Schemes 4.5 and 4.6). Bond reorganization of the last intermediate leads to the final compound, **47**. Hydrolysis of the $SiMe_3$ groups occurs during the process, probably in the alkyne to vinylidene isomerization [42–49].

The reactions of the same iridacycle with terminal alkynes proceed differently, as shown in Scheme 4.22 for the cases of HC≡CPh and $HC{\equiv}CCH_2CH_2OH$. Once more, two molecules of alkyne incorporate into the iridacycle, the first of them performing a typical 1,2-insertion of alkyne, but the second transforms, presumably, into a vinylidene ligand [41].

In comparison with all these precedents, the reaction of iridaindene **34** with BTMSA allowed for the isolation of a further iridanaphthalene **49**, formed in *ca.* 55% yield in admixture with derivative **50** of Scheme 4.23 [50]. The metallaaromatic ring of **49** is identical to that of compound **35** formed by the reaction of the same Ir precursor and ethylene, but instead, the coordination sphere is completed with a $-C{\equiv}CSiMe_3$ ligand.

It is evident that formation of **49** requires two molecules of BTMSA and the hydrogens of the methyl substituent in the metallacycle come from the hydrolysis of three

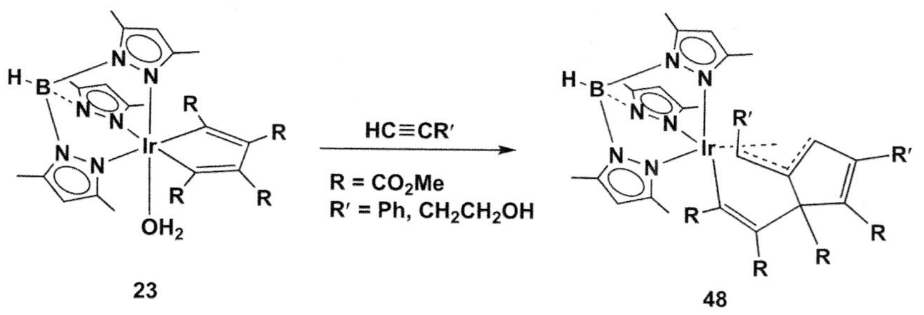

Scheme 4.22

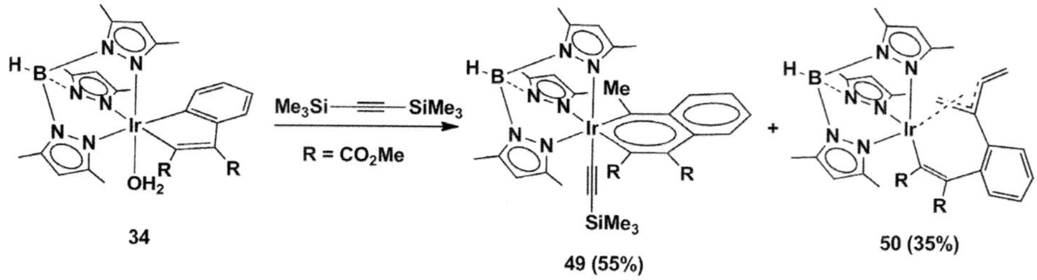

Scheme 4.23

$SiMe_3$ groups. The mechanism of formation of **49** was proposed to proceed by the steps detailed in Scheme 4.24.

BTMSA in intermediate **C24** could coordinate as a vinylidene, as proposed in the scheme, or as η^2-alkyne, to transfer a $SiMe_3$ moiety to the newly formed Ir–C=CH_2 unit of the iridacycle, with eventual hydrolysis to generate the Ir–C(Me) moiety.

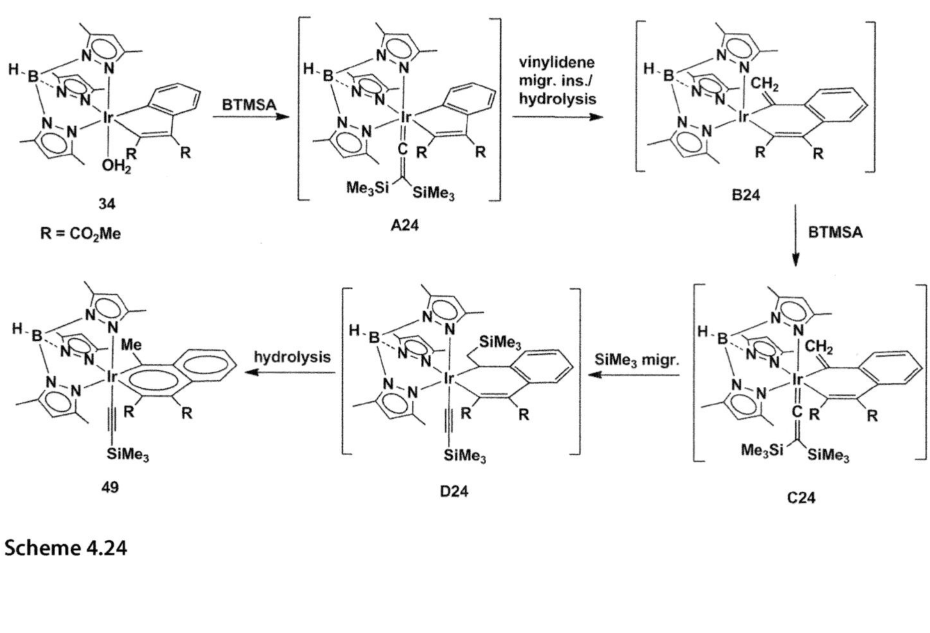

Scheme 4.24

The second product obtained in the reaction of Scheme 4.23 (**50**) has a related precedent in the interaction mentioned above of the iridacycle **23** with the same alkyne (Scheme 4.21). It could be possible that a small amount of an iridabenzene related to **49** forms as well from this precursor, but this was not observed in the reaction mixture or isolated during the column chromatography when **47** was purified. Nevertheless, if any such species formed, it should have been in very small amounts, in view of the high yield (75%) with which product **47** forms.

4.2.2.3.2 Coupling of Two Molecules of Acetylene and a Further C_1 Fragment in a TpIr Precursor

Reaction of the Ir-ethylene derivative $TpIr(C_2H_4)_2$ (**51**), which contains the unsubstituted parent Tp ligand, with acetylene, in different solvents, has resulted in the isolation of further examples of iridabenzenes. Most notably, the reaction in CH_2Cl_2 leads to the isolation of the first parent metallabenzene, i.e. with no substituents on the metallaaromatic ring (Scheme 4.25) [51].

Compound **52** is formed upon treatment of a dichloromethane solution of the starting material $TpIr(C_2H_4)_2$ (**51**) with acetylene (1 atm) from −60°C to room temperature. The reaction conditions shown in Scheme 4.25 are optimized for the best yields of this compound, which is never formed above 50% (spectroscopic yield). A simple change of the temperature at which the reaction mixture is saturated with acetylene, for instance from −60 to −20°C, decreases the yield of compound **52** (in this case, to *ca.* 30%).

The mechanistic investigations (in particular NMR monitoring of the reactions) suggested that formation of the iridabenzene takes place once more by the insertion of a carbene ligand into the Ir–C bonds of an iridacyclopentadiene, i.e. a further case of the ring expansion synthetic method of iridaaromatics.

Monitoring the reaction by variable temperature NMR spectroscopy showed initial formation of the mixed species $TpIr(C_2H_4)(C_2H_2)$ (**54**) and free ethylene (Scheme 4.26). Above 10°C, the second ethylene ligand is displaced by acetylene, with formation of the iridacyclopentadiene TpIr(CH=CH–CH=CH)(L) (**55**). The iridacycle of this compound was conveniently identified in the reaction mixtures by NMR spectroscopy, although the nature of the labile ligand L could not be ascertained, being probably a molecule of solvent (CD_2Cl_2, H_2O, or even acetone, from the commercial acetylene).

Experimental evidence is in agreement with an equilibrium between these two detected intermediate species (**54** and **55**). They are stable up to 10°C, above which temperature they start to transform into the final compounds, the already mentioned iridabenzene **52** and the ethylene-vinyl derivative of composition $TpIr(H)(C_2H_4)$ (CH=CHCl) (**53**). The latter is in fact the first compound to appear, and in accord with

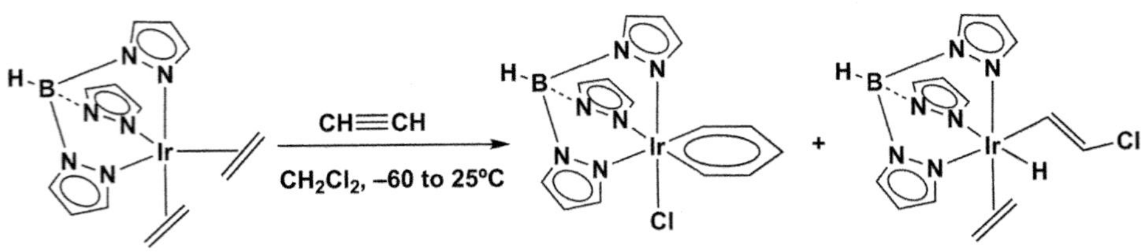

51 | 52 (50%) | 53 (50%)

Scheme 4.25

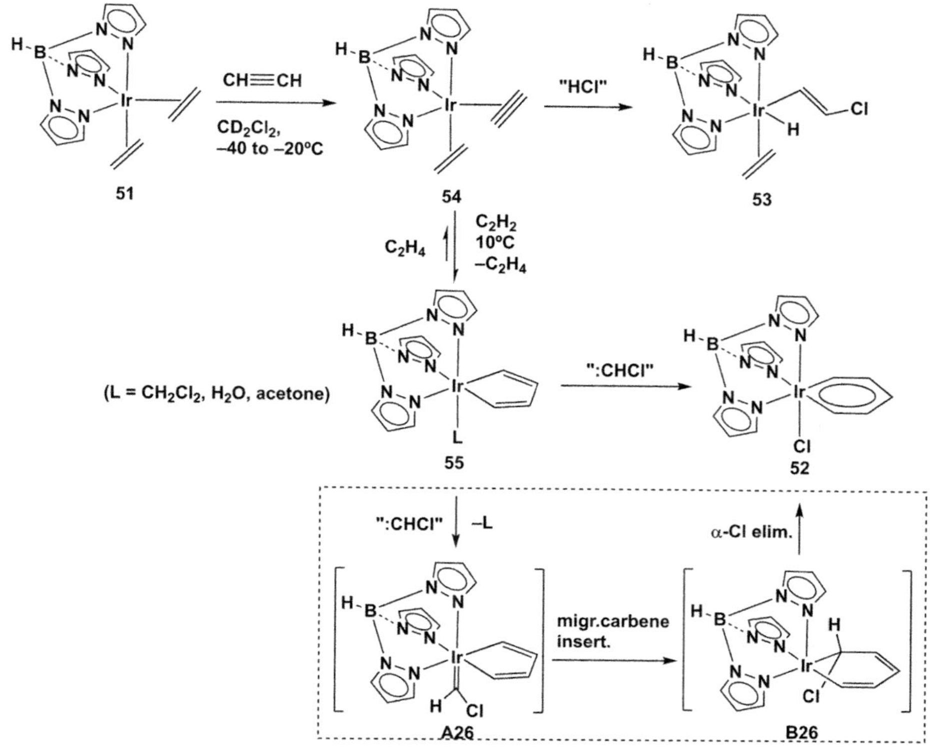

Scheme 4.26

this we proposed that intermediate $TpIr(C_2H_4)(C_2H_2)$ (**54**) is responsible for the fragmentation of CH_2Cl_2, abstracting “H” and “Cl” and generating a carbene :CHCl. The effectiveness of iridacyclopentadiene **55** to trap this carbene under each set of experimental conditions would account for the yield of iridabenzene **52**. Scheme 4.26 shows these transformations, with the proposed intermediate species for the formation of iridabenzene **52** shown at the bottom.

This type of CH_2Cl_2 cleavage mediated by an Ir(III) center has already been observed in a doubly metallated $Tp^{Ms''}Ir(N_2)$ derivative ($Tp^{Ms''}$ = doubly metallated tris(3-mesitylpyrazolyl)borate), which forms a compound of composition $Tp^{Ms'}Ir(Cl)(=CHCl)$ ($Tp^{Ms'}$ = monometallated tris(3-mesitylpyrazolyl)borate) [52]. It is worth mentioning that formation of **53** is not the result of the reaction of **54** with free HCl, since the addition of HCl(aq) to the reaction mixture when **54** has already been formed yields a different product, of composition $TpIr(Cl)(\eta^3\text{-}C_3H_4Me)$, analogous to the corresponding Tp^{Me2} derivative published in the literature [2].

A different iridabenzene, **56**, is obtained when the reaction of $TpIr(C_2H_4)_2$ (**51**) and acetylene is carried out in Et_2O. Under these conditions iridabenzene **56** is formed (also in *ca.* 50% spectroscopic yield) with a methyl substituent in the α position of the iridacycle (Figure 4.3) [51]. The formation of **56** requires clearly the participation of ethylene (released from the starting material) and it can be explained by the metal-induced isomerization of ethylene to ethylidene, “: $C(H)CH_3$” in an analogous process to that

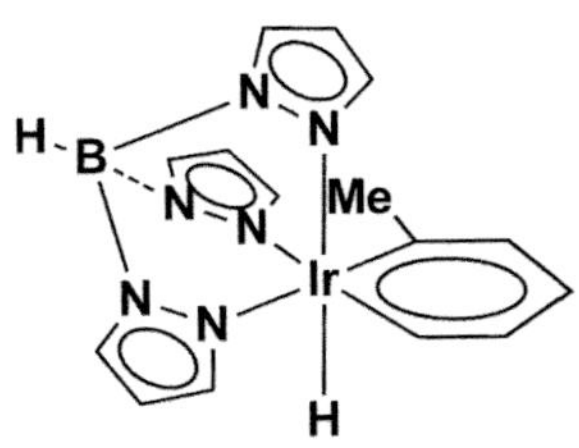

56

Figure 4.3 Structure of Tp-iridabenzene **56**.

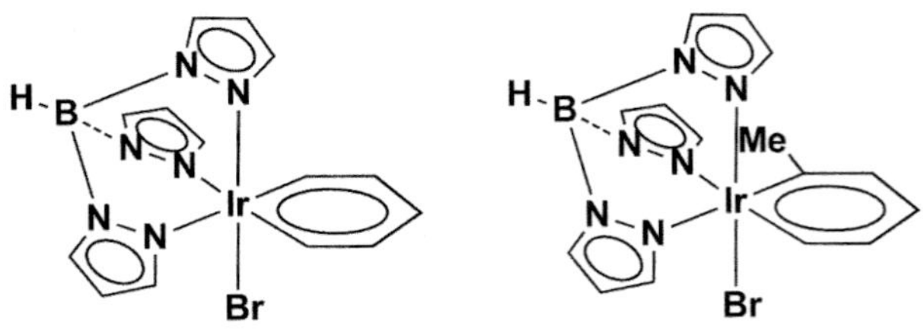

52-Br 56-Br

Figure 4.4 Structure of Tp-iridabenzenes **52-Br** and **56-Br.**

described in Scheme 4.26. In this case, ethylene would coordinate to **55** and isomerize to ethylidene prior to insertion. Remarkably this is a second example of ethylene isomerization to an alkylidene in a Tp′Ir system (Tp′ stands for any type of tris(pyrazolyl) borate ligand, in this case Tp and Tp^{Me2}). In the large majority of reactions where ethylene is involved, it experiences insertion processes into M–C bonds.

It is also worth noting that diethylether, in contrast to CH_2Cl_2, is not activated in this reaction (in spite of known precedents of ether activation in other $Tp^{Me2}Ir$ systems [53–57]) and in its place it is ethylene which perplexingly reacts with **55**, although free ethylene is always present in all these reaction mixtures. Nevertheless, we cannot discard its participation in other parallel reactions leading to other (not isolated) products in this reaction (note that **56** is formed in only 50% yield).

Finally, two different bromide-iridabenzenes have been formed. When using CH_2Br_2 as the solvent for the reaction shown in Scheme 4.25, **52-Br** is formed, and **56-Br** is formed on treatment of **56** with $CHBr_3$ through bromination of the hydride. (See Figure 4.4.)

4.3 Reactivity of Iridaaromatics with Supporting Tris(pyrazolyl)borate ligands

Reactivity of benzene-derived organic compounds is normally driven by their aromatic character, the more straightforward transformations typical of aromatic rings being electrophilic aromatic substitutions. These types of reactions have also been observed for metallaaromatics. From the early studies on the bromination and nitration of an

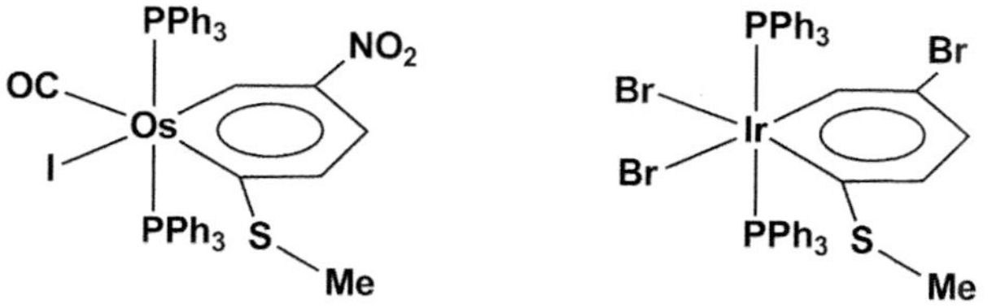

Figure 4.5 Products resulting from electrophilic aromatic substitutions of metallabenzenes.

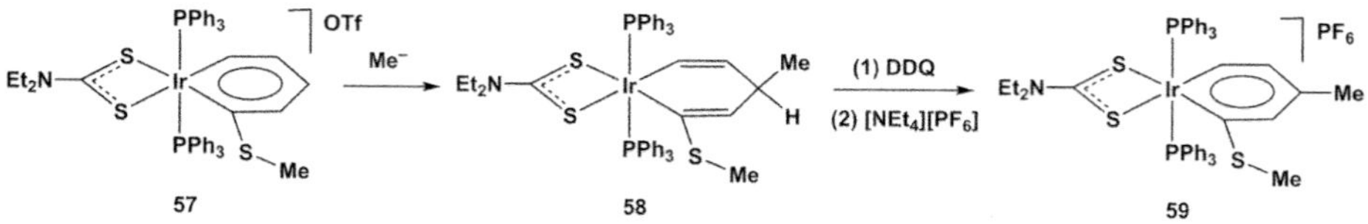

Scheme 4.27

osmabenzene [58], related substitutions have been observed with iridabenzenes of similar composition [59]. Figure 4.5 shows the results for two of these processes, nitration of an osmabenzene and bromination of an analogous iridabenzene ring. In addition, a fused-ring iridabenzofuran, with four unsubstituted C atoms in the iridabenzene ring, has been proved to be an appropriate substrate for electrophilic aromatic substitutions [26].

Apart from these substitutions, many other types of reactions have been observed with metallabenzenes, as for instance nucleophilic aromatic substitution of hydrogen, which can be exemplified by the reaction of cationic iridabenzene **57**, which reacts with nucleophiles (BH_4^-, CH_3^-, OEt^-) to form iridacyclohexa-1,4-dienes, as the result of the selective addition of the nucleophile to the 3-position of the ring (Scheme 4.27). These species are related to the intermediates formed during the nucleophilic aromatic substitution reactions of benzene. The full substitution process can be forced by addition of oxidants, with rearomatization of the metallacycle [60].

Most Tp^{Me2}-iridaaromatics are heavily substituted, and consequently they would not be expected to experience substitution reactions. Nevertheless, some of the few studies carried out on the reactivity of this type of iridaaromatic match the first part of the reactions just mentioned, and constitute a simple nucleophilic addition process. Thus, when α-iridanaphthalene **8**, which contains bulky and electron-withdrawing substituents (CO_2Me), is purified by silica gel column chromatography, a new bicyclic non-aromatic compound **60**, which contains a tertiary alcohol functionality O-bonded to the Ir(III) center, is eluted from the column and isolated [10]. Scheme 4.28 shows this transformation along with a plausible mechanism for it involving a hydroxo intermediate complex from which complex **60** is formed upon nucleophilic attack of the OH to the C_γ of the benzene ring. In that way this last species can be viewed as an internal neutral Jackson–Meisenheimer complex. Aromaticity of the ring is lost with formation of a bicyclic iridacyclohexa-1,4-diene derivative. Interestingly, this reaction can be reversed by treatment of **60** with the acid chloride $ClC(=O)CO_2Me$ and this process presumably takes place by the intermediate formation of an ester species which, as suggested before, decomposes by C–O bond rupture (intermediate **B2** is the same

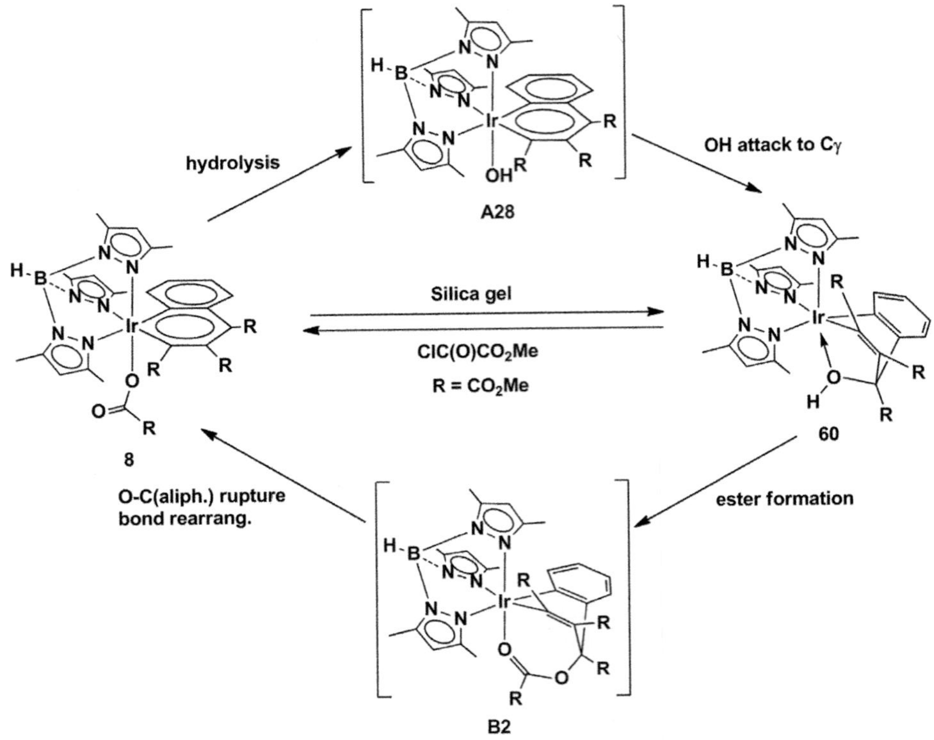

Scheme 4.28

intermediate proposed in Scheme 4.2 to be the immediate precursor of compound **8** during its synthesis).

Hydrolysis of the fully CO_2Me-substituted iridabenzene **7** with silica gel is not feasible, but a compound related to **60** is obtained, although in very low yield, by treatment of a THF-H_2O solution of **7** with *p*-$MeC_6H_4SO_3H$ (*p*-toluenesulfonic acid; Scheme 4.29). This reaction can also be reverted and treatment of **61** with ClC(=O)CO_2Me regenerates complex **7**. When the acid hydrolysis of **7** is carried out in MeOH, an ether derivative **62** is obtained (Scheme 4.29).

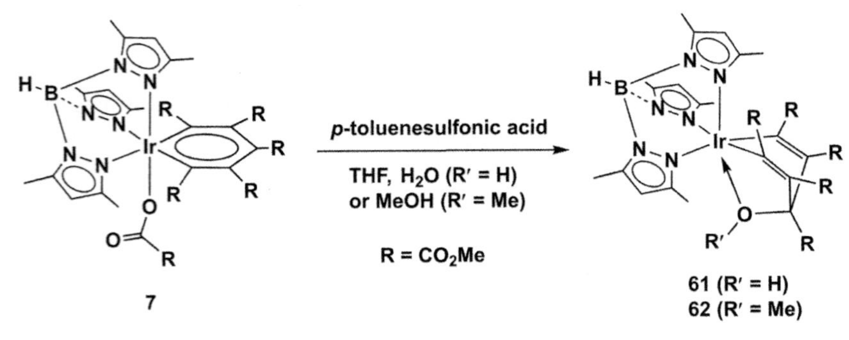

Scheme 4.29

As mentioned above, these reactions do not constitute nucleophilic aromatic substitutions but simply an addition process, since the CO_2Me group on the same C atom is not a suitable leaving group.

Metallaaromatic compounds could experience other types of reactions motivated by the presence of the L_nM fragment in the ring. First, the ancillary ligand can be substituted with a different one. Ancillary ligands normally play a relevant role in organometallic chemistry and their substitution or transformation can induce changes on the properties of the metallaaromatic ring, including reactivity, stability, etc. Many reactions have been described in the literature about this type of transformation [61], and also we have already described some examples along the preceding sections (see, for example, the preparation of the brominated species **52-Br** and **56-Br**). In addition, the latest reactions mentioned could also be included in this category, since they are considered triggered by the substitution of the methyl oxalate ligand with a hydroxide one.

Second, these ancillary ligands can interact with the metallaaromatic ring under the appropriate conditions. The following reactions can be considered a type of these ancillary ligand-induced transformations of metallaaromatics.

Interesting results are obtained upon reaction of iridabenzenes **25** and **32** with NCMe [30]. As represented in Scheme 4.30, iridabenzene **25** reacts with neat NCMe at 20°C to give a 6:1 kinetic mixture of the expected isomeric metallacyclohexadienes **63** and **64**, i.e. the products of the stereospecific migration of the hydride ligand onto the two different Ir-bonded C atoms followed by coordination of NCMe, with **64** being the thermodynamic product of the reaction (**63** transforms cleanly and almost quantitatively into **64** in C_6D_6 at 90°C, $t_{1/2}$ = 4 h). Interestingly, the kinetic **63**/**64** ratio in neat NCMe was found to vary with the temperature and from these data the values

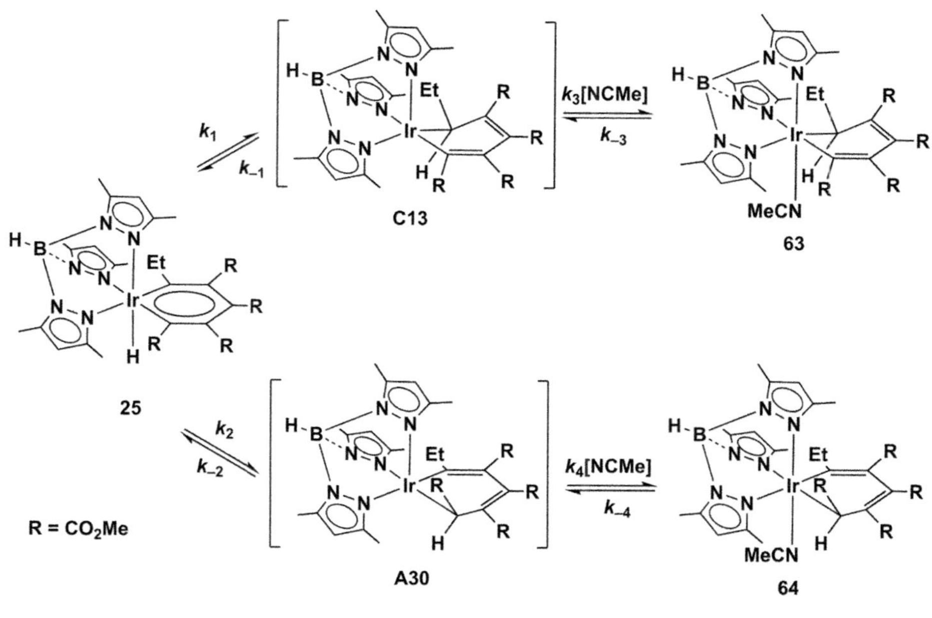

Scheme 4.30

of $\Delta\Delta H^{\neq}$ = 3.75 kcal mol^{-1} and $\Delta\Delta S^{\neq}$ = 5 cal mol^{-1} K^{-1} can be computed, the latter favoring the formation of **64** at high temperatures.

In principle, the transformation of **63** into **64** could be the result of a direct 1,5-suprafacial shift of the hydrogen atom between the two C_α positions. However, several indications are more in accord with a mechanism in which iridabenzene **25** is a true intermediate of this rearrangement, which means that, as shown in Scheme 4.30, all reaction steps are reversible, and the rate determining step is the dissociation of the acetonitrile ligand of **63** (k_{-3}). Intermediate **C13** (See Scheme 4.13) was proposed as the last step on the formation of compound **25** from iridacyclopentadiene **23** and propene. Thus, derivative **63** can be viewed as the result of the trapping of this intermediate and indicates the reversibility of this step. Very likely, some more, if not all, of the Tp′-iridaaromatics with a hydride ligand would participate in this type of H-migration to carbene reactions, although these studies have not been pursued further.

This reaction of migration of a hydride ligand into an Ir-bonded C atom of the metallabenzene is probably occurring in other systems of the literature, as could be the case with iridabenzene **65** of Scheme 4.31, which reacts with hydrogen to form a hydrido-iridacyclohexadiene **66**. This reaction has been proposed to occur by phosphine dissociation, oxidative addition of hydrogen, migration of a hydride ligand to the C_α of the iridabenzene, and coordination of phosphine [62].

Interesting and unpredictable differences were found when studying the reactivity of the methyl-iridabenzene **32** with NCMe (Scheme 4.32). Thus, **32** does not react with neat acetonitrile at 20°C but requires 100°C to transform, and under these conditions it reverts to the iridacyclopentadiene of origin, **24**, isolated in the form of the NCMe adduct. This clearly demonstrates that, despite requiring C–C bond formation and

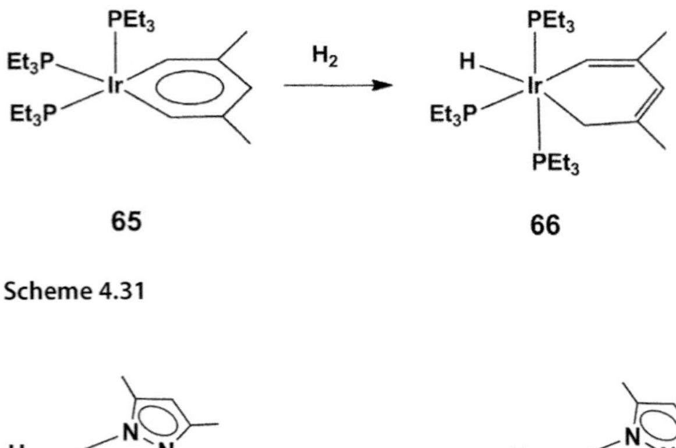

65 66

Scheme 4.31

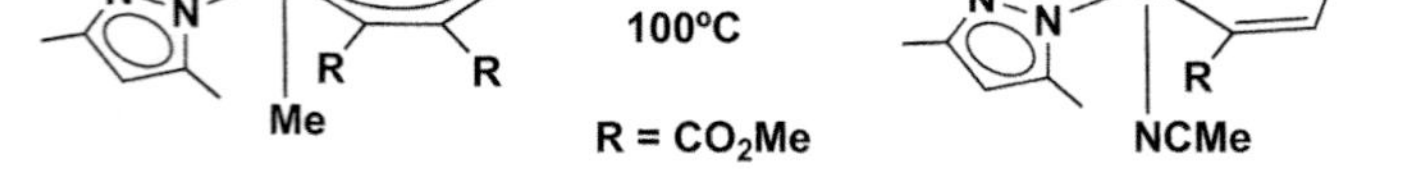

32 24·NCMe (80%)

Scheme 4.32

cleavage, the whole reaction of the iridacyclopentadiene with propene is a reversible process. Furthermore, this scheme represents an unusual fragmentation of a metallabenzene species.

A further closely related example of the disruption of aromaticity in iridabenzenes is the transformation proposed in the last step of Scheme 4.17, where the purported iridanaphthalene intermediate **D17** accepts an H atom from the Ir–CH_2OEt ligand, to allow the generation of a more stable Fischer carbene Ir=CHOEt.

A final comment is pertinent about the stability of the Tp′-iridaaromatics described in this chapter. A widely observed chemical transformation is the decomposition of iridabenzenes into cyclopentadienyl derivatives [63, 64]. The intermediary role of unobserved metallaaromatic species has been proposed in the generation of Cp′M products [65, 66], but also some isolated metallabenzenes have also been observed to undergo this transformation [67]. In particular, a series of iridium derivatives of composition $Ir[C(R')=C(R'')-CH_2CH_2CH_2](CO)(PR_3)_2$ have been observed to transform into Cp′Ir derivatives $Ir(Cp')(CO)(PR_3)$ ($Cp' = C_5H_3(R')(R'')$) at variable rates depending upon the nature of the R′ and R″ groups and the trisubstituted phosphine. The bulkier the phosphine, the easier the transformation of the corresponding iridabenzene into a Cp′-derived compound [68–70]. Iridanaphthalenes of type **9** (Figure 4.1) have also shown to decompose to η^3-indenyl derivatives, upon heating [18, 71].

For the case of the Tp′-iridaaromatics described in this chapter, we have never observed this type of decomposition, although derivatives with Cp′ and Tp′ ligands simultaneously in the coordination sphere of Ir are known (normally, Cp′ coordinates η^5 and Tp′ is forced to adopt a κ^2 bonding mode) [72].

4.4 Structural Data for Iridaaromatics with Supporting Tris(pyrazolyl)borate Ligands

In contrast to the planarity associated with bond conjugation in organic aromatic molecules, non-planarity is a characteristic often observed for metallaaromatics due to the participation of d(π) orbitals from the metal in addition to p(π) orbitals of the C atoms. A detailed theoretical study [73] has dealt with the analysis of the structures of a number of metallaaromatics, and it has been established that the degree of non-planarity depends on different factors, including electronic and steric effects, thus the structures of metallabenzenes can be almost planar or deviate considerably from planarity. In particular, an unsymmetrical ligand environment above and below the metallaaromatic ring exerts different steric pressure at both sides of the ring and enhances the non-planarity. In a similar way, a crowded environment caused by the presence of bulky co-ligands close to the metallacycle, or by substituents on the C atoms of the metallaaromatic ring, can also cause this effect. This is the case for the Tp′Ir metallaaromatic compounds described in this chapter.

A considerable number of iridabenzenes containing tris(pyrazolyl)borate ligands have been characterized by single-crystal X-ray studies. Selected bond distances, as well as the deviation of the metal center from the C_5 plane and the sum of all internal angles within the metallacycle, are collected in Table 4.1. Figure 4.6 shows the numbering scheme for the compounds included in the table.

Table 4.1 Selected bond distances (Å) and other structural data for Tp′-iridabenzenes.

Compound number	7	12	25	32	52
M–C1	1.957(6)	1.939(4)	1.922(4)	1.936(2)	1.955(5)
M–C5	1.921(6)	1.945(4)	1.979(4)	1.957(2)	1.947(5)
C1–C2	1.359(7)	1.397(6)	1.426(6)	1.428(3)	1.380(7)
C2–C3	1.441(8)	1.392(7)	1.386(6)	1.392(3)	1.397(9)
C3–C4	1.393(8)	1.401(7)	1.434(6)	1.415(3)	1.392(9)
C4–C5	1.401(8)	1.388(6)	1.372(5)	1.396(3)	1.382(8)
Deviation of Ir from C_5 plane (Å)	0.74	0.50	0.57	0.70	0.29
Sum of internal angles	704.70	713.34	709.90	706.75	717.8

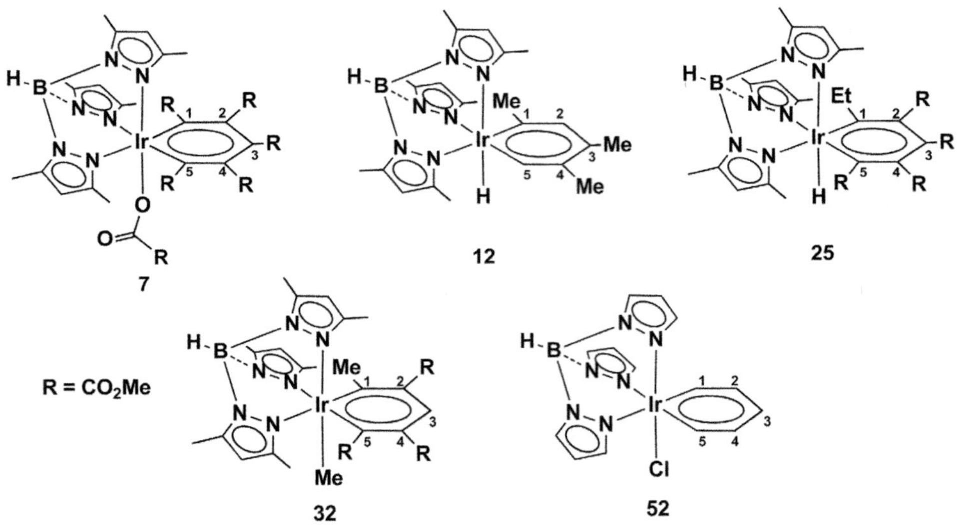

Figure 4.6 Numbering scheme for the compounds referred to in Table 4.1.

The data included in Table 4.1 for the Tp′-iridabenzenes indicate some common trends:

1) C–C bond lengths within the iridacycle have values between single and double C–C bonds (1.46 and 1.34 Å, respectively).
2) The two Ir–C distances have an average value of 1.946 Å, intermediate between Ir–C single and double bonds.
3) In all these iridabenzenes, the iridium atom shows a considerable deviation with respect to the mean plane defined by the five carbon atoms that compose the metallacyclic structure (and which remain essentially coplanar). The iridium atom lies between 0.29 (for **52**) and 0.74 (for **7**) Å out of this C_5 plane. These values are in agreement with the reasons stated above: compound **7**, which presents the largest deviation form planarity, has a crowded environment, with a methyl oxalate ligand and all the carbons of the cycle substituted with CO_2Me groups.

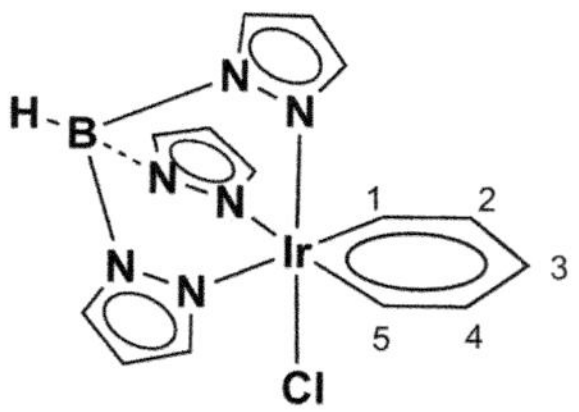

52

D1 = −14.74

D2 = +15.30

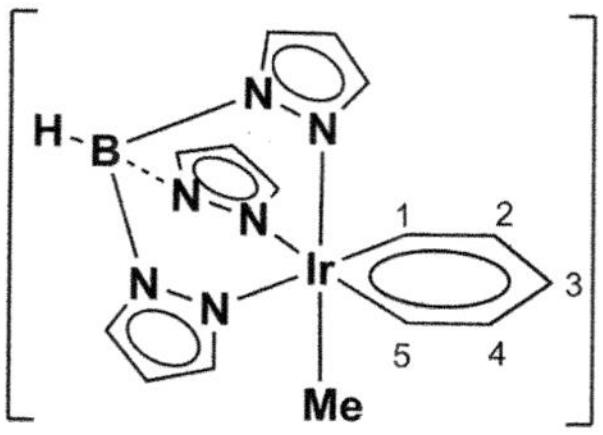

52-Me$_C$

D1 = −14.7

D2 = +14.7

Figure 4.7 D1 and D2 values for compound **52** and the methyl species calculated **52-Me$_C$**. D1 and D2 stand respectively for the dihedral angles (deg) of C2–C1–Ir–C5 and C4–C5–Ir–C1.

Compound **52** represents the simplest possible case in this system, without substituents on the metallacycle and the Tp ligand. Nevertheless, the Ir atom still deviates from planarity (although to a lesser extent than in the other compounds). The lack of symmetry above and below the metallacycle, because of the *fac* κ^3 geometry of the Tp ligand, accounts for the still-present non-planarity in this parent species. In fact, the geometry of a model related to **52** (with a Me ligand instead of a coordinated chloride, species **52-Me$_C$**) was calculated [73] even before **52** was prepared, the values obtained for this species matching very well those measured for compound **52** (see Figure 4.7 for a comparison of these parameters).

Finally, the sum of the internal angles of the six-membered iridacycles (average 710.5°, quite different from the ideal value in benzene of 720°) also reflects the non-planarity of these Tp′-iridaaromatic derivatives.

The corresponding structural data for the Tp^{Me2}-iridanaphthalenes characterized by X-ray crystal structure determinations are collected in Table 4.2. Figure 4.8 shows the numbering scheme for the compounds included in the table.

Table 4.2 Selected bond distances (Å) and other structural data for Tp^{Me2}-iridanaphthalenes.

Compound number	8	35	36	39	49
M–C1	1.930(6)	1.987(6)	2.008(2)	2.026(5)	2.016(3)
M–C5	1.981(6)	1.900(6)	1.898(2)	1.907(5)	1.920(4)
C1–C2	1.402(8)	1.357(9)	1.360(3)	1.343(7)	1.246(5)
C2–C3	1.396(8)	1.465(9)	1.471(4)	1.479(7)	1.471(5)
C3–C4	1.430(8)	1.437(9)	1.425(4)	1.396(7)	1.426(5)
C4–C5	1.445(8)	1.442(8)	1.470(3)	1.475(7)	1.437(5)
Deviation of Ir from C_5 plane (Å)	0.76	0.70	0.67	0.80	0.73
Sum of internal angles	704.2°	706.9°	706.1°	700.1°	705.9°

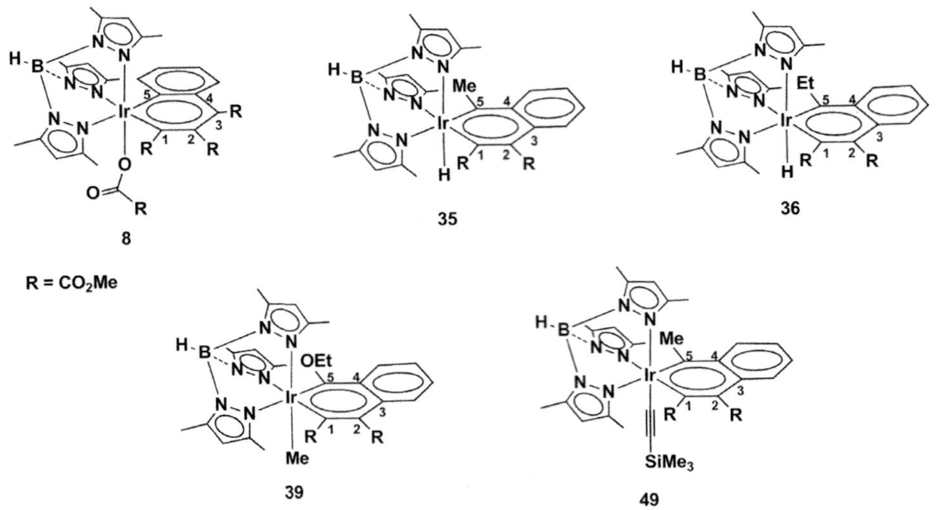

Figure 4.8 Numbering scheme for the compounds referred to in Table 4.2.

In these cases, the differences between bond lengths within the iridacycle rings are greater, reflecting not only the trends encountered in naphthalene but also the preferred aromatic character of the benzene-fused ring in comparison with the iridacycle moiety. Figure 4.9 compares the distances found in **8**, an example of α-iridanaphthalene, **35**, an example of β-iridanaphthalene, and naphthalene.

Thus, while in the α-metallanaphthalene **8** the distance M–C5 is larger than M–C1, for the β-metallanaphthalenes **35**, **36**, **39**, and **49** the trend is the opposite: distances M–C1 are larger than those of M–C5, in accord with a higher double-bond character of these Ir–C5 linkages.

With respect to the distances C–C within the metallacycles, and in agreement with the latter fact, in these β-metallanaphthalenes the distances C1–C2 are always shorter than C2–C3 and C4–C5, as expected for a less delocalized ring.

Again, the six membered rings are non-planar, as the iridium atom deviates significantly (*ca.* 0.7–0.8 Å) from the main plane formed by the five C atoms of these rings.

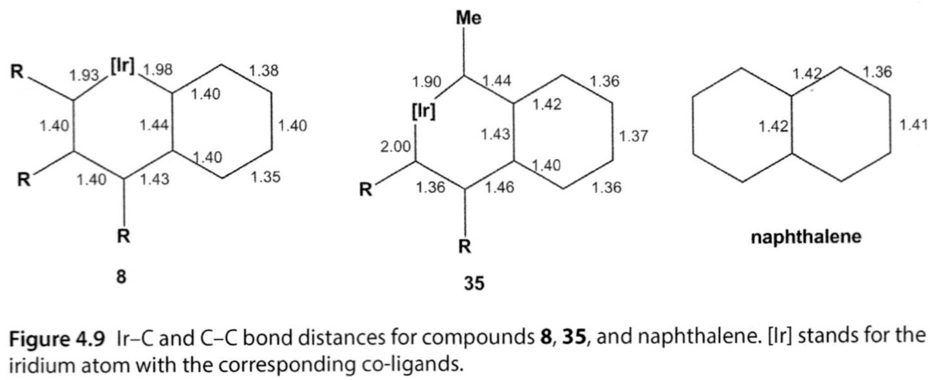

Figure 4.9 Ir–C and C–C bond distances for compounds **8**, **35**, and naphthalene. [Ir] stands for the iridium atom with the corresponding co-ligands.

All the deviations are at the highest limit of the values included in Table 4.1 for Tp′-iridabenzenes, in agreement with the crowded situation in these compounds, with fused-benzene rings and all other C atoms of the iridacycles substituted with different groups. Comparing the data for complex **7** in Table 4.1 and complex **8** in Table 4.2, the deviation of the metal from the C_5 plane is quite similar in both cases (0.74 Å for **7** and 0.76 Å for **8**), indicating that the substitution of two adjacent CO_2Me groups with a fused benzene does not make a big difference.

4.5 Spectroscopic Data for Iridaaromatics with Supporting Tris(pyrazolyl)borate Ligands

NMR data usually clearly reflect the structural characteristics of compounds, and for the case of metallaaromatics, deshielded chemical shifts are expected for the nuclei involved in the ring. Tables 4.3 and 4.4 collect the ^{1}H and ^{13}C NMR data for selected nuclei in these iridabenzenes and iridanaphthalenes, respectively. Figure 4.10 and Figure 4.11 show respectively the numbering scheme for the compounds included in each of these tables.

Table 4.3 Selected NMR chemical shifts (δ, ppm) for Tp′-iridabenzenes.

Compound number	7	12	25	26	27	28	32	33	52	52-Br	56	56-Br
C1	229.3	235.8	274.2	272.8	269.0	256.2	272.7	283.4	230.6	229.8	241.2	254.9
C2	143.0	135.7	140.1	140.3	141.3	137.8	139.5	139.9	131.5	130.7	130.0	135.2
C3	163.2	163.0	151.7	151.1	151.6	153.2	153.7	152.8	162.9	162.8	153.0	158.5
C4	143.0	134.3	131.1	130.8	131.6	132.2	132.0	131.0	131.5	130.7	127.0	131.4
C5	229.3	209.0	202.1	201.5	201.8	206.4	217.6	210.7	230.6	229.8	213.1	211.9
H1	—	—	—	—		—	—	—	14.67	14.55	—	—
H2	—	6.86	—	—		—	—	—	7.52	7.54	6.94	7.00
H3	—	—	—	—		—	9.15	8.95	8.24	8.27	8.06	8.08
H4	—	—	—	—		—	—	—	7.52	7.54	7.23	7.28
H5	—	14.01	—	—		—	—	—	14.67	14.55	14.52	13.56

Table 4.4 Selected NMR chemical shifts (δ, ppm) for Tp^{Me2}-iridanaphthalenes.

Compound number	8	35	36	37	38	39	49
C1	255.0	156.0	155.4	155.4	155.5	152.3	148.6
C2	141.8	133.7	134.2	134.9	134.2	133.9	138.1
C3	163.8	139.9	140.5	140.6	140.9	139.7	141.2
C4	137.0	153.9	152.2	154.4	151.1	142.7	153.6
C5	177.9	294.2	302.1	295.3	285.9	256.4	300.7

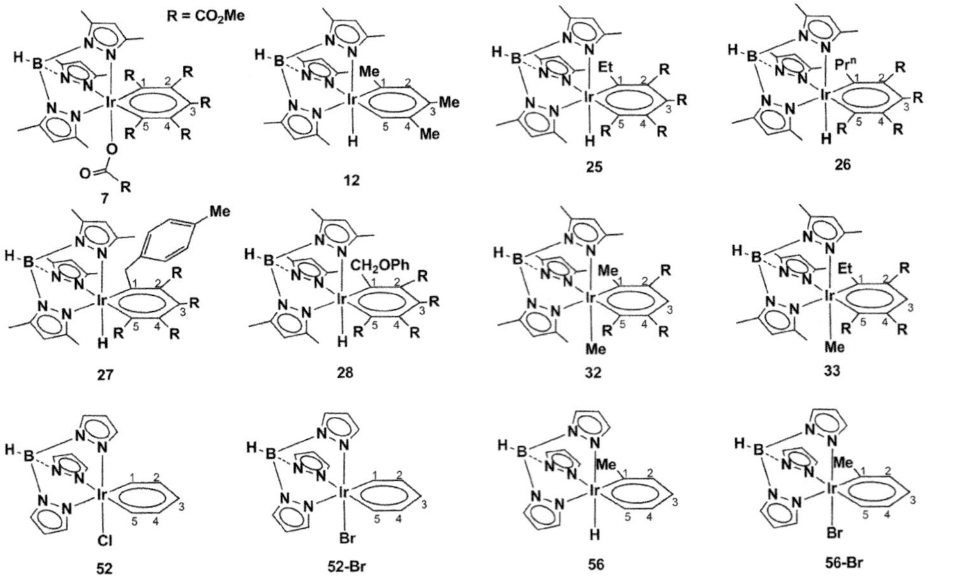

Figure 4.10 Numbering scheme for the compounds referred to in Table 4.3.

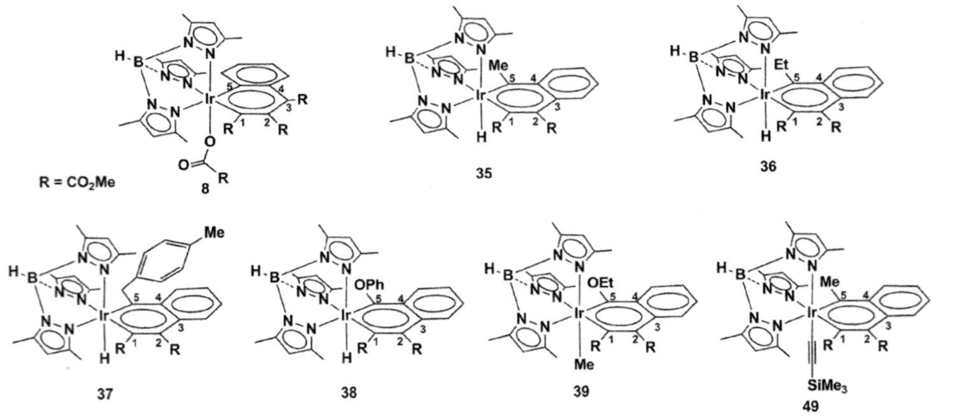

Figure 4.11 Numbering scheme for the compounds referred to in Table 4.4.

In the first case, all the Ir–C resonances (both C1 and C5) are observed at low fields (between 200 and 280 ppm). These values are intermediate between typical Ir–C single bonds and Ir=C double bonds of carbenes. This indicates that the two canonical forms considered for benzene are also participating here, with a larger contribution of one of them when the substituents in carbons 1 and 5 of the cycle have different electron donating properties. Thus, the highest values of chemical shifts for Ir–C nuclei corresponds to those with the more electron-donating groups (methyl, ethyl, and propyl groups for compounds **25**, **26**, **32**, and **33**) on the corresponding carbon atom.

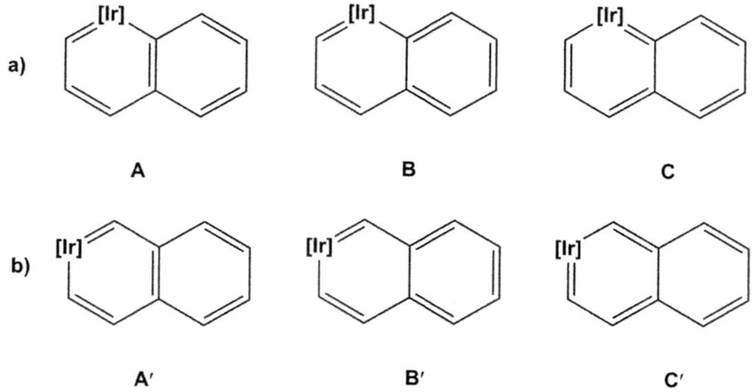

Figure 4.12 Resonance forms for α-iridanaphthalenes and β-iridanaphthalenes. [Ir] stands for the iridium atom with the corresponding co-ligands.

The ring carbon atoms not bonded to the iridium center (C2–C4) appear between 127 and 163 ppm, a typical range for aromatic compounds, with C3 being in the most down-field position.

^{1}H NMR data are also in accord with a partial double-bond character of the M–C units, and thus those protons attached to the metal-bound carbon atoms are detected at very low field values (~14 ppm), corresponding to the highest field chemical shift within the typical region for the proton in carbene functionalities (*ca.* 14–25 ppm). The resonances of the other ring protons have typical aromatic values (6.8–8.2 ppm), with the H3 resonance of **32** and **33** slightly low-field shifted to *ca.* 9 ppm.

For the case of the iridanaphthalenes, Table 4.4 shows very low-field resonances for one of the Ir–C moieties (C1 for **8** and C5 for **35–39** and **49**) in accord with a higher double-bond character of these moieties.

For **8**, the Ir–*C*(phenylic) resonance appears at 177.9 ppm, while for **35–39** and **49** the signals for Ir–*C*(R) appear in the 149–156 ppm range. These values are in accordance with a greater contribution of the resonance forms **A**/**B** and **A′**/**B′** (with the benzo ring fully aromatic) for α- and β-iridanaphthalenes, respectively, as previously commented (Figure 4.12).

4.6 Conclusions

Ligands of type tris(pyrazolyl)borate have proved to be good agents for the stabilization of iridaaromatics. The results summarized in this chapter show that a variety of precursors and reagents have allowed the isolation of iridabenzenes as well as two types of iridanaphthalenes (α- and β-, respectively). These compounds range from the parent TpIr(C_5H_5)(Cl) **52**, with no substituents on the iridabenzene carbons, to heavily substituted derivatives like compound $Tp^{Me2}Ir[C_5(CO_2Me)_5](OC(O)CO_2Me)$ **7**, with five CO_2Me substituents on the iridabenzene, through a variety of complexes with different combinations of functional groups and CH moieties on the iridaaromatic rings. Both electron-withdrawing and electron-donating substituents have proved useful for the stabilization of these iridaaromatics. The rich variety of types of reactions experienced by Tp′Ir derivatives and the *fac* disposition to which the co-ligands are constricted in these compounds are probably responsible for the isolation of this large variety of iridaaromatics.

Several different procedures have been used for the preparation of Tp′-iridaaromatics. Very importantly, the ability of these systems to generate carbene ligands not only by unusual olefin-to-alkylidene rearrangements but also by the fragmentation of halogenated solvents to produce a halocarbene Ir=C(H)(X) (X = Cl, Br) has turned out to be an excellent synthetic tool, provided these functionalities can be generated in the coordination sphere of iridacyclopentadienes. Furthermore, the studies carried out with an α-iridaindene enable the conclusion to be drawn that the presence of the benzo moiety in the iridacyclopentadiene ring favors the olefin-to-alkylidene isomerization in the complex and also drives insertion of the carbene toward the Ir–C(phenylic) bond. This has resulted in the isolation of the first examples of β-iridanaphthalenes of a transition element.

Structural and spectroscopic data of all the Tp′-iridaaromatics reported, as well as the reactions experienced by these derivatives, are in agreement with the delocalized nature of the metallacyclic rings and the aromatic character of the compounds. Finally, in agreement with theoretical studies reported in the literature, all these compounds show non-planarity of the metallaaromatic ring, even in the case of the parent compound **52**, which, while exhibiting the smallest deviation in this system, still lies in an unsymmetrical environment, due to the geometry of the Tp ligand.

References

1 Pettinari, C. (2008) Scorpionates II: *Chelating Borate Ligands*, London: Imperial College Press.
2 Alvarado, Y., Boutry, O., Gutiérrez, E., *et al.* (1997) Formation of Hydrido-η^3-Allyl Complexes of Ir(III) by Sequential Olefinic C–H Bond Activation and C–C Coupling of Alkenyl and Olefin Ligands. *Chemistry: A European Journal*, **3**(6): 860–873.
3 Paneque, M., Posadas, C. M., Poveda, M. L., *et al.* (2007) Reactivity of the Iridium(I) Alkene/Alkyne Complex $Tp^{Me_2}Ir(C_2H_4)(MeO_2CC{\equiv}CCO_2Me)$. *Organometallics*, **26**(13): 3120–3129.
4 Gutiérrez-Puebla, E., Monge, A., Paneque, M., *et al.* (1999) Activation of Aldehydes by the Ir-2,3-Dimethylbutadiene Complex $Tp^{Me_2}Ir(CH_2C(Me)C(Me)CH_2)$. *Journal of the American Chemical Society*, **121**(1): 248–249.
5 Alías, F. M., Daff, P. J., Paneque, M., *et al.* (2002) C–C Bond-Forming Reactions of Ir(III)-Alkenyls and Nitriles or Aldehydes: Generation of Reactive Hydride- and Alkyl-Alkylidene Compounds and Observation of a Reversible 1,2-H Shift in Stable Hydride–Ir(III) Alkylidene Complexes. *Chemistry: A European Journal*, **8**(22): 5132–5146.
6 Paneque, M., Poveda, M. L., Rendón, N., and Mereiter, K. (2004) Isolation of a Stable 1-Iridabicyclo[3.2.0]hepta-1,3,6-triene and Its Reversible Transformation into an Iridacycloheptatriene. *Journal of the American Chemical Society*, **126**(6): 1610–1611.
7 Roa, A. E., Salazar, V., López-Serrano, J., *et al.* (2012) Coupling of Aromatic Aldehydes with CO_2Me-Substituted $Tp^{Me_2}Ir(III)$ Metallacyclopentadienes. *Organometallics*, **31**(8): 3185–3198.
8 Schleyer P. R. (2001) Introduction: Aromaticity. *Chemical Reviews*, **101**(5): 1115–1118.

9 Álvarez, E., Gómez, M., Paneque, M., *et al.* (2003) Coupling of Internal Alkynes in TpMe2Ir Derivatives: Selective Oxidation of a Noncoordinated Double Bond of the Resulting Iridacycloheptatrienes. *Journal of the American Chemical Society*, **125**(6): 1478–1479.
10 Paneque, M., Posadas, C. M., Poveda, M. L., *et al.* (2007) Metallacycloheptatrienes of Iridium(III): Synthesis and Reactivity. *Organometallics*, **26**(14): 3403–3415.
11 Elliott, G. P., Roper, W. R., and Waters, J. M. (1982) Metallacyclohexatrienes or "metallabenzenes": Synthesis of osmabenzene derivatives and X-ray crystal structure of [$Os(CSCHCHCHCH)(CO)(PPh_3)_2$]. *Journal of the Chemical Society: Chemical Communications*, **14**: 811–813.
12 Bleeke, J. R., Xie, Y. F., Peng, W. J., and Chiang, M. (1989) Metallabenzene: Synthesis, structure, and spectroscopy of a 1-irida-3,5-dimethylbenzene complex. *Journal of the American Chemical Society*, **111**(11): 4118–4120.
13 Jacob, V., Weakley, T. J. R., and Haley, M. M. (2002) Metallabenzenes and Valence Isomers: Synthesis and Characterization of a Platinabenzene. *Angewandte Chemie: International Edition*, **41**(18): 3470–3473.
14 Gilbertson, R. D., Weakley, T. J. R., and Haley, M. M. (1999) Direct Synthesis of an Iridabenzene from a Nucleophilic 3-Vinyl-1-cyclopropene. *Journal of the American Chemical Society*, **121**(11): 2597–2598.
15 Thorn, D. L. and Hoffmann, R. (1979) Delocalization in metallocycles. *Nouveau Journal de Chimie*, **3**(1): 39–45.
16 Xia, H., He, G., Zhang, H., *et al.* (2004) Osmabenzenes from the Reactions of $HC{\equiv}CCH(OH)C{\equiv}CH$ with $OsX_2(PPh_3)_3$ (X = Cl, Br). *Journal of the American Chemical Society*, **126**(22): 6862–6863.
17 Zhang, H., Xia, H., He, G., *et al.* (2006) Synthesis and Characterization of Stable Ruthenabenzenes. *Angewandte Chemie: International Edition*, **45**(18): 2920–2923.
18 Talavera, M., Bolaño, S., Bravo, J., *et al.* (2013) Formation of Indanone from an Iridanaphthalene Complex. *Organometallics*, **32**(15): 4058–4060.
19 Liu, B., Xie, H., Wang, H., *et al.* (2009) Selective Synthesis of Osmanaphthalene and Osmanaphthalyne by Intramolecular C-H Activation. *Angewandte Chemie: International Edition*, **48**(30): 5461–5464.
20 O'Connor, J. M., Closson, A., and Gantzel, P. (2002) Hydrotris(pyrazolyl)borate Metallacycles: Conversion of a Late-Metal Metallacyclopentene to a Stable Metallacyclopentadiene–Alkene Complex. *Journal of the American Chemical Society*, **124**(11): 2434–2435.
21 Paneque, M., Poveda, M. L., Salazar, V., *et al.* (2000) Synthesis of $\eta^2{:}\sigma^2$-Diene Complexes of Iridium(III) by the Reaction of $\eta^4{:}\pi^2$-Diene Iridium(I) Species with Lewis Bases. *Organometallics*, **19**(16): 3120–3126.
22 Paneque, M., Posadas, C. M., Poveda, M. L., *et al.* (2007) Oxidative Coupling in the Reaction of $Tp^{Me_2}Ir$(2,3-dimethylbutadiene) with Diphenylacetylene. *Organometallics*, **26**(8): 1900–1906.
23 Paneque, M., Posadas, C. M., Poveda, M. L., *et al.* (2007) Investigations on the Coupling of Ethylene and Alkynes in [$IrTp^{Me_2}$] Compounds: Water as an Effective Trapping Agent. *Chemistry: A European Journal*, **13**(18): 5160–5172.
24 Ilg, K., Paneque, M., Poveda, M. L., *et al.* (2006) Vinylidene Compounds from the Reactions of $Me_3SiC{\equiv}CSiMe_3$ with $Tp^{Me_2}Ir$ Precursors. Protonation to Alkylidene and Iridabenzene Structures. *Organometallics*, **25**(9): 2230–2236.

25 Chin, C. S. and, Lee, H. (2004) New Iridacyclohexadienes and Iridabenzenes by [2+2+1] Cyclotrimerization of Alkynes and Facile Interconversion between Iridacyclohexadienes and Iridabenzenes. *Chemistry: A European Journal*, **10**(18): 4518–4522.

26 Clark, G. R., Johns, P. M., Roper, W. R., *et al.* (2011) Regioselective Mono-, Di-, and Trifunctionalization of Iridabenzofurans through Electrophilic Substitution Reactions. *Organometallics*, **30**(1): 129–138.

27 Clark, G. R., Johns, P. M., Roper, W. R., and Wright, L. J. (2006) An Osmabenzofuran from Reaction between Os(PhC≡CPh)(CS)$(PPh_3)_2$ and Methyl Propiolate and the C-Protonation of this Compound to Form a Tethered Osmabenzene. *Organometallics*, **25**(7): 1771–1777.

28 Vivancos, Á., Hernández, Y. A., Paneque, M., *et al.* (2015) Formation of β-Metallanaphthalenes by the Coupling of a Benzo-Iridacyclopentadiene with Olefins. *Organometallics*, **34**(1): 177–188.

29 Álvarez, E., Paneque, M., Poveda, M. L., and Rendón, N. (2006) Formation of Iridabenzenes by Coupling of Iridacyclopentadienes and Alkenes. *Angewandte Chemie: International Edition*, **45**(3): 474–477.

30 Paneque, M., Poveda, M. L., Rendón, N., *et al.* (2007) The Synthesis of Iridabenzenes by the Coupling of Iridacyclopentadienes and Olefins. *European Journal of Inorganic Chemistry*, **18**: 2711–2720.

31 Freundlich, J. S., Schrock, R. R., and Davis, W. M. (1996) Synthetic and Mechanistic Investigations of Trimethylsilyl-Substituted Triamidoamine Complexes of Tantalum that Contain Metal–Ligand Multiple Bonds. *Journal of the American Chemical Society*, **118**(15): 3643–3655.

32 Ozerov, O. V., Watson, L. A., Pink, M., and Caulton, K. G. (2003) A π-Basic Rhenium Center that Effects Cyclohexene Isomerization to a β-Agostic Carbene Ligand. *Journal of the American Chemical Society*, **125**(32): 9604–9605.

33 Kuznetsov, V. F., Abdur-Rashid, K., Lough, A. J., and Gusev, D. G. (2006) Carbene vs Olefin Products of C–H Activation on Ruthenium via Competing α- and β-H Elimination. *Journal of the American Chemical Society*, **128**(44): 14388–14396.

34 Castro-Rodrigo, R., Esteruelas, M. A., Fuertes, S., *et al.* (2009) Olefin–Alkylidene Equilibrium of 2-Vinylpyridine in Osmium- and Ruthenium-Hydrido-Tris(pyrazolyl) borate and Osmium-Cyclopentadienyl Complexes. *Organometallics*, **28**(20): 5941–5951.

35 Conejero, S., Paneque, M., Poveda, M. L., *et al.* (2010) C–H Bond Activation Reactions of Ethers that Generate Iridium Carbenes. *Accounts of Chemical Research*, **43**(4): 572–580.

36 Ferré, K., Poignant, G., Toupet, L., and Guerchais, V. (2001) Versatile behaviour of iron–arylcarbene complexes towards alkoxides: C–Cl and C–C bond activation reactions. *Journal of Organometallic Chemistry*, **629**(1–2): 19–27.

37 Albrecht, M. and van Koten, G. (2001) Platinum Group Organometallics Based on "Pincer" Complexes: Sensors, Switches, and Catalysts. *Angewandte Chemie: International Edition*, **40**(20): 3750–3781.

38 Rybtchinski, B. and Milstein, D. (1999) Metal Insertion into C–C Bonds in Solution. *Angewandte Chemie: International Edition*, **38**(7): 870–883.

39 Vivancos, Á., Vattier, F., López-Serrano, J., *et al.* (2015) A Diels–Alder Reaction Triggered by a [4 + 3] Metallacycloaddition. *Journal of the American Chemical Society*, **137**(12): 4074–4077.

40 Chin, C. S., Lee, H., and Eum, M.-S. (2005) Iridabenzenes from Iridacyclopentadienes: Unusual C–C Bond Formation between Unsaturated Hydrocarbyl Ligands. *Organometallics*, **24**(20): 4849–4852.
41 Paneque, M., Poveda, M. L., Rendón, N., and Mereiter, K. (2009) Reaction of the Iridacyclopentadiene $Tp^{Me_2}Ir(C(R)=C(R)C(R)=C(R))(H_2O)$ ($R = CO_2Me$) with Alkynes. *Organometallics*, **28**(1): 172–180.
42 Werner, H. (1994) Organometallic chemistry of alkenes and alkynes. *Journal of Organometallic Chemistry*, **475**(1–2): 45–55.
43 Werner, H., Baum, M., Schneider, D., and Windmueller, B. (1994) Reactions of Alkynylsilanes with $[RhCl(P^iPr_3)_2]$: The Synthesis of Four-Coordinate (Alkyne)-, Alkynyl-, and Vinylidenerhodium(I) and Five- and Six-Coordinate Alkynylhydridorhodium(III) Complexes from $RC{\equiv}CSiMe_3$, $Me_3SiC{\equiv}CSiMe_3$, and $HC{\equiv}CSiMe_3$. *Organometallics*, **13**(4): 1089–1097.
44 Werner, H., Lass, R. W., Gevert, O., and Wolf, J. (1997) Alkynes and Diynes as Precursors for Organoiridium Complexes Containing Iridium–Carbon Single and Double Bonds. *Organometallics*, **16**(19): 4077–4088.
45 Jiménez, M. V., Sola, E., Lahoz, F. J., and Oro, L. A. (2005) Reactions of Diamidonaphthalene-Bridged Diiridium Tetrahydrides with Alkynes: Hydrogenation, Vinylidene Formation, and Catalytic C–C Coupling. *Organometallics*, **24**(11): 2722–2729.
46 Bustelo, E., Carbó, J. J., Lledós, A., *et al.* (2003) First X-ray Characterization and Theoretical Study of π-Alkyne, Alkynyl-Hydride, and Vinylidene Isomers for the Same Transition Metal Fragment $[Cp^*Ru(PEt_3)_2]^+$. *Journal of the American Chemical Society*, **125**(11): 3311–3321.
47 De Angelis, F., Sgamellotti, A., and Re, N. (2004) Acetylene to vinylidene rearrangements on electron rich d^6 metal centers: A density functional study. *Dalton Transactions*, **20**: 3225–3230.
48 Singh, V. K., Bustelo, E., de los Ríos, I., *et al.* (2011) Internal Alkyne Isomerization to Vinylidene versus Stable π-Alkyne: Theoretical and Experimental Study on the Divergence of Analogous Cp*Ru and TpRu Systems. *Organometallics*, **30**(15): 4014–4031.
49 Ikeda, Y., Mutoh, Y., Imai, K., *et al.* (2013) Reactivities of Indenylruthenium Complex toward Internal Alkynes: Formation of Disubstituted Vinylidene Complexes and Indenyl–Alkyne Coupling. *Organometallics*, **32**(15): 4353–4358.
50 Vivancos-Ureña, Á. (2015) *Iridaciclos y Compuestos Iridaaromáticos: Síntesis y Reactividad.* Seville, Spain: Consejo Superior de Investigaciones Científicas and Universidad de Sevilla.
51 Vivancos, Á., Paneque, M., Poveda, M. L., and Álvarez, E. (2013) Building a Parent Iridabenzene Structure from Acetylene and Dichloromethane on an Iridium Center. *Angewandte Chemie: International Edition*, **52**(38): 10068–10071.
52 Lopez, J. A., Mereiter, K., Paneque, M., *et al.* (2006) Unusual fragmentation of CH_2Cl_2 by an Ir(III) centre bonded to a doubly metalated Tp^{Ms} Ligand (Tp^{Ms} = hydrotris(3-mesitylpyrazol-1-yl)borate). *Chemical Communications*, **37**: 3921–3923.
53 Lara, P., Paneque, M., Poveda, M. L., *et al.* (2006) Formation and Cleavage of C–H, C–C, and C–O Bonds of *ortho*-Methyl-Substituted Anisoles by Late Transition Metals. *Journal of the American Chemical Society*, **128**(11): 3512–3513.
54 Álvarez, E., Paneque, M., Petronilho, A. G., *et al.* (2007) Activation of Aliphatic Ethers by TpMe2Ir Compounds: Multiple C–H Bond Activation and C–C Bond Formation. *Organometallics*, **26**(5): 1231–1240.

55 Paneque, M., Poveda, M. L., Santos, L. L., *et al.* (2008) Generation of Metallacyclic Structures from the Reactions of Vinyl Ethers with a $Tp^{Me_2}Ir(III)$ Compound. *Organometallics*, **27**(23): 6353–6359.
56 Lara, P., Paneque, M., Poveda, M. L., *et al.* (2009) Experimental and Computational Studies on the Iridium Activation of Aliphatic and Aromatic C–H Bonds of Alkyl Aryl Ethers and Related Molecules. *Chemistry: A European Journal*, **15**(36): 9034–9045.
57 Santos, L. L., Mereiter, K., and Paneque, M. (2013) Reaction of 2-Methylanisole with $Tp^{Me_2}Ir(C_6H_5)_2(N_2)$: A Comprehensive Set of Activations. *Organometallics*, **32**(2): 565–569.
58 Rickard, C. E. F., Roper, W. R., Woodgate, S. D., and Wright, L. J. (2000) Electrophilic Aromatic Substitution Reactions of a Metallabenzene: Nitration and Halogenation of the Osmabenzene $[Os\{C(SMe)CHCHCHCH\}I(CO)(PPh_3)_2]$. *Angewandte Chemie: International Edition*, **39**(4): 750–752.
59 Clark, G. R., Johns, P. M., Roper, W. R., and Wright, L. J. (2008) A Stable Iridabenzene Formed from an Iridacyclopentadiene Where the Additional Ring-Carbon Atom Is Derived from a Thiocarbonyl Ligand. *Organometallics*, **27**(3): 451–454.
60 Clark, G. R., Ferguson, L. A., McIntosh, A. E., *et al.* (2010) Functionalization of Metallabenzenes through Nucleophilic Aromatic Substitution of Hydrogen. *Journal of the American Chemical Society*, **132**(38): 13443–13452.
61 Dalebrook, A. F. and Wright, L. J. (2012) Chapter Three: Metallabenzenes and metallabenzenoids. In: F. H. Anthony and J. F. Mark (eds), *Advances in Organometallic Chemistry*. New York: Academic Press: 93–177.
62 Bleeke, J. R., Xie, Y. F., Bass, L., and Chiang, M. Y. (1991) Metallacyclohexadiene and metallabenzene chemistry: 5: Chemical reactivity of metallabenzene. *Journal of the American Chemical Society*, **113**(12): 4703–4704.
63 Iron, M. A., Martin, J. M. L., and van der Boom, M. E. (2003) Metallabenzene versus Cp Complex Formation: A DFT Investigation. *Journal of the American Chemical Society*, **125**(43): 13020–13021.
64 Iron, M. A., Lucassen, A. C. B., Cohen, H., *et al.* (2004) A Computational Foray into the Formation and Reactivity of Metallabenzenes. *Journal of the American Chemical Society*, **126**(37): 11699–11710.
65 Clark, G. R., Lu, G.-L., Roper, W. R., and Wright, L. J. (2007) Stepwise Reactions of Acetylenes with Iridium Thiocarbonyl Complexes to Produce Isolable Iridacyclobutadienes and Conversion of These to either Cyclopentadienyliridium or Tethered Iridabenzene Complexes. *Organometallics*, **26**(9): 2167–2177.
66 Schrock, R. R., Pedersen, S. F., Churchill, M. R., and Ziller, J. W. (1984) Formation of cyclopentadienyl complexes from tungstenacyclobutadiene complexes and the x-ray crystal structure of an η^3-cyclopropenyl complex, $W[C(CMe_3)C(Me)C(Me)]Me_2NCH_2CH_2NMe_2)Cl_3$. *Organometallics*, **3**(10): 1574–1583.
67 Johns, P. M., Roper, W. R., Woodgate, S. D., and Wright, L. J. (2010) Thermal Rearrangement of Osmabenzenes to Osmium Cyclopentadienyl Complexes. *Organometallics*, **29**(21): 5358–5365.
68 Wu, H.-P., Lanza, S., Weakley, T. J. R., and Haley, M. M. (2002) Metallabenzenes and Valence Isomers: 3: Unexpected Rearrangement of Two Regioisomeric Iridabenzenes to an (η^5-Cyclopentadienyl)iridium(I) Complex. *Organometallics*, **21**(14): 2824–2826.
69 Wu, H.-P., Ess, D. H., Lanza, S., *et al.* (2007) Rearrangement of Iridabenzvalenes to Iridabenzenes and/or η^5-Cyclopentadienyliridium(I) Complexes: Experimental and

Computational Analysis of the Influence of Silyl Ring Substituents and Phosphine Ligands. *Organometallics*, **26**(16): 3957–3968.

70 Wu, H.-P., Weakley, T. J. R., and Haley, M. M. (2005) Regioselective Formation of β-Alkyl-α-phenyliridabenzenes via Unsymmetrical 3-Vinylcyclopropenes: Probing Steric and Electronic Influences by Varying the Alkyl Ring Substituent. *Chemistry: A European Journal*, **11**(4): 1191–1200.

71 Talavera, M., Bravo, J., Castro, J., *et al.* (2014) Electronic effects of substituents on the stability of the iridanaphthalene compound [IrCp*{=C(OMe)CH=C(*o*-C_6H_4)(Ph)}(PMe_3)]PF_6. *Dalton Transactions*, **43**(46): 17366–17374.

72 Carmona, E., Cingolani., A., Marchetti, F., *et al.* (2003) Synthesis and Structural Characterization of Mixed-Sandwich Complexes of Rhodium(III) and Iridium(III) with Cyclopentadienyl and Hydrotris(pyrazolyl)borate Ligands. *Organometallics*, **22**(14): 2820–2826.

73 Zhu, J., Jia, G., and Lin, Z. (2007) Understanding Nonplanarity in Metallabenzene Complexes. *Organometallics*, **26**(8): 1986–1995.

5

Chemistry of Metallabenzynes and Rhenabenzenes

Guochen Jia

Department of Chemistry, The Hong Kong University of Science and Technology, Clear Water Bay, Kowloon, Hong Kong, P. R. China

5.1 Introduction

In 1979, Thorn and Hoffman reported the first theoretically examined transition metal containing metallabenzenes, namely manganabenzene and rhodabenzenes [1]. In 1982, Roper and his workers reported the first isolated transition metal containing metallabenzenes, which are metallabenzenes of the group 8 metal osmium [2]. Since then, the chemistry of metallabenzenes has attracted considerable attention both experimentally and theoretically [3]. Remarkable progress has been made in the chemistry of metallabenzenes, especially in the areas of syntheses and properties of metallabenzenes of group 8–10 transition metals. We have recently started to explore the chemistry of metallabenzenes of middle and early transition metals. One of the themes of this chapter is to summarize our work in the development of chemistry of rhenabenzenes, metallabenzenes containing a group 7 metal.

Compounds closely related to metallabenzenes are metallabenzynes, metallacycles derived from formal replacement of a carbon atom or a CH group in benzyne (**1**) with an isolobal transition metal fragment [4, 5]. The formal replacement can in principle give three classes of metallabenzynes, depending on the location of the transition metal fragment in the six-membered ring as shown by the structures **2–4** in Scheme 5.1. Class I (**2**) contains a formal M≡C bond, and classes II (**3**) and III (**4**) contain a formal M=C bond. While metallabenzynes of classes II and III are still unknown, the first transition metal-containing metallabenzyne of class I, an osmabenzyne, was reported in 2001 [6]. Subsequently, a series of metallabenzynes of class I of osmium and rhenium have been isolated and well characterized. There has also been much interest in defining the properties of these interesting metallacycles. Another theme of this chapter is to summarize our progress in the development of the chemistry of metallabenzynes. Related works in this area from other labs are also briefly described.

Metallabenzenes: An Expert View, First Edition. Edited by L. James Wright.

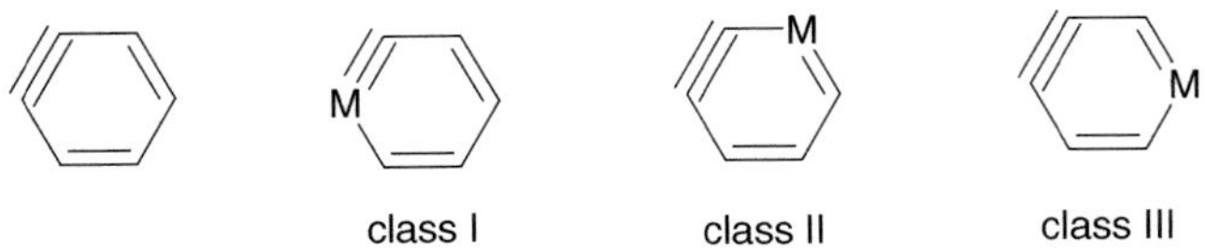

1 2 3 4

Scheme 5.1 Possible structures of metallabenzynes.

5.2 Chemistry of Metallabenzynes

5.2.1 Routes to Construct Metallabenzyne Rings

5.2.1.1 Reactions of Terminal Alkynes with $OsCl_2(PPh_3)_3$

The original route to make metallabenzynes involves the reactions of the osmium complex $OsCl_2(PPh_3)_3$ (**5**) with alkynes [6]. As shown in Scheme 5.2, the reaction of $OsCl_2(PPh_3)_3$ (**5**) with $HC{\equiv}CSiMe_3$ in wet benzene produced the osmabenzyne complex **6** along with other products including the trichloro, methylcarbyne complex **7** and the vinyl, vinylidene complexes **8** and **9**. This method has limited applicability as attempts to make osmabenzynes from reactions of $OsCl_2(PPh_3)_3$ (**5**) with other alkynes such as *t*-butylacetylene [7] and $HC{\equiv}CC(OH)Ph_2$ [8] were unsuccessful.

5.2.1.2 Reactions of Allenylcarbene and Vinylidene Complexes with Acetylides or Alkynes

A general method to prepare osmabenzynes is to use the reactions of allenylcarbene complexes with gold(I) acetylide complexes in the presence of $HNEt_3Cl$ [9]. For example, the reaction of the allenylcarbene complex **10** with $(PPh_3)AuC{\equiv}CPh$ and $HNEt_3Cl$ produced the osmabenzyne **13** (Scheme 5.3). The reaction may proceed by first a metathesis reaction of **10** with $(PPh_3)AuC{\equiv}CPh$ to give the alkynyl-allenylcarbene intermediate **11**, which undergoes a cycloaddition reaction to give the metallacyclic intermediate **12**, followed by protonation.

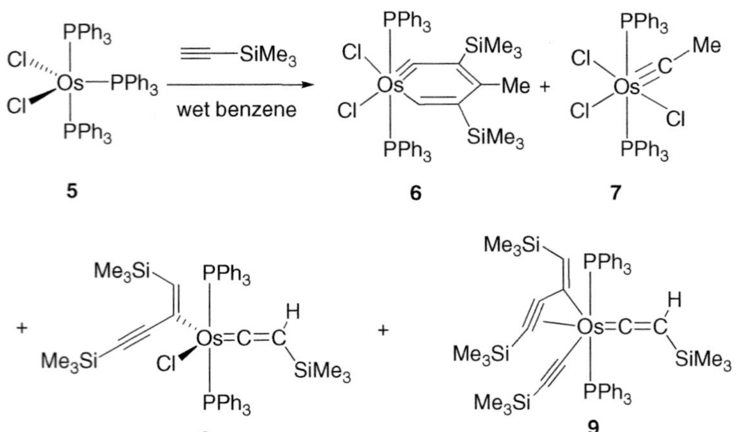

Scheme 5.2 Synthesis of the first metallabenzyne **6**.

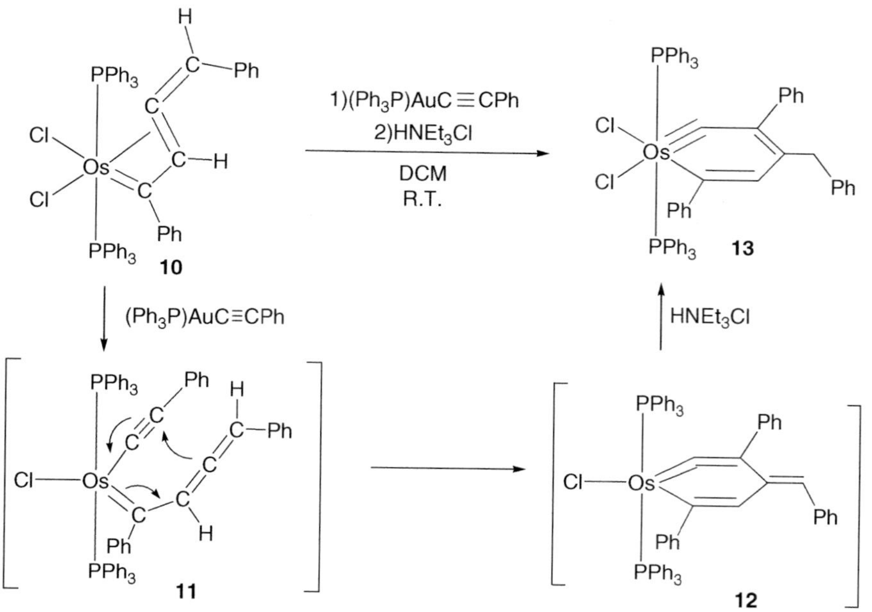

Scheme 5.3 Synthesis of the osmabenzyne **13** from the allenylcarbene complex **10**.

The method has been used to prepare other osmabenzynes. For instance, osmabenzynes with a tolyl (**14**), a trimethylsilyl (**15**), and an *n*-butyl (**16**) group can be similarly prepared by reacting the allenylcarbene complex **10** with the corresponding gold acetylide complexes in the presence of $HNEt_3Cl$ (Scheme 5.4). The method is limited by the availability of allenylcarbene complexes.

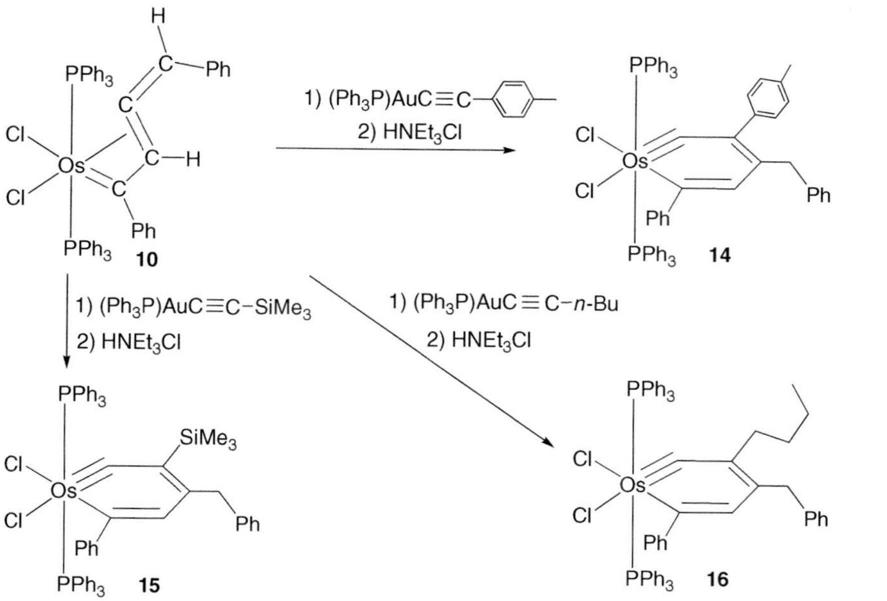

Scheme 5.4 Synthesis of osmabenzynes **14–16** from the allenylcarbene complex **10**.

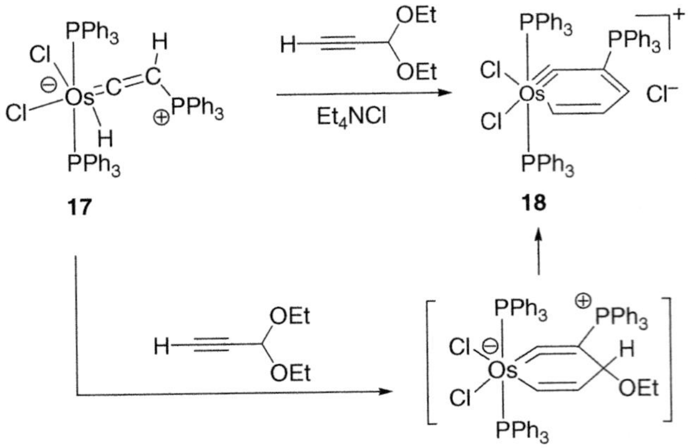

17

18

19

Scheme 5.5 Synthesis of the osmabenzyne **18**.

In related work, Xia *et al.* discovered that the PPh_3-substituted osmabenzyne complex **18** could be made from the reaction of $HC{\equiv}CCH(OEt)_2$ with the osmium hydrido, vinylidene complex **17** (Scheme 5.5) [10]. The reaction was proposed to proceed through the isoosmabenzene intermediate **19**.

5.2.1.3 Oxidative Addition Reactions of Vinyl Carbyne Complexes

Another general route to make osmabenzynes involves oxidative addition reactions of properly functionalized low-valent osmium vinylcarbyne complexes. The method was firstly employed to prepare the osmanaphthalyne **22** (Scheme 5.6) [11], which was obtained by the reaction of the carbyne complex **20** with zinc in THF. The reaction also produced the indenyl complex **25** as a side product. The complex **22** is presumably formed by initial reduction of **20** with zinc to give the $16e^-$ vinylcarbyne complex **21**, followed by a C–Cl oxidative addition reaction. The complex **25** was presumably formed by a C–H oxidative addition reaction of **21** to give the hydrido-osmanaphthalyne intermediate **23** followed by two consecutive migratory insertion reactions via the osmanaphthalene intermediate **24**.

The method has been used to prepare other osmabenzynes. For instance, the *tert*-butyl-substituted osmabenzynes **28** and **29** were obtained from the reactions of zinc with the carbyne complexes **26** and **27** respectively (Scheme 5.7); the adamantyl-substituted osmabenzynes **32** and **33** were obtained from the reactions of zinc with the carbyne complexes **30** and **31** respectively [12].

As an additional example of preparation of osmabenzynes from osmium vinylcarbyne complexes, Xia *et al.* found that the osmanaphthalyne **35** can be prepared by refluxing the hydrido, carbyne complex **34** in $ClCH_2CH_2Cl$ under an O_2 atmosphere for 15 h or by heating a solid sample of **34** in air at 120 °C for 5 h (Scheme 5.8), presumably through the hydrido-osmanaphthalene intermediate **36** formed by C–H activation of **34** [13]. The osmanaphthalyne **35** could also be generated by refluxing the binuclear osmanaphthalene **37** in $ClCH_2CH_2Cl$ under an O_2 atmosphere in the presence of HCl (1 equiv.), HBF_4 (1 equiv.) and excess PPh_3.

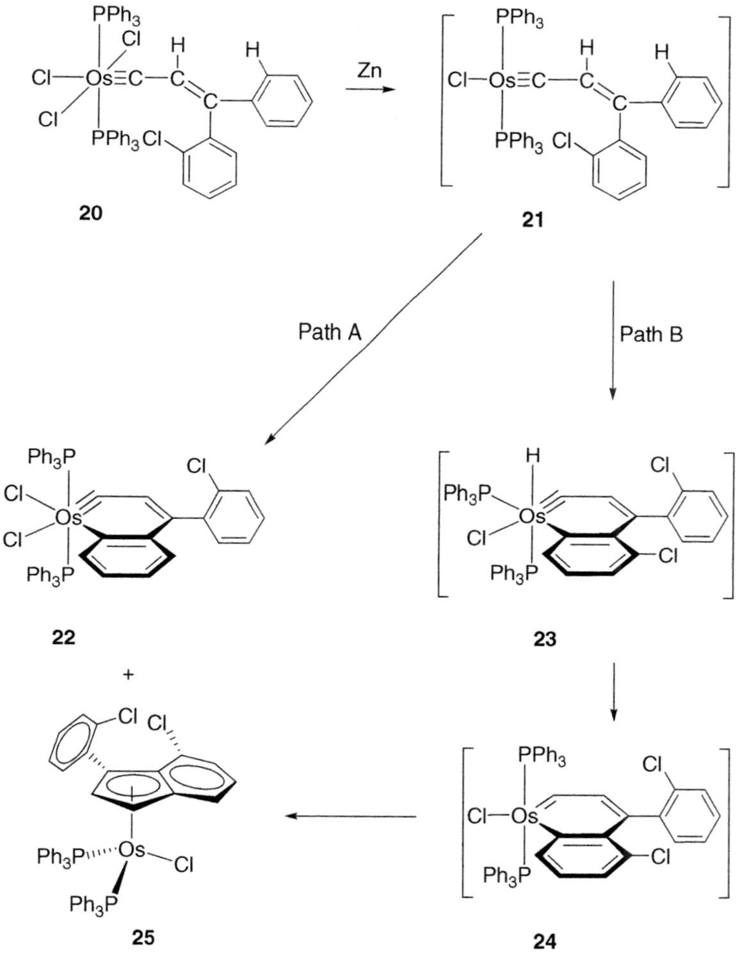

Scheme 5.6 Synthesis of the osmanaphthalyne **22** via zinc reduction of the vinylcarbyne complex **20**.

5.2.1.4 Isomerization of Alkyne-Carbene Complexes

This method has been used for the synthesis rhenabenzynes [14]. For example, the alkyne-carbene complex (**41**), which can be prepared from the vinylcarbyne complex Re{≡CCH=C(CMe$_3$)C≡CSiMe$_3$}Cl$_2$(PMe$_2$Ph)$_3$ (**38**), slowly isomerizes to the *t*-butyl substituted rhenabenzyne **42** (Scheme 5.9). Similarly, the analogous rhenabenzynes **43** (with an isopropyl group) and **44** (with an adamantyl group) were obtained starting from the carbyne complexes Re{≡CCH=C(R)C≡CSiMe$_3$}Cl$_2$(PMe$_2$Ph)$_3$ (R = isopropyl, and adamantly).

5.2.2 Chemical Properties of Metallabenzynes

5.2.2.1 Reactions Not Involving Carbons of the Metallabenzyne Ring

Metallabenzynes can undergo ligand substitution reactions. For example, the two PPh$_3$ ligands of the osmabenzyne **6** can be replaced by PCy$_3$ to give the PCy$_3$-containing

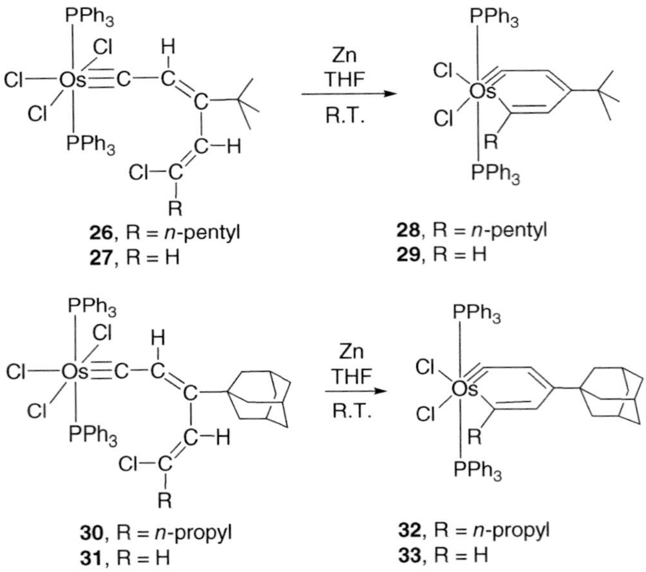

Scheme 5.7 Synthesis of osmabenzynes via zinc reduction of osmium vinylcarbyne complexes.

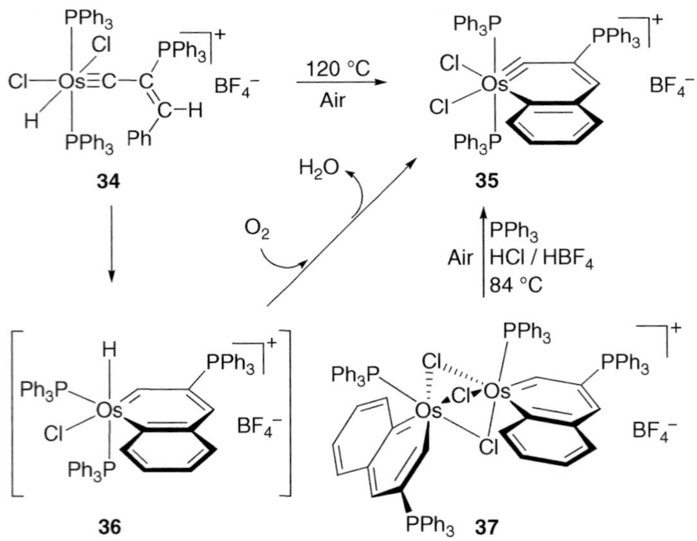

Scheme 5.8 Synthesis of osmanaphthalyne **35** via oxidation of **34**.

osmabenzyne **45** (Scheme 5.10) [15], and the two chloride ligands of the osmabenzyne **6** can be replaced by bipyridine to give the dicationic osmabenzyne **46** [16]. When the osmabenzyne **6** was treated with HBF_4 in the presence of water, the monocationic osmabenzyne **47** was obtained [17]. Similarly, the osmabenzyne **48** reacted with HBF_4 in the presence of water to give the osmabenzyne **49**, which can be converted back to **48** upon treatment with NaCl [17].

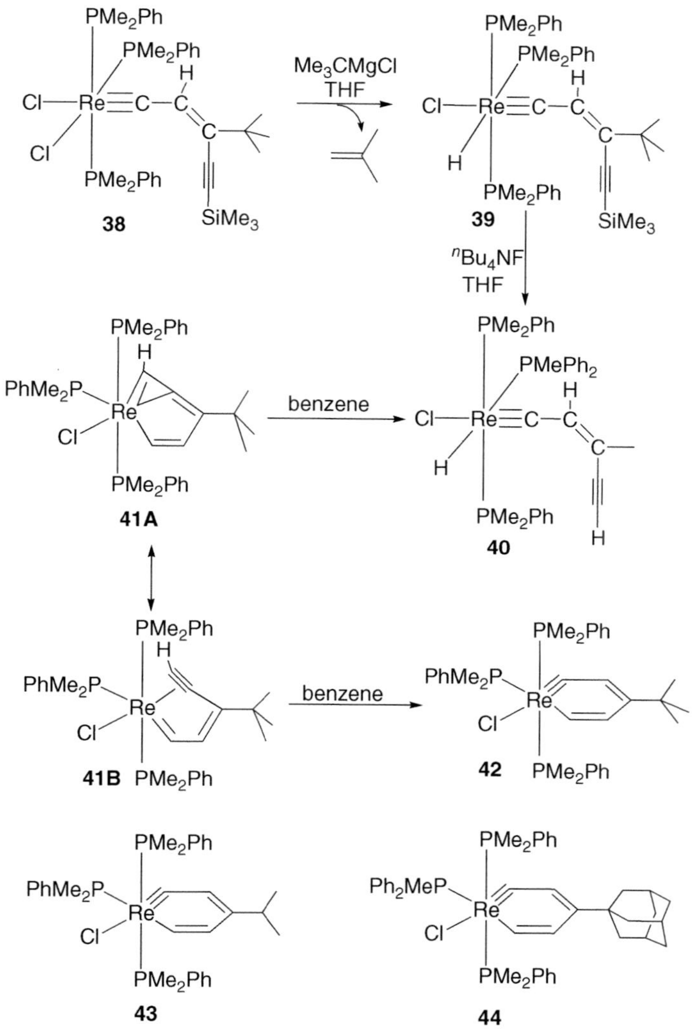

Scheme 5.9 Synthesis of rhenabenzyne complexes.

5.2.2.2 Electrophilic Substitution Reactions

Metallabenzynes can undergo aromatic electrophilic substitution reactions at the carbons *meta* to the metal. For example, the silyl-substituted osmabenzyne **6** reacted with a variety of electrophilic reagents, such as HBF_4, Br_2, $NOBF_4$, NO_2BF_4, and HCl/H_2O_2 to give new osmabenzyne derivatives (Scheme 5.11) [17, 18].

The osmabenzyne complex Os(≡CCH=C(CH$_3$)CH=CH)Cl$_2$(PPh$_3$)$_2$ (**48**), which has no trimethylsilyl groups, can also undergo electrophilic substitution reactions, including H/D exchange with DOTf, nitration, bromination, and chlorination at the carbons *meta* to the metal (Scheme 5.12) [17, 18]. The results imply that the regioselectivity of the electrophilic substitution reactions is not related to the C–Si bond. Theoretical studies indicate that the regioselectivity of the electrophilic substitution reactions is controlled by the HOMO (see discussion in Section 5.2.5).

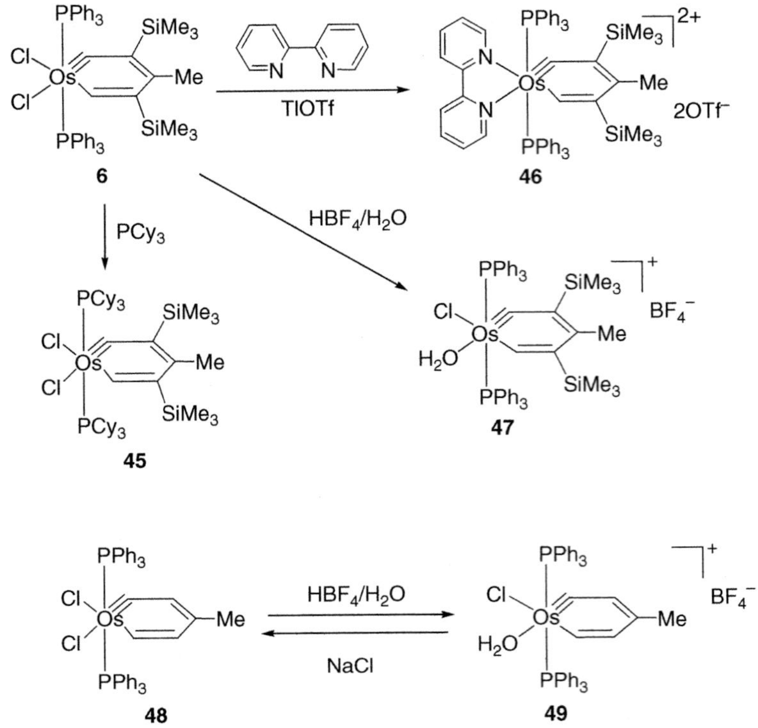

Scheme 5.10 Ligand substitution reactions of metallabenzyne complexes.

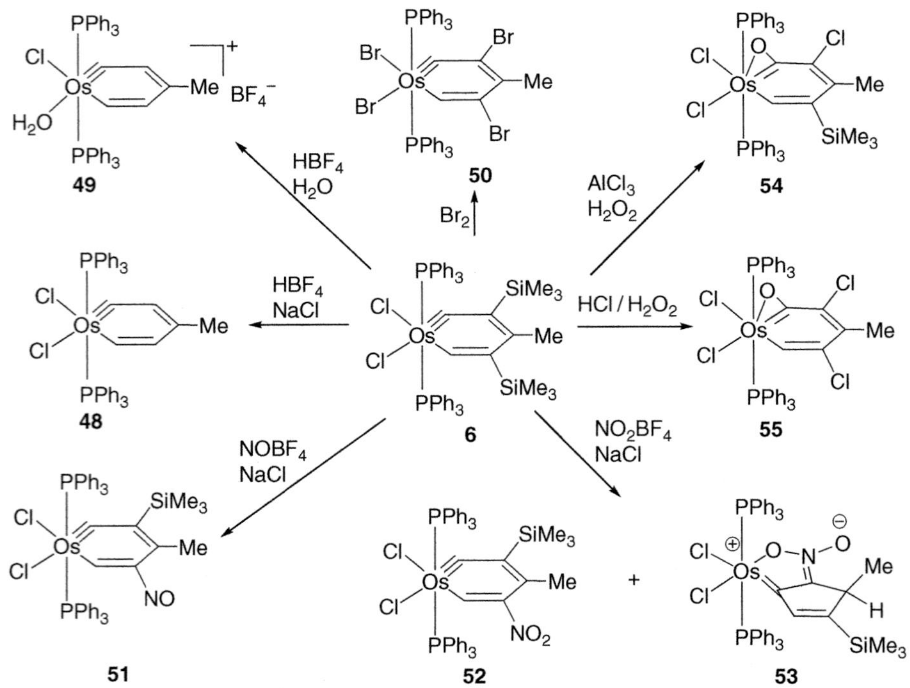

Scheme 5.11 Electrophilic substitution reactions of the osmabenzyne complex **6**.

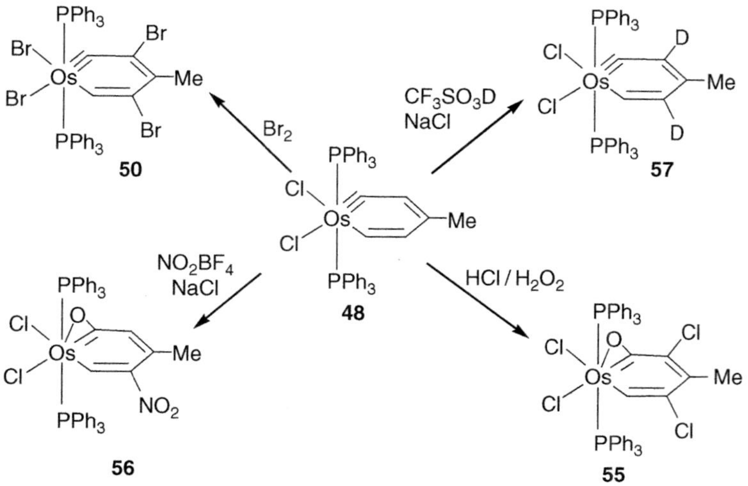

Scheme 5.12 Electrophilic substitution reactions of the osmabenzyne complex **48**.

5.2.2.3 Nucleophilic Addition Reactions

Nucleophiles can add to metallabenzynes. The reactivity and regioselectivity of these reactions are dependent on the ligands as well as the substituents on the metallacycle.

The neutral osmabenzyne $Os(\equiv CC(SiMe_3)=C(CH_3)C(SiMe_3)=CH)Cl_2(PPh_3)_2$ (**6**) is unreactive toward nucleophiles such as $NaBH_4$, amines, and phosphines. On the other hand, the dicationic osmabenzyne $[Os(\equiv CC(SiMe_3)=C(CH_3)C(SiMe_3)=CH)(bipy)(PPh_3)_2][OTf]_2$ (**46**) readily undergoes nucleophilic addition reactions with nucleophiles regioselectively at the carbyne carbon (C1) to give osmabenzenes. For example, complex **46** reacts with methanol in the presence of K_2CO_3 to give the OMe-substituted osmabenzene **58** (Scheme 5.13) [16]. Even water can react with **48** to give the OH-substituted osmabenzene **59**, which is hydrolytically unstable and reacts further with water to give the carbonyl complex **60**. When the osmabenzyne **46** was treated with $NaBH_4$, the cyclopentadienyl complex **62** was produced, as a result of migratory insertion reaction of the metallabenzene intermediate **61**. The regioselectivity of the nucleophilic addition reaction appears to be LUMO-controlled (see discussion in Section 5.2.5).

Nucleophiles could also add to the C3 carbon (*para* to metal) of the metallacycle. For example, Xia *et al.* reported that treatment of the osmabenzyne complex **18** with EtLi and NaSMe produced the isoosmabenzenes **63** and **64**, respectively (Scheme 5.14) [10]. When primary amines $NH_2(n\text{-Bu})$ and $NH_2CH_2C\equiv CH$ were used, the reactions produced the corresponding ring-opened products **65** and **66**. It was suggested that the regioselectivity is related to both electronic and steric effects.

5.2.2.4 Migratory Insertion Reactions

Like hydrido or alkyl, carbyne complexes $L_nM(R)\equiv CR'$ [19], metallabenzynes could undergo migratory insertion reactions to give carbene complexes involving the two α-carbon atoms of the metallabenzyne ring. Experimental and theoretical studies suggest that the thermodynamics and kinetics of the rearrangement reactions are strongly affected by the metal, the ligands, and the substituent on the metallacycle.

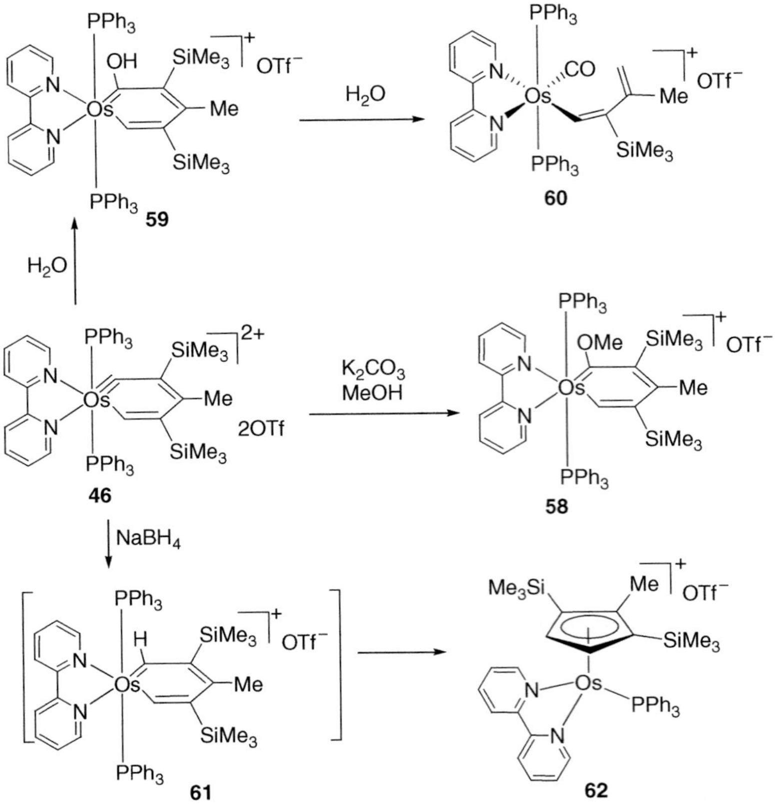

Scheme 5.13 Nucleophilic addition reactions of the osmabenzyne **46**.

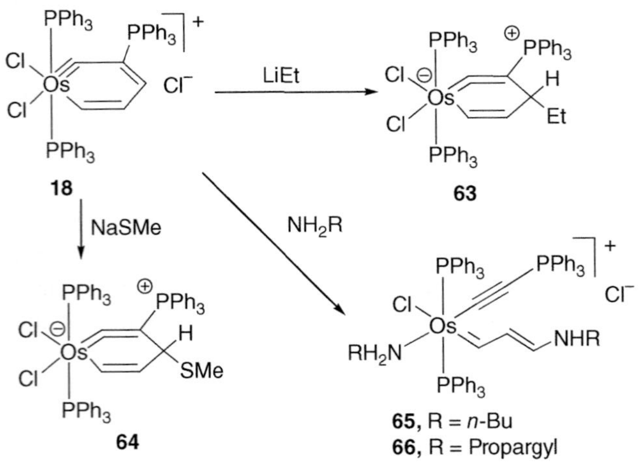

Scheme 5.14 Nucleophilic addition reactions of osmabenzyne **18**.

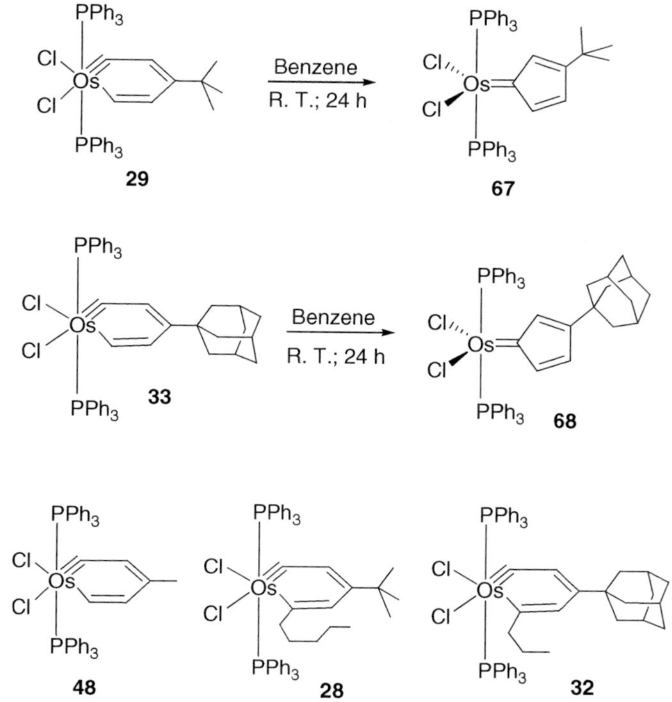

Osmabenzynes that are thermally stable at room temperature.

Scheme 5.15 Substituent effect on the conversion of osmabenzynes to carbene complexes.

The effect of the substituent on the metallacycle on the rearrangement reaction is illustrated by the following experimental observations. The *t*-butyl-containing osmabenzyne **29** and the adamantyl-containing osmabenzyne **33**, which can be generated *in situ* and characterized spectroscopically, undergo migratory insertion reactions at room temperature to form the carbene complexes **67** and **68**, respectively (Scheme 5.15) [12]. In contrast, samples of the related osmabenzynes Os{≡C–CH=CMeCH=CH}$Cl_2(PPh_3)_2$ (**48**), Os{≡C–CH=C(CMe_3)CH=CH(*n*-pentyl)}$Cl_2(PPh_3)_2$ (**28**) and Os{≡C–CH=C(1-adamantyl)CH=CH(*n*-propyl)}$Cl_2(PPh_3)_2$ (**32**) are stable and their solid samples can be stored for months without changes.

Theoretical studies suggest that the difference in the reactivity of these osmabenzynes is mainly of thermodynamic origin [12]. As illustrated in Figure 5.1 by the energy profiles calculated for the rearrangement of metallabenzynes **28**, **29**, and **48** to the corresponding carbene complexes, all the reactions have a similar activation barrier (in the range of 34.7–37.1 kcal/mol). However, the conversion of **29** to the carbene complex **67** is thermodynamically favored by 0.6 kcal/mol, while similar rearrangement reactions of complexes **48** and **28** to give the corresponding carbene complexes are thermodynamically disfavored by 1.5 and 3.0 kcal/mol, respectively, in agreement with experimental observations.

The substituent effect can be explained in terms of steric effects. As shown in Figure 5.2, the optimized structure of the osmabenzyne **29** has short nonbonding H...H

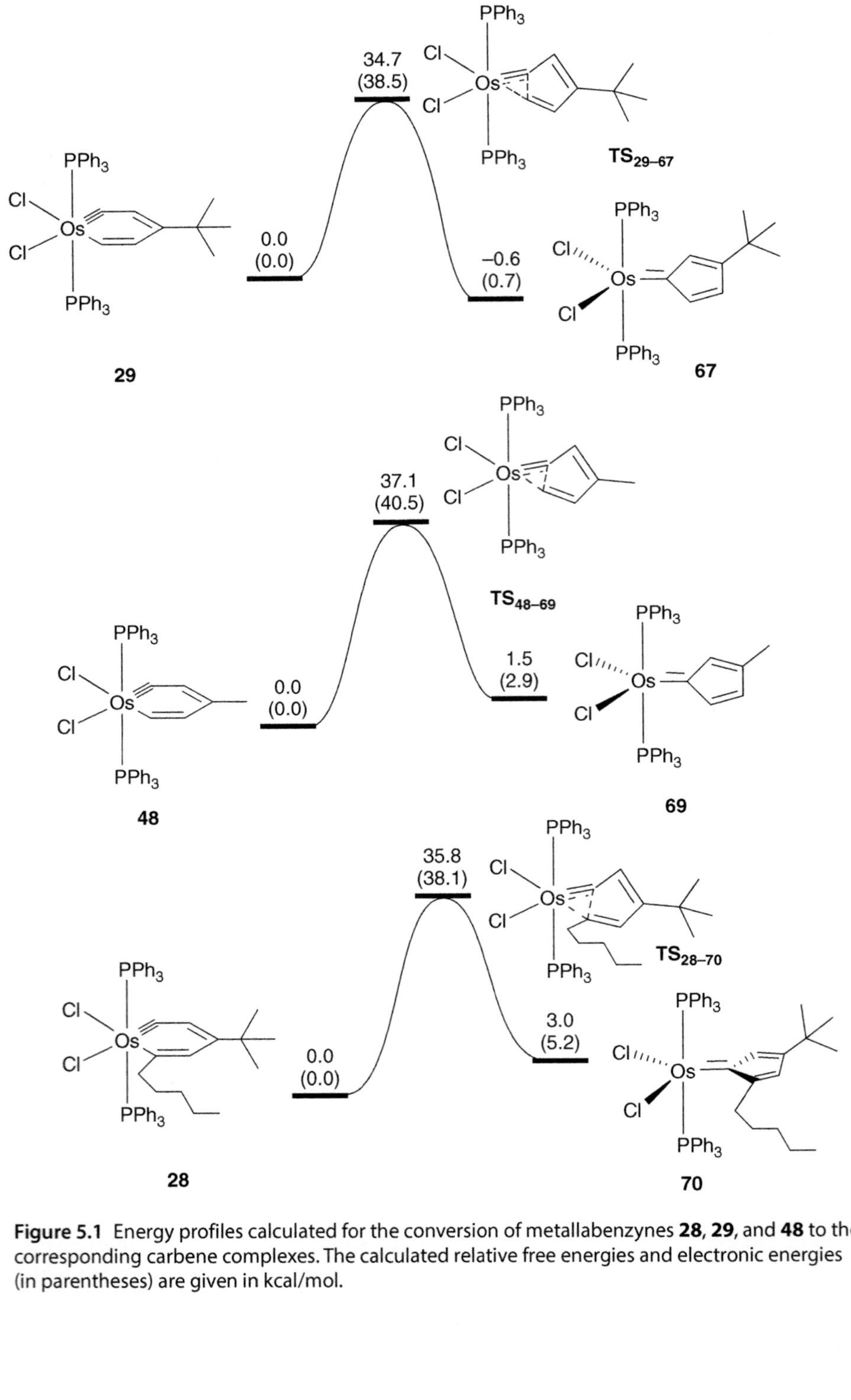

Figure 5.1 Energy profiles calculated for the conversion of metallabenzynes **28**, **29**, and **48** to the corresponding carbene complexes. The calculated relative free energies and electronic energies (in parentheses) are given in kcal/mol.

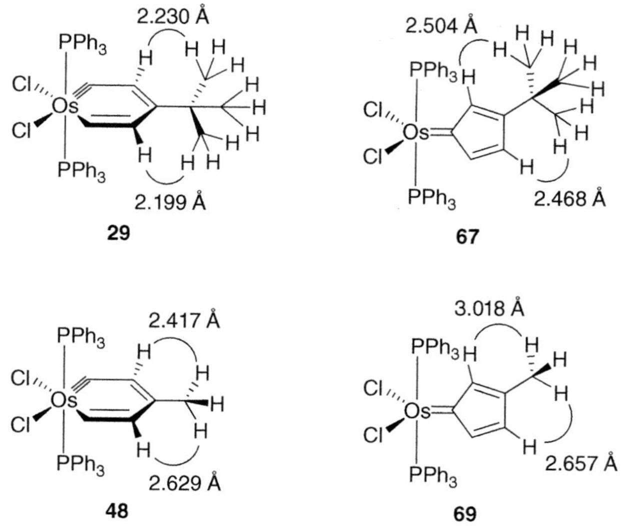

Figure 5.2 Short H...H contacts (in Å) in optimized structures of **29**, **67**, **48**, and **69**.

contacts of 2.199 and 2.230 Å involving the *t*-butyl group and H atoms of the metallacycle at the carbons *meta* to the metal. The distances are shorter than the sum of the van der Waals radius of two hydrogen atoms (2.40 Å), implying the presence of steric repulsive interaction in the osmabenzyne **29**. On the other hand, the shortest nonbonding H...H distances in the structures of the carbene complex **67** (the rearranged product of **29**), the osmabenzyne **48**, and the carbene complex **69** (the rearranged product of **48**) are all longer than the sum of the van der Waals radius of two hydrogen atoms (2.40 Å), implying the lack of significant steric repulsive interaction in these complexes. Therefore, the conversion of the osmabenzyne **29** to the carbene complex **67** is thermodynamically more favorable than the conversion of osmabenzyne **48** to the carbene complex **69**, as conversion of **29** to the carbene complex **67** will release the steric congestion. The conversion of the osmabenzyne **28** to the carbene complex **70** is not favored because the carbene product **70** is forced to adopt an electronically unfavorable conformation to avoid the severe steric repulsive interaction of the *n*-pentyl group with the PPh_3 ligands.

The conversion of the osmanaphthalyne **71** to the indenylidene complex **72** is thermally more favorable than the analogous reactions of osmabenzynes **28**, **29**, and **48**, as revealed by the computational work of Zhu and his co-workers [20]. They found that the reaction has a free energy change of −4.8 kcal/mol with an activation barrier (34.4 kcal/mol) close to those of the osmabenzyne complexes **28**, **29**, and **48** (34.7–37.1 kcal/mol) (Scheme 5.16).

The substituent effect on the rearrangement reactions of osmanaphthalynes with substituents at different positions to give the corresponding indenylidene complexes has been reported by Zhu and his co-workers [20]. They found that a substituent at the C5 or C6 positions has no or minimum effect on the relative stability of the osmanaphthalyne and indenylidene isomers. On the other hand, the relative stability of the two isomers can be strongly influenced by a substituent at the C2, C3, C4, or C7 positions.

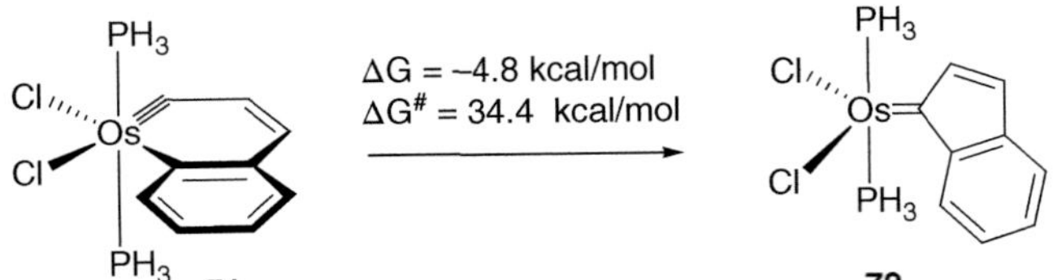

71 **72**

Scheme 5.16 The reaction free energy change and barrier calculated for the rearrangement of the osmanaphthalyne **71** to the indenylidene complex **72**.

The substituent effect is especially pronounced when the substituent is attached to the C2 (in metallabenzynes **73R**) or C3 carbons (in metallabenzynes **75R**), as shown in Figure 5.3. In general, with respect to the conversion of osmanaphthalynes to the corresponding indenylidene isomers, a sterically demanding group will have a stabilizing effect if it is attached to C2 and a destabilizing effect if it is attached to C3; an electron-donating group will have a destabilizing effect if it is attached to C2 and a stabilizing effect if it is attached to C3, while an electron-withdrawing group has the opposite effect. The substituent could also affect the kinetic barrier for the conversion of osmanaphthalynes to indenylidene complexes, although the effect is not as large as that for the thermodynamics. Similar substituent effects were observed for substituted ruthenanaphthalynes.

The ligand effect on the migratory insertion reactions involving the two α-carbon atoms of the metallabenzyne ring can be illustrated by the following experimental

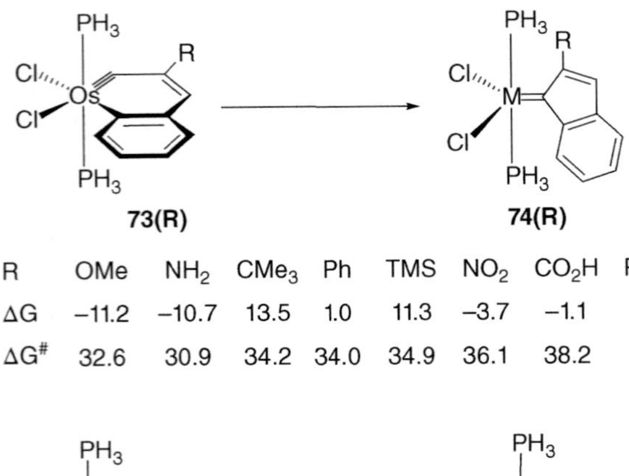

73(R) **74(R)**

R	OMe	NH_2	CMe_3	Ph	TMS	NO_2	CO_2H	PMe_3^+
ΔG	−11.2	−10.7	13.5	1.0	11.3	−3.7	−1.1	−2.3
$\Delta G^\#$	32.6	30.9	34.2	34.0	34.9	36.1	38.2	31.3

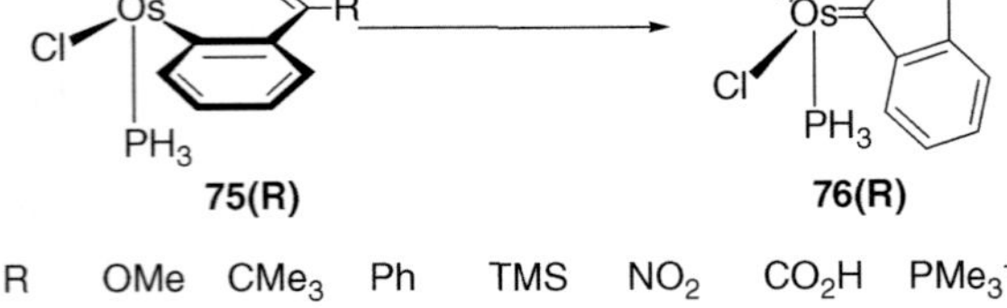

75(R) **76(R)**

R	OMe	CMe_3	Ph	TMS	NO_2	CO_2H	PMe_3^+
ΔG	−0.3	−17.0	−9.9	−12.6	−13.3	−14.8	−23.3

Figure 5.3 Substituent effect on the free energy changes and barriers (in kcal/mol) for the conversion of substituted osmanaphthalynes to the corresponding indenylidene complexes.

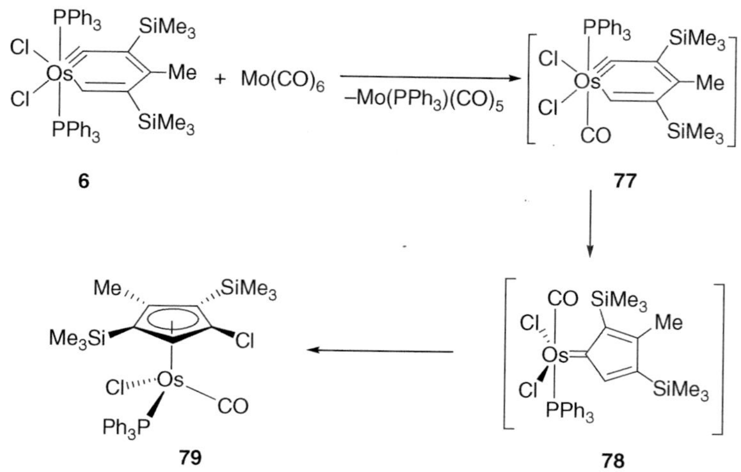

Scheme 5.17 Reaction of complex **6** with $Mo(CO)_6$.

observation [15]. The complex $Os(\equiv CC(SiMe_3)=C(CH_3)C(SiMe_3)=CH)Cl_2(PPh_3)_2$ (**6**) is a thermally stable species that can be stored at room temperature for months without appreciable change. Interestingly, the complex **6** reacted with $Mo(CO)_6$ to produce the chlorocyclopentadienyl complex **79** (Scheme 5.17). The reaction presumably proceeds by initial formation of the CO-containing osmabenzyne complex $Os(\equiv CC(SiMe_3)=C(CH_3)C(SiMe_3)=CH)Cl_2(CO)(PPh_3)$ (77), which rearranges first to the carbene intermediate **78** and then to the chlorocyclopentadienyl complex **79**. The experimental observation implies that π-accepting ligands such as CO can facilitate the migratory insertion reactions of osmabenzyne complexes to form carbene or chlorocyclopentadienyl complexes.

The ligand effect on the relative stability of metallabenzynes $Os(\equiv C–CH=CHCH=CH)Cl_2(L)_2$ (**80**) and the corresponding isomeric carbene complexes $Os\{=C(–CH=CHCH=CH–)\}Cl_2(L)_2$ (**81**) as well as the chlorocyclopentadienyl complexes $Os(\eta^5\text{-}C_5ClH_4)Cl(L)_2$ (**82**) is illustrated in Figure 5.4. For complexes supported with NMe_3 (**a**), PPh_3 (**b**), and PMe_3 (**c**) ligands, the chlorocyclopentadienyl isomer (**82**) is least stable, while the osmabenzyne isomer (**80**) is most stable. Complexes supported with π-accepting ligands such as PCl_3 (**d**) and CO (**e**) have the opposite trend. For complexes supported with pyridine (**f**), the carbene isomer (**81**) is most stable, while the chlorocyclopentadienyl isomer (**82**) is least stable. Complexes supported with the metal fragment $OsCl_2(CO)(PR_3)$ ($PR_3 = PPh_3, PMe_3$) have the same trend in relative stability of the three isomers as those of analogous complexes supported with the metal fragment $OsCl_2(CO)_2$.

A similar ligand effect was revealed by Zhu and his co-workers for the rearrangement reactions of osmanaphthalynes to give the corresponding indenylidene complexes (Figure 5.5) [20]. With respect to the conversion to the corresponding indenylidene isomers, the metallanaphthalynes are stabilized by electron-donating ligands such as PMe_3, and destabilized by π-accepting ligands, such as CO and CH_3CN.

The metal effect on migratory insertion reactions involving the two α-carbon atoms of the metallabenzyne ring have been confirmed by computational studies. Parameswaran and his co-workers studied the conversion of the metallabenzynes

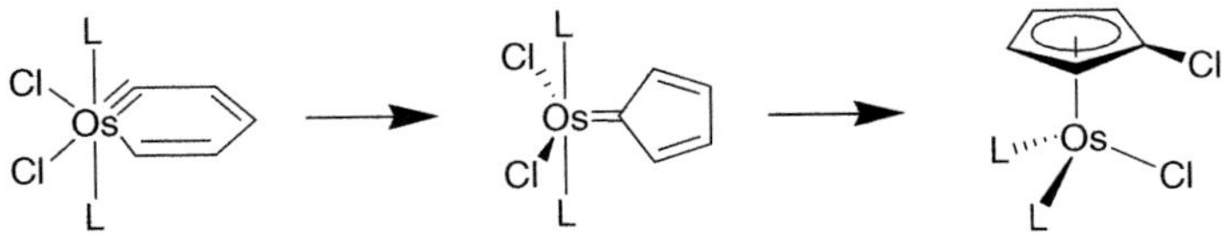

	80	81	82
a, L = NMe_3		6.1(7.9)	23.3(23.2)
b, L = PPh_3		1.9(1.9)	8.5(4.9)
c, L = PMe_3		8.3(9.0)	12.7(11.8)
d, L = PCl_3		−6.0(−5.8)	−21.6(−23.7)
e, L = CO		−14.7(−14.4)	−32.7(−35.0)
f, L = Pyridine		−2.5(−2.4)	12.2(12.0)

Figure 5.4 Ligand effect on the relative free energies and energies (in parenthesis) of osmabenzynes Os(≡CH=CHCH=CH)Cl_2(L)$_2$ (**80**, 0.0 kcal/mol) and their corresponding carbene (**81**) and chlorocyclopentadienyl (**82**) complexes.

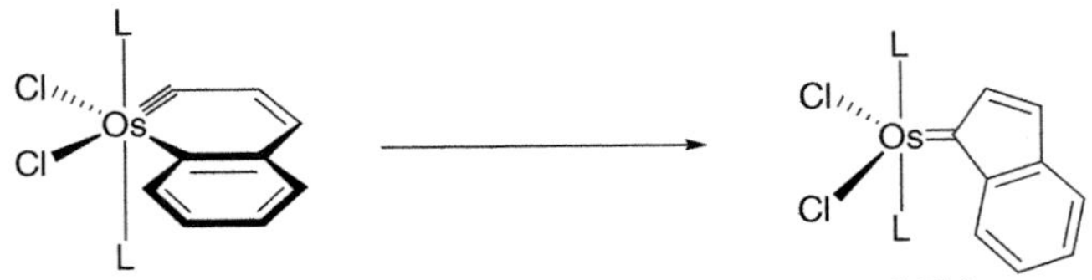

83(L) → **84(L)**

L	CO	NCMe	PMe_3
ΔG	−19.9	−17.3	−1.0

Figure 5.5 Ligand effect on the free energy changes (in kcal/mol) for the conversion of osmanaphthalynes (**83(L)**) to their corresponding carbene isomers (**84(L)**).

M(≡C–CH=CHCH=CH)Cl_2(L)$_2$ (**85(M)**, M = Fe, Ru, and Os) to the corresponding carbene isomers (Figure 5.6) [21]. They found that the iron-containing metallabenzyne **85(Fe)** is less stable than the carbene isomer **86(Fe)** by 19.1 kcal/mol, and the ruthenium-containing metallabenzyne **85(Ru)** is less stable than the carbene isomer **86(Ru)** by 9.7 kcal/mol, while the osmium-containing metallabenzyne **85(Os)** has an energy comparable to that of the carbene isomer **86(Os)** (Figure 5.6). The reaction barriers are also strongly affected by the metals and are 1.8, 21.5, and 34.2 kcal/mol for **85(Fe)**, **85(Ru)** and **85(Os)**, respectively. The results clearly indicate that an analogous metallabenzyne with a lighter metal in the same group has a higher tendency to rearrange to the corresponding carbene isomers for both kinetic and thermodynamic reasons.

A similar metal effect has been reported by Zhu and his co-workers for the conversion of metallanaphthalynes of groups 8 and 9 metals supported by two PH_3 and two chloride ligands to the corresponding indenylidene isomers (Figure 5.7) [20]. For analogous metallanaphthalynes with metals in the same group, a metallanaphthalyne of a

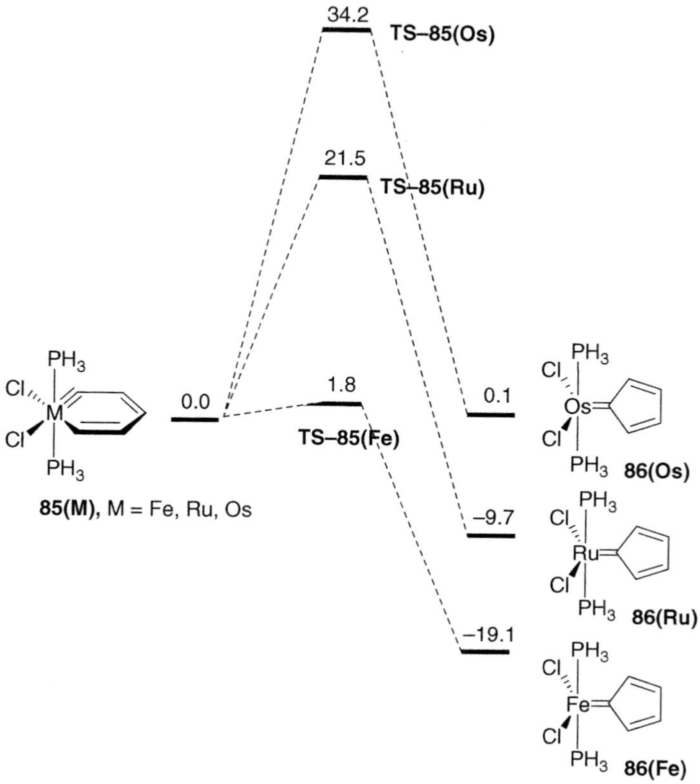

Figure 5.6 Metal effect on the energy changes and barriers (in kcal/mol) calculated for the conversion of metallabenzynes of group 8 metals to the corresponding carbene complexes.

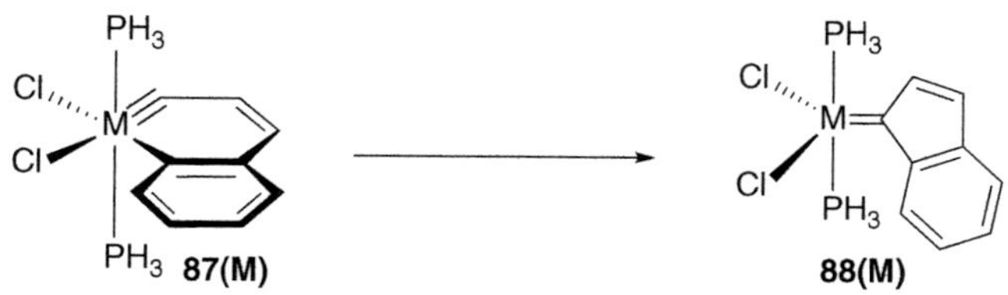

M	Fe	Ru	Co^+	Rh^+	Ir^+
ΔG	–39.9	–25.6	–53.3	–44.9	–23.3

Figure 5.7 Metal effect on the free energy changes (in kcal/mol) for the conversion of metallanaphthalynes of group 8 and 9 metals to the corresponding indenylidene complexes.

lighter metal is thermodynamically less stable than the corresponding indenylidene isomer. Their work also points out that a metallabenzyne of a group 9 metal is in general thermodynamically less stable than that of a group 8 metal in the same period, with respect to the formation of corresponding indenylidene isomers.

Migratory insertion reactions could also occur involving a hydride or alkyl ligand *cis* to the carbyne carbon of the metallabenzyne ring. The reaction has been suggested to

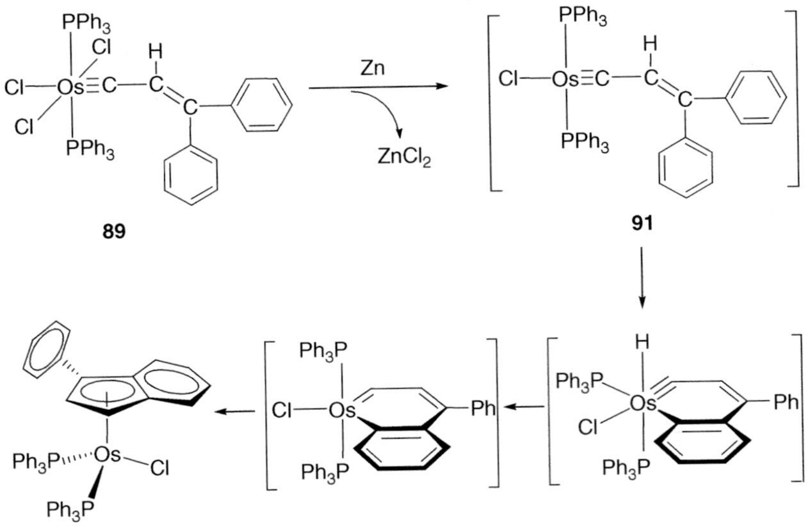

Scheme 5.18 Formation of the indenyl complex **90** via osmanaphthalene intermediate **92**.

account for the formation of the indenyl complex Os{η^5-C_9H_6(Ph)}Cl(PPh_3)$_2$ (**90**) from the reaction of the carbyne complex **89** with zinc powder (Scheme 5.18) [11]. It was proposed that the reaction involves the hydrido-osmanaphthalyne intermediate **92**, generated from cyclometallation reaction of the 16e$^-$ carbyne complex **91**, which undergoes migratory insertion of H to the carbyne carbon to give the osmanaphthalene **93**, which then rearranges to the indenyl complex **90**.

5.2.3 Structural Properties of Metallabenzynes

The structures of 15 osmabenzynes and one rhenabenzyne have been determined by X-ray diffraction. The structural data indicate that the metallacycle of metallabenzynes has a delocalized structure. To illustrate the structural features generally observed for the metallabenzyne rings, selected structural parameters of the osmabenzynes **6** and **48** are presented in Figure 5.8.

All the complexes contain an essentially planar six-membered metallacycle. The M–C(carbyne)–C angles are around 150°, which are significantly smaller than 180° expected for carbyne complexes. The bond angles around the other α-carbon are appreciably larger than the 120° expected for an sp^2 hybridized carbon. The M–C(carbyne) bond distances are at the high end of those observed for typical carbyne complexes and at the low end of those observed for typical vinylidene complexes. The M–C5 bond distances are within the range of M–C (vinyl) bond distances and at the high end of those observed for typical carbene complexes. The ring C–C distances are intermediate between single and double carbon=carbon bonds. Excluding the triple bonds, similarity in the C–C bond distances and the bond angles around the carbons in the osmabenzyne complexes and benzyne [22, 23] can be found, as illustrated in Figure 5.8.

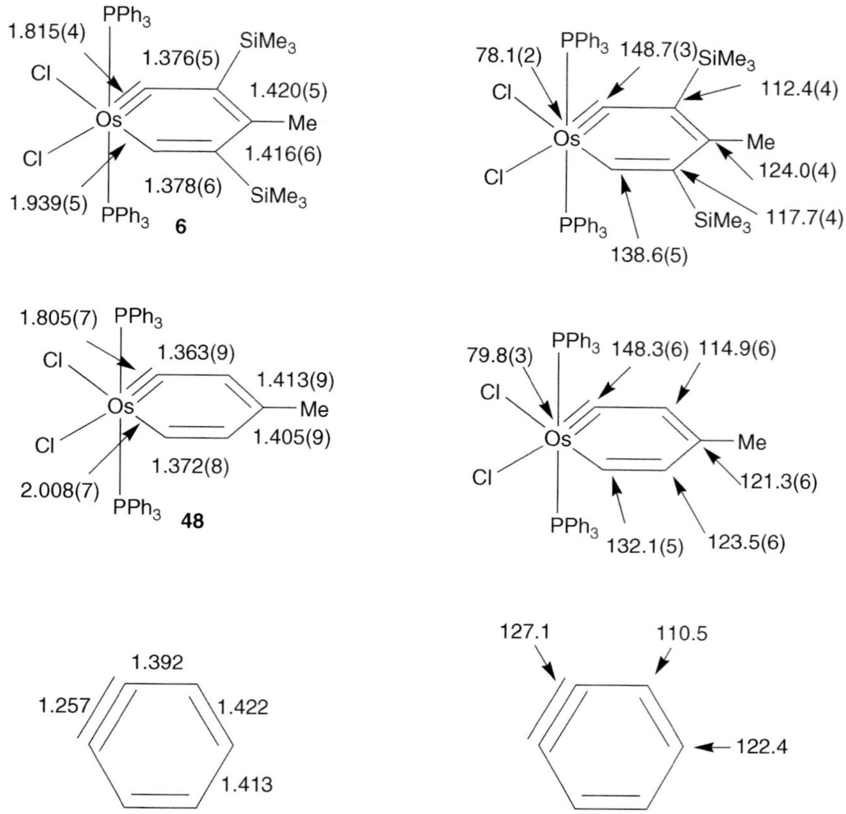

Figure 5.8 Selected bond distances [Å] and angles [degree] of osmabenzynes **6** and **48**, and benzyne.

5.2.4 Ring Strain of Metallabenzynes

It is well established that benzyne and its derivatives have large ring strain caused by bending the formal C≡C triple bond in the six-membered ring. The ring strain of benzyne was estimated to be 53.60 kcal/mol based on the calculated relative energies of optimized $CH_2{=}CHCH{=}CHC{\equiv}CCH{=}CHCH{=}CH_2$ and its deformed form with a geometry similar to that of benzyne [22], or 51.8 kcal/mol based on the calculated relative energies of optimized 2-butyne (**94**) and its deformed form (**95**) (Figure 5.9) [23].

The bending of the angles around the α-carbons of metallabenzyne complexes could also result in ring strain. The ring strain of the osmabenzyne with a geometry similar to that of **6** caused by angle bending at the carbyne carbon was estimated to be only 9.6 kcal/mol by calculating the energy difference between the optimized model carbyne complex $Os({\equiv}CCH_3)(CH_3)Cl_2(PH_3)_2$ (**95**) and its deformed form (**97**) with a geometry similar to that of the osmabenzyne **48** (Figure 5.9). The strain caused by angle bending at the other α-carbon was found to be insignificant. Thus, the ring strain of metallabenzynes is much smaller than that of benzyne. The difference is understandable as significant angle bending (from the ideal value 180° to 127°) occurs at two carbons in benzyne, but only a relatively small angle bending (from 180° to 148.3°) occurs at the carbyne carbon in metallabenzynes.

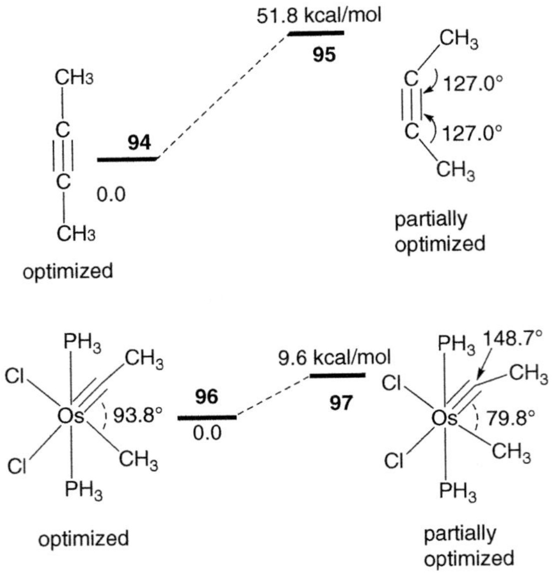

Figure 5.9 Relative energies of optimized 2-butyne (**94**) and the carbyne complex Os(≡CCH_3)(CH_3)$Cl_2(PH_3)_2$ (**96**) and their deformed forms.

5.2.5 Electronic Structures of Metallabenzynes

The electronic structures of group 8 metallabenzyne complexes have been studied by Yang [24], Parameswaran [21] and us [23]. Group 8 metallabenzyne complexes M(≡CCH=CHCH=CH))$Cl_2(PH_3)_2$ can be viewed as d^6 octahedral complexes formed by combination of the d^6 metal fragment $MCl_2(PH_3)_2$ and the carbon fragment C_5H_4. The π-bonding in the metallacycle is illustrated in Figure 5.10, which involves the orbital interaction between the "t_{2g}" orbitals of the metal fragment $OsCl_2(PH_3)_2$ with the π-orbitals of the carbon fragment C_5H_4. In the center of the Figure, the MO1, MO2, MO4, MO5, and MO6 orbitals correspond to the out-of-plane π bonding and are derived from the orbital interactions among out-of-plane π orbitals of C_5H_4 and d_{xz} and d_{yz} of the "t_{2g}" orbitals of the metal fragment. The MO3 orbital corresponds to the in-plane π bonding of the Os≡C triple bond and is derived from the interaction of the in-plane π orbital of C_5H_4 and $d_{x^2-y^2}$ of the "t_{2g}" orbitals of the metal fragment. When the in-plane π bonding is excluded, the bonding in metallabenzynes is similar to that of metallabenzenes [3a].

The first six π-molecular orbitals calculated for the model osmabenzyne complex Os(≡CCH=C(CH_3)CH=CH)$Cl_2(PH_3)_2$ (**98**) are shown in Figure 5.11. For complex **98**, the MO5 and MO6 are the HOMO and LUMO respectively. It is noted that the HOMO (MO5) has significant contribution from the p_π-orbitals at C2 and C4 carbons, the Os(d_{yz}) orbital, and the p_π-orbitals from the chloride ligands; and that the LUMO (MO6) has significant contribution from the p_π-orbitals at C1, C3, and C5 carbons and the d_{xz}(Os) orbital. In other words, the HOMO and LUMO are not related to the in-plane π bonding of the Os≡C triple bond. Interestingly, both the HOMO and LUMO of benzyne are associated with the triple bond and are mainly composed of the in-plane p_π-orbitals of the alkyne carbons [5c].

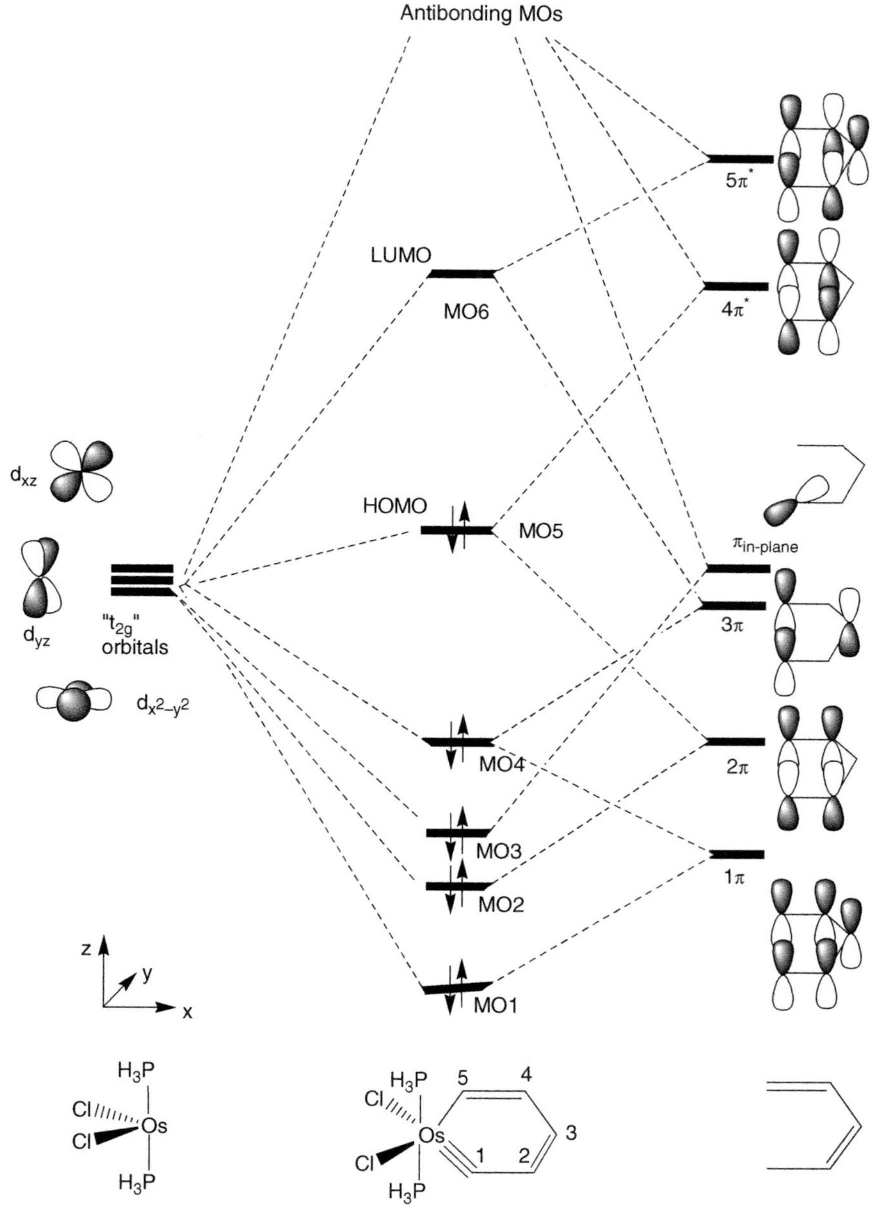

Figure 5.10 Schematic orbital correlation diagram showing the π-interaction for the osmabenzyne complex $Os(\equiv CCH{=}CHCH{=}CH)Cl_2(PH_3)_2$ (**85(Os)**).

5.2.5.1 Aromatic Properties of Metallabenzynes

As mentioned previously, metallabenzynes have a delocalized planar structure. Their electronic structures are similar to those of metallabenzenes. They can undergo electrophilic substitution reactions at the carbon *meta* to the metal. These observations may suggest that metallabenzynes have aromatic character.

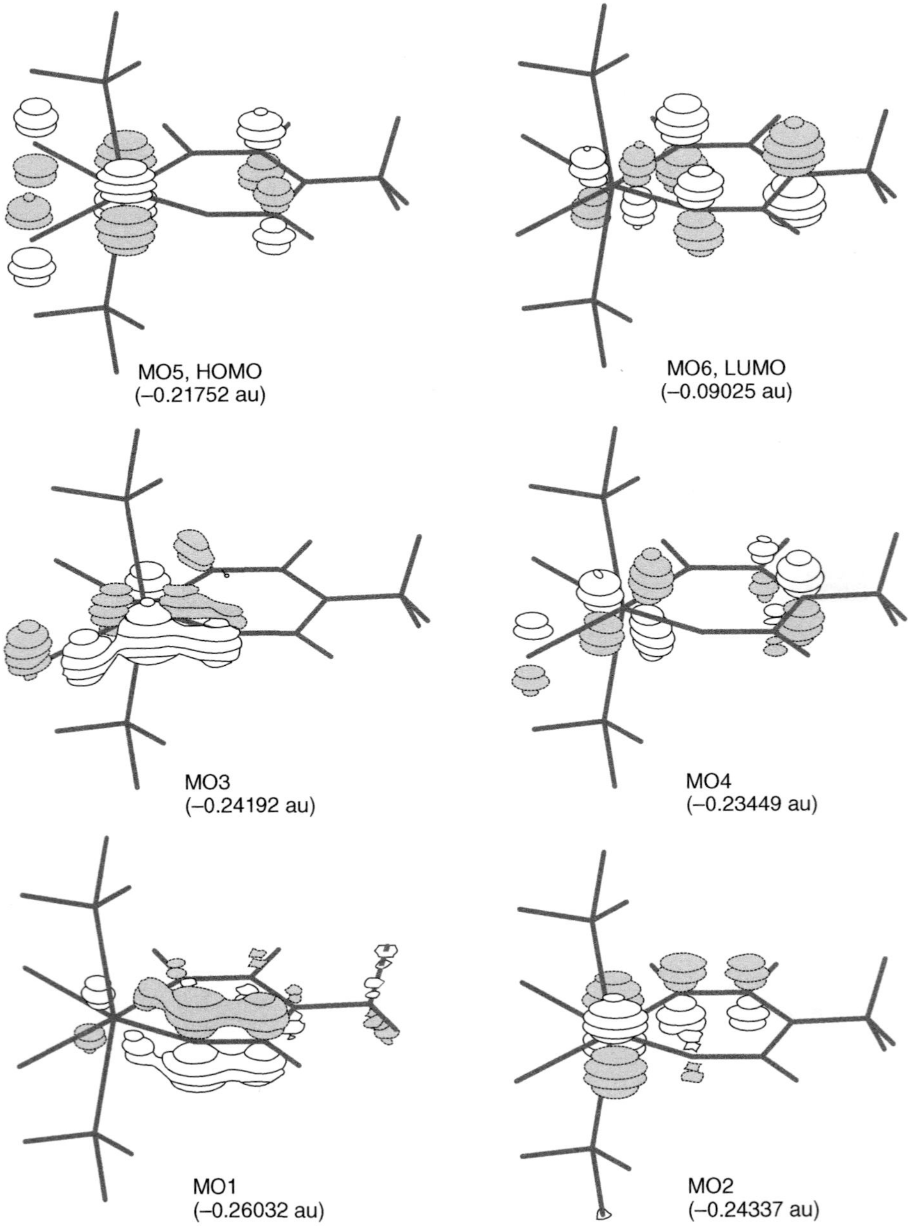

Figure 5.11 Spatial plots of the MO1–MO6 for the model complex Os(≡CCH=C(CH_3)CH=CH) $Cl_2(PH_3)_2$ (**98**).

Energy considerations also support that metallabenzynes are aromatic metallacycles. With the isomerization method introduced by Schleyer *et al.* [25] the aromatic stabilization energy (ASE) values of the model osmabenzyne **98** and the rhenabenzyne complex **105** were estimated to be 19.5 and 14.3 kcal/mol, respectively [14b] (Figure 5.12), while those of benzyne and benzene were found to be 26.8 and 33.9 kcal/mol, respectively.

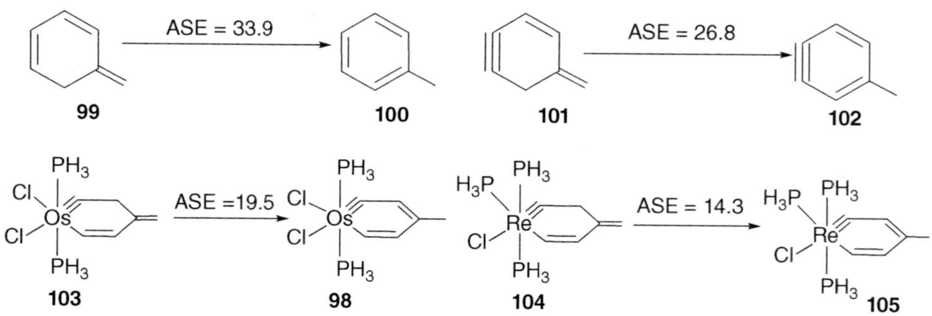

Figure 5.12 ASE values (in kcal/mol) of benzene, benzyne, and metallabenzynes **98** and **105** derived from the isomerization method.

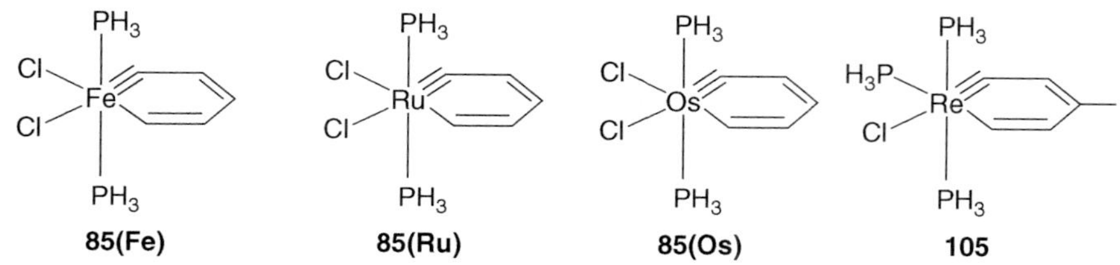

	85(Fe)	85(Ru)	85(Os)	105
NICS(0)	–6.90	–6.77	–7.05	–5.8
NICS(1)	–6.79	–6.33	–6.41	–6.0

Figure 5.13 Calculated NICS values of selected metallabenzynes.

Thus, metallabenzynes indeed have ASE, although the ASE values are less than those of benzene and benzyne.

The aromatic character is also reflected by their nuclear independent chemical shift (NICS) [26] values. Calculations confirm that the model group 8 metallabenzyne complexes **85(M)** [21] as well as the model rhenabenzyne complex **105** [14b] have negative NICS values (Figure 5.13).

5.3 Chemistry of Rhenabenzenes

5.3.1 Synthesis of Rhenabenzenes

Rhenabenzenes [27] and rhenaphenanthrenes [28], have been previously proposed as intermediates for the formation of cyclopentadienyl complexes. We have recently successfully isolated rhenabenzenes, based on the strategy of stabilization by alkoxy groups [29]. These rhenabenzenes were synthesized from the reactions of HC≡COEt with rhenacyclobutadienes **107** (Scheme 5.19). The reactions in general produced a mixture of species, from which rhenabenzene complexes **108** and vinyl-substituted rhenacyclobutadiene complexes **109** could be isolated.

Density functional theory (DFT) calculations suggest that the rhenabenzenes are most likely formed through the mechanism shown in Scheme 5.20. HC≡COEt first attacks the C(OR) carbene carbon of the rhenacyclobutadienes **107** to produce the zwitterionic complexes **110**, which rearrange to the five-membered metallacycles **111**

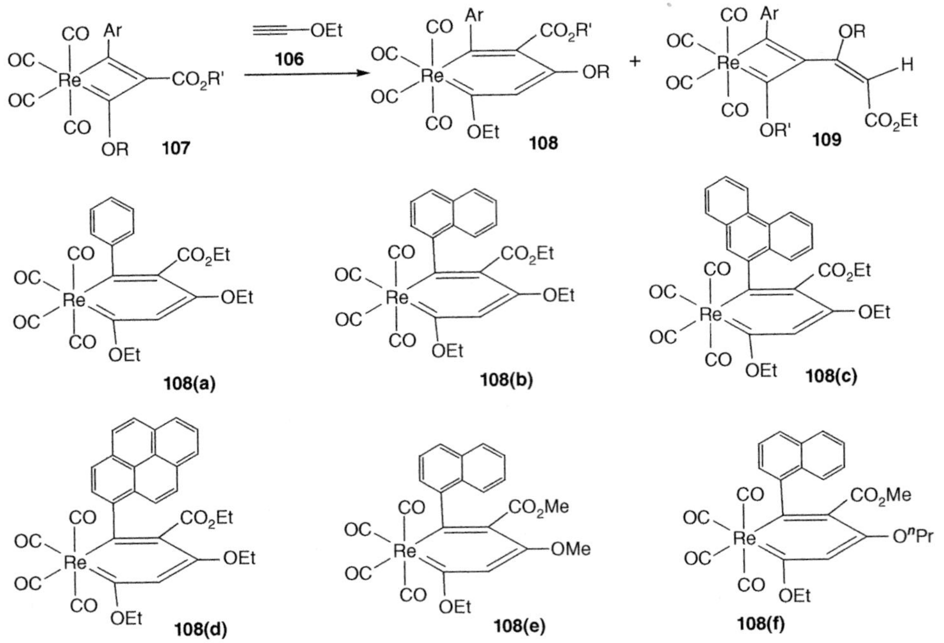

Scheme 5.19 Preparation of rhenabenzenes.

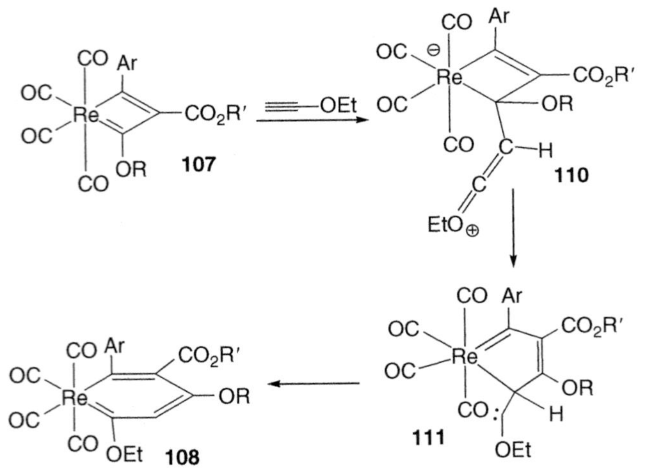

Scheme 5.20 A proposed mechanism for the formation of rhenabenzenes.

containing a free-carbene group stabilized by an OEt group. **111** then evolves to the metallabenzene complexes **108**.

5.3.2 Structural and Aromatic Properties of Rhenabenzenes

X-ray diffraction studies reveal that these complexes contain an essentially planar six-membered metallacycle with the Re–C(Ar) bond being slightly longer than the Re–C(OEt) bond. The Re–C bond distances are within the ranges of Re–C(vinyl) bond

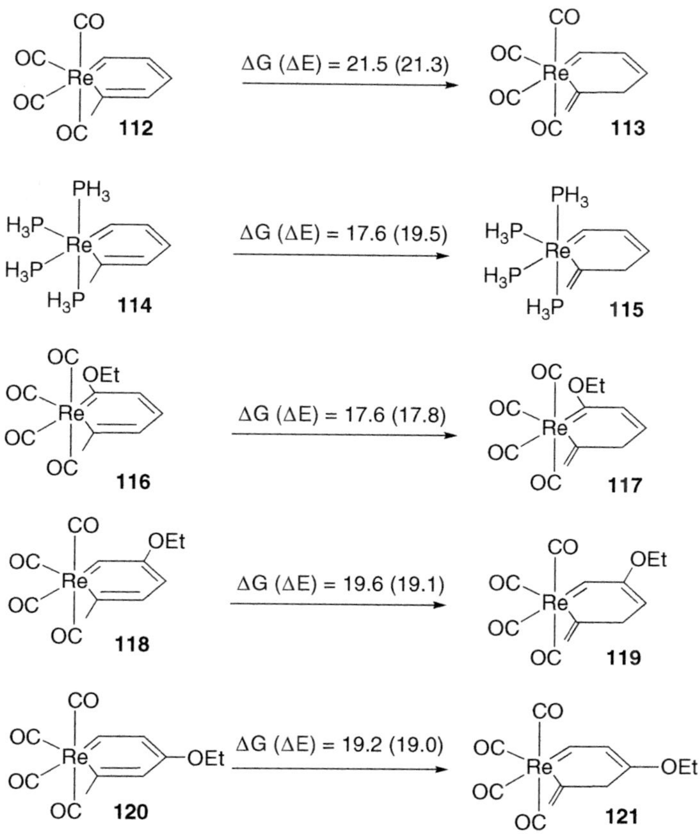

Figure 5.14 ASE values (in kcal/mol) of rhenabenzenes derived from the isomerization method.

distances and at the high end of Re=C(carbene) bond distances. The ring C–C distances are intermediate between single and double carbon–carbon bonds. The structural data suggest that these low-valent rhenabenzenes have a delocalized structure.

A computational study suggests that low-valent model rhenabenzenes have substantial ASE. By using the isomerization method [25], the ASE value of the model rhenabenzene complex **112** was estimated to be 21.5 kcal/mol (Figure 5.14). The value is about 2/3 of that of toluene (33.7 kcal/mol), suggesting that **112** is aromatic. Analogous rhenabenzenes with phosphine ligands (**114**) or an OMe substituent at different positions of the metallacycle (**116, 118, 120**) have similar or slightly lower ASE values.

The calculated ASE values of these low-valent rhenabenzenes are not much different from those (see Figure 5.15) of related model metallabenzenes of Pt, Ir, and Os obtained with the same computational method [30].

5.3.3 Rearrangement of Low-valent Rhenabenzene to η^5-Cyclopentadienyl Complexes

Rearrangement of metallabenzenes to cyclopentadienyl complexes is one of the common reactivities of metallabenzenes. Computational study indicates that the substituent on the metallacycle and ligands around the metal center can significantly influence the

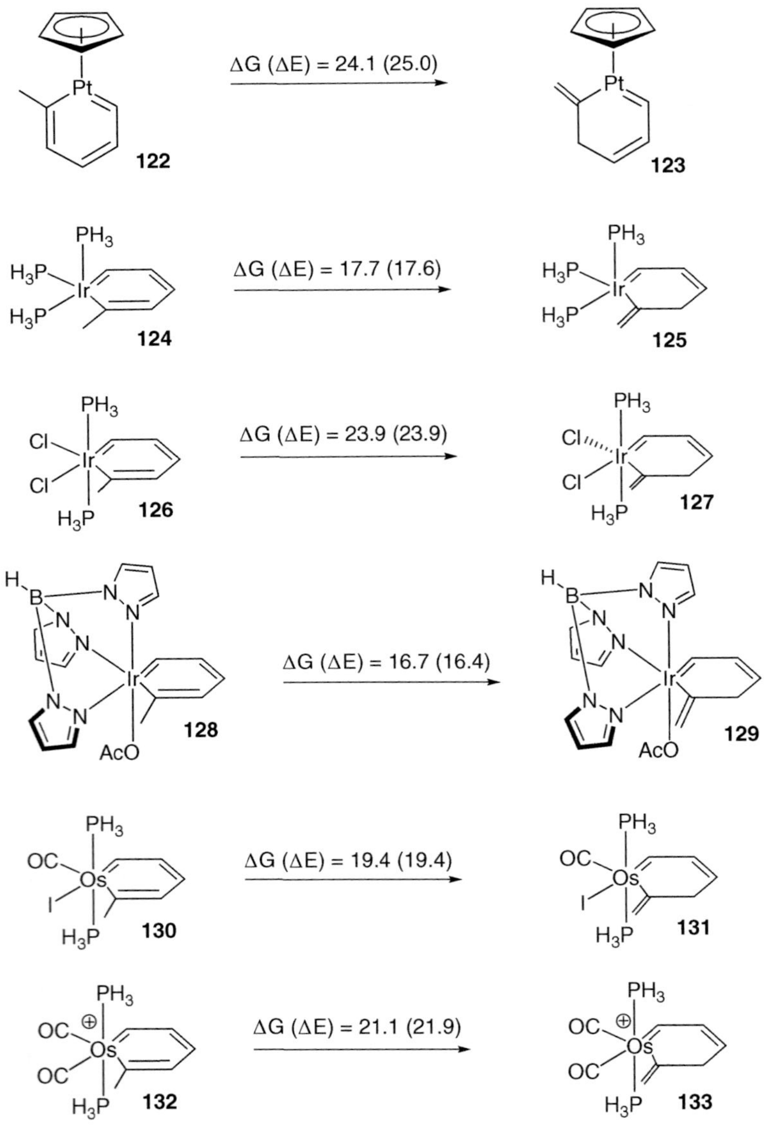

Figure 5.15 ASE values (in kcal/mol) of metallabenzenes derived from the isomerization method.

thermodynamics of the rearrangement of low-valent rhenabenzenes to give η^5-cyclopentadienyl complexes [31, 32].

The effect of the π-donating substituent OMe on the thermodynamics of the rearrangement reactions of rhenabenzenes supported with the metal fragment $Re(CO)_4$ is shown in Figure 5.16. The substituent will significantly increase the stability of the metallabenzenes with respect to formation of cyclopentadienyl complexes if it is at the *ortho* or *para* position on the metallacycle, and decrease slightly the stability of the metallabenzenes if it is at the *meta* position. The effect is so dramatic that the

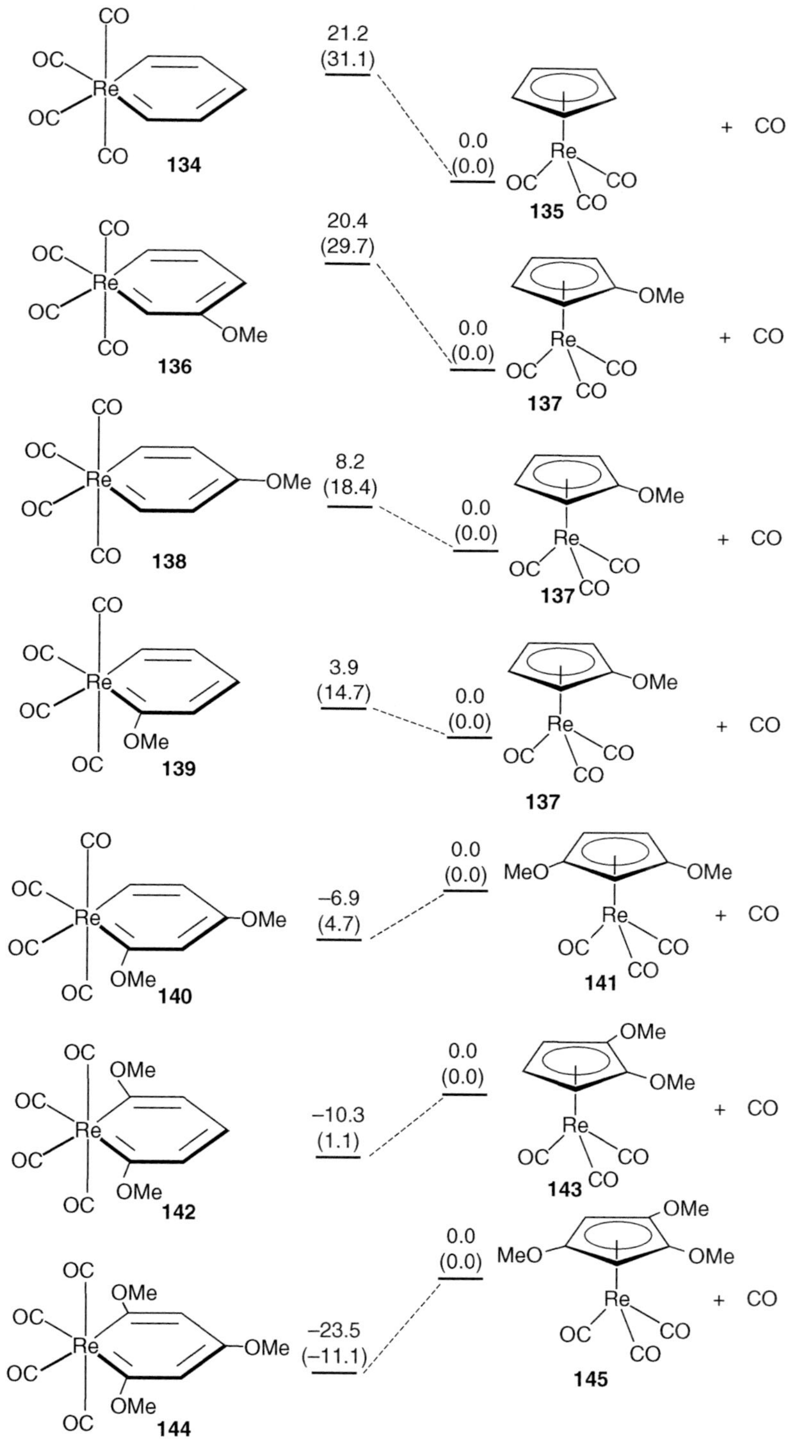

Figure 5.16 Effect of OMe substituent on reaction energies for the conversion of rhenabenzenes to cyclopentadienyl complexes. The relative electronic energies and Gibbs free energies at 298 K (in parentheses) are given in kcal/mol.

unsubstituted rhenabenzene **134** and rhenabenzenes with one OMe group (**136**, **138**, and **139**) are less stable than the corresponding η^5-cyclopentadienyl complexes, while the rhenabenzenes with two or three OMe groups at *ortho* and *para* positions (**140**, **142**, and **144**) are more stable than the corresponding η^5-cyclopentadienyl complexes.

The π-donating groups NMe_2 and SMe have an effect similar to OMe, although with subtle differences. The stabilizing effect of the three substituents is in the order of OMe > NMe_2 > SMe when they are at the *ortho* position, and in the order of NMe_2 > OMe > SMe when they are at the *para* position.

The π-accepting substituents such as CN and NO_2 have almost no effect on the thermodynamics of the rearrangement reactions if they are at the *para* position, but a stabilization effect if they are at the *ortho* or *meta* positions, as illustrated by the calculated free energy changes for the reactions of CN-substituted rhenabenzenes shown in Figure 5.17.

It is expected that the ancillary ligands will also affect the thermodynamics of the rearrangement reactions. Our calculated results indicate that the ligand effect is much

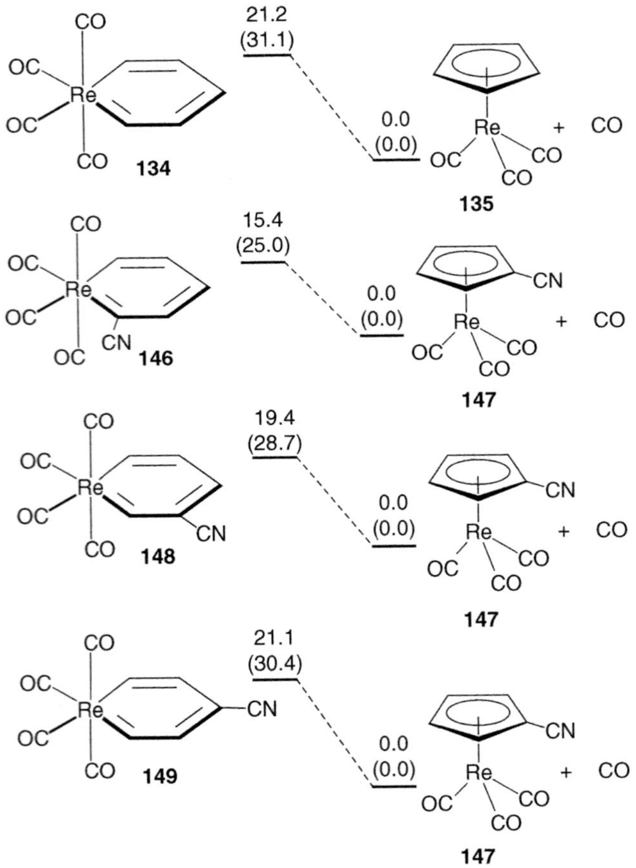

Figure 5.17 Effect of the electron-withdrawing substituent CN on reaction energies for the conversion of rhenabenzenes to cyclopentadienyl complexes. The relative electronic energies and Gibbs free energies at 298 K (in parentheses) are given in kcal/mol.

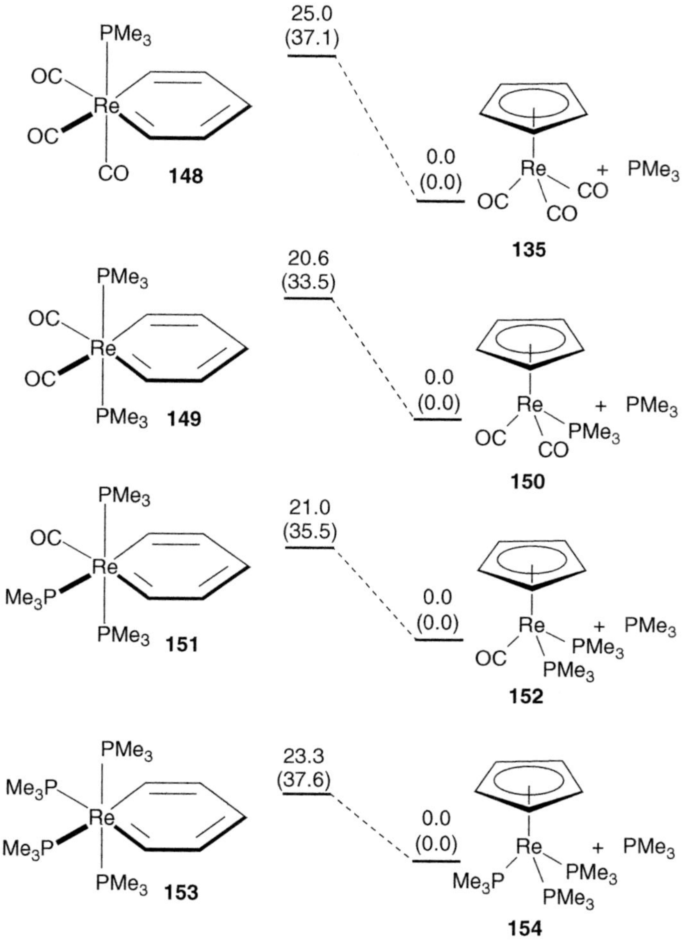

Figure 5.18 Ligand effect on reaction energies for the conversion of rhenabenzenes to cyclopentadienyl complexes. The relative electronic energies and Gibbs free energies (in parentheses) are given in kcal/mol.

smaller than the substituent effect. As shown in Figure 5.18, complexes with different number of phosphine ligands have similar reaction energy changes in the rearrangement reactions.

With respect to the formation of cyclopentadienyl complexes, the rhenabenzene Re(=CHCH=CHCH=CH)$(CO)_4$ (**134**) was calculated to be thermodynamically more stable than the analogous metallabenzene complexes M{=CHCH=CHCH=CH}$(CO)_4$ (M = Tc, Mn) and [Mo(=CHCH=CHCH=CH)$(CO)_4]^-$, but is less stable than the metallabenzene complex [W(=CHCH=CHCH=CH)$(CO)_4]^-$. The relative stability of these complexes are in the order of W > Re > Tc > Mo > Mn (Figure 5.19).

Computationally, studies suggest that the rearrangement reactions of low-valent rhenabenzenes supported with $Re(CO)_4$ to give η^5-cyclopentadienyl complexes generally proceed by initial formation of an η^3-cyclopentadienyl complex, followed by ligand

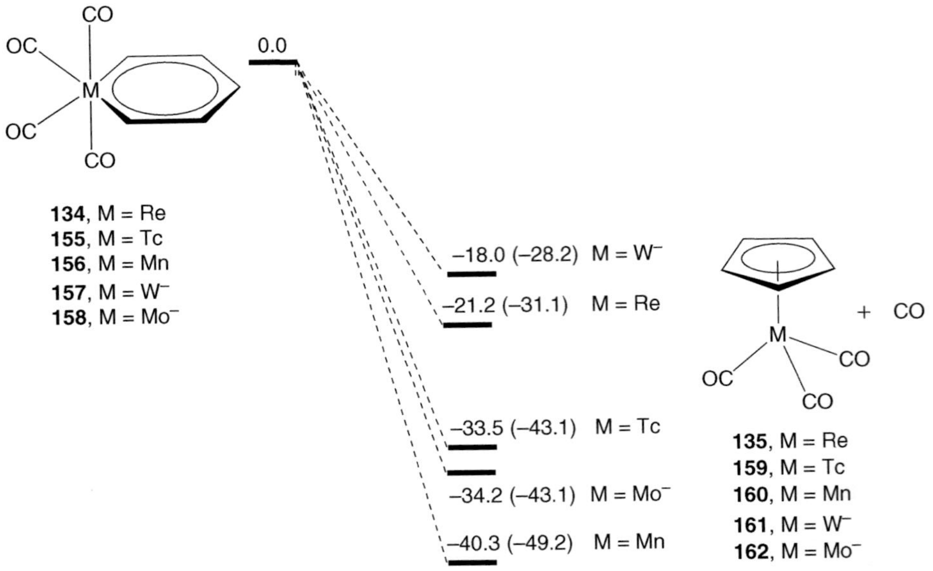

Figure 5.19 Metal effect on reaction energies for the conversion of metallabenzenes to cyclopentadienyl complexes. The relative electronic energies and Gibbs free energies (in parentheses) are given in kcal/mol.

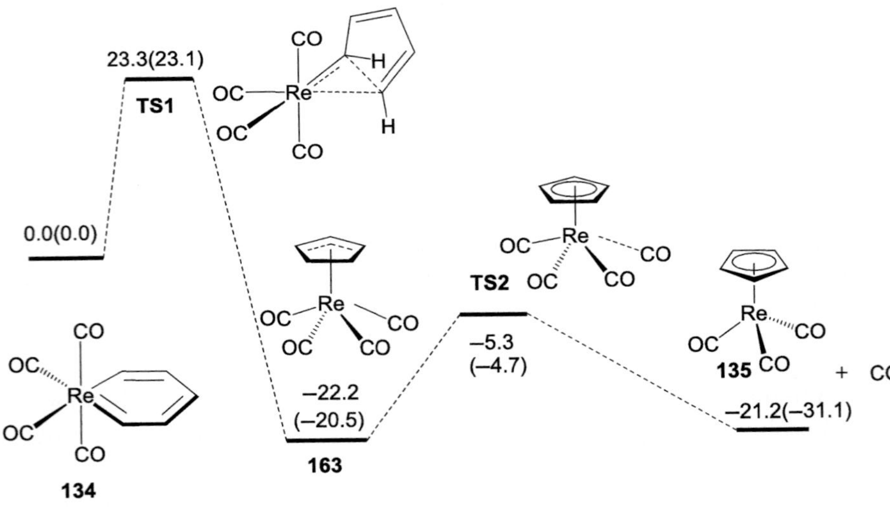

Figure 5.20 Energy profiles calculated for the conversion of **134** to **135**. The relative electronic energies and Gibbs free energies at 298 K (in parentheses) are given in kcal/mol.

dissociation, as illustrated in Figure 5.20 by the reaction of Re{=CHCH=CHCH=CH}$(CO)_4$ (**134**). The rate-determining step is the initial formation of the η^3-cyclopentadienyl complexes, which can be regarded as occurring via nucleophilic attack of the vinyl carbon on the carbene carbon.

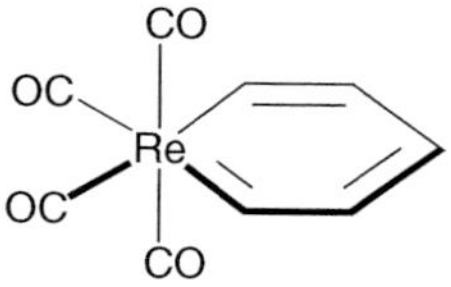

134: 23.3 (23.1)

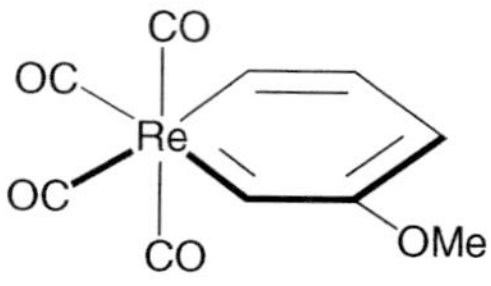

136: 22.6 (22.8)

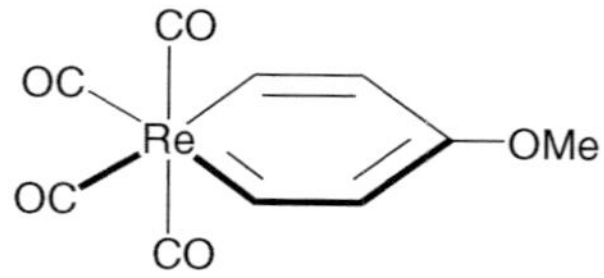

138: 35.7 (34.0)

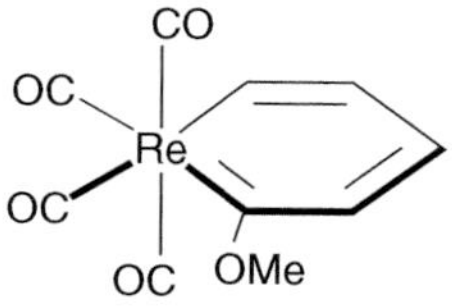

139: 26.6 (26.2)

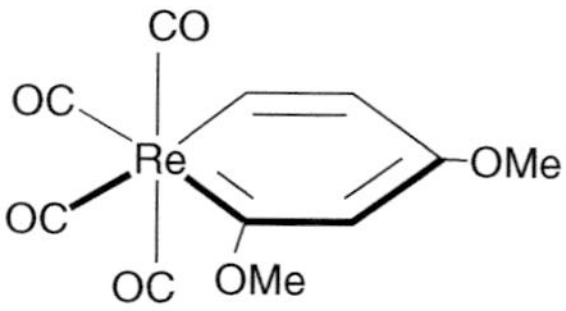

140: 41.0 (39.2)

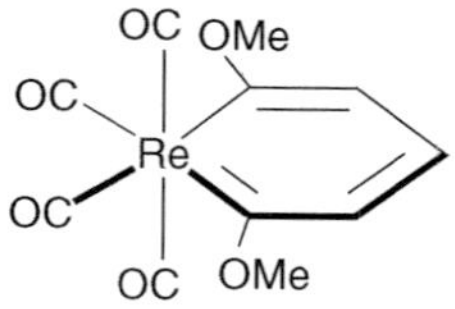

142: 38.7 (38.9)

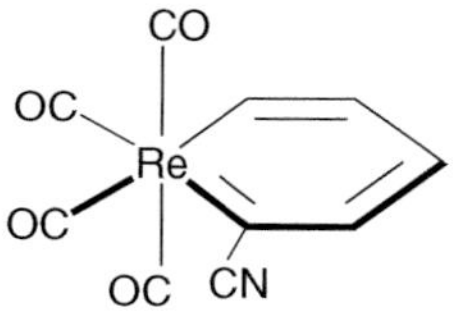

146: 22.6 (22.8)

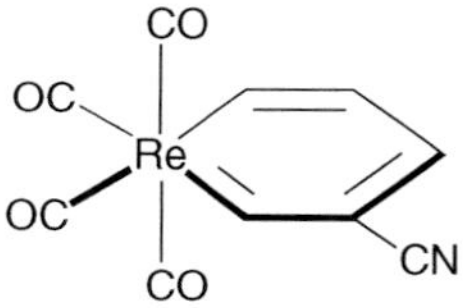

148: 21.9 (23.0)

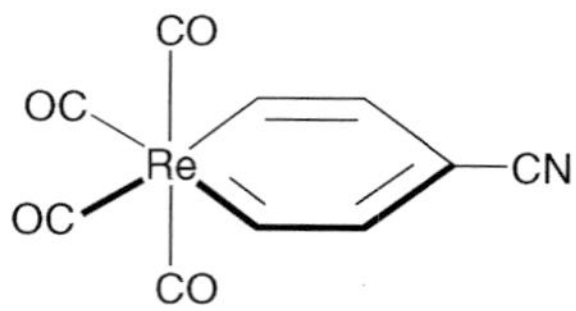

149: 21.9 (23.0)

Figure 5.21 Calculated reaction barriers for the rearrangement reactions of rhenabenzenes. The relative electronic energies ΔE and Gibbs free energies ΔG^0 at 298 K (in parentheses) are given in kcal/mol.

As illustrated in Figure 5.21, the substituents can also affect the reaction barrier of rhenabenzenes supported with the fragment $Re(CO)_4$. The reaction barrier is significantly increased when a π-donating substituent is present at the *para* or *ortho* position of the metallacycle, and decreased slightly when a π-donating substituent is present at the *meta* position. Interestingly, the reaction barrier is not altered significantly by the electron-withdrawing CN substituent. The substituent effect appears to be related to the ability of the substituent to increase or decrease the electron density on the vinyl carbon. It is also noted that the reaction barriers are not related to the relative stability of the rhenabenzenes and the corresponding η^3- or η^5-cyclopentadienyl isomers.

5.4 Summary

A series of metallabenzynes with Os or Re have been isolated. These metallacycles can be regarded as aromatic compounds based on the criteria of electronic structure, geometry, reactivity, ASE, and magnetic properties. They show reactivity of both aromatic systems (e.g. they can undergo electrophilic substitution reaction) and organometallic compounds (e.g. they can undergo migratory insertion reactions to form carbene complexes). Further advances in this area are expected, especially in the syntheses of metallabenzynes with other metals or metallabenzynes of classes II and III, and exploration of their properties and applications.

Several low-valent rhenabenzenes have been isolated based on the strategy of stabilization with π-donating substituent. It remains to be seen whether it is possible to isolate high-valent rhenabenzenes and metallabenzenes with other middle and early transition metals.

References

1 Thorn, D. L. and Hoffmann, R. (1979) Delocalization in metallocycles. *Nouv. J. Chim.*, **3**, 39–45.

2 Elliott, G. P., Roper, W. R., and Waters, J. M. (1982) Metallacyclohexatrienes or "metallabenzenes": Synthesis of osmabenzene derivatives and X-ray crystal structure of [Os(CSCHCHCHCH)(CO)(PPh_3)$_2$]. *J. Chem. Soc. Chem. Commun.*, 811–813.

3 For reviews, see: (a) Fernandez, I., Frenking, G., and Merino, G. (2015) Aromaticity of metallabenzenes and related compounds. *Chem. Soc. Rev.*, **44**, 6452–6463; (b) Frogley, B. J. and Wright, L. J. (2014) Fused-ring metallabenzenes. *Coord. Chem. Rev.*, **270**, 151–166; (c) Cao, X. Y., Zhao, Q., Lin, Z., and Xia, H. (2014) The chemistry of aromatic osmacycles. *Acc. Chem. Res.*, **47**, 341–354; (d) Dalebrook, A. F. and Wright, L. J. (2012) Metallabenzenes and metallabenzenoids. *Adv. Organomet. Chem.*, **60**, 93–177; (e) Paneque, M., Poveda, M. L., and Rendón, N. (2011) Synthesis and reactivity of iridacycles containing the $Tp^{Me2}Ir$ moiety. *Eur. J. Inorg. Chem.*, 19–33; (f) Landorf, C. W. and Haley, M. M. (2006) Recent advances in metallabenzene chemistry. *Angew. Chem. Int. Ed.*, **45**, 3914; (g) Wright, L. J. (2006) Metallabenzenes and metallabenzenoids. *Dalton Trans.*, 1821–1827; (h) Bleeke, J. R. (2001) Metallabenzenes. *Chem. Rev.*, **101**, 1205–1227.

4 Roper, W. R. (2001) First metallabenzenes and now a stable metallabenzyne. *Angew. Chem. Int. Ed.*, **40**, 2440–2441.

5 For reviews, see: (a) Jia, G. (2004) Progress in the Chemistry of metallabenzynes. *Acc. Chem. Res.*, **37**, 479–286; (b) Chen, J. and Jia, G. (2013) Recent development in the chemistry of transition metal-containing metallabenzenes and metallabenzynes. *Coord. Chem. Rev.*, **257**, 2491–2521; (c) Jia, G. (2013) Our Journey to the Chemistry of Metallabenzynes. *Organometallics*, **32**, 6852–6866.

6 Wen, T. B., Zhou, Z. Y., and Jia G. (2001) Synthesis and characterization of a metallabenzyne. *Angew. Chem. Int. Ed.*, **40**, 1951–1953.

7 Wen, T. B., Yang, S. Y., Zhou, Z. Y., *et al.* (2000) Unexpected formation of osmium carbyne and vinylidene complexes from the reaction of $OsCl_2(PPh_3)_3$ with $HC{\equiv}CCMe_3$. *Organometallics*, **19**, 3757–3761.

8 Wen, T. B., Zhou, Z. Y., Lo, M. F., *et al.* (2003) Allenylidene, and carbyne complexes from the reactions of [$OsCl_2(PPh_3)_3$] with $HC{\equiv}CC(OH)Ph_2$. *Organometallics*, **22**, 5217–5225.

9 Wen, T. B., Hung, W. Y., Sung, H. H. Y., *et al.* (2005) Syntheses of metallabenzynes from an allenylcarbene complex. *J. Am. Chem. Soc.*, **127**, 2856–2857.

10 Zhao, Q., Zhu, J., Huang, Z. A., *et al.* (2012) Conversions of osmabenzyne and isoosmabenzene. *Chem. Eur. J.*, **18**, 11597–11603.

11 He, G., Zhu, J., Hung, W. Y., *et al.* (2007) A metallanaphthalyne complex from zinc reduction of a vinylcarbyne complex. *Angew. Chem. Int. Ed.*, **46**, 9065–9068.

12 Chen, J., Shi, C., Sung, H. H. Y., *et al.* (2011) Conversion of metallabenzynes into carbene complexes. *Angew. Chem. Int. Ed.*, **50**, 7295–7299.

13 Liu, B., Xie, H., Wang, H., *et al.* (2009) Selective synthesis of osmanaphthalene and osmanaphthalyne by intramolecular C-H activation. *Angew. Chem. Int. Ed.*, **48**, 5461–5464.

14 (a) Chen, J., Sung, H. H. Y., Williams, I. D., *et al.* (2011) Synthesis and characterization of a rhenabenzyne complex. *Angew. Chem. Int. Ed.*, **50**, 10675–10678; (b) Chen, J., Shi, C., Sung, H. H. Y., *et al.* (2012) Synthesis and characterization of rhenabenzyne complexes. *Chem. Eur. J.*, **18**, 14128–14139.

15 Chen, J., Lee, K. H., Wen, T. B., *et al.* (2015) Rearrangement of metallabenzynes to chlorocyclopentadienyl Complexes. *Organometallics*, **34**, 890–896.

16 Hung, W. Y., Zhu, J., Wen, T. B., *et al.* (2006) Osmabenzenes from the reactions of a dicationic osmabenzyne complex. *J. Am. Chem. Soc.*, **128**, 13742–13752.

17 Wen, T. B., Ng, S. M., Hung, W. Y., *et al.* (2003) Protonation and bromination of an osmabenzyne: Reactions leading to the formation of new metallabenzynes. *J. Am. Chem. Soc.*, **125**, 884–885.

18 Hung, W. Y., Liu, B., Shou, W., *et al.* (2011) Electrophilic substitution reactions of metallabenzynes. *J. Am. Chem. Soc.*, **133**, 18350–18360.

19 See, for example: (a) Spaivak, C. J. and Caulton, K. G. (1988) Hydrogen migration and HCl elimination promoted by metal-centered nucleophilic attack on osmium hydrido-carbynes. *Organometallics*, **17**, 5260–5266; (b) Lee, J. H., Pink, M., and Caulton, K. G. (2006) Triple benzylic dehydrogenation by osmium in an amide ligand environment. *Organometallics*, **25**, 802–804; (c) Bolano, T., Castarlenas, R., Esteruelas, M. A., *et al.* (2005) Hydride-alkenylcarbyne to alkenylcarbene transformation in bisphosphine-osmium complexes. *J. Am. Chem. Soc.*, **127**, 11184–11195.

20 Fan, J., An, K., Wang, X., and Zhu, J. (2013) Interconversion of metallanaphthalynes and indenylidene complexes: A DFT prediction. *Organometallics*, **32**, 6271–6276.

21 Anusha, C., De, S., and Parameswaran, P. (2013) Ring contraction of six-membered metallabenzynes to five-membered metal–carbene complexes: A comparison with organic analogues. *Dalton Trans.*, **42**, 14733–14741.

22 Jiao, H., Schleyer, P. R., Beno, B. R., *et al.* (1997) Theoretical studies of the structure, aromaticity, and magnetic properties of *o*-benzyne. *Angew. Chem. Int. Ed. Engl.*, **36**, 2761–2764.

23 Ng, S. M., Huang, X., Wen, T. B., *et al.* (2003) Theoretical studies on the stabilities of metallabenzynes. *Organometallics*, **22**, 3898–3904.

24 Yang, S. Y., Li, X. Y., and Huang, Y. Z. (2002) Electronic reason for the stabilization of osmabenzyne. *J. Organomet. Chem.*, **658**, 9–14.

25 Schleyer, P. R. and Pühlhofer, F. (2002) Recommendations for the evaluation of aromatic stabilization energies. *Org. Lett.*, **4**, 2873–2876.

26 Chen, Z. F., Wannere, C. S., Corminboeuf, C., *et al.* (2005) Nucleus-independent chemical shifts (NICS) as an aromaticity criterion. *Chem. Rev.*, **105**, 3842–3888.

27 (a) Plantevin, V. and Wojcicki, A. (2004) Chemistry of Fischer-type rhenacyclobutadiene complexes: II: Reactions with alkynes and sulfonium ylides, and rearrangements induced by nitriles and pyridine. *J. Organomet. Chem.*, **689**, 2013–2024; (b) Ferede, R., Hinton, J. F., Korfmacher, W. A., *et al.* (1985) Possible formation of rhenabenzene: Lithium-halogen exchange reactions in η^1-4-bromo-1,4-diphenyl-1,3-butadienyl ligands to form rhenabenzene and ferrabenzene. *Organometallics*, **4**, 614–618.

28 Mike, C. A., Ferede, R., and Allison, N. T. (1988) Evidence for rhenaphenanthrene formation and its conversion to fluorenone. *Organometallics*, **7**, 1457–1459.

29 (a) Poon, K. C., Liu, L., Guo, T., *et al.* (2010) Synthesis and characterization of rhenabenzenes. *Angew. Chem. Int. Ed.*, **49**, 2759–2762; (b) Lin, R., Lee, K. H., Poon, K. C., *et al.* (2014) Synthesis of rhenabenzenes from the reactions of rhenacyclobutadienes with ethoxyethyne. *Chem. Eur. J.*, **20**, 14885–14899; (c) Lin, R., Lee, K. H., Sung, H. H. Y., *et al.* (2015) Rhenabenzenes and unexpected coupling products from the reactions of rhenacyclobutadienes with ethoxyethyne. *Organometallics*, **34**, 167–176.

30 For previous work on the ASEs of other model metallabenzene complexes evaluated by the isomerization methods, see: (a) De Proft, F. and Geerlings, P. (2004) Relative hardness as a measure of aromaticity. *Phys. Chem. Chem. Phys.*, **6**, 242–248; (b) Mauksch, M. and Tsogoeva, S. B. (2010) Demonstration of "Möbius" aromaticity in planar metallacycles. *Chem. Eur. J.*, **16**, 7843–7851. For ASE values of metallabenzenes evaluated with homodesmic reaction: (c) Huang, T. Z., Yang, S. Y., and Li, X. Y. (2004) An investigation of the aromaticity of transition metal heterocyclic complexes by conventional criteria and indices of aromaticity. *J. Organomet. Chem.*, **689**, 1050–1056. For ASE values of metallabenzenes evaluated with energy decomposition analysis (EDA) calculations: (d) Fernández, I. and Frenking, G. (2007) Aromaticity in metallabenzenes. *Chem. Eur. J.*, **13**, 5873–5884.

31 Shi, C., Guo, T., Poon, K. C., *et al.* (2011) Theoretical study on the rearrangement of metallabenzenes to cyclopentadienyl complexes. *Dalton Trans.*, **40**, 11315–11320.

32 For theoretical studies on the transformation of other metallabenzenes to the corresponding cyclopentadienyl complexes, (a) Iron, M. A., Lucassen, A. C. B., Cohen, H., *et al.* (2004) A computational foray into the formation and reactivity of metallabenzenes. *J. Am. Chem. Soc.*, **126**, 11699–11710; (b) Iron, M. A., Martin, J. M. L., and van der Boom, M. E. (2003) Metallabenzene versus Cp complex formation: A DFT investigation, *J. Am. Chem. Soc.*, **125**, 13020–13021.

6

Metallabenzenoid Compounds Bearing Phosphonium Substituents

Hong Zhang and Haiping Xia

Department of Chemistry, Xiamen University, Xiamen, P. R. China

Metallabenzenoid compounds that contain phosphonium substituents are an attractive part of the much larger family of compounds classified as metallaaromatics, not only because of their special structural features but also because they exhibit interesting chemical and physical properties. In this chapter, the syntheses, structures, physical properties and reactions of phosphonium-substituted metallabenzenoid compounds are discussed. A diverse array of polycyclic metallabenzenes produced from monocyclic metallabenzene as a starting material is also summarized.

Metallabenzene chemistry encompasses a wide variety of metallabenzenoid compounds and their reactions: monocyclic compounds containing π-electron delocalized six-membered metallacycles, many polycyclic compounds containing fused metallabenzene rings and reactions that in some cases possess similarities to well-known aromatic reactions and in other cases are significantly different [1–14]. In addition to their intrinsically interesting nature, many metallabenzenoid compounds exhibit special properties and have promising prospects for application. Among the archetypal metallabenzenes, phosphonium-substituted metallabenzenoid compounds are particularly attractive for their high thermal stability resulting from the protective effect of their phosphonium substituents.

Over the past thirty years, the synthesis of metallabenzenes has been achieved by the reactions of simple transition metal compounds with unsaturated hydrocarbons and the transformations of various metallacycles. Among all of the synthetic methods, the ring-forming reactions, as shown in Scheme 6.1, stand out for the formation of a series of metallabenzenoid compounds bearing phosphonium substituents. The strategy can be outlined in three steps: the initial coordination of the alkynes to the transition metal centre, the nucleophilic addition of PPh_3 to the coordinated alkyne and the final cyclization reaction through cycloaddition, cyclometallation or coordination.

Phosphonium-substituted metallabenzenoid compounds possess distinctive properties compared with other metallabenzenoid compounds, not only with regard to their structure and bonding characteristics but also in terms of their chemistry. In this chapter, consideration is restricted to the synthesis, structure and reaction chemistry of metallabenzenoid compounds bearing phosphonium substituents. Their physical

Metallabenzenes: An Expert View, First Edition. Edited by L. James Wright.

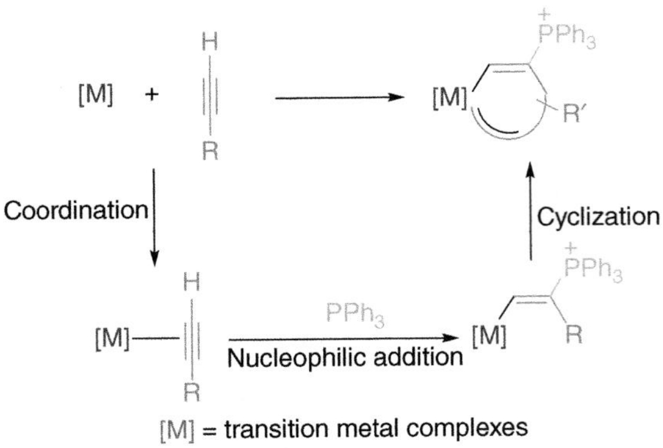

Scheme 6.1

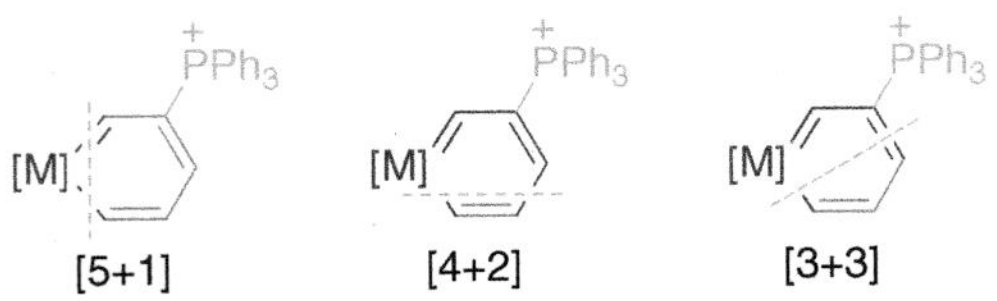

Scheme 6.2

properties including their outstanding thermal stability, photoelectronics, magnetism and stereo-electronics are also discussed in this chapter.

6.1 Synthesis

Public concern for the environment has led to the rise of green chemistry, with the objective of developing environmentally friendly synthetic routes in the chemical industry. One strategy is atom economy, in which reactions are chosen that minimize the formation of by-products or unreacted starting materials. Accordingly, the straightforward construction of a six-membered metallabenzene ring from two components might be achieved by three possible routes, [5+1], [4+2], or [3+3], as illustrated by the retrosynthetic analysis shown in Scheme 6.2. This section will be divided into discussions of the three synthetic routes that can be used for phosphonium-substituted metallabenzenoid compounds and their derivatives.

6.1.1 [5+1] Synthesis

In these syntheses, the organic carbon chains must contain the five carbon atoms of the future metallabenzene skeleton. To further classify these reactions, the following examples are presented successively according to the starting organic carbon chain.

6.1.1.1 Syntheses Starting from HC≡CCH(OH)C≡CH

The first metallabenzene with phosphonium substituents was obtained from the reaction of $OsCl_2(PPh_3)_3$ (**1**) with the readily accessible carbon chain HC≡CCH(OH)C≡CH

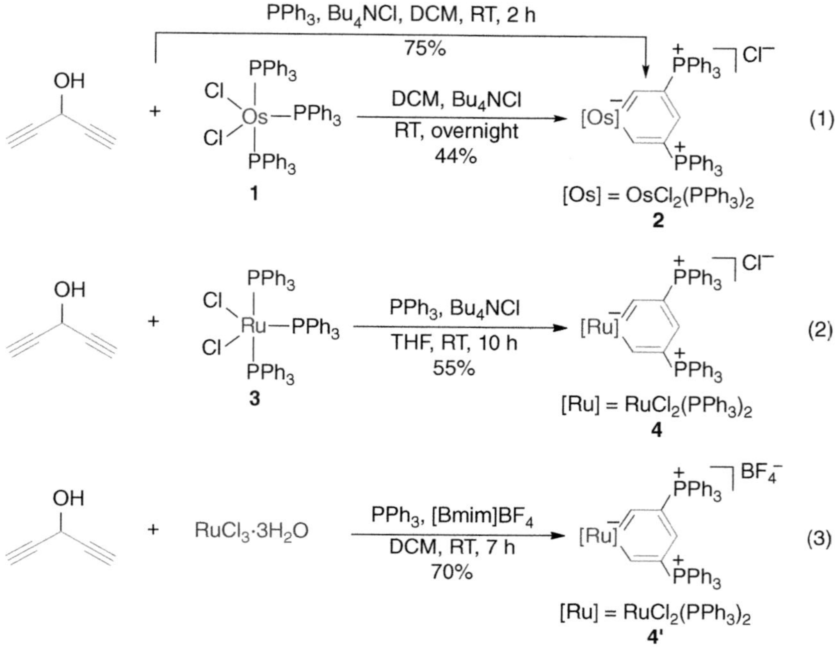

Scheme 6.3

(Scheme 6.3, Equation (1)) [15]. The osmabenzene **2** can be isolated as a green solid in 44% yield when the reaction is performed in CH_2Cl_2 at room temperature (RT) for approximately 12 h. The addition of PPh_3 can increase the overall yield, resulting in a 75% isolated yield of **2** [16]. In a striking extension of this work, the rare thermally stable ruthenabenzene was prepared by the same strategy [17]. As shown in Scheme 6.3, Equation (2), the treatment of $RuCl_2(PPh_3)_3$ (**3**) with HC≡CCH(OH)C≡CH, PPh_3, and Bu_4NCl in tetrahydrofuran (THF) leads to the formation of the ruthenabenzene **4**, which can be isolated as a green solid in 55% yield.

Studies have shown that **4** can even be obtained from the one-pot reaction of the inorganic salt $RuCl_3$ with PPh_3 and HC≡CCH(OH)C≡CH in $CHCl_3$, although the yield is low (*ca.* 20%) [17]. When the one-pot reaction is carried out in a mixed solution of ionic liquid [Bmim]BF_4 (1-butyl-3-methylimidazolium tetrafluoroborate)/CH_2Cl_2 (1:10, v/v), the yield of ruthenabenzene **4**′ can be up to 70% (Scheme 6.3, Equation (3)) [18]. The synthesis of ruthenabenzenes starting from $RuCl_3$ represents the first example of constructing metallabenzenes from inorganic salts.

Scheme 6.4 illustrates the proposed mechanism for the formation of the cationic metallabenzenes **2** and **4** from the one-pot reaction of HC≡CCH(OH)C≡CH with $MCl_2(PPh_3)_3$. As a 16-electron complex, $MCl_2(PPh_3)_3$ can initially react with HC≡CCH(OH)C≡CH to produce the π-alkyne intermediate **A**, which then undergoes the nucleophilic attack of PPh_3 at the coordinated alkyne to generate intermediate **5**. The subsequent attack of another PPh_3 on the coordinated alkyne may form intermediate **B**, which can finally eliminate an OH^- from the γ-carbon to yield metallabenzenes bearing two phosphonium substituents [15]. One of the proposed intermediates,

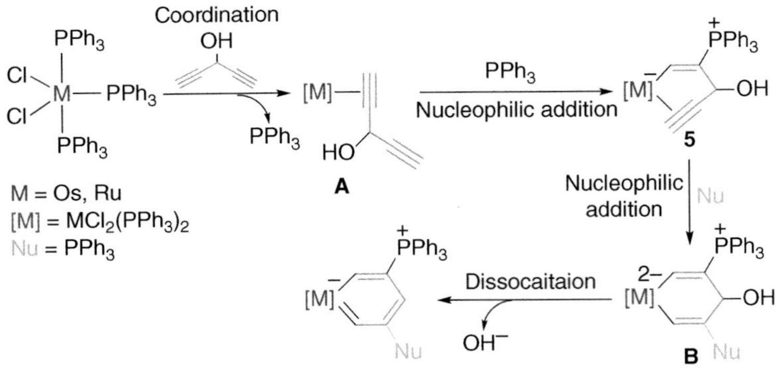

Scheme 6.4

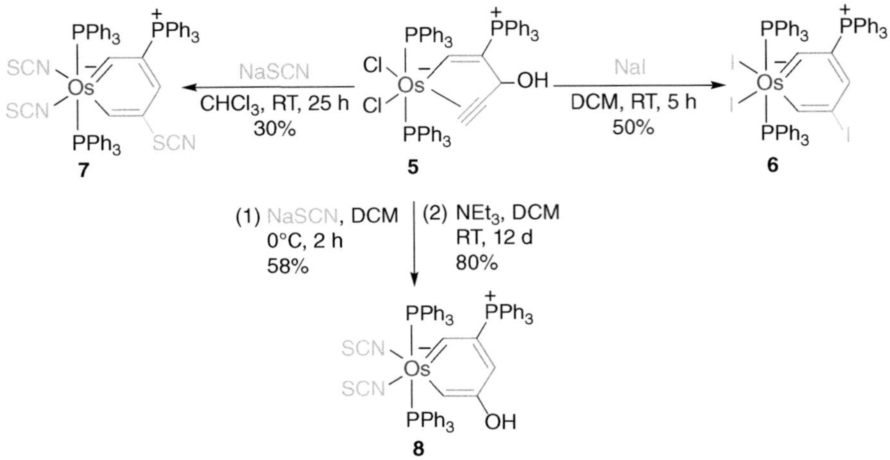

Scheme 6.5

complex **5**, was precipitated as a yellow solid and isolated in 64% yield when the reaction was carried out in THF. Complex **5** is labile in solution at room temperature, resulting in the formation of the osmabenzene **2** as the major product along with other unidentified species. In agreement with the proposed mechanism shown in Scheme 6.4, the treatment of **5** with PPh_3 for 2 h produced **2** in 82% isolated yield.

Complex **5** is also reactive to other nucleophiles (Scheme 6.5) [16]. The addition of NaI to the solution of **2** afforded the iodo-osmabenzene **6**, which is a rare example of a halogen-substituted metallabenzene. Similarly, the addition of NaSCN yielded the thiocyanate-substituted osmabenzene **7**. In the presence of excess triethylamine, complex **5** is converted to form the first *m*-metallaphenol **8**, in which the hydroxyl group takes up a position *meta* to the metal centre [19]. Presumably, the trace amount of water in the solution initiates the reaction. Moreover, the addition of excess triethylamine may facilitate the dehydration process.

This synthetic method offers many possibilities for achieving six-membered metallabenzenoid compounds with phosphonium substituents. An extension of this [5+1]

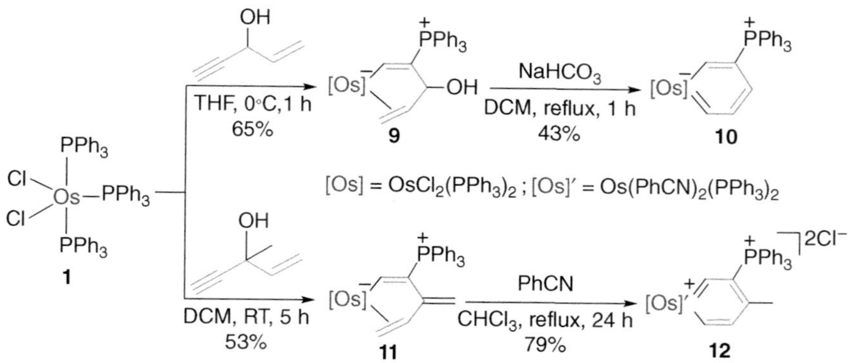

Scheme 6.6

method to metallabenzene synthesis starting with other unsaturated organic carbon chains is also demonstrated below.

6.1.1.2 Syntheses Starting from HC≡CCH(OH)CH=CH$_2$ and HC≡CCCH$_3$(OH)CH=CH$_2$

The carbon chain HC≡CCH(OH)CH=CH$_2$, with a lower degree of unsaturation, was successively considered a starting material for the construction of six-membered metallabenzene rings. The reaction of **1** with HC≡CCH(OH)CH=CH$_2$ in THF leads to the formation of the η^2-allyl alcohol complex **9** (Scheme 6.6) [20]. Heating **9** in dichloromethane (DCM) in the presence of NaHCO$_3$ can give the mono-phosphonium substituted osmabenzene **10**. Scheme 6.6 also shows another example, in which commercially available HC≡CCCH$_3$(OH)CH=CH$_2$ reacts with **1** to form the η^2-allyl complex **11** [21]. Similarly, the expected osmabenzene **12** can be isolated in 79% yield from the transformation of **11**.

As shown in Scheme 6.7, the organic substrate HC≡CCR(OH)CH=CH$_2$ acts very similarly to HC≡CCH(OH)C≡CH. The reactions were also initiated with the coordination of the terminal alkyne to the metal centre and the nucleophilic attack of PPh$_3$ on the coordinated alkyne. For HC≡CCH(OH)CH=CH$_2$, the η^2-allyl alcohol intermediate **9** may undergo the dissociation of OH$^-$ and then deprotonation to form osmabenzene **10**. Presumably, the addition of NaHCO$_3$ can promote the deprotonation step. In the absence of a base, the yield of the transformation to osmabenzene significantly decreased. For HC≡CCCH$_3$(OH)CH=CH$_2$, the dehydration affords the η^2-allyl alcohol intermediate **11**, which then replaces two Cl ligands by two benzonitriles. As a result of ligand substitution, the electron density of the osmacycle was significantly decreased, which should facilitate the hydrogen-transfer process for the formation of the η^2-allene intermediate **13**. The subsequent isomerization of **13** can generate the osmabenzene **12**.

6.1.1.3 Syntheses Starting from HC≡CCH(OH)Ph

This [5+1] synthesis is not limited to that of propynol with an alkynyl or alkenyl. A particularly important case involving HC≡CCH(OH)Ph constitutes another example of the [5+1] synthesis of metallabenzenes, as shown in Scheme 6.8 [22]. The reaction of **1** with HC≡CCH(OH)Ph produces the cyclic complex **14**, which can form the hydrido, alkenylcarbyne complex **15** in the presence of acid. Heating **15** in dichloroethane (DCE)

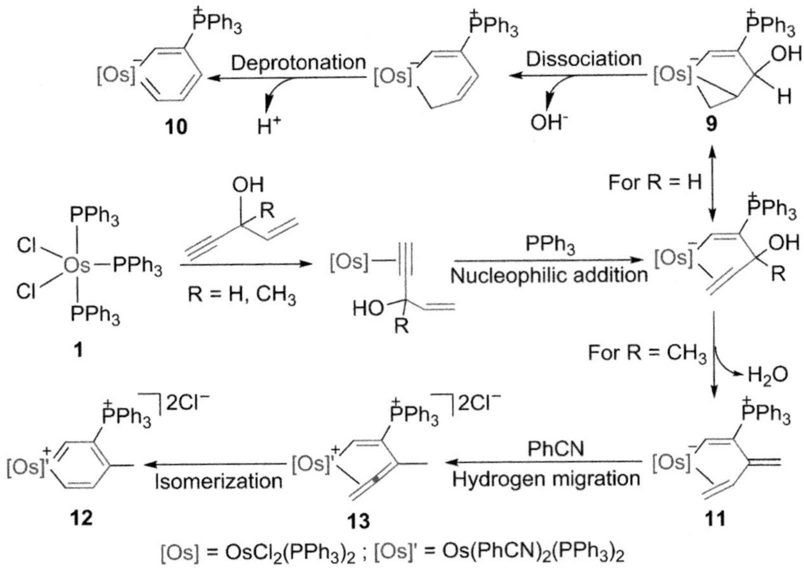

Scheme 6.7

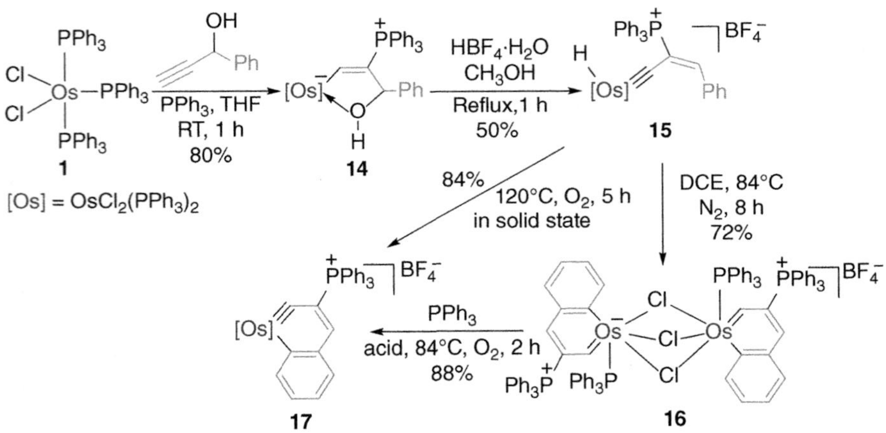

Scheme 6.8

under a N_2 atmosphere yields the (μ-Cl)$_3$-bridged bisosmanaphthalene **16**, whereas, under an O_2 atmosphere, the osmanaphthalyne **17** was obtained. Alternatively, **17** could even be obtained by directly heating a solid sample of **16** in air. In terms of the produced six-membered metallacycles, the transformation from **16** to **17** represents the first example of the transformation from metallabenzene to metallabenzyne.

The likely pathway for the formation of **16** and **17** includes a similar migration of a hydride ligand from the osmium centre to the carbyne carbon atom and the subsequent *ortho* C–H bond activation of a phenyl ring (Scheme 6.9). Assistance by an O_2 atmosphere is essential to the formation of the osmanaphthalyne **17** from the hydride

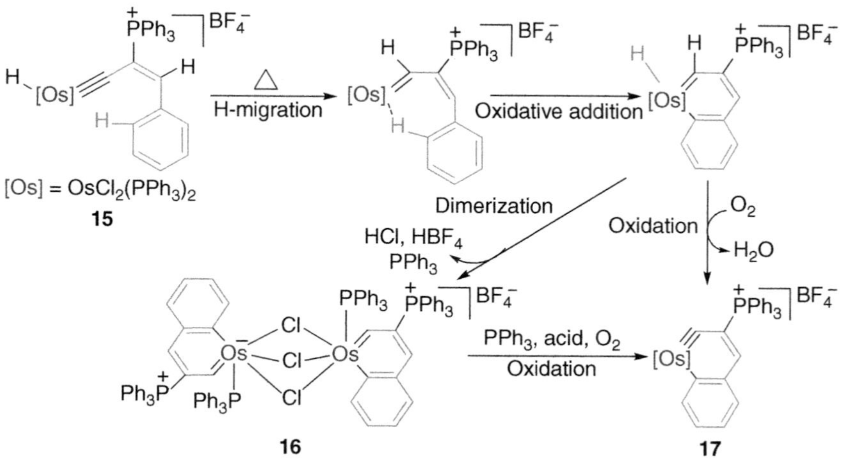

Scheme 6.9

osmanaphthalene intermediate. Dimerization may directly occur under a N_2 atmosphere to yield the bisosmanaphthalene **16**. This analysis is consistent with the conversion of **16** to **17** in the presence of an acid, PPh_3 and O_2.

6.1.1.4 Synthesis Starting from HC≡CCH(OH)Et

Typical [5+1] synthetic routes for six-membered metallabenzenoid compounds with phosphonium substituents differ slightly from those organic substrates used. Scheme 6.10 shows an example of synthetic routes using HC≡CCH(OH)Et as the starting organic material. Here, the hydrido, alkenylvinylidene complex **19** and the

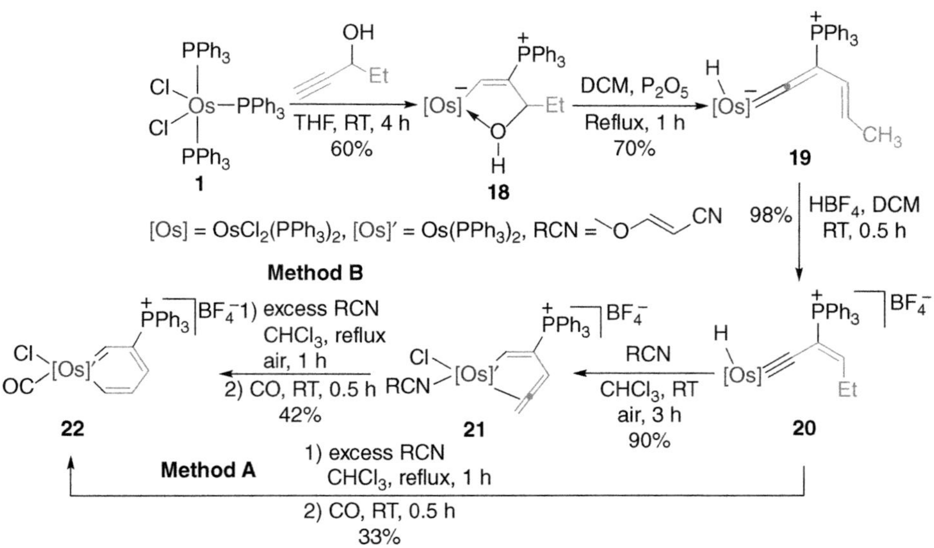

Scheme 6.10

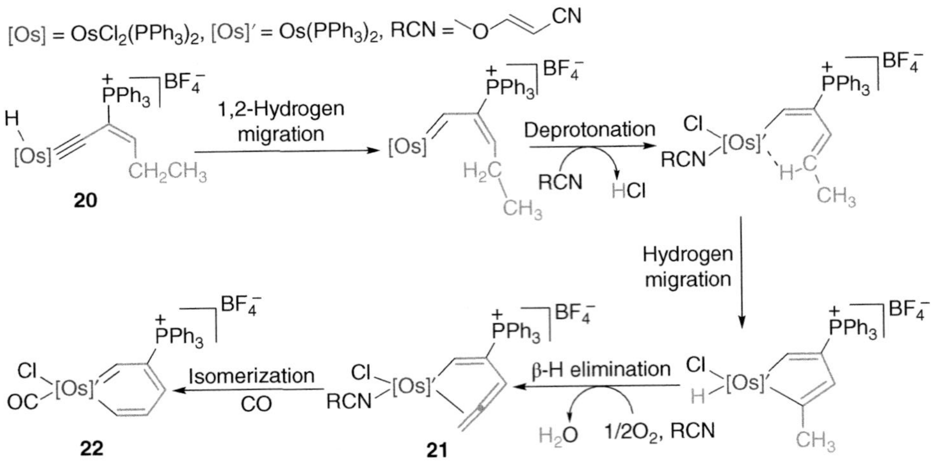

Scheme 6.11

hydrido, butenylcarbyne complex **20** form when the readily accessible HC≡CCH(OH)Et reacts with complex **1** [23]. Note that the final osmabenzene **22** is generated from **20** via a triple hydrogen elimination.

Scheme 6.11 depicts a possible mechanism for the formation of the osmabenzene **22**. The mechanism probably involves a five-step process. The hydrido, butenylcarbyne complex **20** may undergo a 1,2-hydrogen shift to form the butenylcarbene intermediate. The subsequent deprotonation and coordination of a nitrile ligand to the metal centre generates an agostic intermediate. As the C–H bond approaches the metal centre, the dissociation of the nitrile ligand occurs followed by hydrogen migration, resulting in the formation of the osmacyclopentadiene intermediate. Under an air atmosphere, β-H elimination and coordination of the nitrile produce the η^2-allene complex **21**, which tends to isomerize to give the osmabenzene **22**. Experimental evidence and computational analysis have demonstrated that the lability of the nitrile ligand is crucial for this conversion of a hydrido, butenylcarbyne complex into a metallabenzene.

6.1.2 [4+2] Synthesis

The Diels–Alder reaction is particularly important in synthetic organic chemistry as an excellent cycloaddition method for forming six-membered systems. Diels–Alder reactions occur between a conjugated diene and an alkene, usually called the dienophile (Scheme 6.12, Equation (1)). The dienophile component in the Diels–Alder reaction can be an alkene or an alkyne with various types of substituents. This wonderful reaction has also been applied to other π-systems, such as nitriles, to assemble the corresponding six-membered heterocycles, known as the hetero-Diels–Alder reaction (Scheme 6.12, Equation (2)). According to the retrosynthetic analysis shown in Scheme 6.2, the incorporation of heteroatoms or related groups into six-membered metallabenzene systems should be achieved by a similar [4+2] cycloaddition.

An example of [4+2] synthesis is illustrated by the reaction of the styrylcarbyne complex **15** with acetonitrile (Scheme 6.12, Equation (3)). The alkenylcarbyne unit of **15** is used as the precursor of the diene in the reaction with acetonitrile as a dienophile. The

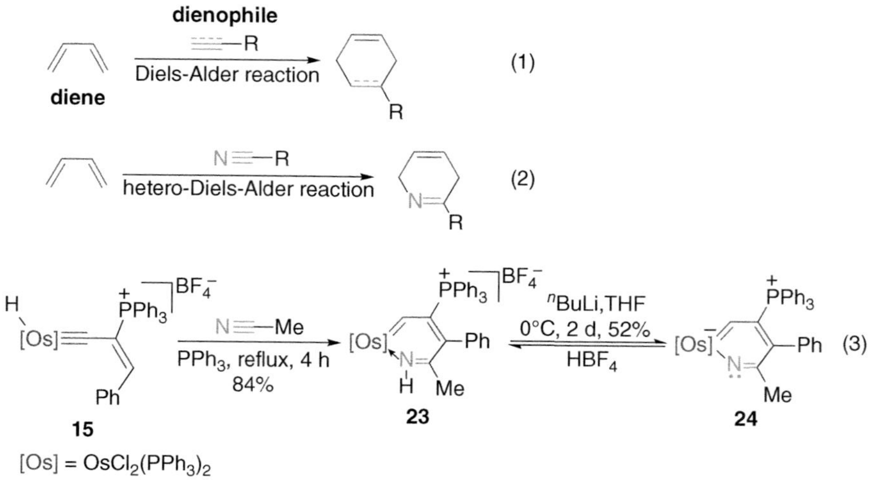

Scheme 6.12

cycloaddition product **23** is formed in good yield as predicted [24, 25]. The deprotonation of **23** by *n*-BuLi produces osmapyridine **24**, which can be protonated to regenerate **23** in the presence of an acid. The interconversion between **23** and **24** resembles the interconversion between pyridinium and pyridine.

The osmapyridine **24** is recognized as the first aza-metallabenzene with aromaticity (the structure is discussed in Section 6.2), because the first metallapyridine (tantalapyridine) does not adopt a π-electron delocalized structure [26]. The mechanism provided for the formation of the osmapyridinium **23** and the osmapyridine **24** shown in Scheme 6.13 involves a formal [4+2] cycloaddition step. In this mechanism, the migration of the hydride ligand to the carbyne carbon atom should generate an alkenylcarbene intermediate that acts as a diene. The coordination of the acetonitrile molecule

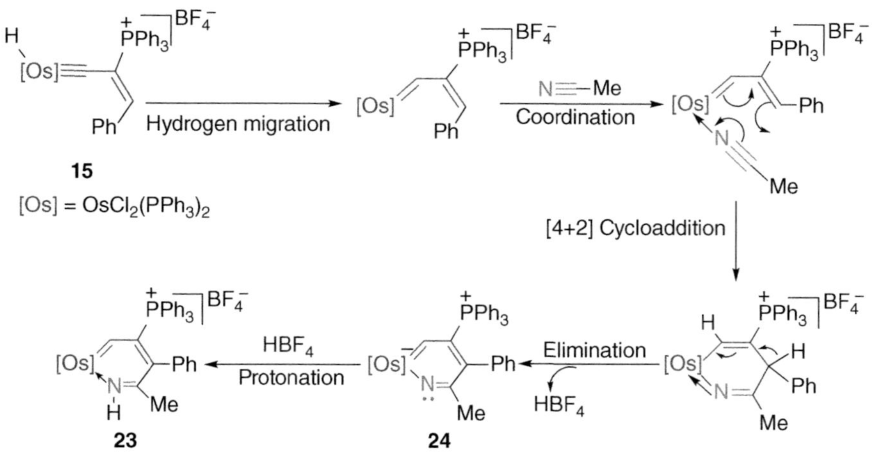

Scheme 6.13

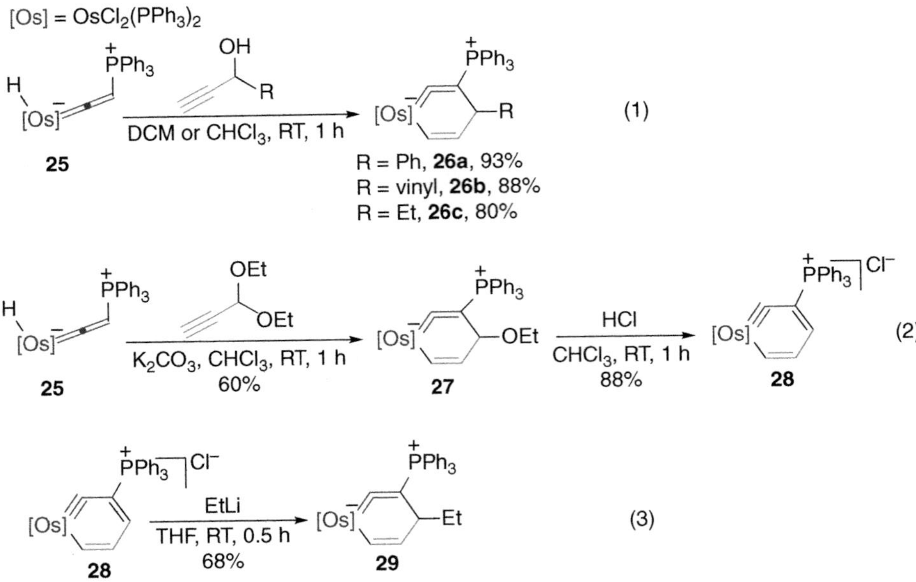

Scheme 6.14

could facilitate the following cycloaddition, resulting in a six-membered metallacyclic intermediate. The subsequent elimination of HBF_4 can produce the aromatic osmapyridine **24**, which can easily undergo protonation by HBF_4 to generate the osmapyridinium **23**.

6.1.3 [3+3] Synthesis

According to Scheme 6.2, vinylidene complexes can serve as substrates of [3+3] synthesis to produce six-membered metallacycles. Two examples are described here. The reaction of osmium hydride vinylidene **25** with alkynols can prepare a series of isoosmabenzenes with different substituents in high yield (Scheme 6.14, Equation (1)) [27]. Similarly, the isoosmabenzene **27** can be obtained from the reaction of **25** with $HC{\equiv}CCH(OEt)_2$ (Scheme 6.14, Equation (2)) [28]. The further conversion of **27** in the presence of acid results in the phosphonium-substituted osmabenzyne **28**. **28** is relatively sensitive to nucleophiles, and it can react with lithium ethide to restore the isoosmabenzene structure, as shown in Scheme 6.14, Equation (3).

A schematic mechanism of these reactions is shown in Scheme 6.15. The insertion of a terminal alkyne into the Os–H bond may generate a vinyl complex, which can undergo C–C coupling and elimination to form the formal [3+3] cycloaddition product **27**. With the aid of acid, the protonation of **27** affords an onium salt intermediate. The subsequent elimination step could allow the formation of osmabenzyne **28**. The nucleophilic addition of a strong nucleophile to the osmabenzyne **28** can restore the isoosmabenzene structure to give **29**.

Note that isoosmabenzenes **26** are very stable. They can remain unchanged after heating in air at 100°C for 5 h in the solid state. The origin of the stability was studied by

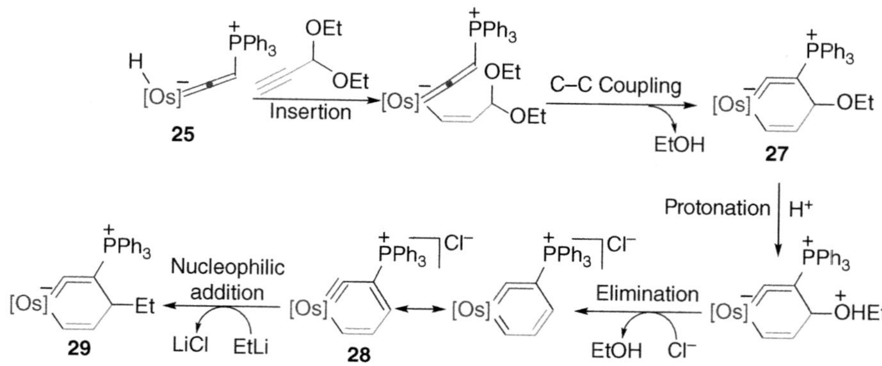

Scheme 6.15

density functional theory (DFT) calculations, which suggest that the phosphonium substituent plays a crucial role in stabilizing isoosmabenzene complexes [10, 27].

The general and convenient approaches to phosphonium-substituted metallabenzenoid compounds have been summarized and exemplified by the above discussion. In these approaches, various species are used as starting materials towards the six-membered metallacycles. The three possible synthetic methods outlined in Scheme 6.2 (i.e. [5+1], [4+2] and [3+3] syntheses) can all be realized in this special metallabenzene system.

6.2 Structure and Bonding

There are several approaches that can help to provide an understanding of the structures of metallabenzenoid compounds bearing phosphonium substituents. Metallabenzene chemistry usually requires X-ray crystallography for characterization. Once a given structure has been determined, it is sometimes possible to use spectroscopic methods to deduce the structures of closely related species. Structures **2A–2F** represent several possible contributing resonance structures of the osmabenzene **2** (Scheme 6.16), as indicated by observation using nuclear magnetic resonance (NMR) spectroscopy and X-ray crystallography characterization [15, 16]. The resonance forms shown in Scheme 6.16 illustrate how phosphonium groups help to stabilize the metallabenzene ring.

The molecular structures of representative metallabenzenoid compounds **2**, **4** and **24** are shown in Figure 6.1 with thermal ellipsoids drawn at 50% probability (the hydrogen

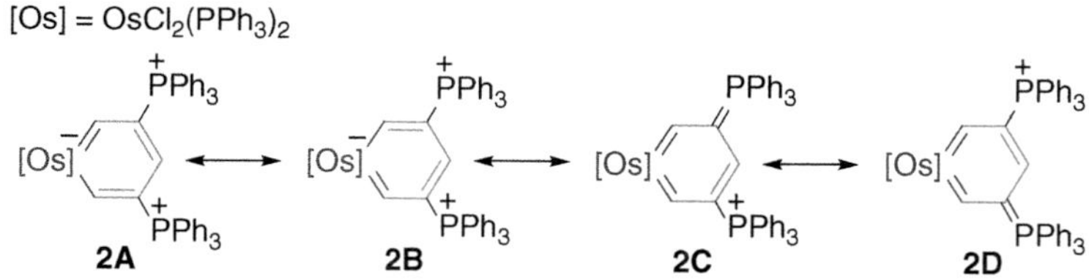

Scheme 6.16

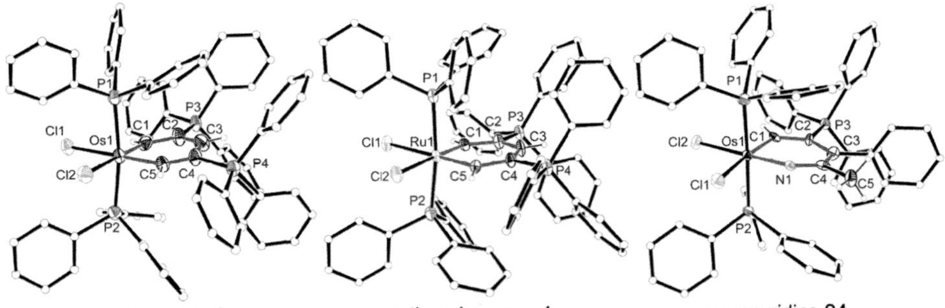

Figure 6.1 Molecular structures of osmabenzene **2**, ruthenabenzene **4** and osmapyridine **24**. The hydrogen atoms of PPh_3 are omitted for clarity. (*See color plate section for the color representation of this figure.*)

atoms of PPh_3 are omitted for clarity), and selected bond lengths and angles are provided in Table 6.1. The X-ray structures clearly show that the complexes have essentially planar six-membered metallacycles with phosphonium substituents. The co-planarity is reflected by the small root mean square (RMS) deviations from the least-squares plane through the six atoms of the metallacycles [15–18, 24]. Notably, the C–C bond distances of the metallacycles are longer than typical C=C double bonds and shorter than typical C–C single bonds, which are very close to the C–C distances of their

Table 6.1 Selected bond lengths (Å) and angles (°) of osmabenzene **2**, ruthenabenzene **4** and osmapyridine **24**.

	Osmabenzene 2 (M = Os, X = C5)	Ruthenabenzene 4 (M = Ru, X = C5)	Osmapyridine 24 (M = Os, X = N1)
Bond lengths (Å)			
M1–C1	1.946(12)	1.922(5)	1.969(8)
C1–C2	1.398(15)	1.382(7)	1.397(11)
C2–C3	1.373(16)	1.395(7)	1.447(11)
C3–C4	1.448(17)	1.378(7)	1.396(12)
C4–X	1.363(15)	1.391(7)	1.326(10)
X–M1	1.971(12)	1.911(5)	1.974(6)
C2–P3	1.775(12)	1.790(5)	1.792(8)
C4–P4	1.782(12)	1.792(5)	—
Bond angles (°)			
M1–C1–C2	128.5(10)	129.5(4)	125.7(6)
C1–C2–C3	123.1(12)	122.2(5)	126.0(8)
C2–C3–C4	126.1(13)	125.2(5)	122.4(8)
C3–C4–X	121.6(11)	123.2(4)	123.7(8)
C4–X–M1	129.5(9)	128.6(4)	131.5(6)

Table 6.2 Selected NMR spectroscopic data of osmabenzene **2**, ruthenabenzene **4** and osmapyridinium **23**.

	osmabenzene 2 (M = Os)	ruthenabenzene 4 (M = Ru)	osmapyridinium 23 (M = Os)
^{1}H NMR (ppm)			
M*CH*	23.13 (d, ^{3}J(PH) = 17.4 Hz)	17.5 (d, ^{3}J(PH) = 18.9 Hz)	48.01 (d, ^{3}J(PH) = 20.3 Hz)
γ-*CH*	8.57 (t, ^{3}J(PH) = 14.5 Hz)	8.2 (t, ^{3}J(PH) = 14.0 Hz)	—
N*H*	—	—	25.38 (s)
CH_3	—	—	1.98 (s)
^{31}P{^{1}H} NMR (ppm)			
M*P*Ph_3	−15.5 (s)	8.1 (s)	11.27 (s)
C*P*Ph_3	20.1 (s)	18.2 (s)	13.56 (s)
^{13}C{^{1}H} NMR (ppm)			
C^1	239.7 (dt, ^{2}J(PC) = 35.9 Hz, ^{2}J(PC) = 8.5 Hz)	284.3 (d, ^{2}J(PC) = 11.6 Hz)	201.2 (t, ^{2}J(PC) = 9.1 Hz)
C^2	112.7 (dd, ^{1}J(PC) = 73.1 Hz, ^{3}J(PC) = 12.6 Hz)	108.3 (dd, ^{1}J(PC) = 73.2 Hz, ^{3}J(PC) = 12.1 Hz)	123.2 (d, ^{1}J(PC) = 69.8 Hz)
C^3	160.5 (t, ^{2}J(PC) = 21.0 Hz)	146.0 (t, ^{2}J(PC) = 21.4 Hz)	147.7 (d, ^{2}J(PC) = 11.8 Hz)
C^4	—	—	234.4 (d, ^{3}J(PC) = 13.8 Hz)

all-carbon aromatic analogues. The transition metal–carbon bond distances are also intermediate between the typical transition metal–carbon single and double bond distances. Even the C–P bond distances between the carbon atoms of the metallacycles with the phosphonium substituents are nearly equivalent. The observations of coplanarity and the lack of significant short/long alternations in the bond distances of the six-membered metallacycles suggest that the phosphonium-substituted metallabenzenoid compounds have electron-delocalized structures.

Support for the electron-delocalized nature of the metallacycles also comes from values obtained from NMR spectroscopy (Table 6.2) and seem to indicate considerable aromatic character (the irregular broad ^{1}H NMR signals of osmapyridine **24** and the unusually low-field ^{1}H NMR signals of osmapyridinium **23** are attributed to their paramagnetism, as is discussed in Section 6.4). In particular, the signals of the CH groups directly attached to the transition metal centre exhibit distinctive very low-field chemical shifts, which are attributed to the influences of the metal and the phosphonium substituents. The resonances of the other CH groups within the metallacycles are observed at values comparable with those of typical arenes.

Taking into account all of these experimental structural observations, aromaticity can be assigned to these metallabenzenoid compounds bearing phosphonium substituents. Meanwhile, all of these metallabenzenoid compounds contain at least one bulky phosphonium substituent that has a protecting effect (see Section 6.1.3). In this context, the high stability of these metallabenzenes is not surprising, although most organometallic compounds are very sensitive to air and moisture. For example, a solid sample of osmabenzene **2** remains nearly unchanged after being heated at 120°C for one day in air [15]. The thermal stability of ruthenabenzene **4** is also remarkable, in that in the solid state it can sustain 100°C for at least 5 h [17]. However, these metallabenzenoid compounds also present rich reactivity in the presence of appropriate reagents, as is classified in the next section.

6.3 Reactions

As predicted by the aromaticity of metallabenzenes, the typical reactivities of arenes are expected for metallabenzenes, although they are not as aromatic as their all-carbon analogues. Reactions of metallabenzene rings are of great interest because some reactions are now available in metallabenzene systems that would be difficult or impossible to achieve using conventional carbon-based aromatic rings. In this section, we will look at some reactions typical of aromatic compounds and some reactions that are significantly different from their all-carbon analogues.

6.3.1 Ligand Substitution Reactions

Transition metal centres of metallabenzenes produce rich reactivity patterns with the usual organometallic ligands. New metallabenzenes can be obtained from metallabenzenoid compounds synthesized using the methods described in the previous section through ligand substitution reactions. Typical examples are shown in Scheme 6.17. The chloride ligands in osmabenzene **2** or ruthenabenzene **4** can be replaced by one or two 1e$^-$ ligands (i.e. SCN^- or $N(CN)_2^-$) [16, 18]. 2e$^-$ ligands, such as pyridine and *tert*-butyl isocyanide (*t*BuNC), can also be used to generate mono-substituted metallabenezenes [2, 4]. Reactions of metallabenzenes with phosphorous ligands leads to the formation of metallabenzenes with three 2e$^-$ ligands binding to the metal centre [16, 18].

Metallabenzenoid compounds with phosphonium substituents can also undergo ligand substitution reactions with bidentate ligands, such as 2,2′-bipyridine, 1,10-phenanthroline and 8-hydroxyquinoline [16, 18, 29, 30]. The reactions are illustrated in Scheme 6.17. New metallabenzene compounds are formed through the replacement of one Cl and one PPh_3 ligand by one bidentate ligand. It is noteworthy that all of the metallabenzene rings of the compounds that are derived from ligand substitution reactions with bidentate ligands deviate significantly from planarity. However, the C–C bond distances of the metallacycles are still intermediate between those of double and single bonds. The lack of significant alternations in the C–C bond distances suggests that these metallabenzenes have a π-delocalized structure, although the metallacycles are not planar. With the aid of DFT calculations, the non-planarity found in these metallabenzene compounds was attributed to electronic and steric factors induced by the unsymmetrical coordination environment of the metal centre [31].

Bismetallabenzenes and their analogues are rare. Ligand substitution reactions allow us to assemble two metallabenzene rings together by bridging the ligands to the other metal centres. Chloric groups usually have a high tendency to form M–Cl–M bridges. Treatment of ruthenabenzene **4** with $AgBF_4$ in CH_2Cl_2 generates binuclear bridged complexes [18]. The structure shown in Scheme 6.17 demonstrates that there are three M–Cl–M bridges. In addition, ligand substitution reactions of osmabenzene **2** with

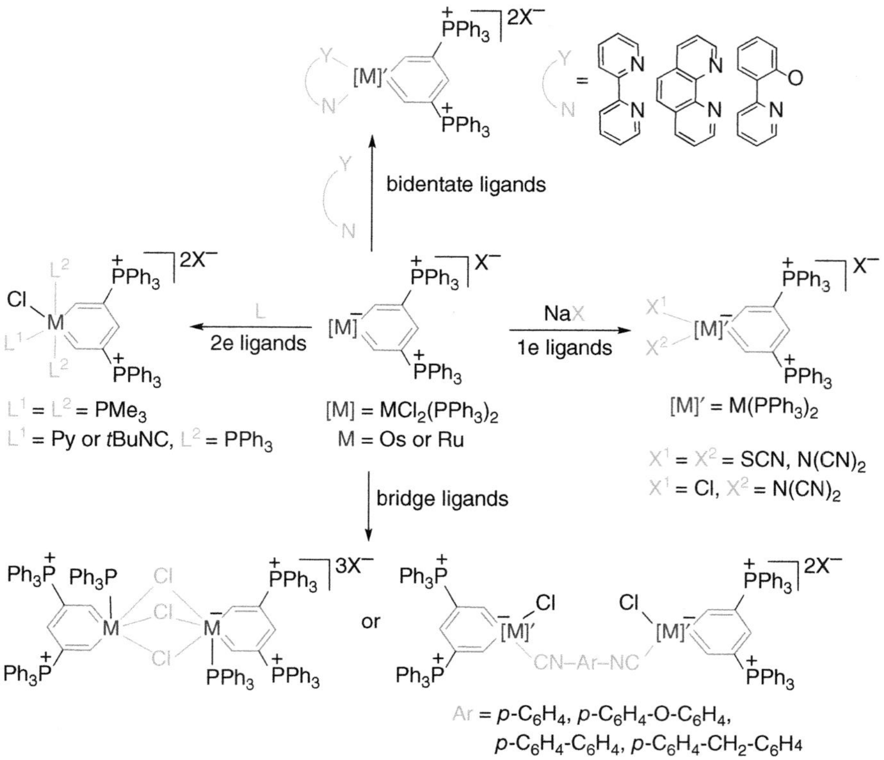

Scheme 6.17

various diisocyanides leads to a series of diisocyanide-bridged bisosmabenzenes (Scheme 6.17) [32]. The electrochemical properties of these bismetallabenzenes have been investigated, revealing that the two metal centres interact with each other through the bridges (see Section 6.4).

6.3.2 Nucleophilic Addition and Nucleophilic Aromatic Substitution Reactions

The substitution reaction is by far the most common type of reaction that benzene rings undergo. Typically, aromatic compounds like benzene tend to undergo substitution reactions with electrophiles, which are called electrophilic aromatic substitutions (S_EAr). Electrophilic substitution is a distinctive feature of aromatic compounds and has generally been regarded as evidence of aromatic character. When strong electron-withdrawing groups are attached to the aromatic rings, the carbons of the aromatic rings are susceptible to nucleophilic attack. The metallacycles of phosphonium-substituted metallabenzenoid compounds are therefore more likely to be attacked by nucleophiles rather than electrophiles. As shown in Scheme 6.18, the phosphorous ligand – i.e. bis(diphenylphosphino)methane (DPPM) – can act not only as a bidentate ligand but also as a nucleophilic reagent [33]. The reaction of osmabenzene **2** with DPPM in a 1:1 molar ratio affords the ligand substitution product **30**, which can react further with DPPM to generate the cyclic osmium η^2-allene complex **31**. For the

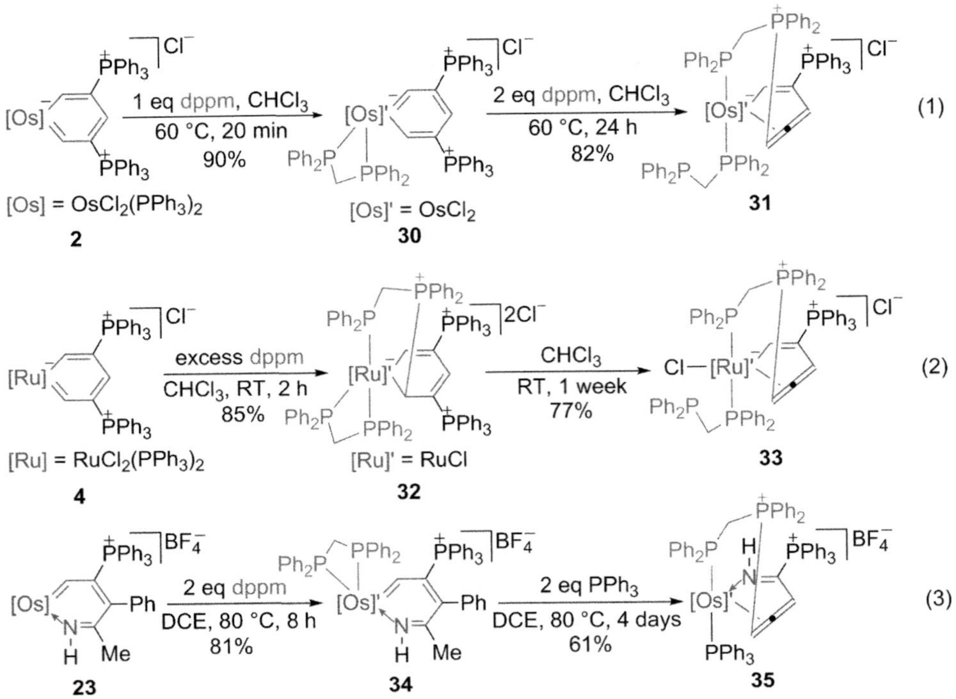

Scheme 6.18

analogous ruthenabenzene **4**, the reaction with DPPM (in a 1:3 molar ratio) leads to the formation of the ruthenacyclohexadiene complex **32**, which can slowly transform in $CHCl_3$ to η^2-allene complex **33**. Formation of the η^2-allene complex was also observed in the reaction of osmapyridinium **23** with DPPM. The complex reacted with DPPM (in 1:2 molar ratio) to produce the substituted osmapyridinium **34**. Heating the osmapyridinium **34** in DCE at 60°C for four days generated the η^2-allene complex **35**.

Scheme 6.19 shows a proposed mechanism for the reactions of phosphonium-substituted metallabenzenes with DPPM. The process involves multiple ligand substation reactions of DPPM to the metal centre. Then, the intramolecular nucleophilic addition of the DPPM to the aromatic metallacycle affords a metallacyclohexadiene, which could be regarded as the typical Meisenheimer complex in nucleophilic aromatic substitution (S_NAr). Probably due to the ring strain of the DPPM-tethered metallacycle, the cleavage of the C–P bond may occur to generate the related η^2-allene complex. The key intermediates of the reactions have been isolated and fully characterized, further supporting this proposed mechanism.

Similar to DPPM, 8-hydroxyquinoline can provide various binding modes, in addition to being a nucleophile. As depicted in Scheme 6.20, the monosubstituted ruthenabenzene **36** is formed through the displacement of one chloride and one triphenylphosphine ligand by 8-hydroxyquinoline [34]. The further reaction of **36** with excess 8-hydroxyquinoline in the presence of CH_3COONa under an air atmosphere yields the fused-ring ruthenabenzene **37**. In the mechanism proposed for the formation of **37**, the reaction

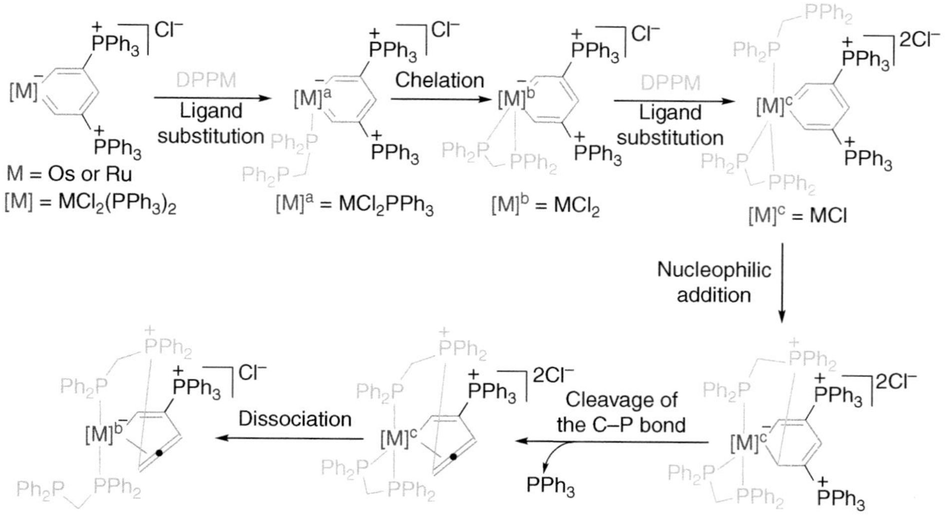

Scheme 6.19

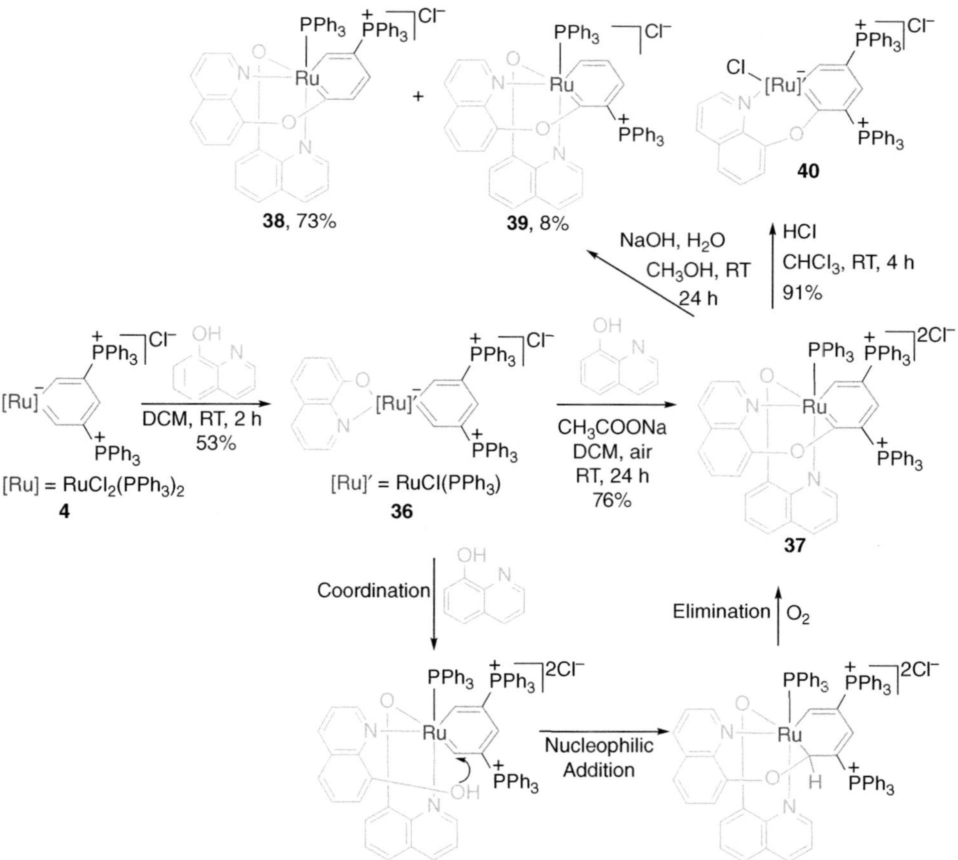

Scheme 6.20

Scheme 6.21

proceeds via the initial coordination of 8-hydroxyquinoline with the metal centre through an N atom and intramolecular S_NAr to form an O–C bond.

The ruthenabenzene **37** can tolerate weak alkali or weak acid in solution. The reaction of **37** with a base gives ruthenabenzenes **38** and **39**, presumably through the cleavage of the metallacycle P–C bond. **37** is also reactive to strong HCl, resulting in a transformation to ruthenabenzene **40** in high yield. Thermal stability experiments demonstrate that these multicyclic metallabenzenes have remarkable stability both in the solid state and in solution under an air atmosphere. **38** and **39** can be heated at 170°C for at least 5 h without appreciable decomposition. **38** and **39** were found to be fluorescent in solution and have similar spectral behaviours to the analogous organic multicyclic compounds (see Section 6.4).

The ruthenabenzene **37** undergoes nucleophilic addition with sodium borohydride to give the first example of a ruthenacyclohexa-1,4-diene in good yield (Scheme 6.21) [35]. Examples of iridacyclohexa-1,4-dienes and osmacyclohexa-1,4-dienes by the nucleophilic addition of BH_4^- to cationic iridabenzenes and osmabenzenes have been documented [36]. Both the iridacyclohexa-1,4-dienes and osmacyclohexa-1,4-dienes oxidize slowly under air to restore the original metallabenzenes. The ruthenacyclohexa-1,4-diene **41** was oxidized much more easily than the iridacyclohexa-1,4-dienes and osmacyclohexa-1,4-dienes. When a solution of **41** in DCM is stirred for 12 h under an oxygen atmosphere at room temperature, **41** was completely consumed to give **37** as an exclusive product. The interconversion of ruthenabenzene **37** to ruthenacyclohexa-1,4-diene **41** has been investigated by means of DFT calculations. A theoretical study showed that the electronic effect, together with the steric effect, contributes to the regioselectivity of the nucleophilic attack at the γ-carbon atom of ruthenabenzene **37**.

The S_NAr reactions with 8-hydroxyquinoline can be extended to osmabenzene **2**, as shown in Scheme 6.22 [29]. In the case of osmabenzene **2**, the 8-hydroxyquinoline monosubstituted osmabenzene **42** converted to η^2-allene-coordinated complex **43** in the presence of excess PPh_3 and NaOH. It was assumed to undergo the cleavage of the P–C bond with the aid of a base. The expected S_NAr reaction product **44** was formed by the treatment of **43** with excess 8-hydroxyquinoline under air. Another S_NAr reaction product **45** was obtained through the reaction of **43** with PPh_3 in the presence of acid. The resulting interconversion of metallabenzenes and cyclic η^2-allene-coordinated complexes produces new access to unsaturated metallacycles, starting from metallabenzenoid compounds with phosphonium substituents.

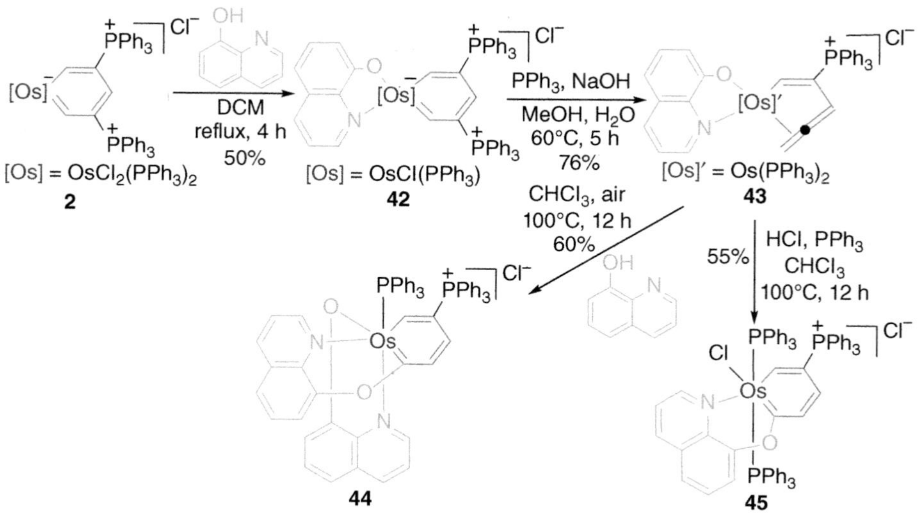

Scheme 6.22

Most aromatic substitution reactions are *ipso*-substitutions (Scheme 6.23, Equation (1)), in which a new substituent is introduced to the same ring position as the departing leaving group. By contrast, other patterns of aromatic substitution, for example *cine*-substitution, are much less common. In *cine*-substitution, the entering group takes up a position adjacent to that previously occupied by the leaving group. The reported substitution reactions of metallabenzenes are limited to *ipso*-substitution, except for the examples of phosphonium-substituted metallabenzenes demonstrated

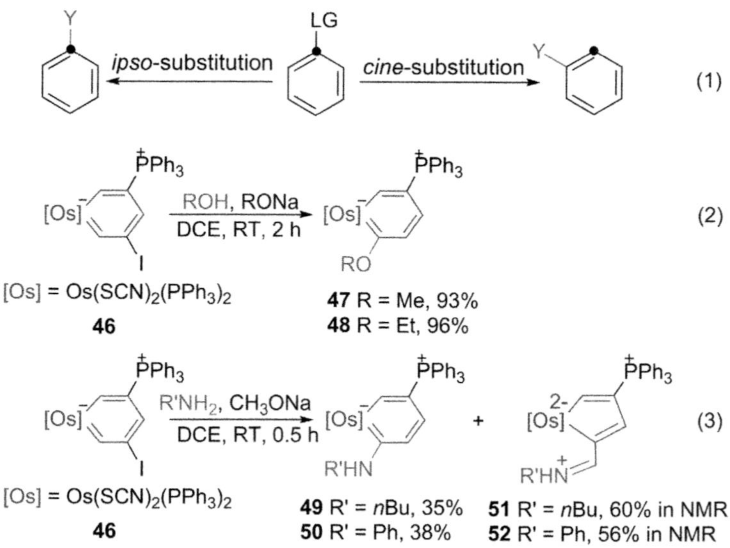

Scheme 6.23

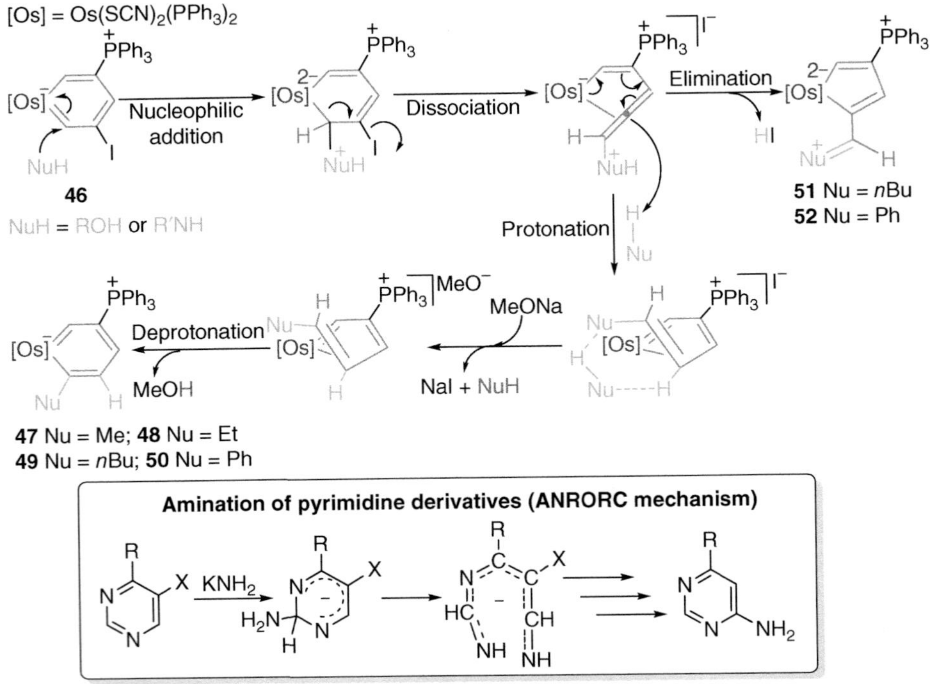

Scheme 6.24

in Scheme 6.23 [37]. The more alkali-resistant osmabenzene **46** is derived from osmabenzene **6** by a ligand substitution reaction. In the presence of a strong alkali, osmabenzene **46** undergoes an S_NAr with methanol or ethanol to give the *cine*-substitution products **47** or **48**. The reactions of **46** with amines, such as *n*-butylamine or aniline, afford the desired *cine*-substitution products **49** and **50**, along with the five-membered metallacyclopentadiene species **51** and **52** under similar reaction conditions. In the absence of sodium methoxide, **46** completely converts to the corresponding metallacyclopentadienes in 5 min.

Detailed mechanisms of the reactions have been studied with the aid of *in situ* NMR experiments and isotopic labelling experiments. As outlined in Scheme 6.24, the *cine*-substitution reactions occur via nucleophilic addition, dissociation of the leaving group, protonation and deprotonation steps, which resemble the classical addition of a nucleophile, ring-opening, ring-closure (ANRORC) mechanism. DFT calculations suggest that the formation of metallacyclopentadiene species in the reaction with alcohols is both kinetically and thermodynamically less favourable. Consistent with the calculation results, only the *cine*-substitution product was observed in the reaction with alcohols. For the analogous reaction with amines, the pathway for the formation of the *cine*-substitution product is kinetically less favourable than the pathway for the formation of the metallacyclopentadiene species, but is much more thermodynamically favourable. These calculation results are supported by the fact that the experimental conversion of **51** to the *cine*-substitution product **49** is observed in an *in situ* NMR experiment carried out for on an isolated pure sample of **51**.

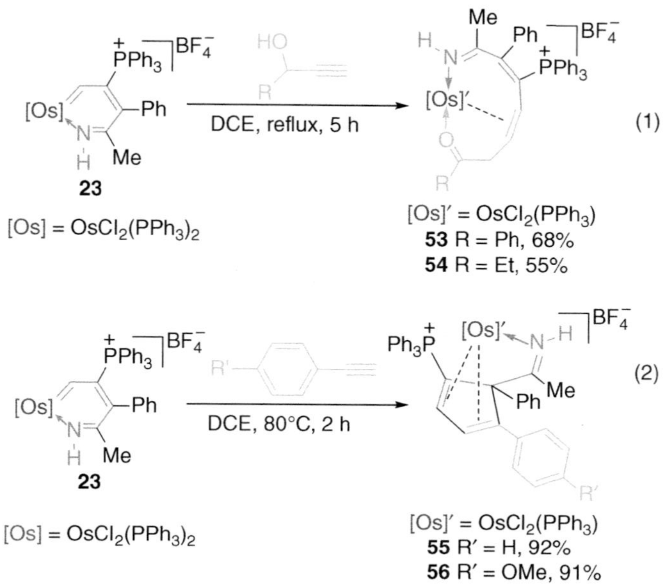

Scheme 6.25

S_NAr reactions are now well established as another good illustration of aromatic character as electron-poor aromatic rings allow substitution reactions with nucleophiles rather than the usual electrophiles. Electron-withdrawing groups *ortho* or *para* to a potential leaving group are crucial for S_NAr reactions. For phosphonium-substituted metallabenzenes, the key step of S_NAr can be facilitated by the phosphonium substituents reducing the electron density of the metallabenzene rings. Therefore, the typical S_NAr reactions can occur, providing strong chemical support for the aromatic character of the metallabenzene ring.

6.3.3 Cycloaddition Reactions

Metallabenzenes can undergo cycloaddition reactions with unsaturated molecules [1–14]. Similar reactions with terminal alkynes also occur in phosphonium-substituted osmapyridinium **23** (Scheme 6.25) [38]. Reactions of osmapyridinium **23** with excess propargyl alcohols in DCM under reflux lead to the formation of the 10-membered osmacycles **53** and **54**, whereas reactions of osmapyridinium **23** with phenylacetylene or substituted phenylacetylene afford the η^4-coordinated cyclopentadiene complexes **55** and **56**.

Scheme 6.26 shows the proposed mechanism of cycloaddition reactions of **23** with terminal alkynes. Coordination of the terminal alkyne to the metal centre in **23** accompanied by the dissociation of the PPh_3 ligand could generate a π-alkyne intermediate. For propargyl alcohols, [2+2] cycloaddition and cycloreversion steps are proposed to form the nine-membered intermediate. The four-membered product of the [2+2] cycloaddition could relieve ring strain by undergoing cycloreversion to form the nine-membered intermediate. The subsequent hydrogen migration and coordination of the carbonyl group to the metal centre could yield the final 10-membered osmacycle **53** or

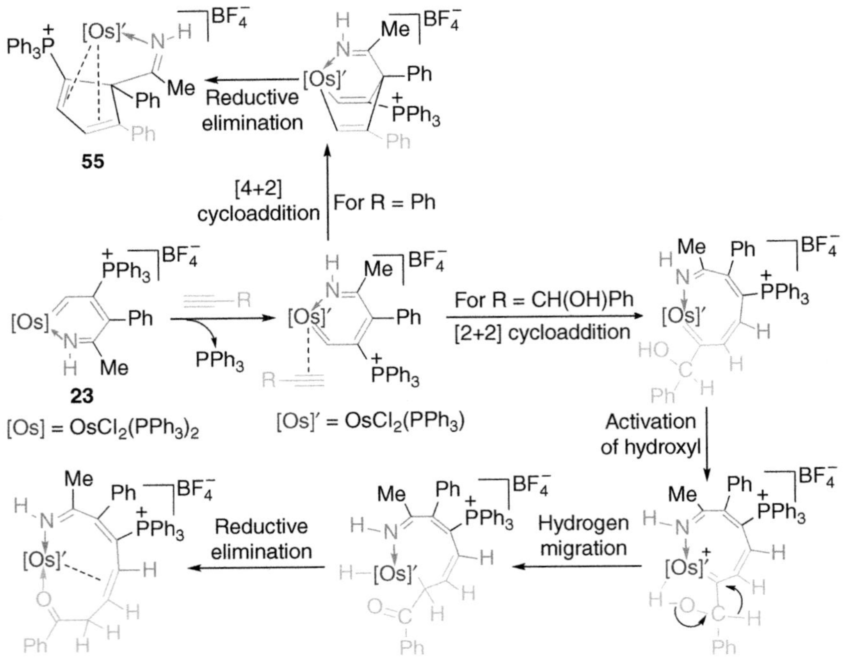

Scheme 6.26

54. For phenylacetylenes, [4+2] cycloaddition of the osmium vinyl carbene fragment with the alkyne is postulated to achieve the metallacyclohexadiene intermediate, which could undergo reductive elimination to generate η^4-coordinated cyclopentadiene complexes **55** or **56**. These cycloaddition reactions of phosphonium-substituted osmapyridinium **23** may be attributed to the carbene bond character of the osmapyridinium **23**.

6.3.4 Formation of Unsaturated Organic Rings

Metallabenzenes are frequently proposed as key intermediates for the formation of η^5- or η^1-cyclopentadienyl complexes. There is firm experimental evidence that a number of conversions involving isolated or characterized metallabenzenes have been observed [8]. Metallabenzenes with phosphonium substituents have been found to convert to free cyclopentadienyl derivatives (Scheme 6.27). Heating a solid sample of ruthenabenzene **4** at 150°C produces cyclopentadienyl anion bearing phosphonium substituents **57** [18]. A similar decomposition occurs when a solid sample of ruthenabenzene **40** is heated at 170°C [34]. The thermal reaction product is characterized as cyclopentadienyl anion derivative **58**. Conversion to the free cyclopentadienyl anion is not common in metallabenzene compounds, but is observed here due to the presence of two bulky phosphonium substituents in the metallabenzene rings.

The transformations to η^5-cyclopentadienyl complexes were also detected in phosphonium-substituted isometallabenzenes (Scheme 6.27) [27]. The rearrangement of **26a** to **59a** was very slow, taking more than one week, whereas the rearrangements of

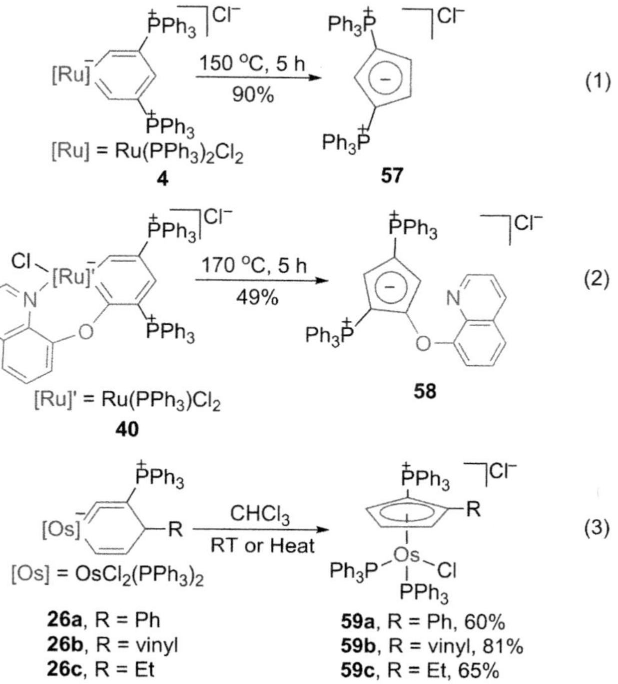

Scheme 6.27

the **26b** and **26c** were faster, which afforded the η^5-cyclopentadienyl complexes **59b** or **59c** within three or two days, respectively. Mechanisms for the isomerization of isometallabenzenes **26** to η^5-cyclopentadienyl complexes **59** were probed via the metallated cyclopentadiene intermediate.

It was found that, on treatment of a DCM solution of ruthenabenzene **4** with nitric acid in DCM for 5 h followed by the addition of $NaHCO_3$, dialdehyde compound **60** can be isolated from solution in 93% yield (Scheme 6.28) [39]. Dialdehyde compound **60** has remarkable thermal stability, with a solid sample remaining virtually unchanged even when heated at 200°C in air for one day. Interestingly, this dialdehyde also has unusual anti-oxidative properties, making it even compatible with nitric acid. Electrochemical experiments indicate the excellent electrochemical anti-oxidative ability of **60**. The

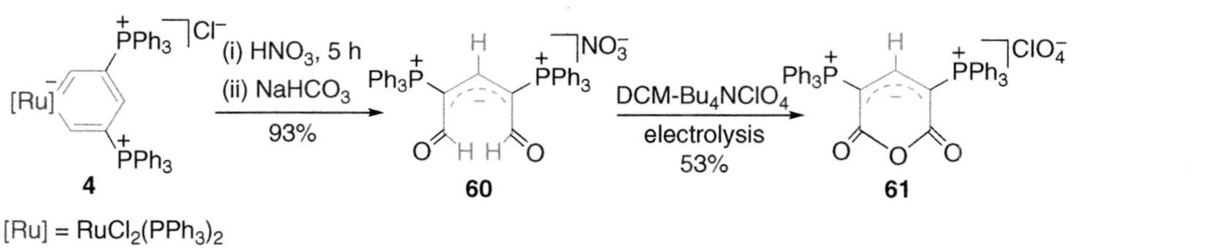

Scheme 6.28

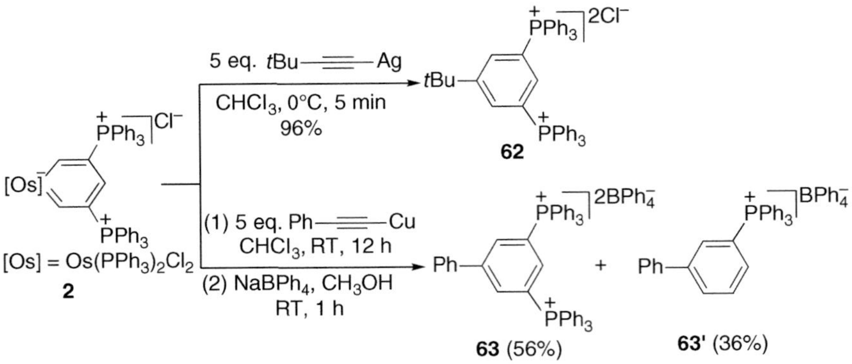

Scheme 6.29

controlled potential electrolysis of **60** was also investigated, affording the isolation of a stable cyclic anhydride **61**.

The conversions of phosphonium-substituted metallabenzene depicted above result in a series of unsaturated organic rings bearing phosphonium substituents. One of the most important phosphonium salts is tetraphenylphosphonium salt, which has gained increasing importance in various fields of chemistry ranging from organic synthesis to material science. The typical synthetic methods for tetraphenylphosphonium salts are palladium or nickel-catalysed reactions of aryl halides or triflates [40–46]. Quaternizations of phosphines are normally employed as metal-free methods, which often require high temperature and/or harsh reaction conditions [47–52]. The conversion of phosphonium-substituted metallabenzene provides a facile synthetic strategy for the preparation of tetraphenylphosphonium salts (Scheme 6.29) [53]. The reaction of osmabenzene **2** with acetylide *t*BuC≡CAg at 0°C produces the dicationic phosphonium salt **62** in high yield. As an alternative transition-metal acetylide, PhC≡CCu reacts with osmabenzene **2** to give a mixture of dicationic phosphonium salt **63** and cationic phosphonium salt **63′**.

A putative mechanism accounting for the transformations of the osmabenzene **2** to the tetraphenylphosphonium salts **62** or **63** is illustrated in Scheme 6.30. As shown in the first four steps, the reaction of osmabenzene **2** with a silver/copper acetylide can produce an acyl intermediate via an alkynyl intermediate and a vinylidene intermediate. The transformations of alkynyl complexes to alkyl-carbonyl complexes with the aid of water molecules are well-established processes [54–57]. Then, acyl decarbonylation and the migration of the alkyl ligand generate the metallacyclohexadiene intermediate. The β-H elimination could generate the η^2-alkenyl-coordinated osmium complex as an intermediate, which would easily convert to the tetraphenylphosphonium salt **62** or **63** and an osmium dihydride complex after a 6-*endo-trig* cyclization. The tetraphenylphosphonium salt **63′** is probably formed by the P–C bond cleavage of **63**, as suggested by the signal of triphenylphosphine oxide in the *in situ* NMR. Further experimental evidence for the proposed reaction mechanism comes from *in situ* experiments. For instance, only a trace amount of tetraphenylphosphonium salt was observed within one week when the reactions of **2** with silver/copper acetylide were carried out in DCM pretreated with calcium hydride (to remove water and acid). In addition, a strong band at 1916.87 cm^{-1} in the infrared (IR) spectrum of the crude product mixture could demonstrate the presence of a carbonyl ligand.

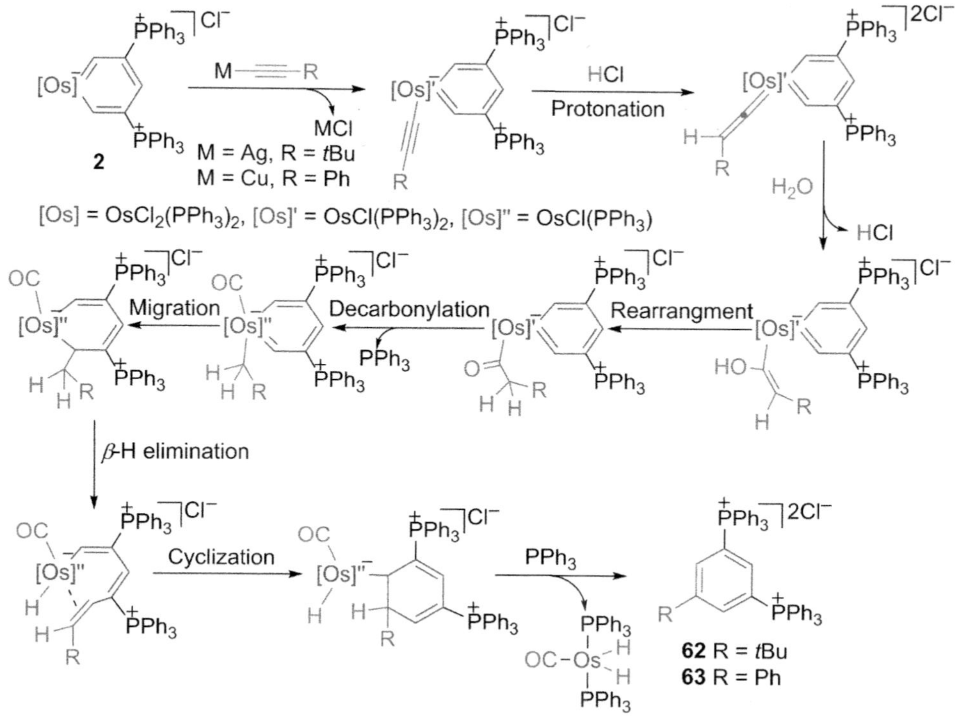

Scheme 6.30

A special group of metallabenzenes that undergo different reactions was explored in this section to produce a diverse array of structures from similar starting materials. Such processes can convert phosphonium-substituted metallabenzenes into a number of different unsaturated molecules, including metallacycles and organic rings. The synthetic possibilities of the reactions of phosphonium-substituted metallabenzenes are promising, but the key to unlocking that potential is a further investigation of the reactivity of these metallabenzenes and an understanding of the rules that determine how these reactions can be achieved.

6.4 Physical Properties

Most organometallic complexes are air-sensitive, which usually means that they react with O_2 and sometimes with water. Because they are so reactive, organometallic complexes are usually stored at low temperature and have to be handled under a dry, inert atmosphere of nitrogen or argon. The main difference of phosphonium-substituted metallabenzenes from usual organometallic compounds is that they exhibit remarkable stability towards air, water and heat (Table 6.3). In some cases, they can even be compatible with bases and acids. The unique stability of phosphonium-substituted metallabenzenes not only facilitates the investigation of their physical properties but also makes it possible to utilize their properties. Unusual physical properties are sometimes exhibited by these special metallabenzenes, as shown in the following discussion.

Table 6.3 Thermal decomposition data of phosphonium-substituted metallabenzenes in solid state.[a]

	100°C	120°C	150°C	170°C	190°C
2	•[b]	•	—	—	—
4	•	△[c]	■$_1$[d]	—	—
8	•	—	—	—	—
16	•	•	—	—	—
17	•	•	—	—	—
23	•	•	—	—	—
36	•	△	■$_1$	—	—
37	•	△	■$_2$[e]	—	—
38	•	•	•	•	■$_3$[f]
39	•	•	•	•	■$_3$[f]
40	•	•	•	■$_4$[g]	—
69	•$_1$[h]	•$_1$	—	—	—
70	•$_1$	•$_1$	—	—	—

a) Most reactions were carried out for 5 h in air unless otherwise stated.
b) • = Stable.
c) △ = partly decomposed.
d) ■$_1$ = completely decomposed, major product is complex **57**.
e) ■$_2$ = completely decomposed, major product is complex **40**.
f) ■$_3$ = completely decomposed, major product is Ph_3PO.
g) ■$_4$ = completely decomposed, major product is complex **58**.
h) •$_1$ = carried out for 5 h in nitrogen.

6.4.1 Electrochemical Properties

The reaction of ruthenabenzene **4** with $AgBF_4$ leads to the formation of bisruthenabenzene **64** (Scheme 6.17), which is composed of several redox active units. The redox behaviour of ruthenabenzene **4** and bisruthenabenzene **64** was probed by cyclic voltammograms collected in DCM containing 0.10 M Bu_4NClO_4 as the supporting electrolyte [18]. When scanned at 100 mV/s in the positive potential direction, **4** undergoes a redox process (**A**) with $E_{p,a} = +0.79$ V, $\Delta E = 112$ mV (see the cyclic voltammogram shown in Figure 6.2). This quasi-reversible oxidation process is followed by a chemical reaction of the oxidized species. The oxidation wave may be assigned to the formation of $[Ru(CHC(PPh_3)CHC(PPh_3)CH)Cl_2(PPh_3)_2]^{2+}$. When voltammetric sweeping is performed in the negative potential direction, two waves for complex **4** appear at −0.75 V (**B**) and −1.02 V (**C**). Wave **B** is considered as one prewave, which means the reduction product is strongly adsorbed on the glassy carbon electrode. Wave **C** is an irreversible process because there is no corresponding oxidation current on the return scan.

In contrast to ruthenabenzene **4**, bisruthenabenzene **64** is more difficult to oxidize, as indicated in the anodic CV scan of **64** (Figure 6.2). The two successive reduction waves (**D** and **E**) of **64** suggest that there is a slight electronic interaction between the two metal centres through the chloric bridges. In addition, bisruthenabenzene **64** exhibits an irreversible reduction process at −1.15 V (**F**), similar to ruthenabenzene **4** at −1.02 V (**C**).

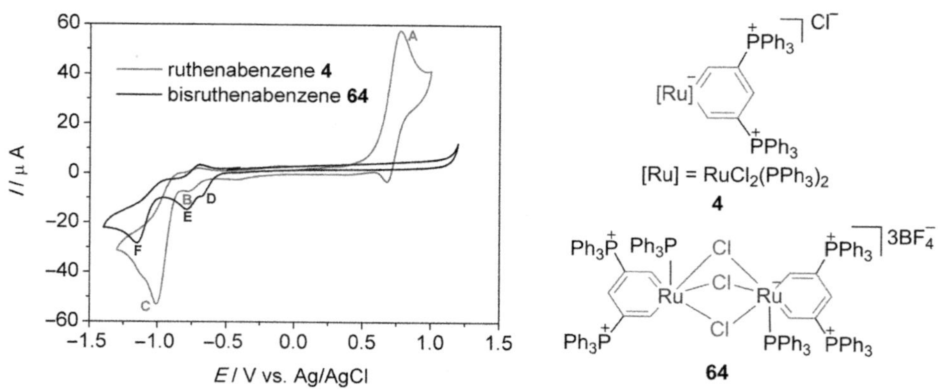

Figure 6.2 Cyclic voltammograms of **4** and **64** measured in DCM with 0.1 M nBu_4NClO_4 as the supporting electrolyte at a scan rate of 0.10 V s^{-1}. (*See color plate section for the color representation of this figure.*)

The electrochemical properties of phosphonium-substituted metallabenzenes are apparently modified on the replacement of different ligands binding to the metal centres. The redox properties of the osmabenzenes **2, 65, 66, 67** and **68** were examined by cyclic voltammetry (Figure 6.3). The replacement of Cl ligand by NCS or $N(CN)_2$ in **2** led to positive shift of oxidative potential. This might be thought that the ligands deplete the electron density of the metallabenzene ring. The cyclic voltammogram of osmabenzene **66** presents only an irreversible oxidation at 1.12 V attributed to the oxidation of free Cl ions. The result suggests that osmabenzene **66** are more difficult to be oxidized

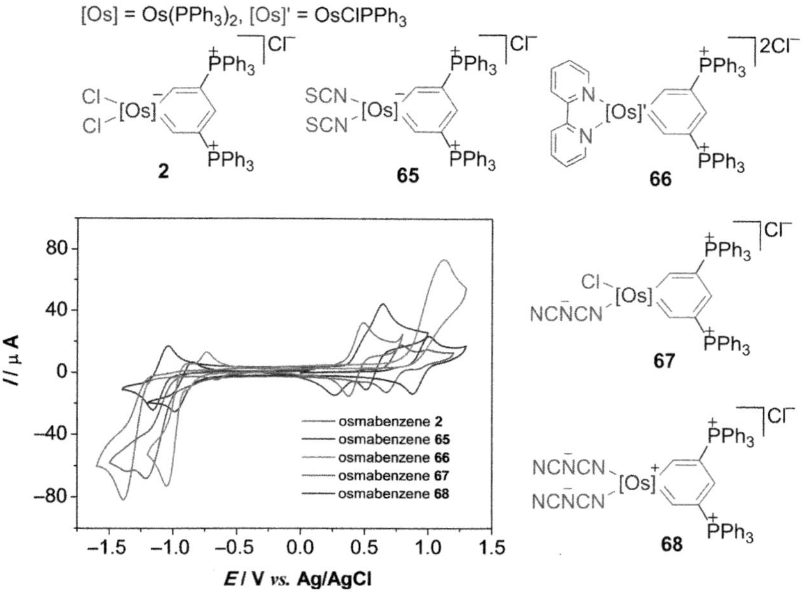

Figure 6.3 Cyclic voltammograms of **2, 65, 66, 67** and **68** measured in DCM with 0.1 M nBu_4NClO_4 as the supporting electrolyte at a scan rate of 0.10 V s^{-1}. (*See color plate section for the color representation of this figure.*)

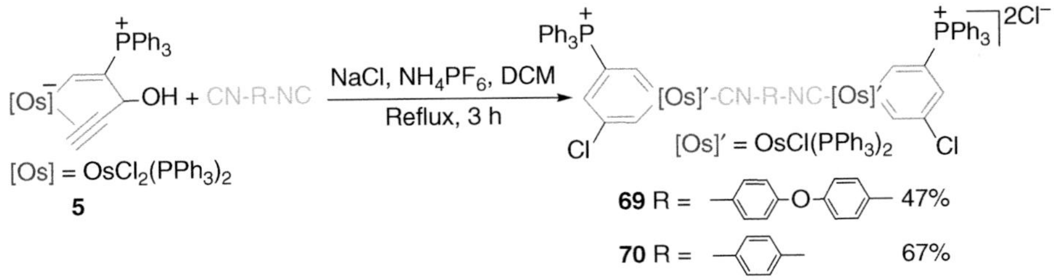

Scheme 6.31

compared with **2**, **65**, **67** and **68**, which may be due to the high valence state and the low electron density of the metal centre of **66**. Additionally, only **65** and **68** can undergo well-behaved, nearly reversible (or a quasi-reversible) reduction process, indicating that two NCS or $N(CN)_2$ ligands can help to stabilize the reduction states.

Diisocyanide-bridged bisosmabenzenes, containing one chloro and one phosphonium substituent on each metallabenzene ring, were obtained from the reactions of complex **5** with corresponding diisocyanides in the presence of NH_4PF_6 and NaCl via nucleophilic addition reactions (Scheme 6.31) [32]. The electrochemical properties of these stable bisosmabenzenes have been investigated (Figure 6.4). Similar to bisruthenabenzene **64**, weak but appreciable electronic interactions between the metal centres can be observed in bisosmabenzene **69**. The two oxidation waves at $E_{p,\,a}$= +0.58 V and $E_{p,\,a}$= +0.77 V are partly overlapping. Compared with **64** and **69**, a stronger interaction between the two metal centres is observed in bisosmabenzene **70**. As shown in Figure 6.3, bisosmabenzene **70** undergoes two consecutive one-electron oxidation processes chemically and electrochemically, resulting in redox waves at $E_{p,}$

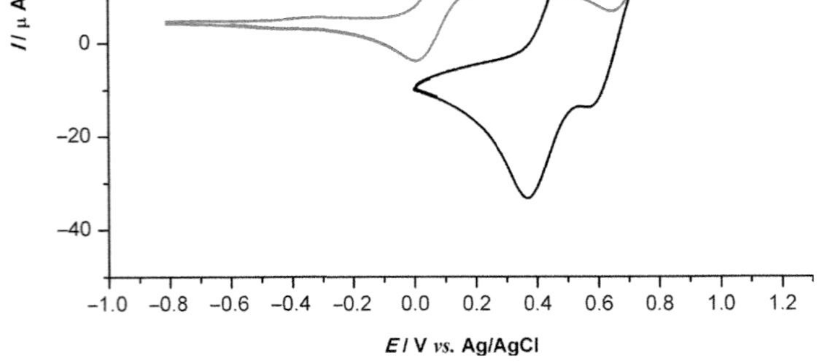

Figure 6.4 Cyclic voltammograms of **69** and **70** measured in DCM with 0.1 M nBu_4NClO_4 as a supporting electrolyte at a scan rate of 0.10 V s^{-1}. (*See color plate section for the color representation of this figure.*)

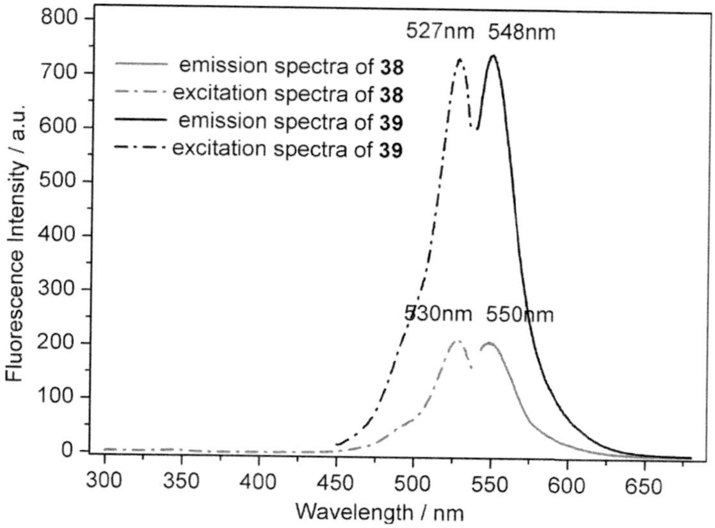

Figure 6.5 Excitation (dot) and emission (solid) spectra of ruthenabenzenes **38** and **39** (1.0×10^{-4} mol L^{-1}) in a methanol-water (1:1, v/v) medium. (*See color plate section for the color representation of this figure.*)

$_{a}$ = +0.16 V and $E_{p,\,a}$= +0.81 V, respectively. According to the X-ray crystallographic study, the impeded communications between the two aromatic osmabenzene rings in **70** may be attributed to the perpendicular orientation between the bridging phenyl ring and the osmabenzene rings.

6.4.2 Optical Properties

In organometallic chemistry, ruthenium complexes are known for their rich coordination stereochemistry, abundant spectral information and bioactivity. A series of ruthenabenzenes incorporated with 8-hydroxyquinolyl groups has been achieved by the reaction of ruthenabenzene **4** with 8-hydroxyquinoline (Scheme 6.20) [34]. The single crystal X-ray structures of fused-ring ruthenabenzenes **37**, **38** and **39** indicate that π-delocalization within the co-planar tetracyclic frameworks occur to some degree. The photoluminescence properties of **37**, **38** and **39** were studied (Figure 6.5), revealing that **38** and **39** are fluorescent in common solvents at room temperature, whereas **37** is almost non-fluorescent. The quantum yields of **38** and **39** were calculated to be 1.0% (25°C) and 4.0% (25°C), respectively, in methanol-water (1:1, v/v), using rhodamine B as the primary standard for calculation. The long-wavelength absorption and emission of **38** and **39** are in accord with the large π-extended system of the co-planar 8-hydroxyquinoline-ruthenabenzene polycyclic systems. Note that only a small Stokes' shift of *ca.* 20 nm has been observed for **38** or **39**, which is comparable to the spectral behaviours of organic multicyclic compounds such as fluorescein and rhodamine.

The ligand substitution reaction of ruthenabenzene **4** with 1,10-phenanthroline produces the ruthenabenzene **71** (Scheme 6.32) [29]. A wide range of biochemical species were investigated in an aqueous solution of **71** at a physiological pH. For L-cysteine, a rapid solution colour change from olive to yellow-brown was observed with a remarkable

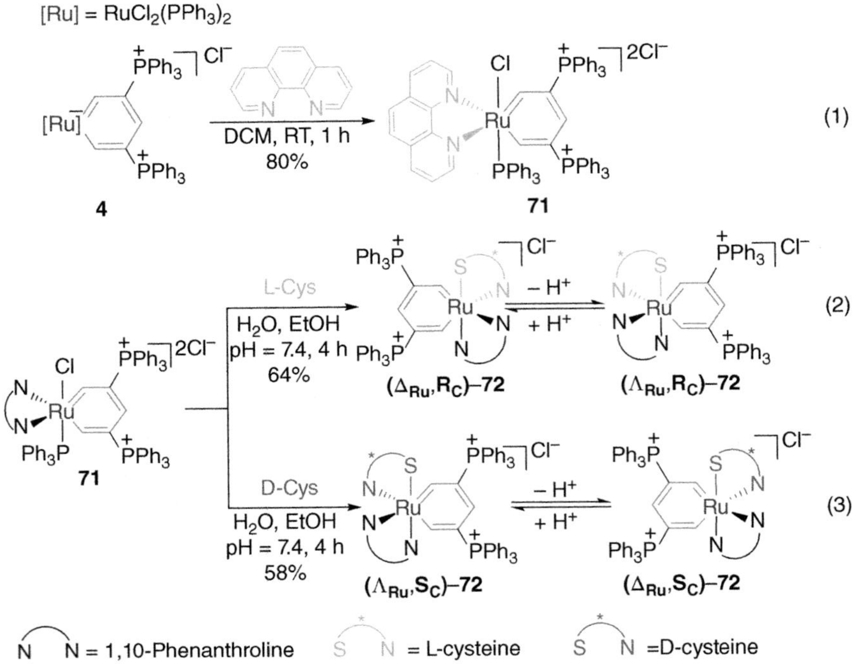

Scheme 6.32

fluorescence enhancement. The X-ray crystal structures of the L-cysteine-**71** adduct reveal that a pair of epimers involved in the L-cysteine-titrated product are (Λ_{Ru}, R_C)-**72** and (Δ_{Ru}, R_C)-**72**, which selectively bind the S atom of L-cysteine as the axial coordination atom (Figure 6.6). It is supposed that the planarity of the metallabenzene ring in **72**

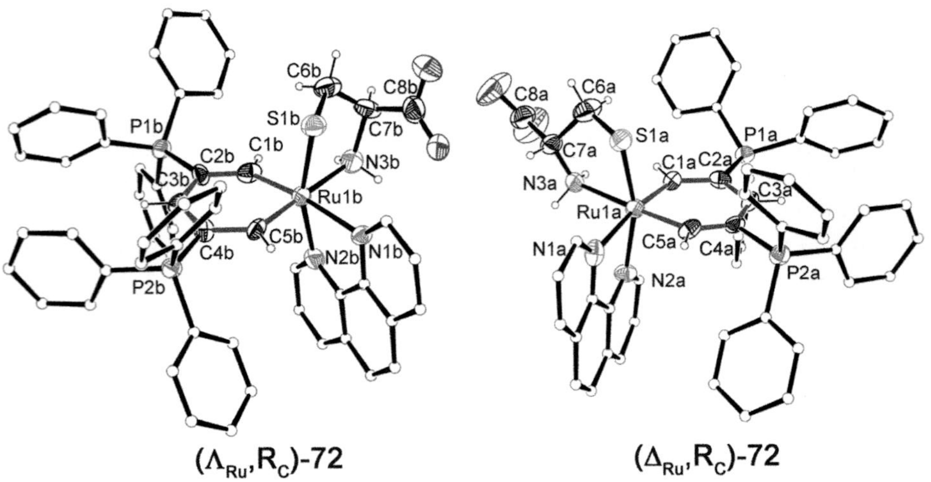

Figure 6.6 Molecular structures of ruthenabenzene (Λ_{Ru}, R_C)-**72** and ruthenabenzene (Δ_{Ru}, R_C)-**72**. The hydrogen atoms of PPh_3 are omitted for clarity. (*See color plate section for the color representation of this figure.*)

may account for the cysteine-induced enhancement of the fluorescence of **71** in aqueous solution because the metallabenzene ring of **71** is non-planar.

As illustrated in Scheme 6.32, the use of D-cysteine instead of L-cysteine leads to an inverted stereodynamic effect. Interestingly, the conformation, and hence the Ru-centred chirality, of the ruthenabenzene **72** can be reversibly tuned between the Δ and Λ forms upon a pH change (Figure 6.7). The dynamic epimerization process of the cysteine-binding product **72** is proposed to be attributed to the significant influence of the intramolecular electrostatic interaction.

It is notable that this is the first investigation of the chemical interaction between metallabenzenes and bioactive molecules. These results provide new understanding to the transmission of chiral information from natural amino acids to metal centres in a bioinorganic manner.

6.4.3 Other Properties

In addition to their electrochemical and optical properties, other physical properties have also been observed in these phosphonium-substituted metallabenzenes. For example, the Os*CH* and *NH* of osmapyridinium **23** exhibit abnormal downfield chemical shifts in the ^{1}H NMR spectrum, whereas the ^{1}H NMR spectrum of osmapyridine **24** shows only several broad signals [24, 25]. Magnetic measurement experiments demonstrate the paramagnetism of both complexes, with the magnetic susceptibility of **24** being higher than that of **23**. The unusual NMR behaviour of the C–H and N–H hydrogen atoms should be attributed to the neighbouring Os^{IV} ($5d^4$) centres in **23** and **24**. Consistent with the experimental results, DFT calculations show that the triplet state of **23** or **24** is more stable than the singlet state. Thus, the NMR behaviour of the neighbouring hydrogen atom could be modified by the net spin populations residing on the metal centre Os^{IV}, together with the tiny paramagnetism of Os-bonded C and N atoms.

These phosphonium-substituted metallabenzenoid compounds, with properties quite unlike those of organic benzene rings and typical organometallic complexes, are finding promising applications in photoelectronics, molecular magnetics, fluorescent molecular probes and stereochemistry. Control of the substituents, metals, ligands and organic fragments may, in turn, provide control of the related physical properties of the final metallabenzenes, which would enable potentially wide applications.

6.5 Polycyclic Metallabenzenoid Compounds Bearing Phosphonium Substituents

A number of syntheses using metallabenzenes as starting materials for making more complex molecules are known [1–14]. The conversion of metallabenzenes to organometallic or organic compounds is particularly important for not only expanding metallabenzene chemistry but also opening up new approaches to complicated compounds. Many reactions of metallabenzenes have been developed into highly useful tools for the synthesis of fused-ring metallabenzenes [12]. The great power of these reactions is that they provide ways to form bonds between groups that are very limited or perhaps impossible in typical organic or organometallic reactions. As exemplified above, metallabenzenes bearing phosphonium substituents exhibit abundant reactivities and

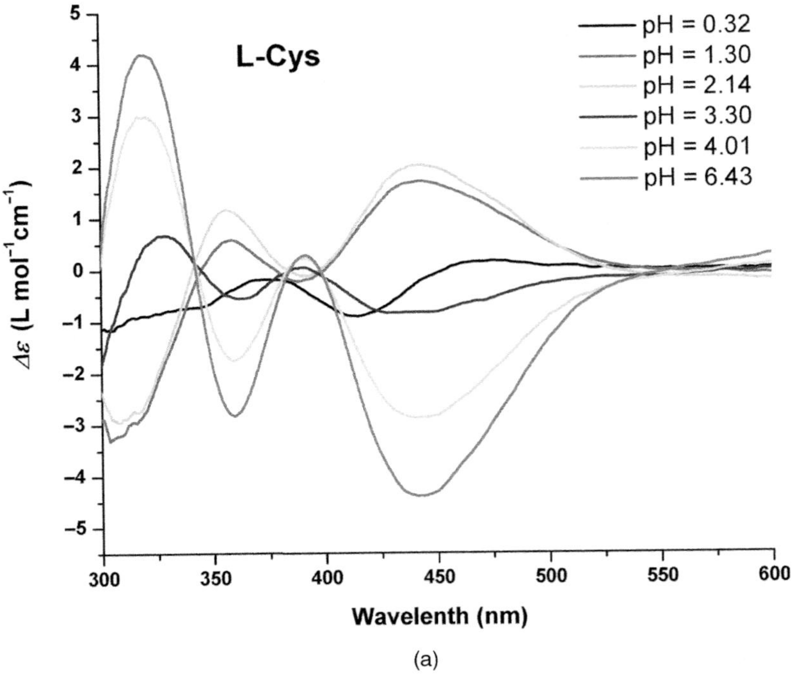

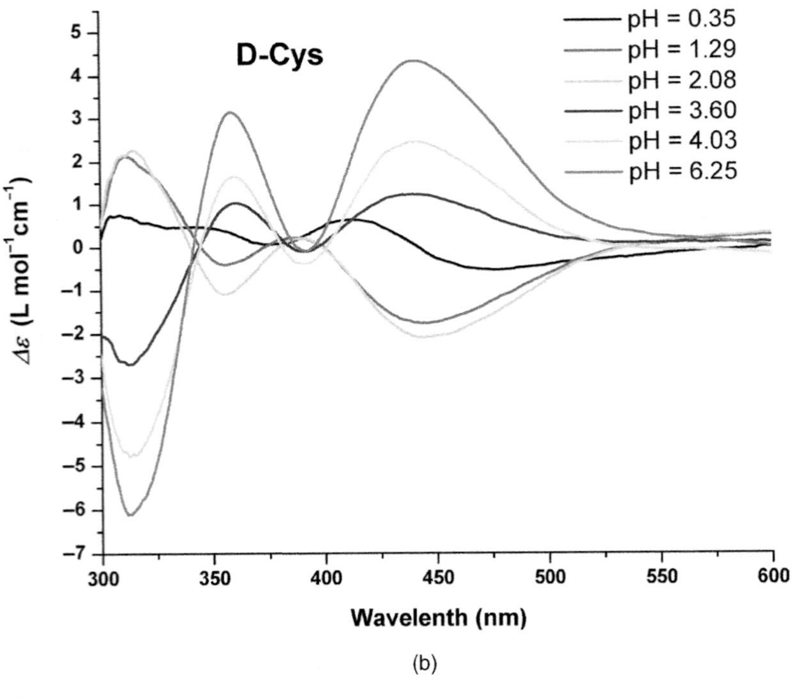

Figure 6.7 Influence of pH on the end CD responses of **71** towards (a) L- and (b) D-cysteine in 10% ethanol/water. [**71**] = 5.86×10^{-4} M. (*See color plate section for the color representation of this figure.*)

properties, which are potentially useful for the synthesis of polycyclic metallabenzenoid compounds bearing phosphonium substituents. The chemistry of phosphonium-substituted polycyclic metallabenzenoid compounds is summarized in this section, especially their synthesis and chemical properties.

6.5.1 Synthesis by Means of S_NAr Reactions

It is extremely difficult to displace a hydride from benzene, even with a strong nucleophile as hydride is a poor leaving group. Nucleophilic substitution reactions of benzene rings occur only when an electronic factor makes the aryl carbon bonded to a good leaving group (such as halogen) susceptible to nucleophilic attack. Any anion-stabilizing (electron-withdrawing) group *ortho* or *para* to a potential leaving group can be used to make a S_NAr reaction possible. The mechanism that operates in these reactions is an addition–elimination mechanism involving the formation of a carbanion with delocalized electrons, called a Meisenheimer intermediate (Scheme 6.33). It is now well established that metallabenzenes can also undergo S_NAr reactions, although they are less common than S_EAr reactions. In accord with classic S_NAr reactions in benzene chemistry, the S_NAr reactions of metallabenzenes take place by an addition–elimination mechanism (Scheme 6.33). Strong evidence for the mechanisms was provided by the isolation of Meisenheimer compounds in the intermolecular S_NAr reactions of cationic metallabenzenes [35, 58, 59].

Electron-withdrawing substituents activate the aromatic ring towards S_NAr, as is certified by the nucleophilic attack in phosphonium-substituted metallabenzenes (Section 6.3.2). Although the major result for S_NAr reactions in metallabenzene chemistry is to make substituted metallabenzenes, the success of this approach has also led to its application in forming polycyclic metallabenzenoid compounds. For example, the fused-ring metallabenzenes bearing phosphonium substituents, ruthenabenzenes **37**,

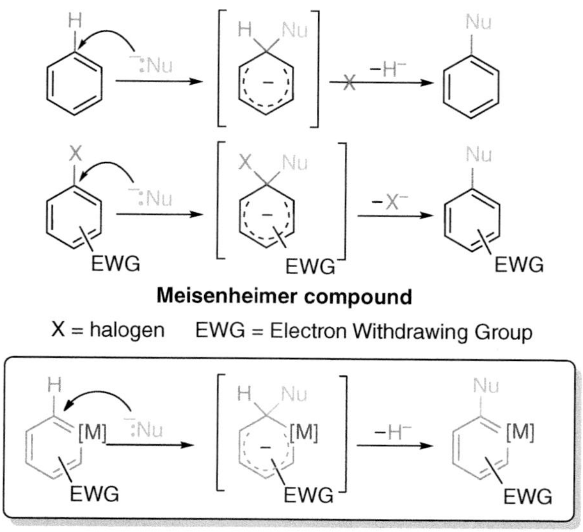

Scheme 6.33

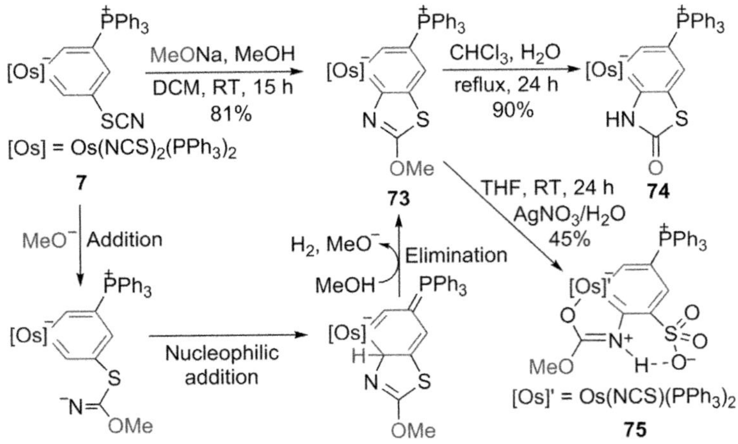

Scheme 6.34

38, 39, 40 and osmabenzenes 44, 45 are depicted in Section 6.3.2. The potential of S_NAr reactions for the construction of polycyclic metallabenzenoid compounds is further addressed by the following examples.

When a methanol solution of excess sodium methoxide was added into the DCM solution of osmabenzene 7, the reaction yielded the first osmabenzothiazole complex (73) in good yield (Scheme 6.34) [60]. It is suggested that the reaction occurs in three steps: the nucleophilic attack of the methoxide at the thiocyanate carbon, the subsequent nucleophilic addition of a nitrogen atom at the metallabenzene ring and the final elimination of the hydride anion. Heating the osmabenzothiazole 73 in wet chloroform at reflux for approximately 24 h leads to the formation of osmabenzothiazolone 74 by hydrolysis of the methoxy group on the thiazole ring. The reaction of 73 with silver nitrate in THF/H_2O produces the osmabenzoxazole sulfonic acid 75.

The structural and spectroscopic data for osmabenzothiazole 73 and osmabenzoxazole 75 indicate electron delocalization within the fused rings. The X-ray crystal structures (Figure 6.8) reveal that the bicyclic rings are essentially co-planar, as is reflected by the maximum deviation from the least-squares plane through the fused-ring atoms (0.0096 Å for 73 and 0.0203 Å for 75). The transition metal–carbon bond distances of 73 or 75 are comparable with those of osmabenzene 2, between those observed for the corresponding M–C single and double bonds (Table 6.4). The carbon–carbon bond distances of the metallabenzene rings in 73 and 75 fall within the range of distances reported for metallabenzenes. The bond distances of the fused five-membered rings are similar to those observed for organic benzothiazole or benzoxazole.

The chemical shifts of the six-membered ring atoms of the osmabenzothiazole 73 and osmabenzoxazole 75 in the ^{1}H and ^{13}C NMR spectra (Table 6.5) appear within the ranges observed for phosphonium-substituted osmabenzenes. In the ^{1}H NMR spectrum of 75, the chemical shift of the hydrogen of N2 occurs at the notably downfield position of 11.9 ppm, which might be interpreted as the intramolecular hydrogen bond (N2–H···O3, 0.86 Å, 2.638(9) Å, 138.6°), as confirmed by single-crystal X-ray diffraction.

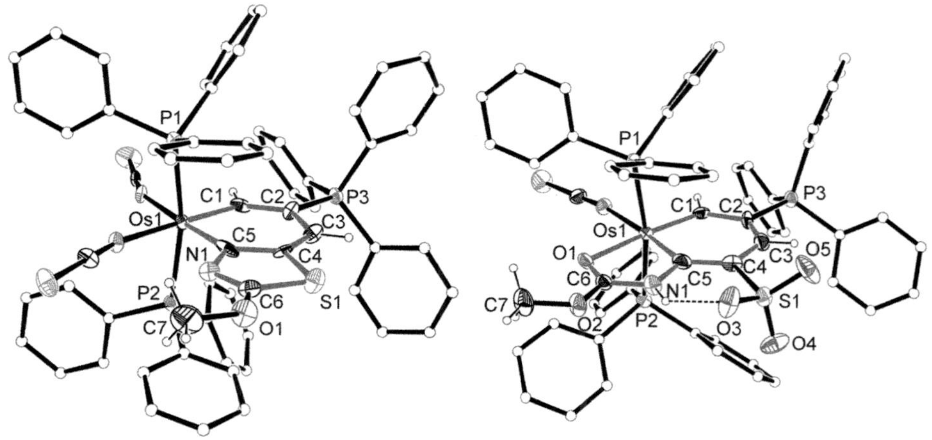

osmabenzothiazole **73** osmabenzoxazole **75**

Figure 6.8 Molecular structures of osmabenzothiazole **73** and osmabenzoxazole **75**. The hydrogen atoms of PPh_3 are omitted for clarity. (*See color plate section for the color representation of this figure.*)

Table 6.4 Selected bond lengths (Å) and angles (°) of osmabenzothiazole **73** and osmabenzoxazole **75**.

	osmabenzothiazole 73	osmabenzoxazole 75
Bond lengths (Å)		
Os1–C1	1.954(7)	1.923(8)
C1–C2	1.366(10)	1.403(10)
C2–C3	1.427(9)	1.402(11)
C3–C4	1.369(10)	1.395(11)
C4–C5	1.438(10)	1.405(11)
C5–Os1	1.953(8)	1.950(9)
C4–S1	1.774(7)	1.791(8)
C5–N1	1.410(8)	1.420(10)
C6–N1	1.327(9)	1.345(10)
C6–X	1.725(8), X = S1	1.235(9), X = O1
Os–O1	—	2.279(5)
C2–P3	1.804(7)	1.786(8)
Bond angle (°)		
C1–Os1–C5	89.1(3)	90.7(3)
Os1–C1–C2	132.5(5)	127.9(6)
C1–C2–C3	122.6(6)	123.7(7)
C2–C3–C4	121.8(7)	127.1(7)
C3–C4–C5	127.6(6)	119.4(7)

(Continued)

Table 6.4 (Continued)

	osmabenzothiazole 73	osmabenzoxazole 75
C4–C5–Os1	126.3(5)	131.1(6)
C5–Y–X	112.7(5), X = S1, Y = C4	78.5(3), X = O1, Y = Os1
Y–C5–N1	109.8(6), Y = C4	113.6(6), Y = Os1
C5–N1–C6	112.7(6)	117.9(7)
N1–C6–X	117.6(6), X = S1	113.5(7), X = O1
C6–X–Y	87.2(4), X = S1, Y = C4	107.1(5), X = O1, Y = Os1

Table 6.5 Selected NMR spectroscopic data of osmabenzothiazole **73** and osmabenzoxazole **75**.

	osmabenzothiazole 73	osmabenzoxazole 75
^{1}H NMR (ppm)		
$C^{1}H$	17.5 (d, ^{3}J(PH) = 22.3 Hz)	16.1 (d, ^{3}J(PH) = 19.4 Hz)
$C^{3}H$	7.9 (d, ^{3}J(PH) = 11.9Hz)	8.2 (d, ^{3}J(PH) = 13.6 Hz)
^{31}P{^{1}H} NMR (ppm)		
OsPPh$_3$	−6.3 (s)	15.8 (s)
CPPh$_3$	19.5 (s)	20.5 (s)
^{13}C{^{1}H} NMR (ppm)		
C1	236.5 (br)	229.7 (br)
C2	105.8 (d, ^{1}J(PC) = 78.2 Hz)	106.1 (d, ^{1}J(PC) = 80.2 Hz)
C3	140.9 (d, ^{2}J(PC) = 23.9 Hz)	153.8 (d, ^{2}J(PC) = 22.7 Hz)
C4	132.3 (d, ^{3}J(PC) = 14.9 Hz)	134.6–121.3
C5	233.6 (br)	231.1 (br)
C6	188.6 (s)	166.7 (s)

A series of bicyclic osmabenzenoid compounds have been prepared starting from the osmaphenol **8** by S_NAr reactions (Scheme 6.35) [19]. The osmaphenol **8** reacts with different isocyanates to form the carbamate intermediates **76**, which then convert to annulation reaction products, osmabenzoxazolone **77**, by an intramolecular S_NAr reaction. Similar annulation reactions can be extended to other unsaturated compounds containing N–C multiple bonds (e.g. isothiocyanates and pyridine), which can produce the corresponding fused osmabenzene complexes **78** and **79**.

The S_NAr reaction is much more restrictive in its applications than the S_EAr reaction in classic benzene chemistry. However, the S_NAr reaction is regarded as one of the general aspects of metallabenzenes, which is often considered the representative step in the mechanisms of reactions involving metallabenzenes, especially those bearing electron-withdrawing substituents. The experimental results suggest that phosphonium-substituted metallabenzenes are very electrophilic and readily react with nucleophiles.

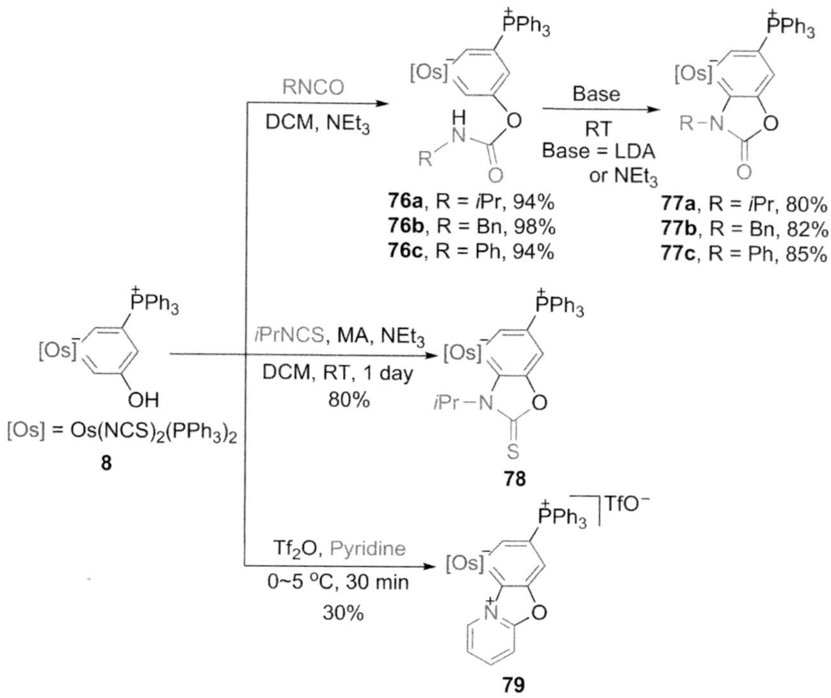

Scheme 6.35

Further advances on the S_NAr reaction in metallabenzene chemistry are expected, especially in applications to the synthesis of polycyclic metallabenzenoid compounds.

6.5.2 Synthesis by Means of Electrophilic Cyclization Reactions

The electrophilic cyclization of alkynes or alkenes mediated by electrophiles such as I_2, ICl or organochalcogen derivatives has been studied intensively in recent years, and it is recognized as an extremely effective method for generating heterocyclic scaffolds [61–65]. The typical process of cyclization reactions involves (1) the addition of an electrophile to activate an unsaturated carbon–carbon multiple bond, (2) the nucleophilic addition of a heteroatom (nitrogen, oxygen or sulfur) to the activated multiple bond and (3) the subsequent removal of the group connected to the heteroatom (Scheme 6.36). In the cases of iodine-mediated electrophilic cyclization reactions, iodonium species **a** have generally been employed as key and common intermediates, although such iodonium intermediates have never been isolated or structurally characterized.

In recent decades, the electrophilic activation of alkynes with I_2 has been applied to produce various fused organic aromatics due to their low cost and eco-friendliness [61–65]. Therefore, this valuable methodology may provide another route to prepare the fused metallabenzene. The potential of the further elaboration of the iodine-mediated electrophilic cyclization in metallabenzene chemistry was demonstrated by the examples of phosphonium-substituted osmabenzenes (Scheme 6.37) [66]. The phenylethynylthio substituent can be introduced to the osmabenzene ring by the reaction of

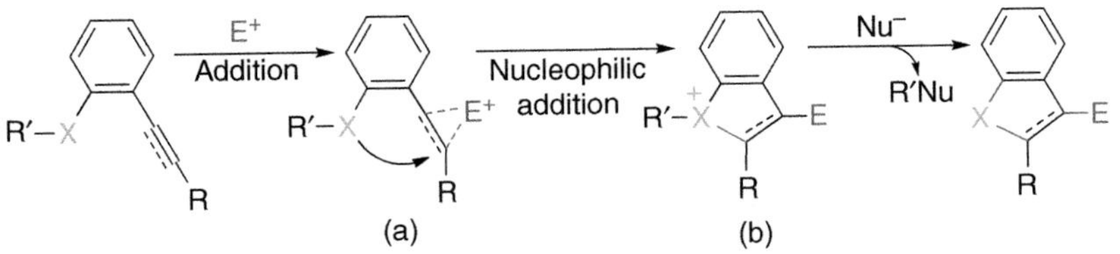

Scheme 6.36

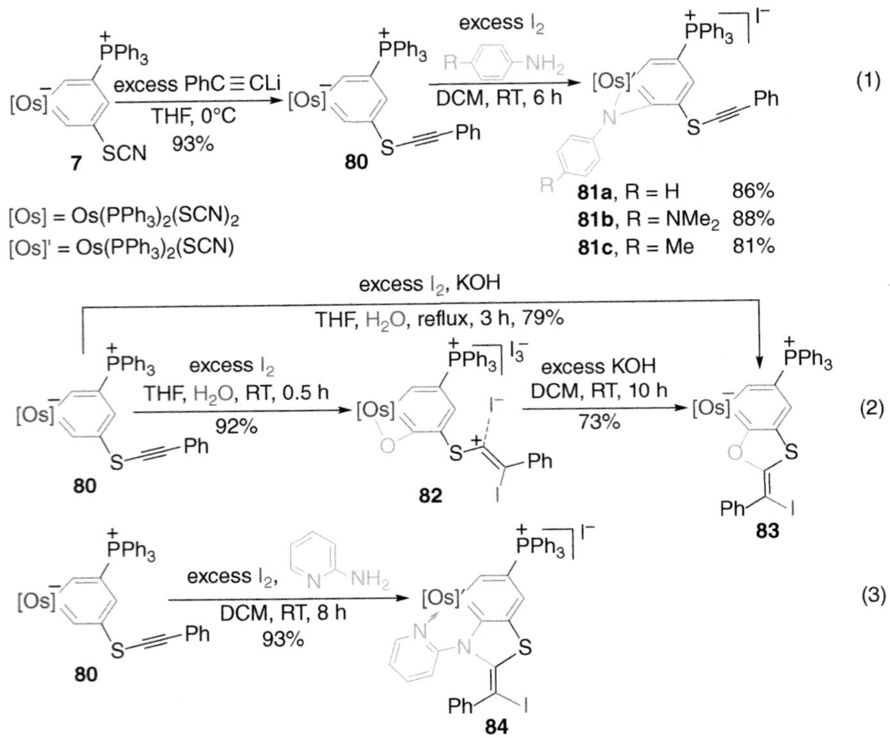

Scheme 6.37

osmabenzene **7** with PhC≡CLi [67]. The treatment of **80** with excess I_2 and nucleophilic amines cannot obtain the desired iodocyclization products, but rather osmabenzenes **81** (R = Ph(**81a**), *p*-tolyl(**81b**), *p*-(dimethylamino)phenyl(**81c**)). In THF/water (30:1, v/v) solution, however, the reaction of **80** with excess I_2 in the presence of a strong base affords the fused metallabenzene **83**. The control experiments demonstrate that water and base are required for the formation of the fused osmabenzene **83**. During the examination of the reaction conditions, intermediate **82** of the cyclization was isolated when the reaction was performed at room temperature in the absence of base. The conversion of **82** to **83** in the presence of KOH provides clear evidence for the active role of **82** in the formation of **83**. When 2-aminopyridine is used as the nucleophile, the corresponding iodocyclization product **84** is obtained in high yield.

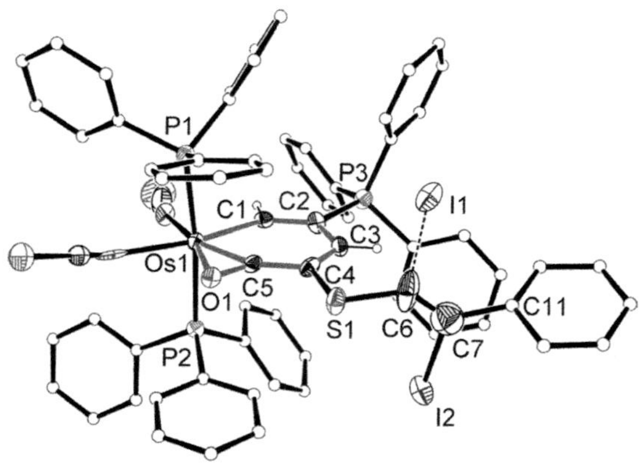

fused osmabenzene **82**

Figure 6.9 Molecular structures of fused osmabenzene **82**. The hydrogen atoms of PPh_3 are omitted for clarity. (*See color plate section for the color representation of this figure.*)

The isolated intermediate **82** has been characterized by X-ray crystallography (Figure 6.9). The structure of **82** contains an essentially planar metallabenzene unit, which is reflected by the mean deviation (0.0134 Å) from the least-squares plane through the six atoms Os1, C1, C2, C3, C4 and C5. The C5–O1 unit is bonded to the osmium centre in an η^2 bonding mode, which affords the fused three-membered ring. An obvious alternation in the C–C distances within the osmabenzene ring is observed, and the Os1–C5 bond is relatively short compared with the Os1–C1 bond (Table 6.6). Similar structural features are observed in the hexa-coordinate metallabenzothiirenes [68–70]. For the iodo-substituted phenylethenylthio substituent, the real structure resembles the intimate ion pair form rather than the iodonium ion form, as shown by the long C6–I2 and short C7–I1 bonds found by X-ray studies. DFT calculations further

Table 6.6 Selected bond lengths (Å) and angles (°) of fused osmabenzene **82**.

Bond Distances (Å)		
Os1–C1 2.061(9)	C3–C4 1.344(15)	C7–C11 1.55(2)
Os1–C5 1.971(9)	C4–C5 1.398(14)	C5–O1 1.229(12)
Os1–O1 2.159(6)	C4–S1 1.768(10)	C7–I1 2.098(17)
C1–C2 1.361(14)	C6–S1 1.713(14)	C6–I2 2.362(19)
C2–C3 1.415(14)	C6–C7 1.17(2)	C2–P3 1.799(10)
Bond Angles (°)		
C1–Os1–C5 78.8(4)		C5–O1–Os1 64.5(5)
C2–C1–Os1 133.4(7)		O1–C5–Os1 81.3(6)
C1–C2–C3 124.1(9)		C5–Os1–O1 34.2(3)
C2–C3–C4 123.5(9)		C4–S1–C6 101.5(6)
C3–C4–C5 117.9(9)		S1–C6–C7 136.9(17)
C4–C5–Os1 142.3(8)		C6–C7–C11 128.7(18)

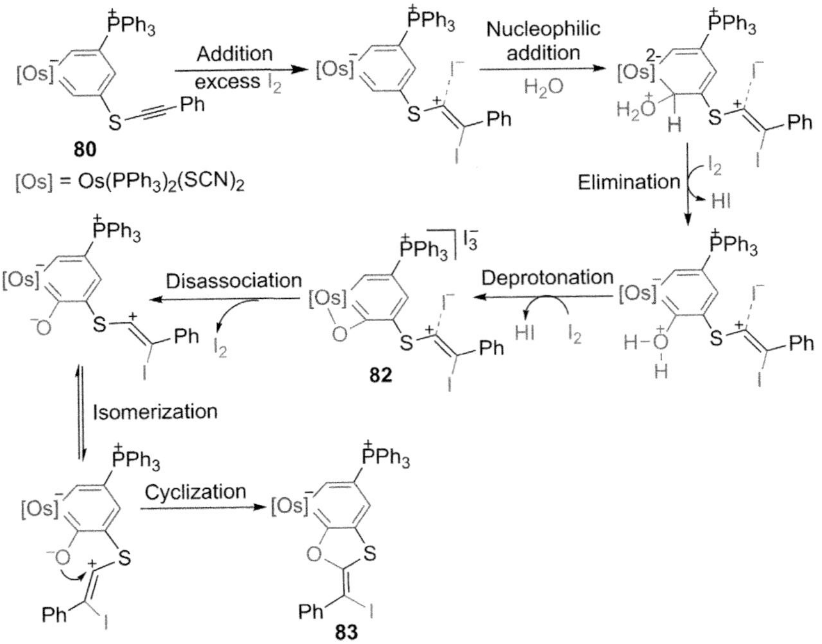

Scheme 6.38

support that the resonance contributor in the intimate ion pair form is the dominant one, although the iodonium ion is generally accepted as the key intermediate in iodine-mediated electrophilic cyclization reactions. In addition, the distance of the iodo-substituted double bond is remarkably short compared with the normal double-bond distances, probably due to the charge effect of carbocation C6.

The proposed mechanism for the formation of **83**, shown in Scheme 6.38, involves the initial activation of the double bond, followed by the intermolecular S_NAr reaction of the metallabenzene ring with water. Then, the oxidative deprotonation and the coordination of the oxygen atom form intermediate **82**. The conversion of **82** to **83** is proposed from the dissociation of iodine to give an Os(II) intermediate, which can undergo the addition of a dissociated oxygen to the ethenylium group. In the absence of base, the dihapto acyl oxygen has insufficient nucleophilicity to carry out the next cyclization step, and so the intermediate **82** can be isolated. As for the reactions with amines, the much stronger Os–N interaction compared with the Os–O interaction in **82** may prevent further iodocyclization and afford only the fused osmabenzenes **81**. However, in the case of 2-aminopyridine, the coordination ability of the pyridyl substituent at the N atom may facilitate the formation of the cyclization product **84**.

As amply demonstrated above, iodine-mediated electrophilic cyclization reactions can act as a new strategy for the construction of complex polycyclic systems. During the course of transformations, reactive species that are difficult to capture in normal organic reactions can be stabilized and even isolated in the phosphonium-substituted metallabenzene system. A fused-ring osmabenzene that incorporates an *m*-metallapyridine segment has been prepared from the osmabenzene **80** under similar iodocyclization

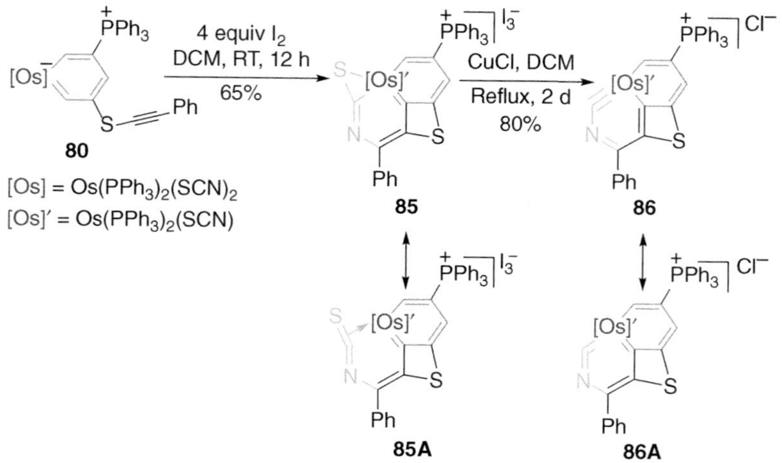

Scheme 6.39

reaction conditions (Scheme 6.39). On treatment of **85** with copper(I) chloride, the fused-ring osmabenzene **86** is formed in 80% yield.

These two fused osmabenzenes are intriguing. The single crystal X-ray structure determination suggests that both of them are rare metal-bridged polycyclic metallabenzenoid compounds, in which the transition metal centre is located in a ring junction position of the two six-membered rings. As shown in Figure 6.10, **85** contains a nearly planar polycyclic system that is composed of one three-membered ring, one four-membered ring and two six-membered rings. The maximum deviation from the least-squares plane

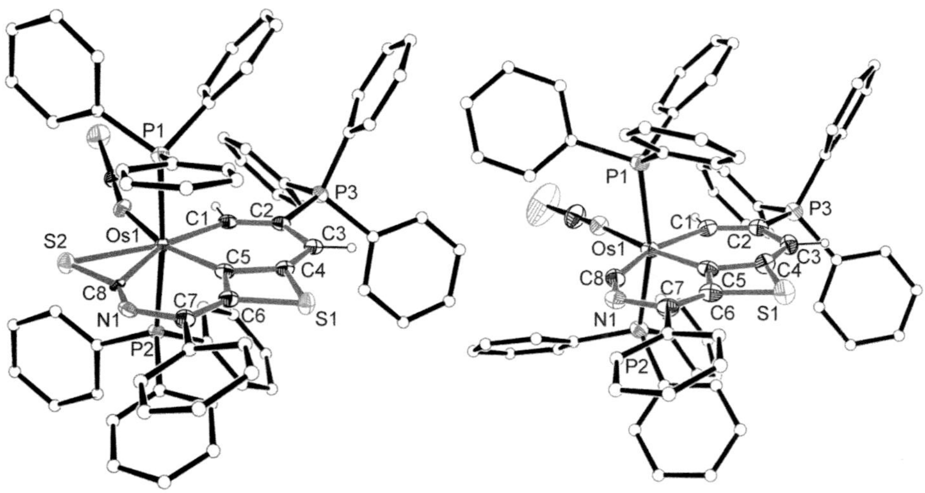

Figure 6.10 Molecular structures of osmaisoquinoline **85** and osmaisoquinolyne **86**. The hydrogen atoms of PPh_3 are omitted for clarity. (*See color plate section for the color representation of this figure.*)

through the 12 atoms (Os1, N1, C1–C8, S1, S2) of the four rings is only 0.0178 Å. The metal centre is shared by the three rings. Compared with the distances of the Os1–C1 and Os1–C5 bonds (Table 6.7), the markedly longer distance of the Os1–C8 bond may be attributed to the competition of the S2 atom and C8 atom for bonding with the metal centre. Thus, the π-bound organoisothiocyanate structure **85A** also contributes to the overall structure of **85**, except for the osmaisoquinoline structure. In the case of osmaisoquinolyne **86**, the co-planarity of three fused rings is even better than in **85**, which is reflected by a very small mean deviation (0.0099 Å) from the least-squares plane through the 10 atoms. The appreciably long distance of the Os–C8 bond (1.894(4) Å) together with the Os–C5 bond length (1.954(4) Å) indicate that two resonance structures **86** and **86A** are accounted for in the real structure, with **86A** being more important. Both osmaisoquinoline **85** and osmaisoquinolyne **86** display pronounced C–C bond distance alternation within the six-membered metallacycles, although the distances still fall within the normal aromatic range.

Table 6.7 Selected bond lengths (Å) and angles (°) for osmaisoquinoline **85** and osmaisoquinolyne **86**.

	osmaisoquinoline 85	osmaisoquinolyne 86
Bond lengths (Å)		
Os1–C1	2.065(8)	2.085(4)
C1–C2	1.333(10)	1.360(5)
C2–C3	1.451(10)	1.456(5)
C3–C4	1.363(10)	1.338(5)
C4–C5	1.458(10)	1.456(5)
Os1–C5	1.996(7)	1.954(4)
C4–S1	1.802(8)	1.794(4)
C5–C6	1.435(10)	1.447(5)
C6–S1	1.796(8)	1.785(4)
C6–C7	1.352(11)	1.355(5)
C7–N1	1.411(10)	1.416(5)
C8–N1	1.230(9)	1.236(5)
Os1–C8	2.199(7)	1.894(4)
Os1–S2	2.477(2)	—
C8–S2	1.609(7)	—
C2–P3	1.791(8)	1.791(4)
Bond angles (°)		
C1–Os1–C5	80.8(3)	83.65(14)
Os1–C1–C2	138.8(6)	132.9(3)
C1–C2–C3	122.7(7)	124.4(3)
C2–C3–C4	116.7(7)	118.6(3)
C3–C4–C5	130.1(7)	128.7(3)

(Continued)

Table 6.7 (Continued)

	osmaisoquinoline 85	osmaisoquinolyne 86
C4–C5–Os1	130.9(6)	131.7(3)
C5–C4–S1	96.1(5)	94.5(2)
C4–C5–C6	94.3(6)	96.1(3)
C5–C6–S1	97.2(5)	95.2(3)
C4–S1–C6	72.3(4)	74.22(17)
C5–C6–C7	129.2(7)	127.1(4)
C6–C7–N1	115.9(7)	114.7(3)
C7–N1–C8	124.9(7)	118.7(3)
N1–C8–Os1	139.2(6)	151.4(3)
C8–Os1–C5	75.9(3)	75.89(15)
Os1–C5–C6	134.7(6)	132.3(3)
Os1–C8–S2	79.5(3)	—
Os1–S2–C8	60.8(3)	—
C8–Os1–S2	39.69(18)	—

6.6 Future Prospects

The synthesis and characterization of stable metallabenzenes constitute two of the central themes of metallabenzene chemistry. The three direct methods for the synthesis of six-membered metallabenzenes (i.e. [5+1], [4+2] and [3+3]) are presented in this chapter. In this methodology, simple transition metal complexes with phosphine ligands can be used as metal sources, and substituted alkynes have proven very effective carbon sources to afford a series of metallabenzenes bearing phosphonium substituents. These cyclization reactions can be performed under relatively mild reaction conditions and tolerate a wide variety of functional groups. Most of the resulting phosphonium-substituted metallabenzenes have excellent stability in air, water and heat and can undergo a variety of reactions and transformations to produce interesting monocyclic or polycyclic metallacycles or organic rings. Some species have special properties providing promising applications in photoelectronics, molecular magnetics, fluorescent molecular probes and stereochemistry. However, the understanding of the chemistry of the phosphonium-substituted metallaaromatics is still in a much more primitive state than classic aromatic chemistry. A number of challenges remain to be taken up, such as the construction of various metallaaromatics with remarkable properties ranging from monocyclic compounds to polycyclic systems, the conversion of metallaaromatics to complicated organic rings difficult to achieve with typical organic reactions and the stabilization of reactive and short-lived intermediates of important reactions by introducing highly stable phosphonium-substituted metallaaromatics. Future efforts and developments will certainly expand metallabenzene chemistry into a rather broad and exciting domain with many opportunities for research and development of applications.

References

1 Bleeke, J. R. (2001) Metallabenzenes. *Chemical Reviews*, **101**(5), 1205–1228.

2 He., G., Xia, H. and Jia, G. (2004) Progress in the synthesis and reactivity studies of metallabenzenes. *Chinese Science Bulletin*, **49**(15), 1543–1553.

3 Jia, G. (2004) Progress in the chemistry of metallabenzynes. *Accounts of Chemical Research*, **37**(7), 479–486.

4 Landorf, C. W. and Haley, M. M. (2006) Recent advances in metallabenzene chemistry. *Angewandte Chemie: International Edition*, **45**(24), 3914–3936.

5 Wright, L. J. (2006) Metallabenzenes and metallabenzenoids. *Dalton Transactions*, **15**, 1821–1827.

6 Bleeke, J. R. (2007) Aromatic iridacycles. *Accounts of Chemical Research*, **40**(10), 1035–1047.

7 Jia, G. (2007) Recent progress in the chemistry of osmium carbyne and metallabenzyne complexes. *Coordination Chemistry Reviews*, **251**(17), 2167–2187.

8 Chen, J. and Jia, G. (2013) Recent development in the chemistry of transition metal-containing metallabenzenes and metallabenzynes. *Coordination Chemistry Reviews*, **257**(17), 2491–2521.

9 Jia, G. (2013) Our journey to the chemistry of metallabenzynes. *Organometallics*, **32**(23), 6852–6866.

10 Zhu, C., Cao, X.-Y. and Xia, H. (2013) Double stabilization of highly strained six-membered rings by phosphonium and transition metal. *Chinese Journal of Organic Chemistry*, **33**(4), 657–662.

11 Cao, X.-Y., Zhao, Q., Lin, Z. and Xia, H. (2014) The chemistry of aromatic osmacycles. *Accounts of Chemical Research*, **47**(2), 341–354.

12 Frogley, B. J. and Wright, L. J. (2014) Fused-ring metallabenzenes. *Coordination Chemistry Reviews*, **270**, 151–166.

13 Fernández, I., Frenking, G. and Merino, G. (2015) Aromaticity of metallabenzenes and related compounds. *Chemical Society Reviews*, **44**, 6452–6463.

14 Dalebrook, A. F. and Wright, L. J. (2012) Chapter Three: Metallabenzenes and metallabenzenoids. In: *Advances in Organometallic Chemistry*, Anthony F. H., Mark J. F. (eds), New York: Academic Press, Elsevier, 93–177.

15 Xia, H., He., G., Zhang, H., *et al.* (2004) Osmabenzenes from the reactions of HC≡CCH(OH)C≡CH with $OsX_2(PPh_3)_3$ (X= Cl, Br). *Journal of the American Chemical Society*, **126**(22), 6862–6863.

16 Zhang, H., Wu, L., Lin, R., *et al.* (2009) Synthesis: Characterization and electrochemical properties of stable osmabenzenes containing PPh_3 substituents. *Chemistry: A European Journal*, **15**(14), 3546–3559.

17 Zhang, H., Xia, H., He., G., *et al.* (2006) Synthesis and characterization of stable ruthenabenzenes. *Angewandte Chemie: International Edition*, **45**(18), 2920–2923.

18 Zhang, H., Feng, L., Gong, L., *et al.* (2007) Synthesis and characterization of stable ruthenabenzenes starting from HC≡CCH(OH)C≡CH. *Organometallics*, **26**(10), 2705–2713.

19 Han, F., Wang, T., Li, J., *et al.* (2014) m-Metallaphenol: Synthesis and reactivity studies. *Chemistry: A European Journal*, **20**(15), 4363–4372.

20 Gong, L., Lin, Y., Wen, T. B., *et al.* (2008) Formation of four conjugated osmacyclic species in a one-pot reaction. *Organometallics*, **27**(11), 2584–2589.

21 Gong, L., Chen, Z., Lin, Y., *et al.* (2009) Osmabenzenes from osmacycles containing an η^2-coordinated olefin. *Chemistry: A European Journal*, **15**(25), 6258–6266.
22 Liu, B., Xie, H., Wang, H., *et al.* (2009) Selective synthesis of osmanaphthalene and osmanaphthalyne by intramolecular C–H activation. *Angewandte Chemie: International Edition*, **48**(30), 5461–5464.
23 Chen, J., Zhang, C., Xie, T., *et al.* (2013) Conversion of a hydrido–butenylcarbyne complex to η^2-allene-coordinated complexes and metallabenzenes. *Organometallics*, **32**(14), 3993–4001.
24 Liu, B., Wang, H., Xie, H., *et al.* (2009) Osmapyridine and osmapyridinium from a formal [4+2] cycloaddition reaction. *Angewandte Chemie: International Edition*, **48**(30), 5430–5434.
25 Liu, B., Zhao, Q., Wang, H., *et al.* (2012) Synthesis of osmapyridiniums by [4+2] cycloaddition reaction between osmium alkenylcarbyne and nitriles. *Chinese Journal of Chemistry*, **30**(9), 2158–2168.
26 Weller, K. J., Filippov, I., Briggs, P. M. and Wigley, D. E. (1998) Pyridine degradation intermediates as models for hydrodenitrogenation catalysis: Preparation and properties of a metallapyridine complex. *Organometallics*, **17**(3), 322–329.
27 Zhao, Q., Gong, L., Xu, C., *et al.* (2011) Stable iso-osmabenzenes from a formal [3+3] cycloaddition reaction of metal vinylidene with alkynols. *Angewandte Chemie: International Edition*, **50**(6), 1354–1358.
28 Zhao, Q., Zhu, J., Huang, Z. A., *et al.* (2012) Conversions of osmabenzyne and isoosmabenzene. *Chemistry: A European Journal*, **18**(37), 11597–11603.
29 Lin, R., Zhang, H., Li., S., *et al.* (2011) pH-Switchable inversion of the metal-centered chirality of metallabenzenes: Opposite stereodynamics in reactions of ruthenabenzene with L- and D-cysteine. *Chemistry: A European Journal*, **17**(8), 2420–2427.
30 Lin, R., Zhao, J., Chen, H., *et al.* (2012) Interconversion of metallabenzenes and cyclic η^2-allene-coordinated complexes. *Chemistry: An Asian Journal*, **7**(8), 1915.
31 Zhu, J., Jia, G., and Lin Z. (2007) Understanding nonplanarity in metallabenzene complexes. *Organometallics*, **26**(8), 1986–1995.
32 Huang, J., Lin, R., Wu, L., *et al.* (2010) Synthesis, characterization, and electrochemical properties of bisosmabenzenes bridged by diisocyanides. *Organometallics*, **29**(13), 2916–2925.
33 Zhang, H., Lin, R., Hong, G., *et al.* (2010) Nucleophilic aromatic addition reactions of the metallabenzenes and metallapyridinium: Attacking aromatic metallacycles with bis(diphenylphosphino)methane to form metallacyclohexadienes and cyclic η^2-allene-coordinated complexes. *Chemistry: A European Journal*, **16**(23), 6999–7007.
34 Lin, R., Zhang, H., Li., S., *et al.* (2011) New highly stable metallabenzenes via nucleophilic aromatic substitution reaction. *Chemistry: A European Journal*, **17**(15), 4223–4231.
35 Zhang, H., Lin, R., Li, J., *et al.* (2014) Interconversion between ruthenacyclohexadiene and ruthenabenzene: A combined experimental and theoretical study. *Organometallics*, **33**(19), 5606–5609.
36 Clark, G. R., Ferguson, L. A., McIntosh, A. E., *et al.* (2010) Functionalization of metallabenzenes through nucleophilic aromatic substitution of hydrogen. *Journal of the American Chemical Society*, **132**(38), 13443–13452.
37 Wang, T., Zhang, H., Han, F., *et al.* (2013) *cine*-Substitution reactions of metallabenzenes: An experimental and computational study. *Chemistry: A European Journal*, **19**(33), 10982–10991.

38 Wei, Y., Zhou, X., Hong, G., *et al.* (2015) Reactions of osmapyridinium with terminal alkynes. *Organic Chemistry Frontiers*, **2**(5), 560–568.

39 Wu, L., Feng, L., Zhang, H., *et al.* (2008) Synthesis and characterization of a novel dialdehyde and cyclic anhydride. *The Journal of Organic Chemistry*, **73**(7), 2883–2885.

40 Marcoux, D. and Charette, A. B. (2008) Nickel-catalyzed synthesis of phosphonium salts from aryl halides and triphenylphosphine. *Advanced Synthesis & Catalysis*, **350**(18), 2967–2974.

41 Marcoux, D. and Charette, A. B. (2008) Palladium-catalyzed synthesis of functionalized tetraarylphosphonium salts. *Journal of Organic Chemistry*, **2**, 590–593.

42 Migita, T., Nagai, T., Kiuchi, K. and Kosugi, M. (1983) Convenient preparation of tetraarylphosphonium halides. *Bulletin of the Chemical Society of Japan*, **56**(9), 2869–2870.

43 Migita, T., Shimizu, T., Asami, Y. *et al.*, (1980) The palladium catalyzed nucleophilic substitution of aryl halides by thiolate anions. *Bulletin of the Chemical Society of Japan*, **53**(5), 1385–1389.

44 Horner, L. and Duda, U.-M. (1970) Phosphororganische verbindungen, 69. Darstellung und eigenschaften einiger vollaromatischer phosphoniumsalze mit *p*-dimethylamino- und *p*-methoxygruppen. *Tetrahedron Letters*, **11**(59), 5177–5181.

45 Horner, L., Mummenthey, G., Moser, H. and Beck, P. (1966) Phosphororganische verbindungen, 51. Die einführung von arylresten in tertiäre phosphine mit hilfe von komplexen der übergangsmetalle (komplexsalzmethode). *Chemische Berichte*, **99**(9), 2782–2788.

46 Hirusawa, Y., Oku, M. and Yamamoto, K. (1957) Neue synthetische methode von tetraphenyl-phosphoniumhalogeniden. *Bulletin of the Chemical Society of Japan*, **30**(6), 667–670.

47 Shevchuk, M. I., Bukachuk, O. M. and Zinzyuk, T. A. (1985) Tetraarylphosphonium salts from *p*-bromobutyrophenone. *Zhurnal Obshchei Khimii*, **55**(2), 346–350.

48 Shevchuk, M. I. and Bukachuk, O. M. (1982) Synthesis and study of formyl-containing tetraarylphosphonium salts. *Zhurnal Obshchei Khimii*, **52**(4), 830–838.

49 Bukachuk, O. M., Megera, I. V. and Shevchuk, M. I. (1980) Synthesis of mono- and bisphosphonium salts containing phenyl and heterocyclic nuclei in phosphorus. *Zhurnal Obshchei Khimii*, **50**(8), 1730–1732.

50 McDonald, R. N. and Campbell, T. W. (1960) Synthesis of hydrocarbon derivatives by the Wittig reaction: III: Wittig reaction as a polymerization method. *Journal of the American Chemical Society*, **82**, 4669–4671.

51 Horner, L. and Hoffmann, H. (1958) Phosphorus organic compounds: VIII: Quaternary arylphosphonium salts by the diazo method. *Chemische Berichte*, **91**, 45–49.

52 Remond, E., Tessier, A., Leroux, F. R., *et al.* (2010) Efficient synthesis of quaternary and *p*-stereogenic phosphonium triflates. *Organic Letters*, **12**(7), 1568–1571.

53 Han, F., Li, J., Zhang, H., *et al.* (2015) Reactions of osmabenzene with silver/copper acetylides: From metallabenzene to benzene. *Chemistry: A European Journal*, **21**(2), 565–567.

54 Esteruelas, M. A., López, A. M. and Oliván, M. (2007) Osmium-carbon double bonds: Formation and reactions. *Coordination Chemistry Reviews*, **251**(5-6), 795–840.

55 Chin, C. S., Chong, D., Maeng, B., *et al.* (2002) New water-soluble alkyl-carbonyl iridium complexes containing both Cp*($C_5Me_5^-$) and $PAr_3(P(m\text{-}C_6H_4SO_3Na)_3)$: Cleavage of C≡C and C=C Bonds with water. *Organometallics*, **21**(8), 1739–1742.

56 Bianchini, C., Casares, J. A., Peruzzini, M., *et al.* (1996) The mechanism of the Ru-assisted CC bond cleavage of terminal alkynes by water. *Journal of the American Chemical Society*, **118**(19), 4585–4594.

57 Chin, C. S. and Lee, H. (2004) New iridacyclohexadienes and iridabenzenes by [2+ 2+ 1] cyclotrimerization of alkynes and facile interconversion between iridacyclohexadienes and iridabenzenes. *Chemistry: A European Journal*, **10**(18), 4518–4522.

58 Paneque, M., Posadas, C. M., Poveda, M. L., *et al.* (2007) Metallacycloheptatrienes of iridium (III): Synthesis and reactivity. *Organometallics*, **26**(14), 3403–3415.

59 Paneque, M., Posadas, C. M., Poveda, M. L., *et al.* (2003) Formation of unusual iridabenzene and metallanaphthalene containing electron-withdrawing substituents. *Journal of the American Chemical Society*, **125**(33), 9898–9899.

60 Wang, T., Li., S., Zhang, H., *et al.* (2009) Annulation of metallabenzenes: From osmabenzene to osmabenzothiazole to osmabenzoxazole. *Angewandte Chemie: International Edition*, **48**(35), 6453–6456.

61 Parvatkar, P. T., Parameswaran, P. S. and Tilve, S. G. (2012) Recent developments in the synthesis of five- and six- membered heterocycles using molecular iodine. *Chemistry: A European Journal*, **18**(18), 5460–5489.

62 Godoi, B., Schumacher, R. F. and Zeni, G. (2011) Synthesis of heterocycles via electrophilic cyclization of alkynes containing heteroatom. *Chemical Reviews*, **111**(4), 2937–2980.

63 Tsuji, H., Ilies, L. and Nakamura, E. (2014) Synthetic strategy for multisubstituted fused furan compounds using main-group metal reagents. *Synlett*, **25**(15), 2099–2110.

64 Jiang, X. and Liu, H. (2014) 4.07 Electrophilic Cyclization. In P. Knochel (ed.), *Comprehensive Organic Synthesis II*, 2nd edn. Amsterdam: Elsevier, 412–494.

65 Mendoza, A., Fananas, F. J. and Rodríguez, F. (2013) Asymmetric halocyclizations of unsaturated compounds: An overview and recent developments. *Current Organic Synthesis*, **10**(3), 384–393.

66 Wang, T., Zhang, H., Han, F., *et al.* (2013) Key intermediates of iodine-mediated electrophilic cyclization: Isolation and characterization in an osmabenzene system. *Angewandte Chemie: International Edition*, **52**(35), 9251–9255.

67 Wang, T., Zhang, H., Han, F., *et al.* (2012) Synthesis and characterization of a metallapyridyne complex. *Angewandte Chemie: International Edition*, **51**(39), 9838–9841.

68 Elliott, G. P., Roper, W. R. and Waters, J. M. (1982) Metallacyclohexatrienes or "metallabenzenes". Synthesis of osmabenzene derivatives and X-ray crystal structure of [Os(CSCHCHCHCH)(CO)$(PPh_3)_2$]. *Journal of the Chemical Society: Chemical Communications*, **14**, 811–813.

69 Rickard, C. E., Roper, W. R., Woodgate, S. D. and Wright, L. J. (2001) Reaction between the thiocarbonyl complex, Os(CS)(CO)$(PPh_3)_3$, and propyne: Crystal structure of a new sulfur-substituted osmabenzene. *Journal of Organometallic Chemistry*, **623**(1), 109–115.

70 Dalebrook, A. F. and Wright, L. J. (2009) Annulation of an iridabenzene through formal cycloaddition reactions with organonitriles. *Organometallics*, **28**(18), 5536–5540.

7

Theoretical Studies of Metallabenzenes: From Bonding Situation to Reactivity

Israel Fernández[1] and Gernot Frenking[2]

[1] *Departamento de Química Orgánica I and Centro de Innovación en Química Avanzada, Facultad de Ciencias Químicas, Universidad Complutense de Madrid, Madrid, Spain*
[2] *Fachberich Chemie, Philipps-Universität Marburg, Marburg, Germany and Donostia International Physics Center (DIPC), Donostia, Euskadi, (Spain)*

7.1 Introduction

Metallabenzenes constitute an important family of organometallic compounds which are characterized by the formal replacement of a CH unit in benzene by an isolobal transition-metal fragment. Since the isolation and full characterization by Roper and co-workers in 1982 of the first metallabenzene, an osmabenzene complex [1], the chemistry of these compounds has experienced a tremendous development [2]. As a result and as shown in the previous chapters in this book, a great number of metallabenzenes including related compounds such as heteroatom-containing analogues (i.e. metallapyridines [3], metallapyryliums [4], metallathiabenzenes [5]), fused-ring metallabenzenes [6–8], metallabenzynes [9], or even species incorporating two transition-metals into the six-membered ring (known as dimetallabenzenes) [10] have been successfully prepared so far.

Despite the large amount of experimental work focused on the chemistry of this family of organometallic compounds, the study of these species using computational tools has received comparatively little attention. This is somehow surprising if we take into account that the origin of metallabenzenes was indeed computational. The seminal report by Thorn and Hoffmann in 1979 [11] predicted that metallabenzenes might be synthesized as stable molecules on the basis of the computed electron delocalization within the six-membered ring. This delocalization was considered the crucial mechanism to stabilize these organometallic complexes and was viewed as an indicator of their aromatic character.

The difficulties associated with the calculations of transition-metal compounds, which were computationally highly demanding until the end of the 1980s [12], are the reason behind the comparatively low number of computational studies on metallabenzenes. The successful employment of density functional theory (DFT) methods together with the use of small-core relativistic pseudopotentials and the tremendous development of the computational facilities and algorithms have enormously contributed to our current understanding of the chemistry of transition-metal complexes in general, and metallabenzenes in particular. In this chapter, we shall summarize the insight gained into the chemistry of

Metallabenzenes: An Expert View, First Edition. Edited by L. James Wright.

metallabenzenes from a computational point of view. Issues such as the bonding situation, aromatic character, synthetic pathways and reactivity will be covered herein by describing the main works reported since the initial prediction by Thorn and Hoffmann.

7.2 Structure and Bonding Situation

7.2.1 Bonding Situation and Molecular Orbitals

As mentioned above, a considerable number of metallabenzenes have been fully characterized by X-ray diffraction techniques [1, 2, 13]. According to these experimental data, the following key structural features are shared by this family of transition-metal complexes: (1) the C–C bond lengths within the six-membered ring are invariably found to be intermediate between double and single bonds (1.34 to 1.46 Å); (2) the average of the four C–C distances is very close to the C–C distances of *ca.* 1.4 Å in benzene [14]; (3) the M–C bond lengths are also intermediate between those found for corresponding M–C (alkyl) and M–C (carbene) bonds, which is consistent with partial multiple bonding between the transition metal and these carbon atoms; and (4) regarding planarity, the five carbon atoms of the metallabenzene ring are essentially co-planar, but the transition metal can either be placed within this plane or may be significantly displaced.

In general, the calculated geometries (mainly by DFT methods) concur quite well with the available experimental data. Thus, the computed complexes reproduce the above commented geometrical features (i.e. bond-length equalization and co-planarity of the carbon framework). In a recent report, Lin and co-workers have computationally explored the origin of the deviation from planarity [15]. According to the correlation diagram depicted in Figure 7.1, metallabenzenes can be considered 8π-electron systems. Whereas the highest lying doubly occupied π-orbital (π_4), which constitutes the HOMO of many metallabenzenes, can be considered as an antibonding molecular orbital (MO), the remaining π-orbitals (π_1, π_2 and π_3) are bonding combinations between the metal centre and the metal-bonded ring-carbon atoms. Planar geometries maximize the bonding interactions in the three bonding MO's while maximizing the antibonding interaction. In contrast, a non-planar geometry reduces the antibonding interaction and the bonding interactions. It is the balance between these opposite forces which mainly controls the degree of ring distortion in metallabenzenes. However, as the three low-lying occupied π-MO's have bonding between the metal centre and the metal-bonded ring-carbon atoms, the electronic driving force for non-planarity is not very large, which implies that other factors also have an influence on the degree of non-planarity in these systems. For instance, strong π-acceptor ligands *trans* to the M–C bonds favour planar structures by withdrawing electron density from the d_{xy} atomic orbital. In contrast, good σ- and π-donor ligands favour non-planar structures. Besides electronic effects, steric effects exerted by the ancillary ligands were found to be decisive in determining the geometry adopted by metallabenzenes as well.

The molecular orbitals of metallabenzenes and particularly their frontier MO's have been discussed in numerous computational studies [11, 16–18], where the emphasis was on the factors leading to the observed C–C bond length equalization. The resemblance between the MO's of metallabenzene complexes and those for benzene becomes evident when comparing the most representative MO's of these species (Figure 7.2). Thus, the typical "doughnut shaped" HOMO-2 of benzene strongly

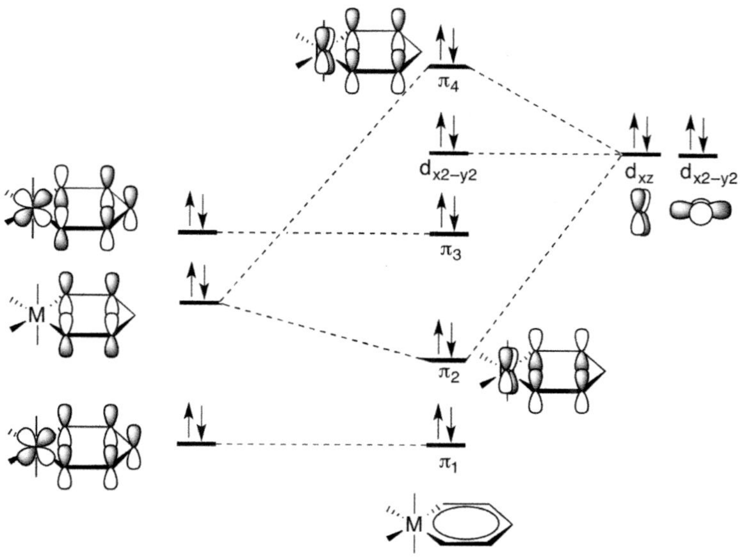

Figure 7.1 MO correlation diagram between the metal centre and the metal-bonded ring-carbon atoms. Figure adapted from reference [15a].

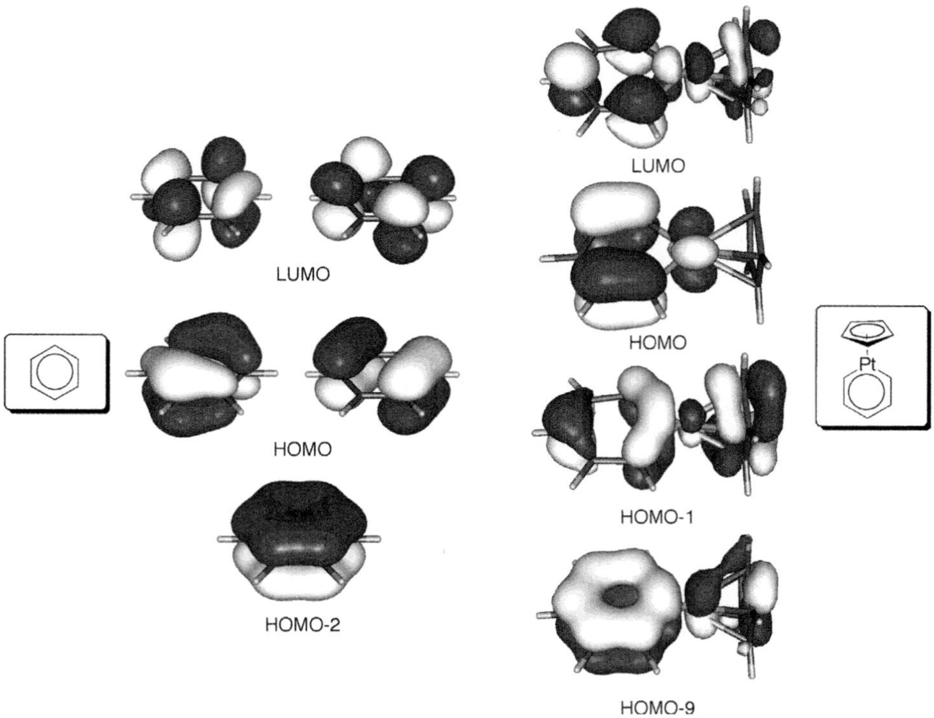

Figure 7.2 Representative MO's computed for benzene (left) and model platinabenzene Pt[C_5H_5](Cp) (right). Figure adapted from reference [17a]. (*See color plate section for the color representation of this figure.*)

resembles that of the HOMO-9 of the model platinabenzene Pt[C_5H_5](Cp). Similarly, the HOMO-1 and HOMO of the complex match the doubly degenerate HOMO of benzene. Of course, the relative energies of the calculated MO's and the degree of participation of the metal d-atomic orbitals depend on the transition metal and its coordination sphere as well as the substituents attached to the C5 moiety [17a]. For instance, the involvement of the d_{z^2} atomic orbital is generally associated with a slight deviation from planarity in order to maximize the orbital overlap.

The total number of π-electrons in metallabenzenes remains debatable. In their original report, Thorn and Hoffmann partitioned the metallabenzene into contributions coming from the four π-electron fragments $[C_5H_5]^-$ and a suitable transition metal fragment $[M]^+$ (with M = Mn, Rh) [11]. They suggested that metallabenzenes possess 6π-electrons, the most important π-bonding contribution coming from the $d_{xz}(M) \rightarrow 3\pi^*(C_5H_5^-)$ π-backdonation, due to the strong π-acceptor character of the vacant 3π* MO (Figure 7.3). Alternatively, Schleyer has suggested that the doubly occupied d_{yz} metal atomic orbital significantly contributes to the π-orbital interactions in metallabenzenes as well [19]. Thus, the interaction with the occupied 2π-orbital of $C_5H_5^-$ yields a pair of bonding and antibonding π-orbitals, whereby the latter becomes stabilized by mixing with the vacant 4π* MO of $C_5H_5^-$. The alternative interpretation of metallabenzenes suggests that they are 8π-electron systems, which agrees with the view by Jia and co-workers [15] considering a completely different fragmentation scheme (see Figure 7.1). However, it should be pointed out that the additional orbital involving the d_{yz} metal AO and the π-orbitals of the $C_5H_5^-$ anion has δ instead of π symmetry, which would mean that they possess Möbius aromaticity (see below).

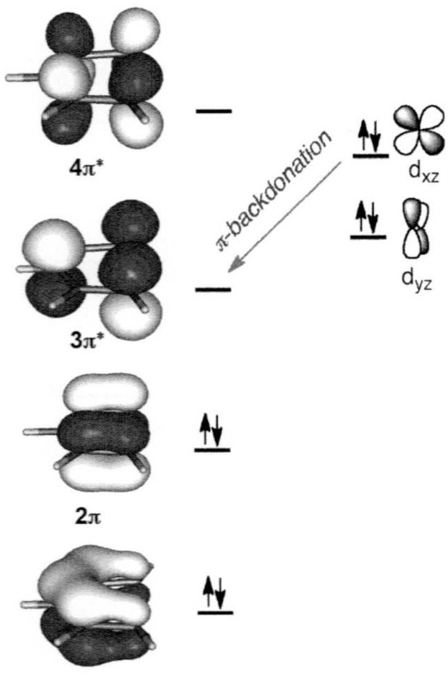

Figure 7.3 Schematic representation of the π-orbital interactions in metallabenzenes, adapted from reference [2a]. (*See color plate section for the color representation of this figure.*)

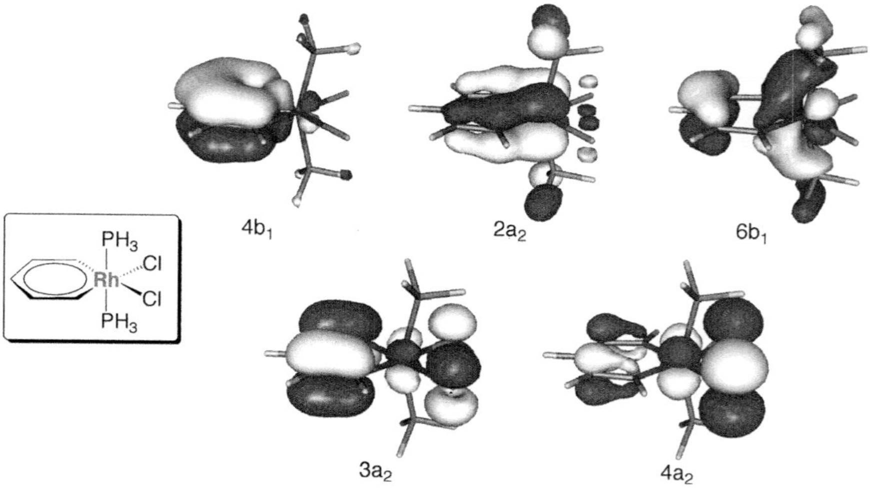

Figure 7.4 π-MO's computed for model rhodabenzene $Rh[C_5H_5](Cl)_2(PH_3)_2$ (figure adapted from reference [20]). (*See color plate section for the color representation of this figure.*)

The π-bonding in the C_{2v} symmetric rhodabenzene $Rh[C_5H_5](Cl)_2(PH_3)_2$, a model compound suggested by Thorn and Hoffmann as a candidate for a stable metallacyclic compound [11], was analysed in detail by us [20]. It was found that this particular metallabenzene has seven occupied π-MO's, where five of them (in the energetic order $4b_1 < 2a_2 < 6b_1 < 3a_2 < 4a_2$; see Figure 7.4) have coefficients in the six-membered ring, therefore indicating that this rhodabenzene is actually a 10π-electron species. Closer examination of these π-MO's suggests that the $4b_1$ MO is the result of the interaction of the 1π fragment orbital of $C_5H_5^-$ (see Figure 7.3) with a vacant orbital involving the transition metal fragment. In addition, the $2a_2$ MO is the result of the bonding contribution between the d_{yz} metal orbital and the 2π-orbital of the $C_5H_5^-$ fragment, whereas the $6b_1$ orbital is clearly identified as the result of the $d_{xz}(M) \rightarrow 3\pi^*(C_5H_5^-)$ π-backdonation suggested by Thorn and Hoffmann [11]. Finally, the two antibonding orbitals $3a_2$ and $4a_2$ arise from the bonding and antibonding combinations between the metal d_{yz} atomic orbital and the chlorine $p(\pi)$ orbitals. The antibonding nature of these MO's is somewhat diminished by mixing with the vacant 4π* of $C_5H_5^-$. This view of metallabenzenes as 10π-electron systems has been confirmed in a recent study by Solà and co-workers, who analysed the relative stabilities of the *ortho*, *meta* and *para* isomers of a series of heterometallabenzenes with the general formula $MClY(XC_4H_4)(PH_3)_2$ (M = Ir, Rh; X = N, P; Y = Cl and M = Os, Ru; X = N, P; Y = CO) [21].

The energy decomposition analysis (EDA) [22] method was also used by us to gain quantitative insight into the strength of the different energy contributions to the metallacyclic bonding in the model rhodabenzene [20]. According to the computed interaction and orbital energies using different partitioning schemes (namely $C_5H_5^-$ and $[M]^+$, $C_5H_5^+$ and $[M]^-$, and C_5H_5 and [M] as neutral fragments), the model rhodabenzene is best described as a result of the interactions between charged $C_5H_5^-$ and $[M]^+$ fragments. Interestingly, the breakdown of the attractive orbital ΔE_{orb} term into contributions from orbitals with different symmetry shows a contribution of 83 and 17% for the total σ- and π-bonding, respectively. The latter contribution comes mainly

Table 7.1 EDA results for selected metallabenzenes C_5H_5 and [TM] as fragments calculated at BP86/TZ2P. Energy values in kcal/mol (data taken from reference [20]).

Molecule	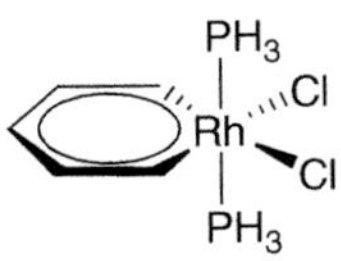	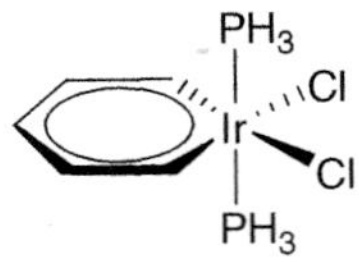	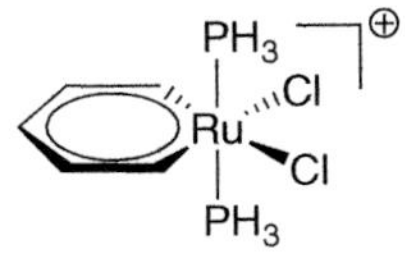	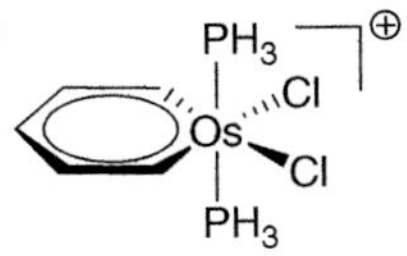
Symmetry	C_{2v}	C_{2v}	C_{2v}	C_{2v}
ΔE_{int}	−209.4	−234.9	−182.6	−210.4
ΔE_{Pauli}	329.6	407.2	238.1	298.4
$\Delta E_{elstat}{}^{a}$	−242.7 (45.0%)	−308.4 (48.0%)	−163.3 (38.8%)	−212.1 (41.7%)
$\Delta E_{Orb}{}^{a}$	−296.2 (55.0%)	−333.7 (52.0%)	−257.4 (61.2%)	−296.7 (58.3%)
$\Delta E\sigma(a_1)^{b}$	−119.2 (40.2%)	−137.6 (41.2%)	−82.0 (31.8%)	−101.7 (34.3%)
$\Delta E\pi(a_2)^{b}$	−10.3 (3.5%)	−12.6 (3.8%)	−8.8 (3.4%)	−11.0 (3.7%)
$\Delta E\pi(b_1)^{b}$	−52.3 (17.7%)	−55.8 (16.7%)	−58.8 (22.8%)	−60.9 (20.5%)
$\Delta E\sigma(b_2)^{b}$	−114.5 (38.6%)	−127.8 (38.3%)	−107.9 (41.9%)	−123.1 (41.5%)

a) The percentage values in parentheses give the contribution to the total attractive interactions $\Delta E_{elstat} + \Delta E_{orb}$.
b) The percentage values in parentheses give the contribution to the total orbital interactions ΔE_{orb}.

from orbitals of b_1 symmetry (−31.0 kcal mol^{-1}), while the a_2 orbitals are comparatively less significant (−14.6 kcal mol^{-1}), which is in agreement with the finding that the model rhodabenzene has one (strongly) binding $2a_2$ orbital but two antibonding $3a_2$ and $4a_2$ orbitals (see Figure 7.4).

The EDA method was also applied to a series of 16- and 18-valence electrons metallabenzenes which include 4d and 5d transition metals [20]. Similarly, it was found that σ-bonding is clearly more important (*ca.* 75%) than π-bonding (see Table 7.1). Moreover, σ-bonding is stronger in 18-electron metallabenzenes than in 16-electron systems, which can be ascribed to the different orbital occupation in both types of complexes. Whereas the highest-lying a_1 σ orbital has a metal–C_5H_5 bonding character in the 18-electron complexes, it is vacant in the 16-electron complexes, which is translated into weaker metal–C_5H_5 σ-bonding. The EDA results also provide a reasonable explanation for the experimentally observed higher stability of metallabenzenes having a third row transition metal compared to complexes of the second transition row [3]. It was found that π-conjugation contributes very little to the overall stronger attractive interaction energies computed for 5d complexes. Instead, the main difference between both types of complexes comes from the electrostatic term, which is clearly higher for 5d complexes. Therefore, the enhanced stability of these complexes is suggested to be related to the more electrostatic (and consequently less covalent) nature of the metal–C_5H_5 bonding.

7.2.2 Aromaticity of Metallabenzenes

The aromatic nature of metallabenzene complexes has attracted the interest of both experimental and theoretical chemists from the very beginning. As commented above,

Thorn and Hoffmann already concluded that the crucial mechanism for the stabilization of these complexes is the electronic delocalization within the six-membered ring. Indeed, the term "aromatic" was coined for these species in their pioneering work [11]. Since then, a large number of experimental and computational studies were reported that focused on assessing the putative aromatic character of this family of organometallic compounds. As the aromaticity of metallabenzenes and heterometallabenzenes has been reviewed quite recently by us [23]; herein, we shall only discuss the most relevant issues associated with this controversial topic.

From an experimental point of view, metallabenzenes satisfy the so-called structural criterion for aromaticity as they generally exhibit planarity and bond-length equalization. In addition, the CH groups in the metallabenzene ring which are not directly attached to the transition metal usually present chemical shifts in the typical "aromatic" range (5.5–8.0 ppm and 120–150 ppm in the ^{1}H and ^{13}C-NMR spectra, respectively). Furthermore, some metallabenzenes undergo electrophilic aromatic substitution reactions, S_EAr [2, 24], the archetypal reaction of aromatic compounds. According to these experimental observations, metallabenzenes fulfil the criteria (i.e. structural, magnetic and reactivity) usually used to characterize aromatic species.

However, and as commented above, many metallabenzenes exhibit significant deviations from planarity. In addition, a significant number of complexes may also engage in reactions which are unusual for classical aromatic systems such as oxidative addition of the halogen substituent to the transition metal [25], cycloaddition reactions [26, 27], nucleophilic aromatic substitution (S_NAr) of hydrogen via the corresponding Meisenheimer intermediates [28] or rearrangement processes typically to cyclopentadienyl complexes [29]. It becomes clear, therefore, that metallabenzene complexes exhibit a rich reactivity which is markedly different from that found for typical organic aromatic compounds. Hence, from an experimental point of view, it can be concluded that the strength of aromaticity of metallabenzenes is different from that of their all-carbon analogues.

Various aromaticity descriptors have been computationally applied to diagnose the aromatic nature of metallabenzenes. In this section, the most relevant descriptors are summarized.

7.2.2.1 [4*n*+2]-rule

The simple inspection and comparison of the MO's of metallabenzene complexes with respect to those of benzene and related aromatic compounds clearly indicate the close relationship between both types of compounds. As commented above, metallabenzenes were initially viewed as 6π-electron system (Figure 7.3), therefore satisfying the [4*n*+2]-rule [30] for Hückel aromatic compounds. The alternative view by Schleyer [19] or Jia and co-workers [15] considers metallabenzenes as 8π-electrons. However, one of these π-orbitals actually possesses actual δ symmetry, which means that metallabenzenes could be considered Möbius aromatic species [31].

7.2.2.2 Absolute Hardness

Absolute hardness (η, defined as half the HOMO–LUMO gap for Hartree–Fock (HF) or Hückel orbitals, i.e. $\eta = (\varepsilon_{LUMO} - \varepsilon_{HOMO})/2$ [32]) is frequently used to estimate the stabilization and reactivity of a molecule [33]. In addition, this parameter has been developed as a quantitative aromaticity measure. Thus, the frontier between aromatic and antiaromatic species in typical organic compounds is defined as $0.2\eta_B$, where η_B is the absolute hardness of benzene [34]. In general, the aromatic strength increases with larger absolute hardness values.

Chamizo and co-workers calculated an absolute hardness of 0.60 eV for iridabenzene $Ir[C_5H_3(Me\text{-}2, 4)](PEt_3)_3$ using the Extended Hückel Method [35]. As this value is clearly lower than that computed for benzene (2.27 eV) or thiophene (2.17 eV), the authors concluded that this particular iridabenzene should not be considered an aromatic molecule. These earlier calculations were revisited at the DFT level by Yang and co-workers, who considered the model osma- and iridabenzenes $Os[C_5H_5](PH_3)_2(CO)(I)$ and $Ir[C_5H_5](PH_3)_3$ [36]. An absolute hardness of 4.43 and 4.22 eV was computed (B3LYP/6-31G(d, p)&LanL2DZ) for these complexes, respectively, which corresponds to *ca.* 65% of the value calculated for benzene at the same level of theory (6.47 eV). These results strongly suggest that the degree of stabilization by aromaticity in metallabenzene complexes is lower than in benzene. It should, however, be noted that the use of absolute hardness values as a quantitative measure of aromaticity must be used with great caution as serious deficiencies have been found when applied to heterocyclic compounds [37].

7.2.2.3 Magnetic Descriptors

The anomalous behaviour of arene ^1H-NMR (and ^{13}C-NMR) chemical shifts are ascribed to the ability of aromatic compounds to sustain an induced diatropic ring current as suggested by Pople's ring current model [38]. Although many computational methods have been developed to probe the magnetic responses of aromatic compounds, both globally (exalted diamagnetic susceptibilities [39] and anisotropies for the entire molecule) or locally by means of induced magnetic field, $\boldsymbol{B}^{ind}{}_z$ [40], values, ring currents [41–43], etc., only a few of which have been applied to metallabenzene complexes.

Among the different magnetic descriptors, nucleus-independent chemical shift (NICS) [44] values are arguably the most popular computational tool to diagnose magnetic aromaticity, both qualitatively (NICS < 0 ppm denotes aromatic character) and quantitatively (by means of the dissected $NICS_{zz}$ values [45] which take into account contributions arising from the zz vector component of the shielding tensor). This is because such evaluations are relatively easy to compute and do not rely on reference compounds. Martin and co-workers calculated (B97/aug-SDB-cc-pVDZ//mPW1K/SDD level) the isotropic NICS(0) and NICS(1) values (computed at the ring centre and 1 Å above, respectively) of a series of metallabenzenes (a representative selection is given in Table 7.2) [17a]. From the data in Table 7.2 it is clear that, although the NICS values are in general negative, the absolute values are not very large, which does not give a definite support for the aromaticity of these species. Furthermore, the computed magnetic susceptibility anisotropy ($\Delta\chi$) values, also summarized in Table 7.2, do not support a clear aromatic character for these complexes either (aromatic molecules typically exhibit large negative $\Delta\chi$ values). According to these results, Martin and co-workers concluded that "based on the above-mentioned methods, it is difficult to state with any certainty whether the metallabenzene complexes are truly aromatic or not" [17a]. This is due to the severe limitations found on the application of these methods to this particular type of organometallic complex because both shielding and magnetic susceptibility tensors are greatly disturbed by the close proximity to ligands on the metal centre.

Table 7.2 Computed NICS(0) and NICS(1) (in ppm), magnetic susceptibility anisotropy ($\Delta\chi$ in cgs ppm) values, [a] and ASE values (in kcal/mol)[b] for representative model metallabenzenes.

	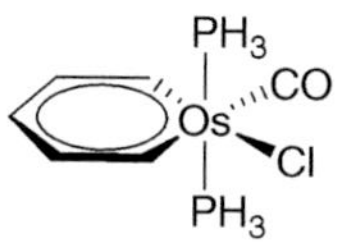	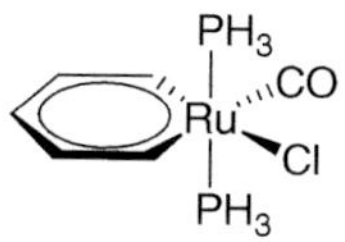	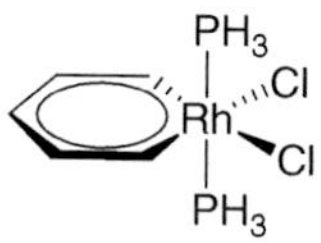	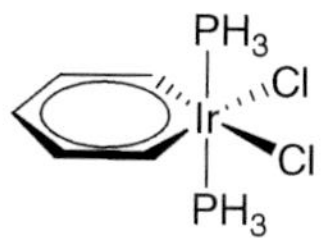	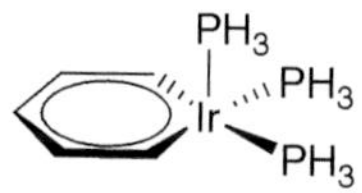			
NICS(0)	+2.5	+3.2	—	+2.8	−3.7	−2.6	—	−7.7
NICS(1)	−3.5	−3.2	—	−3.2	−8.8	−6.4	—	−10.0
$\Delta\chi$	−10.0	−43.4	—	−22.7	+93.9	+43.2	—	−61.7
ASE	18.7	20.0	33.4	33.5	22.4	37.6	32.8	42.5

a) Values taken from reference [17a].
b) Values taken from reference [20].

We want to point out that there is no direct correlation between aromaticity, which is by definition an energy (stability) property of a molecule and its magnetic property. The fact that benzene has a specific magnetic property (magnetic susceptibility anisotropy) does not mean that the absence or occurrence of such magnetic property necessarily indicates the absence or occurrence of energetic stabilization due to delocalization.

A similar finding was observed by Han *et al.* [46], who reported the synthesis of the first example of *m*-osmaphenol and calculated the corresponding NICS(1) value to assess the aromatic character of this species. Again, the computed NICS(1) value was negative, albeit not very large (NICS(1) = −3.1 ppm). Quite similar values were computed by Zhang *et al.* [47] for the ruthenabenzene $[(C_9H_6NO)Ru\{CC\text{-}(PPh_3)CHC(PPh_3)CH\}(C_9H_6NO)(PPh_3)]Cl_2$ (NICS(1) and $NICS(1)_{zz}$ values of −3.2 and −11.3 ppm, respectively, at the B3LYP-GIAO/6-31G(d)&LanL2DZ(f) level). This supports some aromatic character for this species, which agrees with the lack of reactivity observed experimentally when the ruthenabenzene is treated with different nucleophiles and electrophiles.

Mauksch and Tsogoeva [48] studied the aromaticity of a series of metallacycloheptatrienes and metallacyclooctatetraenes. Both type of complexes possess formally eight π-electrons and fulfil the requirement of Möbius aromaticity [31]. Indeed, aromaticity was supported by NICS and harmonic oscillator model of aromaticity (HOMA, which is based on the geometry of the molecule) [49] calculations. Particularly, the dissected $NICS(1)_{zz}$ values range from −1.2 to −65.3 ppm, therefore confirming the aromatic nature of these species.

Ring currents in metallabenzenes have also been analysed computationally. For instance, Herges and co-workers applied the anisotropy of the induced current density (ACID) [43, 50] method to visualize the delocalization of electrons in the model iridabenzene $Ir[C_5H_5](CO)(PH_3)_2$ [50]. The aromatic nature of this species is confirmed by the presence of a clear diatropic (clockwise vectors) circulation within the six-membered ring, which clearly resembles the corresponding ACID plot computed for benzene (Figure 7.5). Periyasamy *et al.* [51] also studied the induced current corresponding to

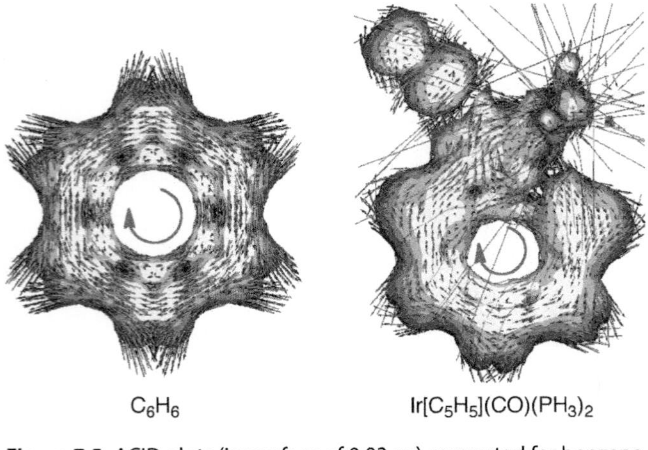

Figure 7.5 ACID plots (isosurface of 0.03 au) computed for benzene and $Ir[C_5H_5](CO)(PH_3)_2$. Aromatic species exhibit clockwise diatropic circulations (arrows). Figure adapted from reference [50]. (*See color plate section for the color representation of this figure.*)

the out-of-plane MO's of a series of metallabenzenes containing Ir, Rh, Os, Ru, Pt and Pd. Despite the fact that the reported electron counting of these complexes is questionable, the authors found either diatropic (aromatic) or paratropic (antiaromatic) ring currents for complexes having the same occupancy of π-MO's. For instance, platina- and palladametallabenzenes are highly aromatic compounds but only when coordinated to the cyclopentadienyl ligand. Similar current maps were found recently by Havenith *et al.* [52], who calculated the ring currents, including relativistic effects for selected systems studied by Periyasamy *et al.* [51].

7.2.2.4 Energetic Descriptors

It is well known that the fundamental property of aromatic compounds is their enhanced thermochemical stability with regard to the acyclic conjugated references. Whereas properties such as NMR chemical shifts and bond length equalization are secondary manifestations of aromaticity (which in many cases can be misleading), the energetic criterion is considered the principal descriptor for aromaticity as it governs the reactivity and much of the chemical behaviour of a molecule.

One method used both experimentally and computationally to obtain the so-called aromatic stabilization energy (ASE), defined as the extra thermodynamic stabilization in a molecule due to aromaticity, is based on the reaction energies of isodesmic equations (where the number and type of bonds is exactly the same on both sides of the equation). As a reference, an ASE value of 28.8 kcal/mol has been computed for benzene using the reaction depicted in Scheme 7.1, Equation (a) [53]. For metallabenzenes, Yang and co-workers proposed the homodesmotic equations (b) and (c) to estimate the ASE values of two model osma- and iridabenzenes [36]. Similarly, the reaction (d) was used to compute the ASE value of benzene at the same level of theory (B3LYP/6-31G(d, p)&LanL2DZ). Although the calculated ASE values confirm a significant aromatic stabilization in both model metallabenzenes (ASE = 18.0 and 14.7 kcal/mol, respectively), they are clearly less aromatic than benzene (ASE = 32.2 kcal/mol).

It has been suggested that in most cases these equations are contaminated by different flaws such as strain, hyperconjugation, "proto"-branching or *syn-anti* effects which make the calculated ASE values not always reliable [54]. To avoid these shortcomings, Schleyer and Pühlhofer introduced the so-called isomerization method (ISE) to evaluate ASE values more accurately [54]. This approach is based on the differences between the total energies computed for only two species: a methyl derivative of the aromatic system and its nonaromatic exocyclic methylene isomer. Using this approach, an ASE value of 33.2 kcal/mol was calculated for benzene (Scheme 7.2, Equation (a)) which is close to the value obtained using the strain-balanced Equation (a) in Scheme 7.1. This methodology was applied by de Proft and Geerlings to estimate the ASE value of the platinabenzene Pt[$C_5H_3Me_2$](Me_2Cp) [55]. The computed ASE value of 23.4 kcal/mol (Scheme 7.2, Equation (b)) is *ca.* two-thirds of the value for benzene (33.8 kcal/mol) at the same level of theory (B3LYP/6-311+G*&LANL2DZ for Pt), which confirms that metallabenzenes are aromatic molecules whose aromaticity strength is weaker than that of typical aromatic compounds like benzene but still noticeable. Similar results were found recently by Lin *et al.*, who compared the ASE-ISE values of a series of model metallabenzenes with Pt, Re, Ir and Os [56]. The calculated values suggest that the strength of aromaticity of the model rhenabenzene (ASE = 21.5 kcal/mol; Scheme 7.2, Equation (c)) is lower than that of platinabenzene and iridabenzene, and comparable to osmabenzene.

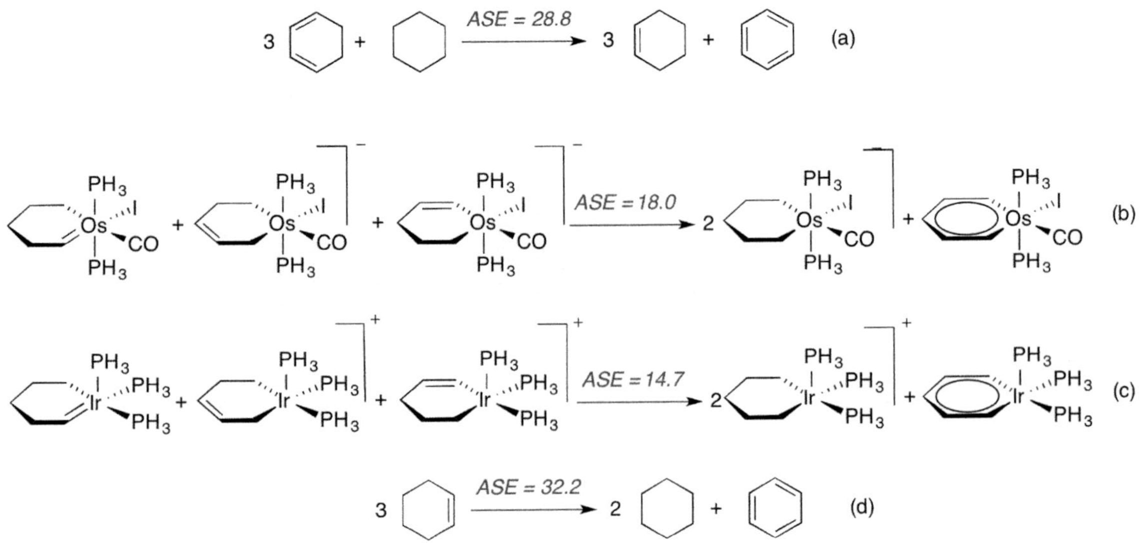

Scheme 7.1 Isodesmic equations to estimate ASE values.

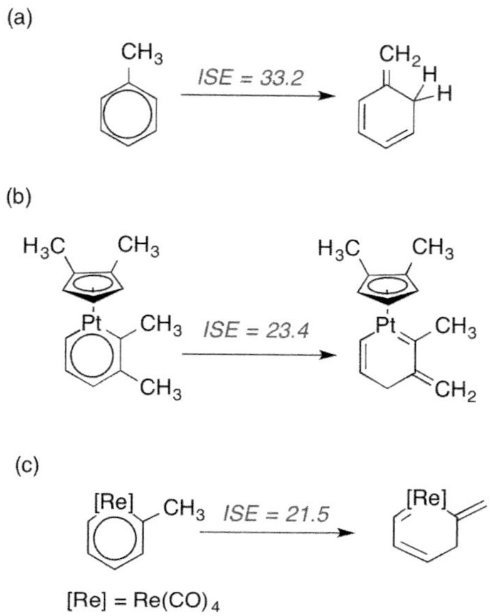

Scheme 7.2 ASE values (in kcal/mol) computed by means of the ISE method.

Using the EDA method [22], we developed a completely different approach to estimate the ASE values [57, 58]. The EDA method is able to consider only the π-orbitals of the interacting fragments in the geometry of the molecule to exclusively estimate π-interactions without recourse to reference molecules. Thus, the orbital contribution derived from the $\Delta E\pi$ values (i.e. the out-of-plane orbital component) serves as a direct estimation of the π-conjugation and hyperconjugation of a molecule, even in complex π-extended systems [59]. Therefore, our approach consists of comparing the π-cyclic conjugation strength in the aromatic species with the π-conjugation of an appropriate acyclic reference system. The difference between the $\Delta E\pi$ (cyclic molecule) and $\Delta E\pi$ (acyclic reference) value provides directly the extra energetic stabilization due to cyclic conjugation, i.e. due to aromaticity (aromatic molecules exhibit ASE > 0, whereas antiaromatic species possess ASE < 0). The ASE values computed in this way for metallabenzenes span from nonaromatic or low-aromatic compounds (ASE = 8.7 kcal/mol for $[Ir(C_5H_5)(PH_3)_2(MeCN)_2]^{+2}$) to highly aromatic species (see, for instance, the series of compounds gathered in Table 7.2). Considering that the ASE value calculated for benzene at the same level of theory is higher (ASE = 42.5 kcal/mol, BP86/TZVP) than for the considered metallabenzenes, it can be concluded once again that the aromatic strength in metallabenzenes is significant but comparatively weaker than in benzene. Within this methodology, no clear correlation between the computed ASE values and the nature of the transition metal fragment (charge, metal, ligands, oxidation state of the metal) was found.

7.3 Computational Studies on Synthetic Pathways towards Metallabenzenes

The routes towards the synthesis of metallabenzenes can be classified into two main groups of reactions, namely (1) the modification of a preformed metallacyclic ring via either ring expansion or contraction and (2) addition reaction of an appropriate molecule or anion having the carbonated framework to a transition metal substrate.

The synthesis of the first metallabenzene complex, the osmabenzene **1**, by Roper and co-workers [1] belongs to the first group of processes. The overall transformation from precursor **2** involves cyclization at the metal centre of two HC≡CH molecules and the CS ligand. Martin and co-workers proposed a plausible reaction mechanism for the synthesis of **1** using a model osmium complex (where the bulky PPh_3 ligands were replaced by PH_3; see Figure 7.6) [17a]. According to the computed data (PCM(toluene)-mPW1K/SDB-cc-pVDZ//mPW1K/SDD level), the transformation begins with the exergonic PH_3 ligand exchange by a molecule of acetylene to produce complex **4**. This species evolves to complex **5** via a migratory insertion of the CS ligand into metal-carbon σ-bond of the osmacyclopropene through transition state $\mathbf{TS_{4-5}}$ ($\Delta G^{\neq}$ = 31.4 kcal/mol). Then, complex **5** isomerizes to **6**, where the coordination of a second molecule of acetylene occurs, yielding complex **7**. A subsequent C–C coupling via $\mathbf{TS_{7-8}}$ ($\Delta G^{\neq}$ = 30.6 kcal/mol) produces complex **8** in a strongly exergonic transformation (ΔG = −40.8 kcal/mol). Finally, intermediate **8** renders the final osmabenzene **1** through a reaction that can be described as either an intramolecular C=S oxidative addition or complete C=S bond coordination through the transition state $\mathbf{TS_{8-1}}$ ($\Delta G^{\neq}$ = 14.8 kcal/mol) in an exergonic process ($\Delta G^{\neq}$ = −18.7 kcal/mol).

Martin and co-workers [17a] also computationally explored the formation of metallabenzenes by reacting metal halide salts with 2, 3-diphenylcyclopropenylvinyllithium **9** (Figure 7.7). To date, this elegant and direct route to metallabenzenes has been only achieved for irida- and platinabenzenes [13, 60]. Two possible reaction mechanisms can be envisaged for this transformation, namely one involving a metallabenzvalene intermediate (**13**) and the other involving a metalla-Dewar benzene-like structure (**14**). According to the computed data, it was suggested that metallabenzenes are formed preferentially from metallabenzvalene intermediates through Ir–C and C–C bond breaking (via transition state $\mathbf{TS_{13-14}}$), since the alternative route through a Dewar-metallabenzene species **14** is significantly higher in energy (via transition state $\mathbf{TS_{14-15}}$).

The formation of osmabenzene **20** by the reaction of osmium(II) complex **16** and 3-methylpent-1-en-4-yn-3-ol was described by Xia and co-workers (Figure 7.8) [61]. In addition, DFT calculations (B3LYP/6-31G(d, p)&LanL2DZ level) were carried out to understand the role of benzonitrile in the transformation. Two alternative reaction pathways are proposed from intermediate **18** which involve complexes **19** and **21**, respectively (Figure 7.8). Using model complexes where the PPh_3 and PhCN ligands were replaced by PH_3 and HCN, respectively, it was found that both processes are feasible to yield the final osmabenzene **20**, which nicely agrees with the experimental findings. Moreover, in both cases the key steps involve C–H activation by the metal- and chloride-assisted proton transfers. As the rate-determining step involves an electrophilic substitution, replacement of the chloride ligands by benzonitrile would increase the electrophilicity of the metal centre, therefore favouring the whole transformation.

In a closely related work, Xia and co-workers reported the synthesis of a similar osmabenzene **27** by reaction of $OsCl_2(PPh_3)_3$ with alkynol HC≡CCH(OH)Et (Figure 7.9) [62]. DFT calculations (PCM-B3LYP/6-31G(d, p)&LanL2DZ level) suggest that the transformation is both kinetically and thermodynamically favourable in the presence of a nitrile ligand, whereas CO inhibits the process by forming the highly stable intermediate complex **29**. This finding is consistent with the experimental

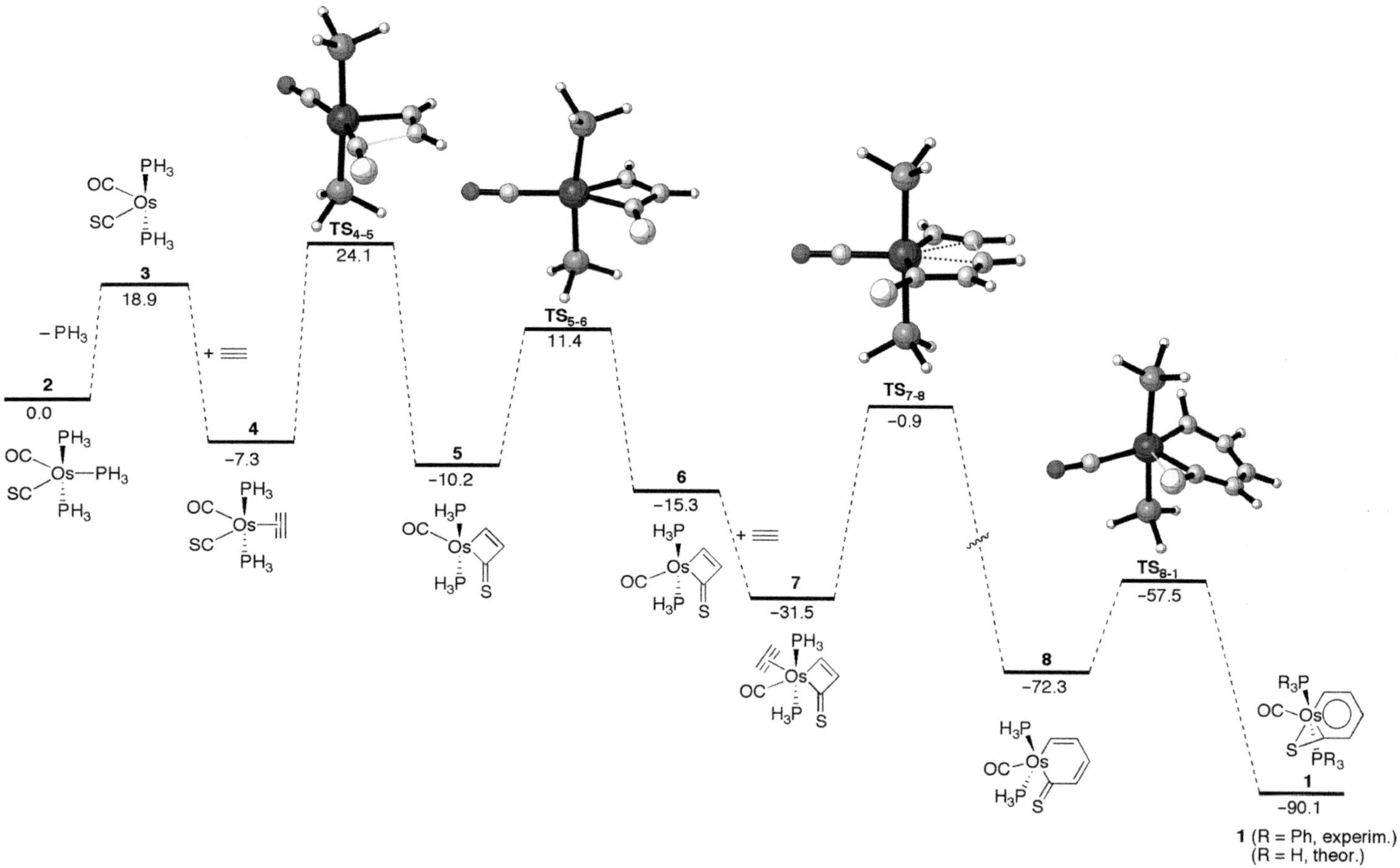

Figure 7.6 Computed reaction profile (PCM(toluene)-mPW1K/SDB-cc-pVDZ//mPW1K/SDD level) for the formation of model osmabenzene **1** (coordinates of the transition states were taken from reference [17a]). Relative free energies (ΔG_{298}, at 298 K) are given in kcal/mol. (*See color plate section for the color representation of this figure.*)

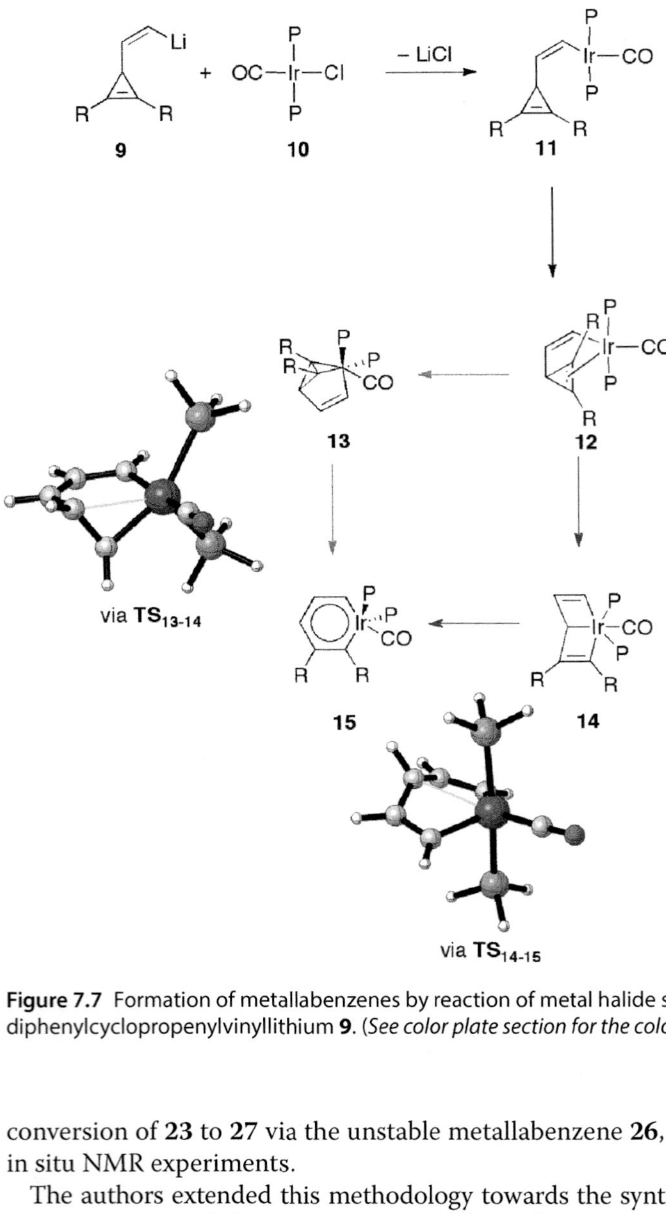

Figure 7.7 Formation of metallabenzenes by reaction of metal halide salts with 2,3-diphenylcyclopropenylvinyllithium **9**. (*See color plate section for the color representation of this figure.*)

conversion of **23** to **27** via the unstable metallabenzene **26**, which was observed using in situ NMR experiments.

The authors extended this methodology towards the synthesis the osmanaphthalene **31** and osmanaphthalyne **32** (Figure 7.10) [63]. The mechanism involved in the transformation starting from carbyne complex **28** was elucidated by means of DFT calculations (B3LYP/6-31g(d)&LanL2DZ level; see Figure 7.10). It was suggested that the 18-electron hydride-alkenylcarbyne complex **28** evolves via $\mathbf{TS_{28-29}}$ ($\Delta E^{\neq}$ = 31.5 kcal/mol) to intermediate **29**, which is stabilized by C_{aryl}–H···Os agostic interactions. This species is then transformed into the naphthalene complex **30** through an *ortho* C–H activation reaction of the phenyl ring via the transition state $\mathbf{TS_{29-30}}$ ($\Delta E^{\neq}$ = 17.6 kcal/mol). It was suggested that the latter intermediate can evolve under nitrogen atmosphere either to the experimentally observed dinuclear species **32** through H–Cl coupling, HCl removal

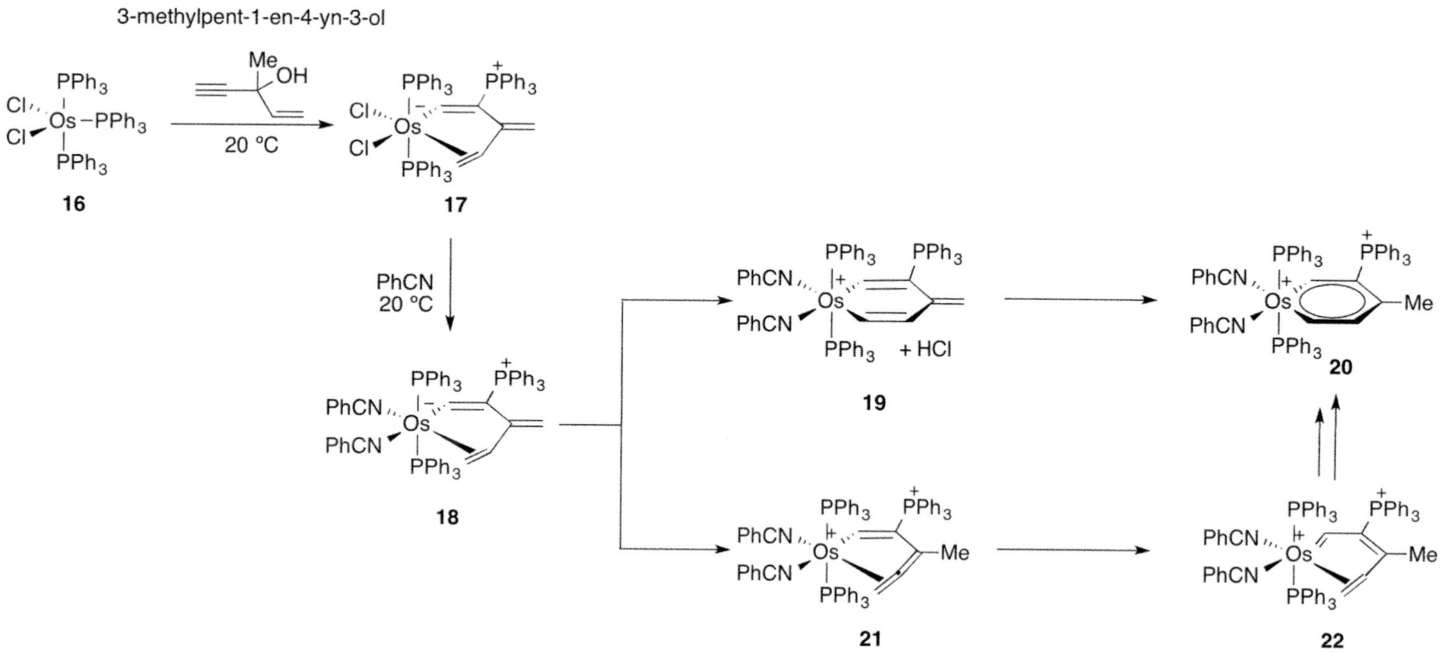

Figure 7.8 Formation of osmabenzene **20** from $OsCl_2(PPh_3)_3$ and 3-methylpent-1-en-4-yn-3-ol.

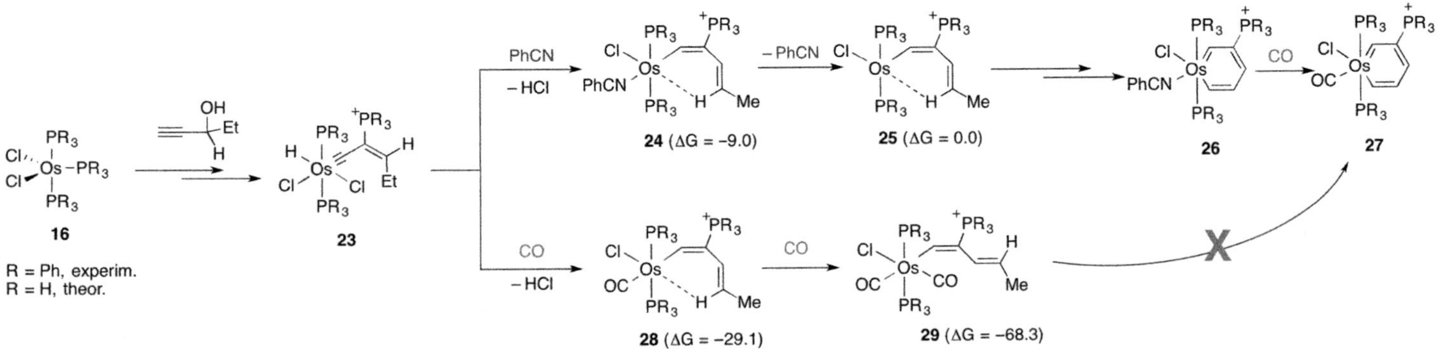

Figure 7.9 Proposed mechanism for the formation of osmabenzene **27** from $OsCl_2(PPh_3)_3$ and alkynol HC≡CCH(OH)Et. Relative free energies (ΔG, computed at the PCM-B3LYP/6-31G(d, p)&LanL2DZ level) are given in kcal/mol.

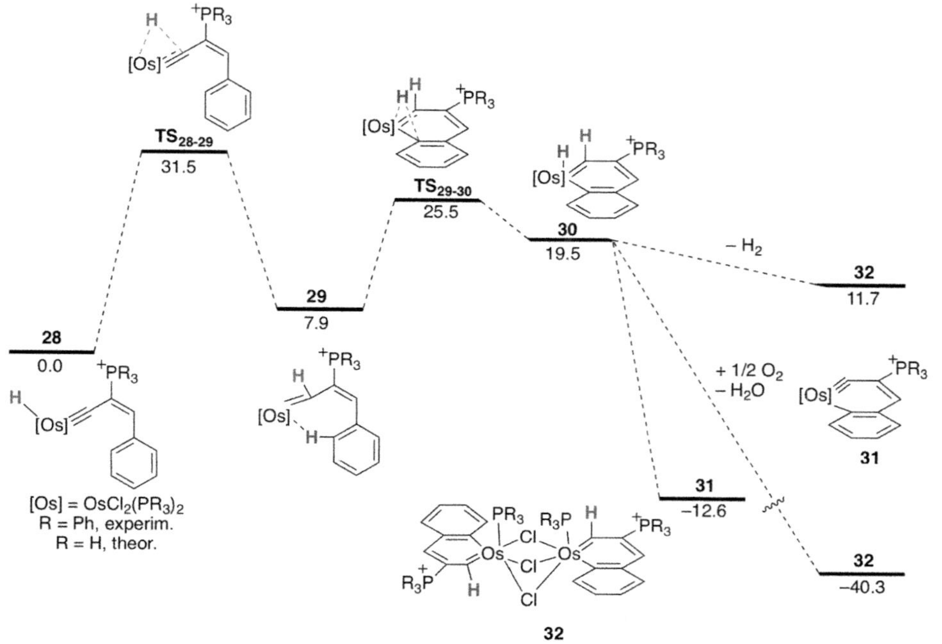

Figure 7.10 Computed (B3LYP/6-31g(d)&LanL2DZ level) reaction profile for the formation of osmanaphthalene **31** and osmanaphthalyne **32** (figure adapted from reference [63]). Relative energies are given in kcal/mol.

and final dimerization, or alternatively to naphthanyle complex **32** through H–H coupling (the latter species being thermodynamically favoured under oxygen atmosphere).

A different way to prepare osmanaphthalenes has been computationally predicted quite recently. Zhu and co-workers carried out DFT calculations (B3LYP/6-31+G(d)& LanL2DZ level) on the rearrangement of osma-indenyl complexes **33** into osmanaphthalenes **37** (Figure 7.11) [64]. According to the computed data, the latter species are in general thermodynamically more stable than the former complexes. The computed mechanism for the transformation involves the initial coordination of the PMe_3 ligand to the metal centre of the η^5-indenyl complex **33** which yields the η^1-indenyl complex **34**. This species then isomerizes to the corresponding osmanaphthalene **37** through the transition state $\mathbf{TS_{34-37}}$ ($\Delta G^{\neq}$ = 32.5 kcal/mol, with respect to the initial complex **33**). Alternatively, it was suggested that the coordination of the PMe_3 ligand to the metal centre occurs after the isomerization of η^5-indenyl **33** to η^1-indenyl complex **35**. In this case, the required energy to reach the corresponding transition state $\mathbf{TS_{35-36}}$ is identical to that for $\mathbf{TS_{34-37}}$. Therefore, both pathways seem equally feasible in view of the exergonicity of the process and the moderate energy barriers. Interestingly, the barrier for the isomerization becomes lower when placing electron donor substituents within the metallacycle. For instance, when the R substituents in **33** are methoxy groups, the computed barriers are 23.9 and 23.4 kcal/mol for the corresponding processes involving $\mathbf{TS_{34-37}}$ and $\mathbf{TS_{35-36}}$, respectively.

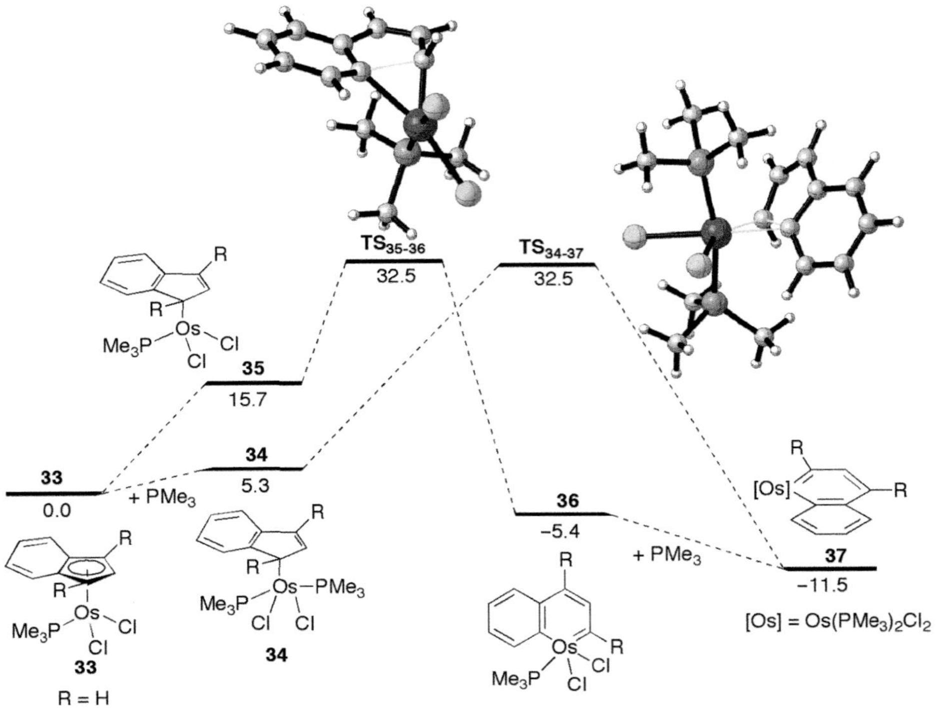

Figure 7.11 Computed (B3LYP/6-31+G(d)&LanL2DZ level) reaction profile for the rearrangement reaction of indenyl-complexes **33** into osmanaphthalenes **37** (coordinates of the transition states were taken from reference [64]). Relative free energies are given in kcal/mol. (*See color plate section for the color representation of this figure.*)

7.4 Computational Studies on the Reactivity of Metallabenzenes

As commented above, metallabenzenes present a rich reactivity which is markedly different from that found for typical organic aromatic compounds. In this chapter, we summarize the computational efforts made to understand the main reactions in which metallabenzenes are involved.

7.4.1 Electrophilic Aromatic Substitution Reactions

As typical organic aromatic molecules, metallabenzenes undergo electrophilic aromatic substitution reactions, such as bromination or nitration reactions where the substitution is directed in the same way as for benzenes by the ring substituents [2, 24].

Condensed Fukui functions (f_k^-) [65] derived from DFT were used to rationalize the regioselectivity of the electrophilic substitution reactions involving the iridabenzofurans shown in Figure 7.12 [66]. The Fukui functions can be used as a reactivity index to assess the most nucleophilic site (i.e. the atom with the largest f_k^- value) and the most electrophilic site (the atom with the largest f_k^+ value) in different systems, including organometallic complexes [67]. This parameter is computed as $f_k^- = q_k(N) - q_k(N{-}1)$,

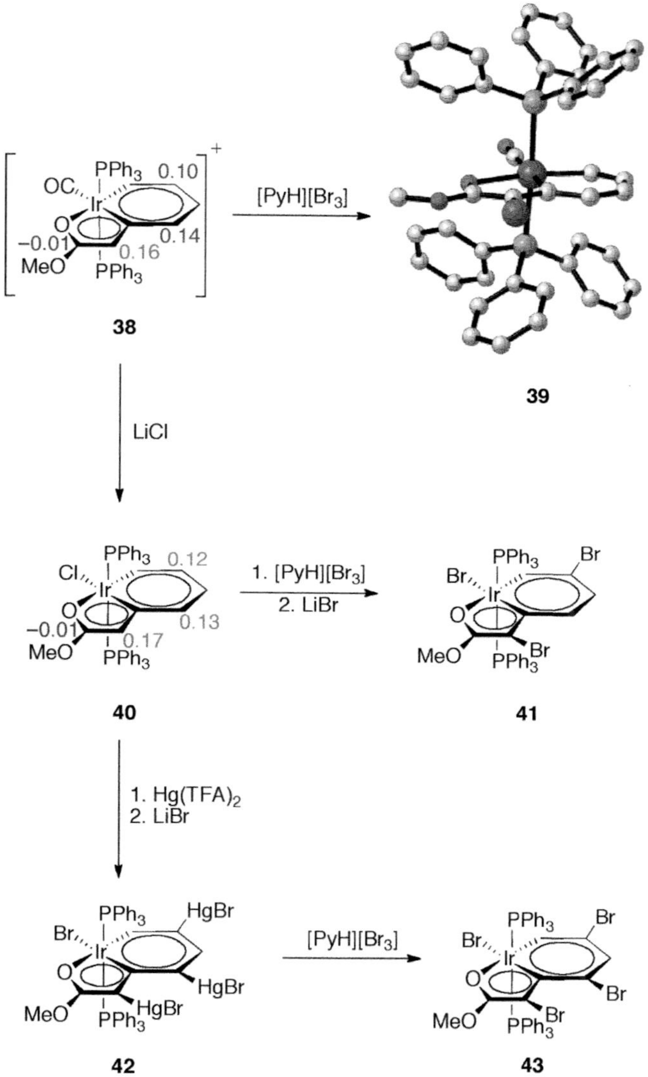

Figure 7.12 Electrophilic substitution reactions on iridabenzofurans. Numbers close to carbon atoms indicate the corresponding computed Fukui functions (f_k^-). Coordinates of complex **39** were taken from reference [66]. (*See color plate section for the color representation of this figure.*)

where q_k are the atomic charges at the kth atomic site and $q_k(N)$ and $q_k(N-1)$ are the electron populations on atom k for the N and $N-1$ electron species [65]. As seen in Figure 7.12, in compound **38** the atom having the largest f_k^- value (0.16) is C6, which indicates that the bromination reaction should occur at this position to produce complex **39**, as experimentally found. Similarly, in compound **40** the largest f_k^- value (0.17) was found again for C6, but, in this case, high f_k^- values have also been computed at the C2 and C4 positions. This agrees with the experimental observation that complex **40** may undergo either dibromination to produce **41** or trimercuration, yielding complex

43. Despite the good agreement between computational and experimental data, the authors noted that the predictive value of this methodology has limitations in situations where steric effects become significant.

7.4.2 Nucleophilic Aromatic Substitution Reactions

S_NAr reactions typically occurs in deactivated aromatic rings and in most cases are characterized by a simple two-step addition/elimination mechanism in which a Meisenheimer σ-adduct is central [68]. Metallabenzenes can also undergo this type of reaction.

Wright and co-workers described a convenient way to functionalize metallabenzenes through nucleophilic aromatic substitution of hydrogen [28]. Thus, treatment of the cationic metallabenzenes **44** and **47** with different nucleophiles ($NaBH_4$, MeLi or NaOEt) results in the formation of metalla-Meisenheimer complexes **45** and **48**, which can then be oxidized to metallabenzenes **46** and **49**, respectively (Figure 7.13). This two-step process proceeds therefore in a regioselective manner, as it occurs exclusively at the metallabenzene ring position γ to the metal. The condensed Fukui functions f_k^+, computed as $f_k^+ = q_k(N+1) - q_k(N)$, were calculated to explain the experimental observation that nucleophiles preferentially attack at the C3 position of the metallabenzenes. In the case of complex **44**, although C1, C3 and C5 exhibit large f_k^+ values (i.e. most electrophilic sites), the largest numerical value (0.18) is found for C3, which nicely agrees with the experimental findings. A similar scenario is found for complex **47**, although in this case the C3 and C5 carbon atoms exhibit identical f_k^+ values (0.17) therefore suggesting that both centres are equally electrophilic. As commented above, this discrepancy may be ascribed to the steric congestion in the nucleophilic addition reaction which is not taken into account when computing the Fukui indices.

A S_NAr mechanism has been invoked to explain the formation of ruthenabenzene **51** from the reaction between complex **50** and 8-hydroxyquinoline (Figure 7.14) [69]. Interestingly, the newly formed metallabenzene **51** can be transformed into the ruthenacyclohexa-1,4-diene **52** by treatment with excess $NaBH_4$ in dichloromethane [47].

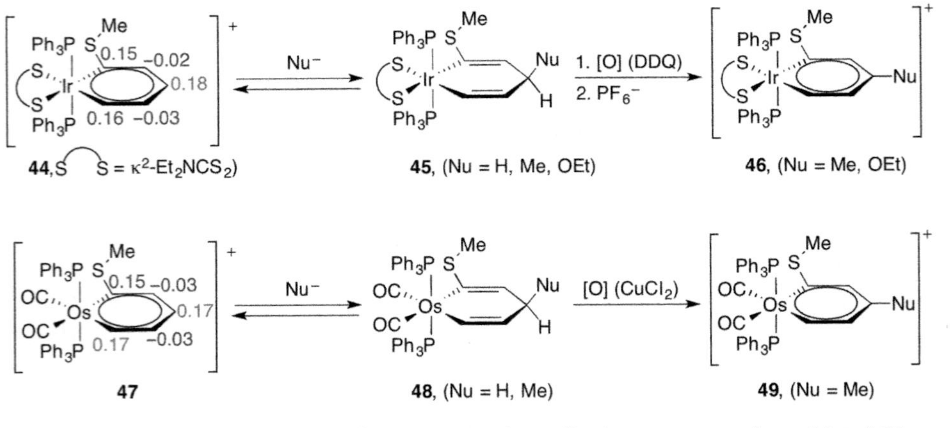

Figure 7.13 Nucleophilic aromatic substitution reactions of hydrogen on complexes **44** and **47**. Numbers close to carbon atoms indicate the corresponding computed Fukui functions (f_k^+, taken from reference [28]).

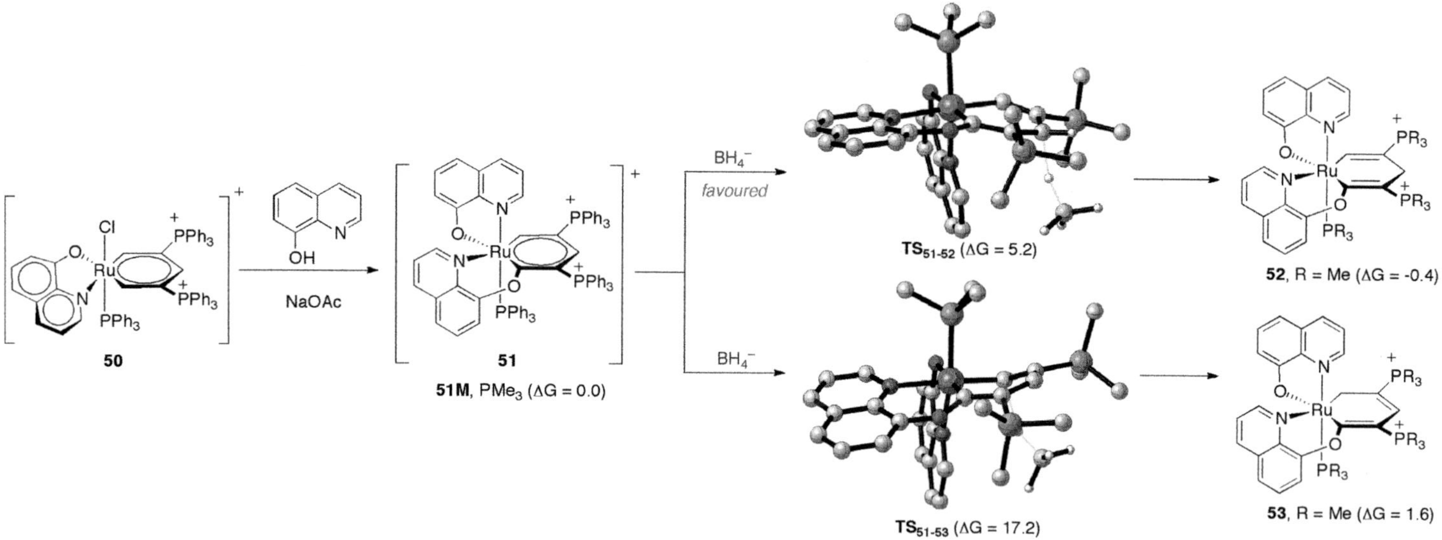

Figure 7.14 Computed (B3LYP/6-31G(d)&LanL2DZ level) reaction profile for the interconversion of ruthenabenzene **51** into ruthenacyclohexa-1,4-diene **52**. Relative free energy values are given in kcal/mol. Coordinates of the transition states were taken from reference [47]. (*See color plate section for the color representation of this figure.*)

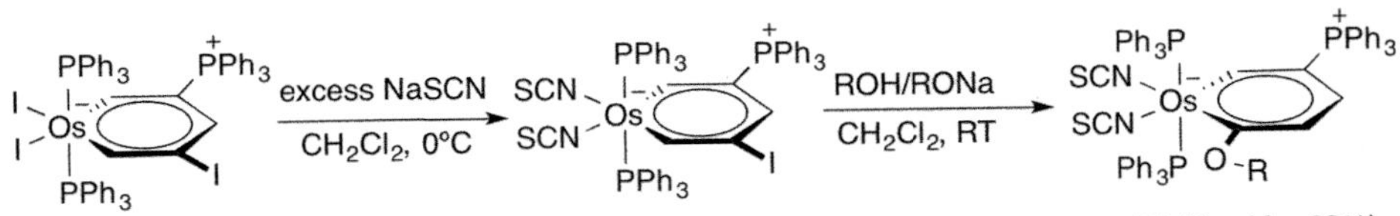

54

55 (89%)

56 (R = Me, 93%)
(R = Et, 93%)

Scheme 7.3 *Cine*-substitution in osmabenzene **55**.

According to DFT calculations (B3LYP/6-31G(d)&LanL2DZ level), the complete regioselectivity of the process takes place mainly under kinetic control in view of the much lower activation energy ($\Delta G^{\neq}$ = 5.2 kcal/mol, via $\mathbf{TS_{51-52}}$) computed for the pathway involving the C3 carbon atom compared to the reaction involving C1 ($\Delta G^{\neq}$ = 17.2 kcal/mol, via $\mathbf{TS_{51-53}}$; see Figure 7.14). In addition, the regioselectivity also occurs under thermodynamic control in view of the higher stability of the 1, 4-cyclohexadiene derivative **52** ($\Delta\Delta G$ = 2.0 and 1.6 kcal/mol, for PMe_3 and PPh_3 ligands, respectively).

Lin, Xia and co-workers recently reported that metallabenzenes can undergo the so-called *cine*-substitution reaction [70]. This process can be considered a particular type of nucleophilic aromatic substitution where the nucleophile takes up a position adjacent to that previously occupied by the leaving group [71]. Thus, when alkali-resistant osmabenzene **55** is reacted with MeOH or EtOH in the presence of strong alkali and at room temperature, the corresponding *cine*-substituted osmabenzene **56** is formed in excellent reaction yields (Scheme 7.3). Under similar reaction conditions, the reaction using amines as nucleophiles produces five-membered ring species in addition to the product derived from the *cine*-substitution. With the help of computational (DFT) and experimental tools, a plausible reaction mechanism has been suggested. According to the computed reaction profile depicted in Figure 7.15 (where the bulky PPh_3 groups were replaced by PH_3, B3LYP/6-31G(d,p)&LanL2DZ(plus polarization)), the model osmabenzene **57** initially evolves to allene-coordinated complex **58** via transition $\mathbf{TS_{57-58}}$ ($\Delta E^{\neq}$ = 27.2 kcal/mol) in a slightly endothermic reaction (ΔE = 1.5 kcal/mol). This saddle point is associated with the concomitant release of the iodide leaving group and nucleophilic addition of MeOH to the adjacent position of this group. Then, complex **58** is transformed into intermediate **59** through the attack of a proton from a new molecule of MeOH onto the central carbon atom of the allene moiety via $\mathbf{TS_{58-59}}$ ($\Delta E^{\neq}$ = 3.2 kcal/mol). Intermediate **59** can be viewed as an ion-pair species that may mix well with the base (MeONa) thus facilitating further reaction. Indeed, anion exchange of MeO^- for I^- is suggested to take place next, followed by deprotonation which leads to the formation of intermediate **60** via $\mathbf{TS_{59-60}}$ ($\Delta E^{\neq}$ = 14.9 kcal/mol). Finally, complex **60** evolves to the final *cine*-substituted osmabenzene **61** through $\mathbf{TS_{60-61}}$ ($\Delta E^{\neq}$ = 1.6 kcal/mol), a saddle point associated with the formation of the Os–C bond accompanied with aromatization in a strongly exothermic transformation (overall reaction energy of –59.6 kcal/mol). This proposed mechanism resembles the classical ANRORC mechanism described for organic heterocyclic compounds [72], thereby proceeding through a sequence that comprises three key steps, namely addition of the nucleophile (AN), ring opening (RO) and final ring closure (RC).

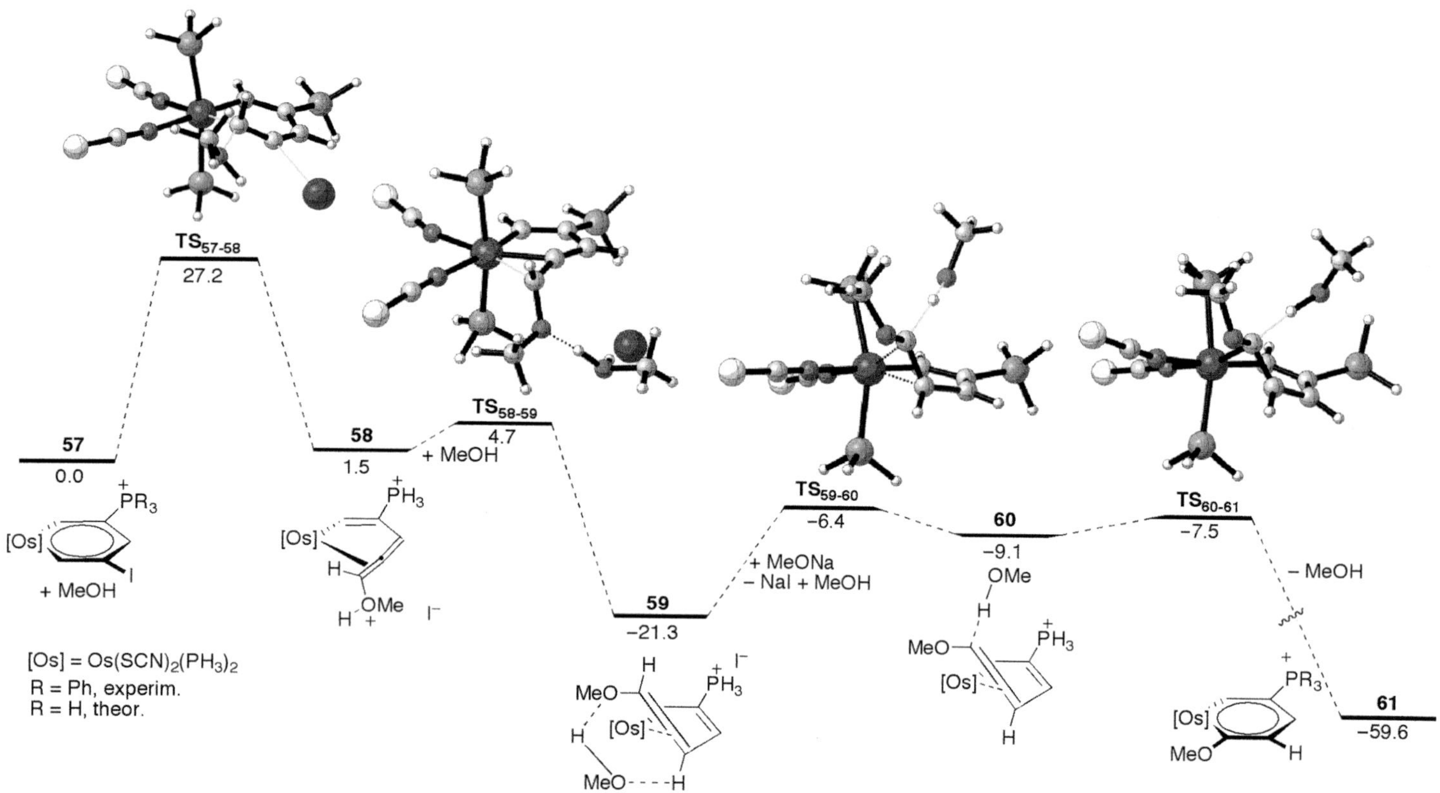

Figure 7.15 Computed reaction profile (B3LYP/6-31G(d, p)&LanL2DZ (plus polarization) level) for the transformation of osmabenzene **57** into **61**. Relative energy values are given in kcal/mol. Coordinates of transition states were taken from reference [70]. (*See color plate section for the color representation of this figure.*)

7.4.3 Addition and Cycloaddition Reactions

As reported by Bleeke and co-workers, metallapyrylium complex **62-O** reversibly adds acetone in a 1,4-fashion to yield the bicyclic complex **63** (Scheme 7.4) [4]. This reaction was not observed for the analogous metallabenzene **62-C** or metallathiabenzene **62-S**. DFT calculations (mPW1K/SDB-cc-pVDZ//mPW1K/SDD level) were carried out to gain more insight into the mechanism of this addition and the reasons behind this differential behaviour [73]. As shown in Figure 7.16, the process for **64-O** begins with the initial exergonic coordination ($\Delta G = -7.9$ kcal/mol) of acetone to produce the

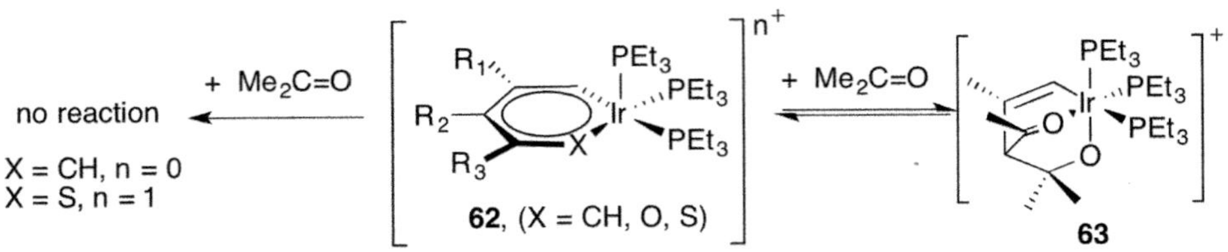

Scheme 7.4 Acetone addition to iridabenzenes **62**.

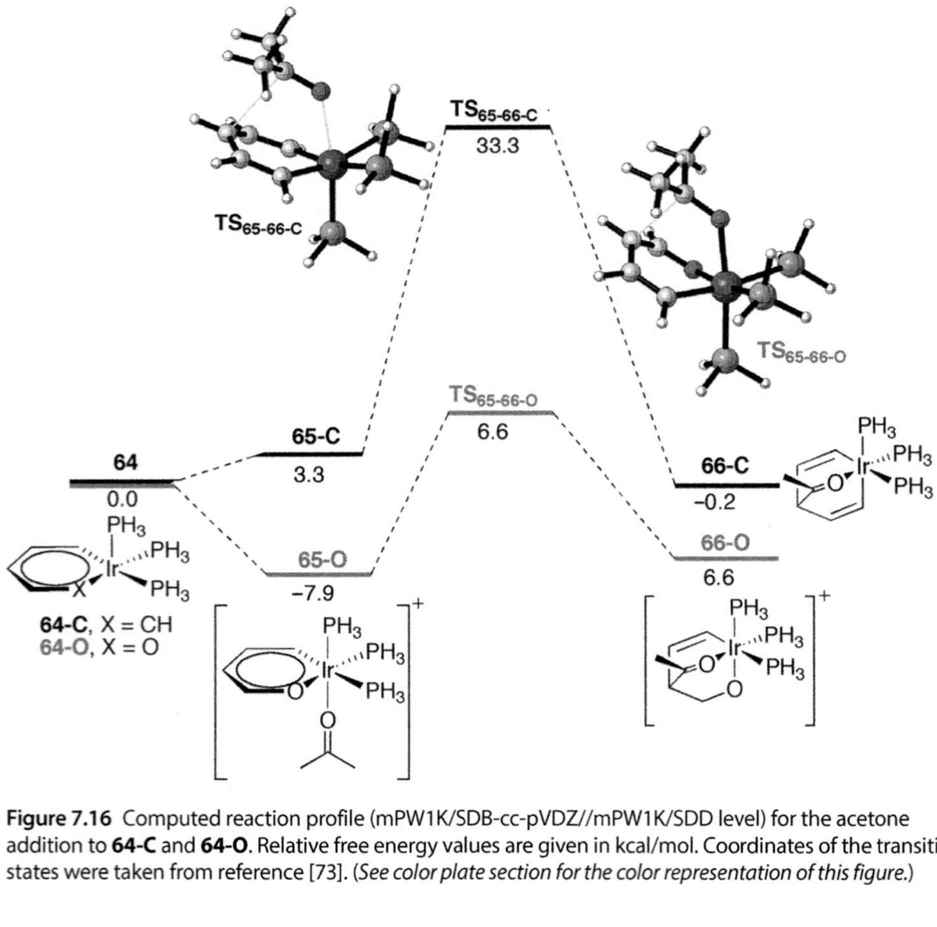

Figure 7.16 Computed reaction profile (mPW1K/SDB-cc-pVDZ//mPW1K/SDD level) for the acetone addition to **64-C** and **64-O**. Relative free energy values are given in kcal/mol. Coordinates of the transition states were taken from reference [73]. (*See color plate section for the color representation of this figure.*)

η^1-complex **65-O**. This species evolves to the final product **66-O** (which exhibits a puckered ring) through the transition state $\mathbf{TS_{65-66-O}}$, which is associated with the formation of the new C–C bond ($\Delta G^{\neq}$ = 14.5 kcal/mol). The fact that the computed overall reaction is only mildly exergonic (ΔG = −6.8 kcal/mol) nicely agrees with the reversible nature of the process. In contrast, the initial acetone coordination for the carbon counterpart **64-C** was not located on the corresponding potential energy surface. Instead, a long-range reactant complex **65-C** was found from which the addition of the acetone occurs in a concerted manner. The computed activation barrier for the process involving $\mathbf{TS_{65-66-C}}$ is much higher ($\Delta G^{\neq}$ = 33.2 kcal/mol), which hampers the addition, even though the overall reaction is thermoneutral ($\Delta G^{\neq}$ = −0.2 kcal/mol, see Figure 7.16). This computational finding matches the experimentally observed lack of reactivity of iridabenzene **62-C** (Scheme 7.4). Although the barrier is not that high for the metallathiabenzene analogue ($\Delta G^{\neq}$ = 23.4 kcal/mol), the formation of the corresponding bicyclic species is endergonic (ΔG = 6.7 kcal/mol) and therefore, the rate of the reverse reaction would be approximately 14 orders of magnitude faster.

Similar stepwise versus concerted reaction mechanisms have been computed by the authors in an extensive computational study on the 1,2- and 1,4-cycloaddition reactions of acetone, CO_2 and CS_2 to isostructural iridabenzene, iridapyrylium and iridathiabenzene complexes, as well as their rhodium analogues [74]. Thus, while complexes of the first type react in a concerted manner, the presence of the heteroatom in the six-membered ring makes the process proceed stepwise. This is mainly ascribed to the lone pair of the heteroatom which allows for an additional resonance structure (as compared to the carbon analogue) whereby an additional ligand can coordinate permitting a stepwise mechanism. For instance, Figure 7.17 shows the computed

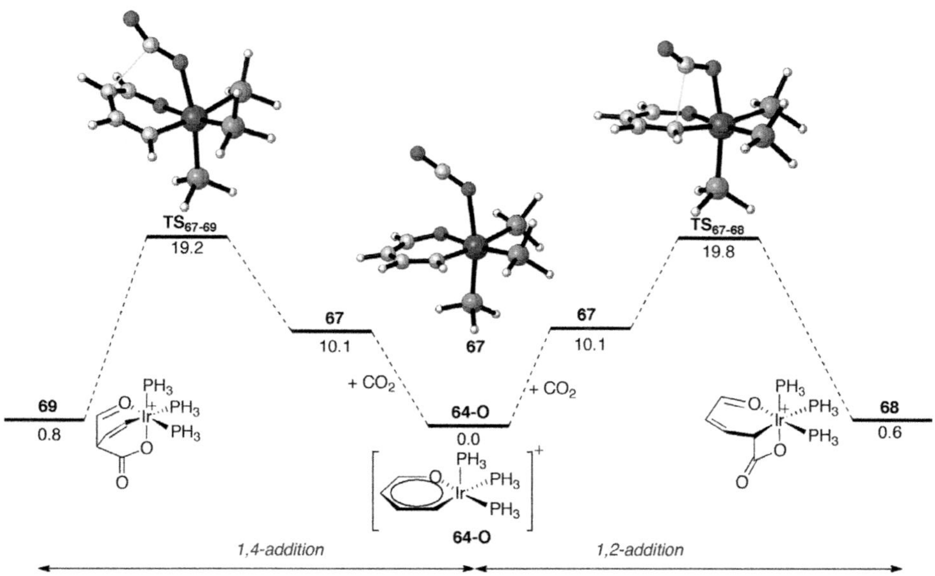

Figure 7.17 Computed reaction profile (PCM/mPW1K/SDB-cc-pVDZ//mPW1K/SDD level) for the cycloaddition reaction of iridapyrylium complex **64-O** and CO_2. Relative free energy values are given in kcal/mol. Coordinates of the transition states were taken from reference [74]. (*See color plate section for the color representation of this figure.*)

reaction profile (PCM/mPW1K/SDB-cc-pVDZ//mPW1K/SDD level) for the 1, 2- versus 1,4-addition of CO_2 to the model iridapyrylium **64-O**. The calculations predict that a mixture of both CO_2 addition products should be experimentally expected in view of the quite similar barrier ($\Delta\Delta G^{\neq}$ = 0.6 kcal/mol) and reaction energies. A nonselective process is also predicted for the analogous iridathiabenzene. However, none of these processes has been studied so far. Interestingly, the barriers for the isostructural iridabenzene are comparatively higher ($\Delta G^{\neq}$ = 37.4 kcal/mol for the 1, 4-addition). Inspection of the corresponding frontier molecular orbitals suggests that the processes involve the electrophilic attack of the substrate (acetone, CO_2 or CS_2) on the aromatic ring. Indeed, the HOMO of the metallacycle (see above) and the LUMO of these substrates (i.e. the π*(C=X) MO) have the appropriate orbital symmetries for 1,4-addition. According to the Woodward–Hoffmann rules [75], the process can be considered as a [4+2] symmetry allowed reaction, whereas the 1,2-addition reaction is symmetry forbidden. Finally, quite similar reaction profiles were computed for the rhodium analogues, therefore indicating that the outcome of the reaction is metal independent.

7.4.4 Rearrangement Reactions

Metallabenzenes exhibit a strong tendency to rearrange into cyclopentadienyl complexes [29]. This rearrangement reaction, which has been identified as the main decomposition route for metallabenzenes, involves the coupling of the two metal-bound carbon atoms in the metallabenzene to form a cyclopentadienyl ligand.

In their seminal reports, Martin and co-workers also studied the rearrangement of a series of model metallabenzenes at the DFT level [17]. This process is suggested to proceed via a three-atom-centred asymmetric transition state (similar to that depicted in Figure 7.18) that couples the C–C bond by pinching two CH groups together. In this

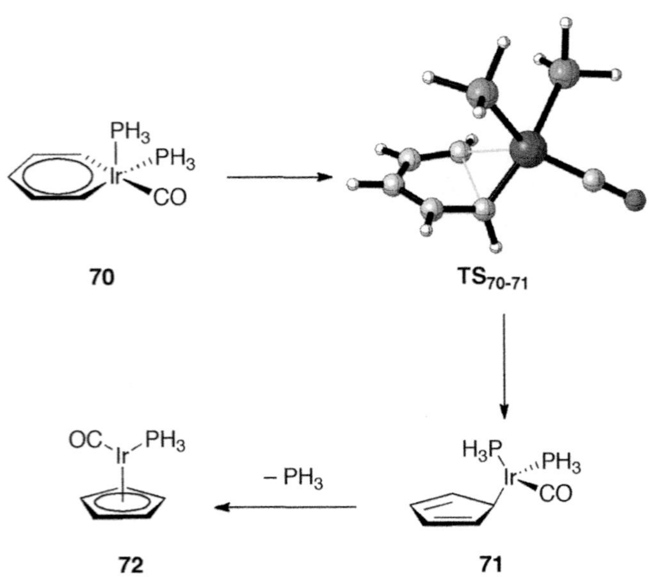

Figure 7.18 Metallabenzene to cyclopentadienyl complex rearrangement. Coordinates of transition state **TS**$_{70-71}$ were taken from reference [17a].

saddle point, both the metal centre and its ligands are distorted from planarity in order to accommodate the C–C bond formation. Therefore, this transformation can be viewed as a carbene-vinyl migratory insertion reaction. In general, the C–C formation is followed by phosphine ligand dissociation to generate the η^5-cyclopentadienyl ligand complex in an exergonic process. With the exception of the model iridabenzene $Ir(C_5H_5)Cl_2(PH_3)_2$, all the metallabenzenes studied in this work were computed to be thermodynamically unstable with respect to this rearrangement.

Interestingly, the activation barrier of the rearrangement strongly depends on the nature of transition metal (including its coordination number and formal oxidation state), as well as the surrounding ligands. For instance, the barrier height is reduced when going from $Ir(C_5H_5)(PH_3)_3$ (44.4 kcal/mol) to $Ir(C_5H_5)Cl_2(PH_3)_2$ (35.2 kcal/mol, at the PCM-mPW1K/SDB-cc-pVDZ//mPW1K/SDD level). Moreover, the substitution on the hydrocarbon ring *para* to the metal fragment also affects the barrier heights. This becomes evident by the very good linear relationships found when plotting the corresponding activation barrier versus the Hammett-σ^+ [76] substituent parameters. These linear correlations support a migratory insertion mechanism, since π-donor groups raise the barrier height for carbene migration, whereas π-acceptors lead to lower activation barriers [17a].

Haley, Houk and co-workers investigated the influence on the rearrangement of different substituents (R = H, Ph, $SiMe_3$) attached to the *ortho* and *meta* positions with respect to the metal fragment [77]. The calculated data (B3LYP/6-311+G(2d, p)&LanL2DZ level) indicate that *ortho*-trimethylsilyl- and *ortho, meta*-bis(trimethylsilyl) iridabenzenes have moderate barriers (22.1 and 15.6 kcal/mol, respectively) for rearrangement due to direct electronic effects at the *ortho* position (i.e. metal-bound carbon atoms). In the case of the bis-trimethylsilyl derivative, there are significant steric repulsions between the adjacent $SiMe_3$ groups, leading to a non-planar geometry in the ground state of **73** which is relieved in the corresponding transition state $\mathbf{TS_{73-74}}$ (Figure 7.19).

Similar three-membered ring asymmetric transition states were also located by Lin, Jia and co-workers for the rearrangement of a series of rhenabenzenes into the corresponding cyclopentadienyl complexes [78]. In an extensive computational study at the DFT level (B3LYP/6-31G&LanL2DZ), it was confirmed that substituents on the metallabenzene ring have a remarkable effect on both the thermodynamics and kinetics of the transformation. For instance, in mono-substituted rhenabenzenes π-donor groups placed at the *ortho* or *para* positions increase the stability of the compound with respect to the cyclopentadienyl complex (no effect was found when placed at the *meta* position). In contrast, π-acceptor groups (such as NO_2 or CN) increase the stability significantly when placed at the *ortho* position. However, a smaller effect is observed when they are placed in the *meta* position, and essentially no effect is seen when they are located in the *para* position. The effect of the ligands in the coordination sphere of the transition metal was found to be comparatively less important than the effect of the substituents in the six-membered ring. In this sense, it was found that di-substituted rhenabenzenes $Re\{C_5H_3(OMe)_2\text{-}1,3\}(CO)_4$ and $Re\{C_5H_3(OMe)_2\text{-}1,5\}(CO)_4$, as well as the tri-substituted rhenabenzene $Re\{C_5H_2(OMe)_3\text{-}1,3,5\}(CO)_4$, are all thermodynamically more stable than their cyclopentadienyl counterparts. Regarding the barrier heights, it was calculated that π-donor groups increase the activation barrier when placed at the *ortho* and *para* positions, while a decrease is observed in the *meta*

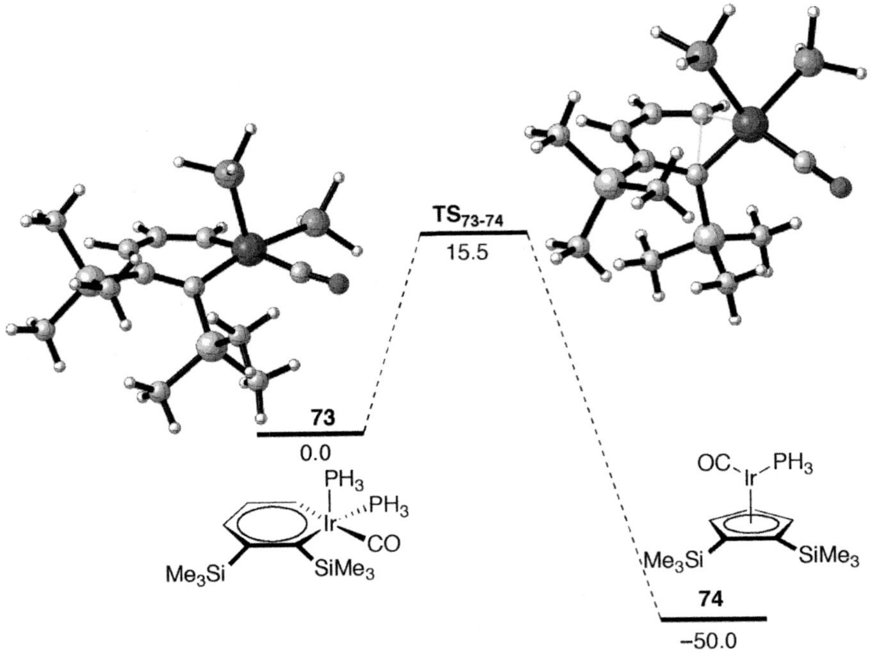

Figure 7.19 Computed reaction profile (B3LYP/6-311+G(2d, p)&LanL2DZ level) for the rearrangement of iridabenzene **73** into cyclopentadienyl complex **74**. Relative energy values are given in kcal/mol. Coordinates of complex **73** and $\mathbf{TS_{73-74}}$ were taken from reference [77].

position. Electron-withdrawing groups (such as CN) have a negligible influence on the barrier regardless of their location on the six-membered ring.

The reverse reaction (i.e. the possible formation of metallabenzenes from cyclopentadienyl complexes) has been computationally explored for a series of metallaphosphabenzenes by Zhu and co-workers [79]. In this report, the effects of the metal centres, ligands and substituents were systematically examined at the DFT level (PCM-B3LYP/6-31+G(d)&LanL2DZ). It was found that the third transition-metal row atom osmium is more prone to produce osmaphosphabenzene complexes than first- and second-row counterparts iron and ruthenium. By varying the substituents in the ring, it is suggested that donor and acceptor groups placed in *para* and *ortho* positions, respectively, favour the formation of the metallacycle. For instance, when the methoxy and PH_3 substituents are considered, the model osmaphosphabenzene **77** is thermodynamically 7.9 kcal/mol more stable than the cyclopentadienyl complex **75**. In addition, the computed activation barrier of 20.4 kcal/mol (via $\mathbf{TS_{76-77}}$) suggests that this reverse rearrangement reaction is even feasible at room temperature (Figure 7.20).

7.5 Concluding Remarks and Outlook

From the contents described in this chapter, it becomes evident that the study of metallabenzenes by means of computational methods has been key to understand and

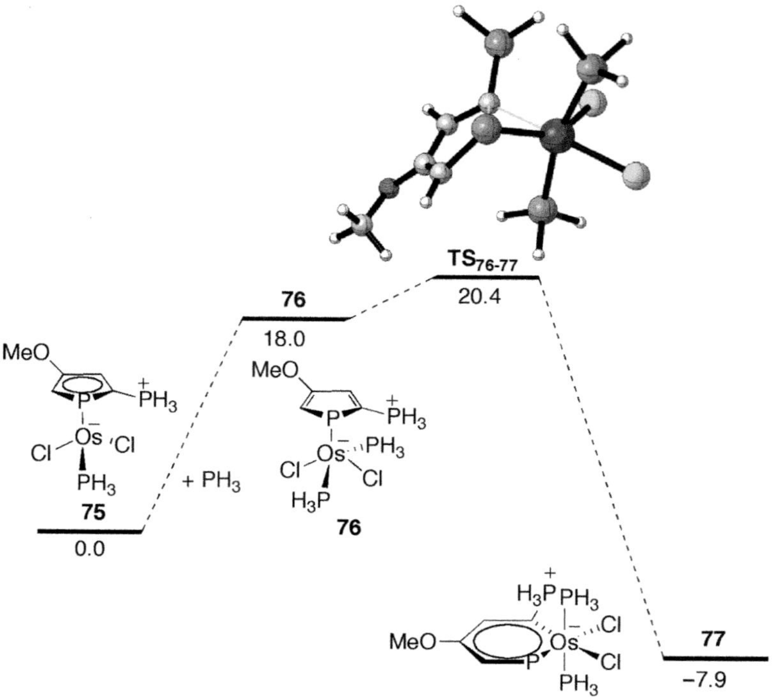

Figure 7.20 Computed reaction profile PCM-B3LYP/6-31+G(d)&LanL2DZ level) for the formation of model osmaphosphabenzene **77**. Relative free energies are given in kcal/mol. Coordinates of transition state $\mathbf{TS}_{76-77}$ were taken from reference [79].

rationalize both the bonding situation and rich reactivity of this interesting family of organometallic compounds. In our opinion, from the initial prediction by Thorn and Hoffmann in 1979, the insight gained by the calculations has significantly contributed to the tremendous and rapid development of the chemistry of these species. In this sense, we do believe that the predictive ability of the state-of-the-art computational methods will contribute enormously to the discovery of not only new metallabenzene derivatives with interesting bonding situations but also to new reactions which will stimulate intensive research of the involved reaction mechanisms from both computational and experimental points of view. Therefore, we think that computational chemistry will be helpful to the future development of the chemistry of metallabenzenes, whose limits are still far from being reached.

7.6 Acknowledgements

The authors gratefully acknowledge financial support from Spanish MINECO-FEDER (grants CTQ2013-44303-P, CTQ2016-78205-P, and CTQ2014-51912-REDC) and the Deutsche Forschungsgemeinschaft. The ORFEO-CINQA network is also acknowledged.

References

1 Elliot, G. P., Roper, W. R. and Waters, J. M. (1982) Metallacyclohexatrienes or "metallabenzenes". Synthesis of osmabenzene derivatives and X-ray crystal structure of $[Os(CSCHCHCHCH)(CO)(PPh_3)_2]$. *J. Chem. Soc. Chem. Commun.*, 811–813.

2 For recent reviews, see: (a) Bleeke, J. R. (2001) Metallabenzenes. *Chem. Rev.*, **101**, 1205–1227; (b) He, G., Xia, H. and Jia, G. (2004) Progress in the synthesis and reactivity studies of metallabenzenes. *Chin. Sci. Bull.*, **49**, 1543–1553; (c) Wright, L. J. (2006) Metallabenzenes and metallabenzenoids. *Dalton Trans.*, 1821–1827; (d) Landorf, W. C. and Haley, M. M. (2006) Recent advances in metallabenzene chemistry. *Angew. Chem. Int. Ed.*, **45**, 3914–3936; (e) Dalebrook, A. F. and Wright, L. J. (2012) Metallabenzenes and Metallabenzenoids. *Adv. Organomet. Chem.*, **60**, 93–177.

3 Selected examples: (a) Weller, K. J., Filippo, I., Briggs, P. M. and Wigley, D. E. (1997) Isolation and characterization of a monomeric metallapyridine complex: Models for pyridine hydrodenitrogenation intermediates. *J. Organomet. Chem.*, **528**, 225–228; (b) Weller, K. J., Filippo, I., Briggs, P. M. and Wigley, D. E. (1998) Pyridine Degradation Intermediates as Models for Hydrodenitrogenation Catalysis: Preparation and Properties of a Metallapyridine Complex. *Organometallics*, **17**, 322–329; (c) Liu, B., Wang, H., Xie, H., *et al.* (2009) Osmapyridine and osmapyridinium from a formal [4+2] cycloaddition reaction. *Angew. Chem. Int. Ed.*, **48**, 5430–5434; (d) Wang, T., Zhang, H., Han, F., *et al.* (2012) Synthesis and Characterization of a Metallapyridyne Complex. *Angew. Chem. Int. Ed.*, **51**, 9838–9841; (e) Liu, B., Zhao, Q., Wang, H., *et al.* (2012) Synthesis of Osmapyridiniums by [4+2] Cycloaddition Reaction between Osmium Alkenylcarbyne and Nitriles. *Chin. J. Chem.*, **30**, 2158–2168.

4 Bleeke, J. R. and Blanchard, J. M. B. (1997) Synthesis and Reactivity of a Metallapyrylium. *J. Am. Chem. Soc.*, **119**, 5443–5444; (b) Bleeke, J. R., Blanchard, J. M. B. and Donnay, E. (2001) Synthesis, Spectroscopy, and Reactivity of a Metallapyrylium. *Organometallics*, **20**, 324–336.

5 (a) Chen, J., Daniels, L. M. and Angelici, R. J. (1990) New modes of thiophene coordination and reactivity: Structures of Cp*Ir(η^2-thiophene), an iridathiabenzene, and Cp*Ir(η^4-thiophene-BH_3). *J. Am. Chem. Soc.*, **112**, 199–204; (b) Chen, J., Daniels, L. M. and Angelici, R. J. (1990) Reactions of the η^2- and η^4-isomers of Cp*Ir(2, 5-dimethylthiophene) with phosphines, CO and H_2. *Polyhedron*, **9**, 1883–1891; (c) Chin, R. M. and Jones, W. D. (1992) Dimerization of Thiophene to Give a Linear $S(CH)_8S$ Fragment with $[(C_5Me_5)Rh(C_2H_4)_2]$. *Angew. Chem. Int. Ed. Engl.*, **31**, 357–358; (d) Bianchini, C., Meli, A., Peruzzini, M. *et al.* (1993) Opening, desulfurization, and hydrogenation of thiophene at iridium: An experimental study in a homogeneous phase. *J. Am. Chem. Soc.*, **115**, 2731–2742; (e) Bianchini, C., Meli, A., Peruzzini, M., *et al.* (1994) HDS Model Systems: Coordination, Opening, and Hydrogenation of Benzo[b]thiophene at Iridium. *J. Am. Chem. Soc.*, **116**, 4370–4381; (f) Bleeke, J. R., Hinkle, P. V. and Rath, N. P. (1999) Synthesis, Structure, Spectroscopy, and Reactivity of a Metallathiabenzene. *J. Am. Chem. Soc.*, **121**, 595–596; (g) Bleeke, J. R., Hinkle, P. V. and Rath, N. P. (2001) Synthesis, Structure, Spectroscopy, and Reactivity of a Metallathiabenzene. *Organometallics*, **20**, 1939–1951.

6 (a) Clark, G. R., Johns, P. M., Roper, W. R. and Wright, L. J. (2006) An Osmabenzofuran from Reaction between $Os(PhC{\vdots}CPh)(CS)(PPh_3)_2$ and Methyl Propiolate and the C-Protonation of this Compound to Form a Tethered Osmabenzene. *Organometallics*,

25, 1771–1777; (b) Clark, G. R., O'Neale, T. R., Roper, W. R., *et al.* (2009) Stable Cationic and Neutral Ruthenabenzenes. *Organometallics*, **28**, 567–572; (c) Frogley, B. J. and Wright, L. J. (2017) A Metallaanthracene and Derived Metallaanthraquinone. *Angew. Chem. Int. Ed.*, **56**, 143–147.

7 Clark, G. R., Lu, G.-L., Roper, W. R. and Wright, L. J. (2007) Stepwise Reactions of Acetylenes with Iridium Thiocarbonyl Complexes To Produce Isolable Iridacyclobutadienes and Conversion of These to either Cyclopentadienyliridium or Tethered Iridabenzene Complexes. *Organometallics*, **26**, 2167–2177.

8 Paneque, M., Posadas, C. M., Poveda, M. L., *et al.* (2003) Formation of Unusual Iridabenzene and Metallanaphthalene Containing Electron-Withdrawing Substituents. *J. Am. Chem. Soc.*, **125**, 9898–9899.

9 For a recent review, see: Jia, G. (2013) Our Journey to the Chemistry of Metallabenzynes. *Organometallics*, **32**, 6852–6866.

10 (a) Profilet, R. D., Fanwick, P. E. and Rothwell, I. P. (1992) 1,3-Dimetallabenzene Derivatives of Niobium or Tantalum. *Angew. Chem. Int. Ed. Engl.*, **31**, 1261–1263; (b) Riley, P. N., Profilet, R. D., Salberg, M. M., *et al.* (1998) "1,3-dimetallabenzene" derivatives of niobium and tantalum. *Polyhedron*, **17**, 773–779.

11 Thorn, D. L. and Hoffmann, R. (1979) Delocalization in Metallocycles. *Nouv. J. Chim.*, **3**, 39–45.

12 Frenking, G. and Fröhlich, N. (2000) The Nature of the Bonding in Transition-Metal Compounds. *Chem. Rev.*, **100**, 717–774.

13 Representative examples: (a) Gilbertson, R. D., Weakley, T. J. R. and Haley, M. M. (1999) Direct Synthesis of an Iridabenzene from a Nucleophilic 3-Vinyl-1-Cyclopropene. *J. Am. Chem. Soc.*, **121**, 2597–2598; (b) Gilbertson, R. D., Weakley, T. J. R. and Haley, M. M. (2000) Synthesis, Characterization, and Isomerization of an Iridabenzvalene. *Chem. Eur. J.*, **6**, 437–441; (c) Jacob, V., Weakley, T. J. R. and Haley, M. M. (2002) Metallabenzenes and Valence Isomers: Synthesis and Characterization of a Platinabenzene. *Angew. Chem. Int. Ed.*, **41**, 3470–3473.

14 Bacon, G. E., Curry, N. A. and Wilson, S. A. (1964) A Crystallographic Study of Solid Benzene by Neutron Diffraction. *Proc. R. Soc. Lond. A.*, **279**, 98–110.

15 (a) Zhu, J., Jia, G. and Lin, Z. (2007) Understanding Nonplanarity in Metallabenzene Complexes. *Organometallics*, **26**, 1986–1995. See also, Lin Z. (2010) Interplay between Theory and Experiment: Computational Organometallic and Transition Metal Chemistry. *Acc. Chem. Res.*, **43**, 602–611.

16 Effertz, U., Englert, U., Podewils, F., *et al.* (2003) Reaction of Pentadienyl Complexes with Metal Carbonyls: Synthetic, Structural, and Theoretical Studies of Metallabenzene π-Complexes. *Organometallics*, **22**, 264–274.

17 (a) Iron, M. A., Lucassen, A. C. B., Cohen, H., *et al.* (2004) A Computational Foray into the Formation and Reactivity of Metallabenzenes. *J. Am. Chem. Soc.*, **126**, 11699–11710; (b) Iron, M. A., Martin, J. M. L. and van der Boom, M. E. (2003) Metallabenzene versus Cp Complex Formation: A DFT Investigation. *J. Am. Chem. Soc.*, **125**, 13020–13021.

18 Lin, M., Li, P. and Cao, Z. (2011) Effects of structural modification on the ground state of metallabenzenes: Singlet versus triplet state. *J. Theor. Comp. Chem.*, **10**, 861–874.

19 Schleyer, P. v. R. and Wang, Z.-X. personal communication cited in reference [2a].

20 Fernández, I. and Frenking, G. (2007) Aromaticity in Metallabenzenes. *Chem. Eur. J.*, **13**, 5873–5884.

21 El-Hamdi, M., El-Farri, O. E. B., Salvador, P., *et al.* (2013) Analysis of the Relative Stabilities of Ortho, Meta, and Para $MClY(XC_4H_4)(PH_3)_2$ Heterometallabenzenes (M = Rh, Ir; X = N, P; Y = Cl and M = Ru, Os; X = N, P; Y = CO). *Organometallics*, **32**, 4892–4903.

22 For recent reviews on the EDA method, see: (a) von Hopffgarten, M. and Frenking, G. (2012) Energy decomposition analysis. *WIREs Comput. Mol. Sci.*, **2**, 43–62; (b) Frenking, G. and Bickelhaupt, F. M. (2014) The EDA Perspective of Chemical Bonding, in G. Frenking and S. Shaik (eds), *The Chemical Bond: Fundamental Aspects of Chemical Bonding*. Weinheim, Germany: Wiley-VCH, 121–157.

23 Fernández, I., Frenking, G. and Merino, G. (2015) Aromaticity of metallabenzenes and related compounds. *Chem. Soc. Rev.*, **44**, 6452–6463.

24 Rickard, C. E. F., Roper, W. R., Woodgate, S. D. and Wright, L. J. (2000) Electrophilic Aromatic Substitution Reactions of a Metallabenzene: Nitration and Halogenation of the Osmabenzene $[Os\{C(SMe)CHCHCHCH\}I(CO)(PPh_3)_2$. *Angew. Chem. Int. Ed.*, **39**, 750–752.

25 Bleeke, J. R., Behm, R., Xie, Y.-F. *et al.* (1997) Synthesis, Structure, Spectroscopy, and Reactivity of a Metallabenzene. *Organometallics*, **16**, 606–623.

26 Bleeke, J. R., Behm, R., Xie, Y.-F. *et al.* (1994) Metallacyclohexadiene and Metallabenzene Chemistry: 10: Cycloaddition Reactions of a Metallabenzene. *J. Am. Chem. Soc.*, **116**, 4093–4094.

27 (a) Bleeke, J. R., Xie, Y. F., Bass, L. A. and Chiang, M. Y. (1991) Metallacyclohexadiene and metallabenzene chemistry: 5: Chemical reactivity of metallabenzene. *J. Am. Chem. Soc.*, **113**, 4703–4704; (b) Bleeke, J. R., Behm, R. and Beatty, A. M. (1997) Oligomerization Chemistry of a Metallabenzene Cycloadduct: Synthesis and Structure of an Organometallic Crown Compound. *Organometallics*, **16**, 1103–1105.

28 Clark, G. R., Ferguson, L. A., McIntosh, A. E., *et al.* (2010) Functionalization of Metallabenzenes through Nucleophilic Aromatic Substitution of Hydrogen. *J. Am. Chem. Soc.*, **132**, 13443–13452.

29 See, for instance: (a) Yang, J., Jones, W. M., Dixon, J. K. and Allison, N. T. (1995) Detection of a Ruthenabenzene, Ruthenaphenoxide, and Ruthenaphenanthrene Oxide: The First Metalla Aromatics of a Second-Row Transition Metal. *J. Am. Chem. Soc.*, **117**, 9776–9777; (b) Wu, H.-P., Lanza, S., Weakley, T. J. R. and Haley, M. M. (2002) Metallabenzenes and Valence Isomers: 3: Unexpected Rearrangement of Two Regioisomeric Iridabenzenes to an (η^5-Cyclopentadienyl)iridium(I) Complex. *Organometallics*, **21**, 2824–2826; (c) Wu, H.-P., Weakley, T. J. R. and Haley, M. M. (2005) Regioselective Formation of β-Alkyl-α-Phenyliridabenzenes via Unsymmetrical 3-Vinylcyclopropenes: Probing Steric and Electronic Influences by Varying the Alkyl Ring Substituent. *Chem. Eur. J.*, **11**, 1191–1200.

30 Hückel, E. (1937) Grundzüge der Theorie ungesättigter und aromatischer Verbindungen. *Z. Elektrochemie*, **43**, 752–788.

31 For an excellent review on Möbius aromaticity, see: (a) Rzepa, H. S. (2005) Möbius Aromaticity and Delocalization. *Chem. Rev.*, **105**, 3697–3715. See also, (b) Mauksch, M., Gogonea, V., Jiao, H. and Schleyer, P. v. R. (1998) Monocyclic $(CH)_9^+$: A Heilbronner Möbius Aromatic System Revealed. *Angew. Chem. Int. Ed.*, **37**, 2395–2397; (c) Herges, R. (2006) Topology in Chemistry: Designing Möbius Molecules. *Chem. Rev.*, **106**, 4820–4842; (d) McKee, W. C., Wu, J. I., Rzepa, H. S. and Schleyer, P. v. R. (2013) A Hückel Theory Perspective on Möbius Aromaticity. *Org. Lett.*, **15**, 3432–3435;

(e) see also, Schleyer, P. v. R., Wu, J. I., Cossío, F. P. and Fernández, I. (2014) Aromaticity in transition structures. *Chem. Soc. Rev.*, **43**, 4909–4921, for additional references.
32 Pearson, R. G. (1986) Absolute electronegativity and hardness correlated with molecular orbital theory. *Proc. Natl. Acad. Sci.*, **83**, 8440–8441.
33 Pearson, R. G. (1989) Absolute electronegativity and hardness: Applications to organic chemistry. *J. Org. Chem.*, **54**, 1423–1430.
34 Zhou, Z. and Parr, R. G. (1988) Absolute hardness as a measure of aromaticity. *Tetrahedron Lett.*, **38**, 4843–4846.
35 Chamizo, J. A., Morgado, J. and Sosa, O. (1993) Organometallic aromaticity. *Organometallics*, **12**, 5005–5007.
36 Huang, Y.-Z., Yang, S.-Y. and Li, X.-Y. (2004) An investigation of the aromaticity of transition metal heterocyclic complexes by conventional criteria and indices of aromaticity. *J. Organomet. Chem.*, **689**, 1050–1056.
37 (a) Bird, C. W. (1997) Absolute hardness as a convenient criterion of heteroaromaticity. *Tetrahedron*, **53**, 3319–3324; (b) Bean, G. J. (1998) Application of Natural Bond Orbital Analysis and Natural Resonance Theory to Delocalization and Aromaticity in Five-Membered Heteroaromatic Compounds. *J. Org. Chem.*, **63**, 2497–2506.
38 Pople, J. A. (1956) Proton Magnetic Resonance of Hydrocarbons. *J. Chem Phys.*, **24**, 1111.
39 Dauben, H. J., Wilson, J. D. and Laity, J. L. (1968) Diamagnetic susceptibility exaltation as a criterion of aromaticity. *J. Am. Chem. Soc.*, **90**, 811–813.
40 (a) Merino, G., Heine, T. and Seifert, G. (2004) The Induced Magnetic Field in Cyclic Molecules. *Chem. Eur. J.*, **10**, 4367–4371; (b) Islas, R. Heine, T. and Merino, G. (2012) The Induced Magnetic Field. *Acc. Chem. Res.*, **45**, 215–228.
41 (a) Gomes, J. A. N. F. and Mallion, R. B. (2001) Aromaticity and Ring Currents. *Chem. Rev.*, **101**, 1349–1384; (b) Lazzeretti, P. (2004) Assessment of aromaticity via molecular response properties. *Phys. Chem. Chem. Phys.*, **6**, 217–223.
42 Fliegl, H., Taubert, S., Lehtonena, O. and Sundholm, D. (2011) The gauge including magnetically induced current method. *Phys. Chem. Chem. Phys.*, **13**, 20500–20518.
43 Herges, R. and Geuenich, D. (2001) Delocalization of Electrons in Molecules. *J. Phys. Chem. A.*, **105**, 3214–3220.
44 (a) Schleyer, P. v. R., Maerker C., Dransfeld, A., *et al.* (1996) Nucleus-Independent Chemical Shifts: A Simple and Efficient Aromaticity Probe. *J. Am. Chem. Soc.*, **118**, 6317–6318; (b) Chen, Z., Wannere, C. S., Corminboeuf, C., *et al.* (2005) Nucleus-Independent Chemical Shifts (NICS) as an Aromaticity Criterion. *Chem. Rev.*, **105**, 3842–3888.
45 Fallah-Bagher-Shaidaei, H., Wannere, C. S., Corminboeuf, C., *et al.* (2006) Which NICS Aromaticity Index for Planar π Rings Is Best? *Org. Lett.*, **8**, 863–866.
46 Han, F., Wang, T., Li, J., *et al.* (2014) *m*-Metallaphenol: Synthesis and Reactivity Studies. *Chem. Eur. J.*, **20**, 4363–4372.
47 Zhang, H., Lin, R., Li, J., *et al.* (2014) Interconversion between Ruthenacyclohexadiene and Ruthenabenzene: A Combined Experimental and Theoretical Study. *Organometallics*, **33**, 5606–5609.
48 Mauksch, M. and Tsogoeva, S. B. (2010) Demonstration of "Möbius" Aromaticity in Planar Metallacycles. *Chem. Eur. J.*, **16**, 7843–7851.
49 (a) Kruszewski, J. and Krygowski, T. M. (1972) Definition of aromaticity basing on the harmonic oscillator model. *Tetrahedron Lett.*, **13**, 3839–3842. For a review, see: (b) Krygowski, T. M., Szatylowicz, H., Stasyuk, O. A., *et al.* (2014) Aromaticity from the

Viewpoint of Molecular Geometry: Application to Planar Systems. *Chem. Rev.*, **114**, 6383–6422.
50 Geuenich, D., Hess, K., Köhler, F. and Herges, R. (2005) Anisotropy of the Induced Current Density (ACID): A General Method to Quantify and Visualize Electronic Delocalization. *Chem. Rev.*, **105**, 3758–3772.
51 Periyasamy, G., Burton, N. A., Hillier, I. H. and Thomas, J. M. H. (2008) Electron Delocalization in the Metallabenzenes: A Computational Analysis of Ring Currents. *J. Phys. Chem. A.*, **112**, 5960–5972.
52 Havenith, R. W. A., De Proft, F., Jenneskens, L. W. and Fowler, P. W. (2012) Relativistic ring currents in metallabenzenes: An analysis in terms of contributions of localised orbitals. *Phys. Chem. Chem. Phys.*, **14**, 9897–9905.
53 (a) Schleyer, P. v. R., Manoharan, M., Jiao, H. and Stahl, F. (2001) The Acenes: Is There a Relationship between Aromatic Stabilization and Reactivity? *Org. Lett.*, **3**, 3643–3646; (b) see Suresh, C. H. and Koga, N. (2002) Accurate Calculation of Aromaticity of Benzene and Antiaromaticity of Cyclobutadiene: New Homodesmotic Reactions. *J. Org. Chem.*, **67**, 1965–1968, for another approach giving ASE = 29 kcal/mol for benzene.
54 Schleyer, P. v. R. and Pühlhofer, F. (2002) Recommendations for the Evaluation of Aromatic Stabilization Energies. *Org. Lett.*, **4**, 2873–2876.
55 De Proft, F. and Geerlings, P. (2004) Relative hardness as a measure of aromaticity. *Phys. Chem. Chem. Phys.*, **6**, 242–248.
56 Lin, R. , Lee, K.-H., Poon, K. C., *et al.* (2014) Synthesis of Rhenabenzenes from the Reactions of Rhenacyclobutadienes with Ethoxyethyne. *Chem. Eur. J.*, **20**, 14885–14899.
57 Fernández, I. and Frenking, G. (2007) Direct estimate of conjugation and aromaticity in cyclic compounds with the EDA method. *Faraday Discuss.*, **135**, 403–421.
58 Fernández, I. (2014) Direct Estimate of Conjugation, Hyperconjugation, and Aromaticity with the Energy Decomposition Analysis Method, in G. Frenking and S. Shaik (eds), *The Chemical Bond: Fundamental aspects of chemical bonding*, Weinheim, Germany: Wiley-VCH, 357–381.
59 Conjugation and Hyperconjugation: (a) Cappel, D., Tüllmann, S., Krapp A. and Frenking, G. (2005) Direct Estimate of the Conjugative and Hyperconjugative Stabilization in Diynes, Dienes, and Related Compounds. *Angew. Chem. Int. Ed.*, **44**, 3617; (b) Fernández, I. and Frenking, G. (2006) Direct Estimate of the Strength of Conjugation and Hyperconjugation by the Energy Decomposition Analysis Method. *Chem. Eur. J.*, **12**, 3617–3620; (c) Fernández, I. and Frenking, G. (2006) Correlation between Hammett Substituent Constants and Directly Calculated π-Conjugation Strength. *J. Org. Chem.*, **71**, 2251–2256; (d) Fernández, I. and Frenking, G. (2006) π-Conjugation in donor-substituted cyanoethynylethenes: An EDA study. *Chem. Commun.*, 5030–5032; (e) Fernández, I. and Frenking, G. (2007) Hyperconjugative Stabilization in Alkyl Carbocations: Direct Estimate of the β-Effect of Group-14 Elements. *J. Phys. Chem. A.*, **111**, 8028–8035; (f) Fernández, I. and Frenking, G. (2007) EDA Study of π-Conjugation in Tunable Bis(gem-diethynylethene) Fluorophores. *J. Org. Chem.*, **72**, 7367–7372.
60 Landorf, C. W., Jacob, V., Weakley, T. J. R. and Haley, M. M. (2004) Rational Synthesis of Platinabenzenes. *Organometallics*, **23**, 1174–1176.
61 Gong, L., Chen, Z., Lin, Y. *et al.* (2009) Osmabenzenes from Osmacycles Containing an η^2-Coordinated Olefin. *Chem. Eur. J.*, **15**, 6258–6266.

62 Chen, J., Zhang, C., Xie, T., *et al.* (2013) Conversion of a Hydrido–Butenylcarbyne Complex to η^2-Allene-Coordinated Complexes and Metallabenzenes. *Organometallics*, **32**, 3993–4001.

63 Liu, B., Xie, H., Wang, H., *et al.* (2009) Selective Synthesis of Osmanaphthalene and Osmanaphthalyne by Intramolecular C–H Activation. *Angew. Chem. Int. Ed.*, **48**, 5461–5464.

64 Fan, J., Wand, X. and Zhu, J. (2014) Theoretical Study on the Stability and Aromaticity of Metallasilapentalynes. *Organometallics*, **33**, 2336–2340.

65 Parr, R. G. and Yang, W. (1989) in *Density-Functional Theory of Atoms and Molecules*. New York: Oxford University Press.

66 Clark, G. R., Johns, P. M., Roper, W. R., *et al.* (2011) Regioselective Mono-, Di-, and Trifunctionalization of Iridabenzofurans through Electrophilic Substitution Reactions. *Organometallics*, **30**, 129–138.

67 See, for instance: (a) Otero, N., Mandado, M. and Mosquera, R. A. (2007) Revisiting the calculation of condensed Fukui functions using the quantum theory of atoms in molecules. *J. Chem. Phys.*, **126**, 234108/1-234108/6; (b) Andrada, D. M., Granados, A. M., Solà, M. and Fernández, I. (2011) DFT Study of Thermal 1, 3-Dipolar Cycloaddition Reactions between Alkynyl Metal(0) Fischer Carbene Complexes and 3H-1,2-Dithiole-3-thione Derivatives. *Organometallics*, **30**, 466–476.

68 For a recent report on the rate-determining factors in S_NAr reactions involving organic compounds, see: Fernández, I., Frenking, G. and Uggerud, E. (2010) Rate-Determining Factors in Nucleophilic Aromatic Substitution Reactions. *J. Org. Chem.*, **75**, 2971–2980.

69 Lin, R., Zhang, H., Li, S., *et al.* (2011) New Highly Stable Metallabenzenes via Nucleophilic Aromatic Substitution Reaction. *Chem. Eur. J.*, **17**, 4223–4231.

70 Wang, T., Zhang, H., Han, F., *et al.* (2013) *cine*-Substitution Reactions of Metallabenzenes: An Experimental and Computational Study. *Chem. Eur. J.*, **19**, 10982–10991.

71 For a review on *cine*-substitution reactions, see: Suwinski, J. and Swierczek, K. (2001) *cine*- and *tele*-Substitution reactions. *Tetrahedron*, **57**, 1639–1662.

72 (a) Buncel, E., Dust, J. M. and Terrier, F. (1995) Rationalizing the Regioselectivity in Polynitroarene Anionic σ-Adduct Formation: Relevance to Nucleophilic Aromatic Substitution. *Chem. Rev.*, **95**, 2261–2280; (b) Chupakhin, O. N., Charushin, V. N. and van der Plas, H. C. (1988) Nucleophilic substitution of hydrogen in azines. *Tetrahedron*, **44**, 1–34; (c) van der Plas, H. C., Wozniak, M. and van den Haak, H. J. W. (1983) Reactivity of Naphthyridines toward Nitrogen Nucleophiles. *Adv. Heterocycl. Chem.*, **33**, 95–146; (d) van der Plas, H. C. (1978) The SN(ANRORC) mechanism: A new mechanism for nucleophilic substitution. *Acc. Chem. Res.*, **11**, 462–468.

73 Iron, M. A., Iron, J. M. L. and Van der Boom, M. E. (2003) Mechanistic aspects of acetone addition to metalloaromatic complexes of iridium: A DFT investigation. *Chem. Commun.*, **1**, 132–133.

74 Iron, M. A., Martin, J. M. L. and Van der Boom, M. E. (2003) Cycloaddition Reactions of Metalloaromatic Complexes of Iridium and Rhodium: A Mechanistic DFT Investigation. *J. Am. Chem. Soc.*, **125**, 11702–11709.

75 Woodward, R. B. and Hoffmann, R. (1969) The Conservation of Orbital Symmetry. *Angew. Chem. Int. Ed. Engl.*, **8**, 781–853.

76 Hammett, L. P. (1935) Some Relations between Reaction Rates and Equilibrium Constants. *Chem. Rev.*, **17**, 125–136; (b) Hammett, L. P. (1970) in *Physical Organic Chemistry*, 2nd edn. New York: McGraw-Hill.

77 Wu, H.-P., Ess, D. H., Lanza, S., *et al.* (2007) Rearrangement of Iridabenzvalenes to Iridabenzenes and/or η^5-Cyclopentadienyliridium(I) Complexes: Experimental and Computational Analysis of the Influence of Silyl Ring Substituents and Phosphine Ligands. *Organometallics*, **26**, 3957–3968.

78 (a) Shi, C., Guo, T., Poon, K. C., *et al.* (2011) Theoretical study on the rearrangement of metallabenzenes to cyclopentadienyl complexes. *Dalton Trans.*, **40**, 11315–11320. See also, (b) Poon, K. C., Liu, L., Guo, T., *et al.* (2010) Synthesis and Characterization of Rhenabenzenes. *Angew. Chem. Int. Ed.*, **49**, 2759–2762.

79 Huang, C., Hao, Y., Zhao, Y. and Zhu, J. (2014) Computations Offer an Unconventional Route to Metallaphosphabenzene from a Half-Phosphametallocene. *Organometallics*, **38**, 817–822.

Index

a

Absolute hardness 101, 273
Addition reactions
  electrophilic, *see* Electrophilic addition
  nucleophilic, *see* Nucleophilic addition
η^2-Allene 223, 226, 233
Allenylcarbene 186
Anisotropy of the induced current density (ACID) 276
Aromaticity
  of metallabenzenes 51, 272
  of metallabenzynes 205
Aromatic stabilization energy
  of metallabenzenes 101, 208, 277
  of metallabenzynes 206
Aromatic substitution
  electrophilic, *see* Electrophilic substitution
  nucleophilic, *see* Nucleophilic substitution

b

Berry-type process 59, 80, 89, 124
Bis(diphenylphosphino)methane (DPPM) 233
Bismetallabenzenes 96, 232
Bromination, *see* Electrophilic substitution

c

cine-substitution 237, 290
η^6-Coordination
  in iridabenzenes 71
  in iridathiabenzenes 86
  in platinabenzenes 131
Cycloaddition
  to give isoosmabenzenes 228
  to give osmapyridines 227
  in iridabenzenes 65
  in iridapyrylium 81
  in iridathiabenzenes 86
  in osmapyridinium 239
  in platinabenzenes 131
  theoretical study 68, 292
Cyclopentadienyl complexes
  via rearrangement of iridabenzenes 116, 122
  via rearrangement of osmabenzenes 34, 193
  via rearrangement of osmabenzynes 199
  via rearrangement of rhenabenzenes 209
  via rearrangement of ruthenabenzenes 38, 240
  theoretical study 199, 294

d

Dewar metallabenzene 120 137
Diamagnetic susceptibility exaltation 100, 274
Diels–Alder reactions 83, 160, 226
Dimethylacetylene diacarboxylate (DMAD) 149, 153
2,4-Dimethyl-5-oxapentadienide 78
2,4-Dimethylpentadienide 54, 57

e

Electrophilic addition
  to iridabenzenes 69

Electrophilic substitution
in metallabenzenes 24
in metallabenzofurans 25
in metallabenzynes 191
theoretical study 26, 286
Energy decomposition analysis (EDA) 101, 271, 279

f
Fukui function 42, 286, 288

h
Hammett plot 135
8-Hydroxyquinoline 232, 236, 288

i
$IrCl(CO)(PPh_3)_2$, *see* Vaska's complex
Iridabenzene
BF_3 addition 70
bromination 25
coordination to metal substrates 71
cycloaddition 65, 292
16-electron 133
electron counting 56
electronic structure 99
electrophilic addition 69
electrophilic substitution 25
hydride migration 169
ligand substitution 8, 60
magnetic properties 100
nucleophilic addition 166
nucleophilic substitution 38
oxidative addition 63
rearrangement reactions 294
spectral data 10, 42, 58, 123, 175
stability 100, 121
structural data 8, 41, 58, 99, 125, 134, 171
synthesis 7, 54, 114, 146, 149, 153, 164
Iridabenzofuran
bromination 25
electrophilic substitution 25, 286
Fukui functions 286
mercuration 25
nitration 28
spectral data 18, 28
structural data 18, 28
synthesis 17
Iridabenzothiazolium cation
spectral data 22
structural data 22
synthesis 21
Iridabenzothiophene
spectral data 21
structural data 20
synthesis 19
Iridabenzvalene
isomerization 120
spectral data 123
structural data 126
synthesis 114
Iridacyclobutadiene 19
Iridacycloheptatriene 146
Iridacyclohexadienone 74
Iridacyclopentadiene
alkyne insertion 17, 162
CS insertion 7
olefin insertion 153
synthesis from iridabenzenes 67, 170
vinylidene insertion 153
Iridaindene 157, 162
Iridanaphthalene
nucleophilic addition 167
spectral data 175
structural data 173
synthesis 146, 157
Iridaphenol 73
Iridapyran 78
Iridapyrylium
cycloaddition 81
electron counting 80
ligand addition 80
spectral data 78
synthesis 78
valence bond structures 80
Iridathiabenzene
coordination to metal substrates 86
cycloaddition 86
ligand addition 86, 91
spectral data 84, 90
structural data 85
synthesis 83, 89
Isometallabenzene
of osmium, *see* Isoosmabenzene
Isoosmabenzene 193, 228

l

Ligand substitution
  in iridabenzenes 7, 57, 60
  in iridathiabenzenes 91
  in metal-coordinated metallabenzenes 94
  in osmabenzenes 232
  in osmabenzynes 189
  in ruthenabenzenes 232

m

Magnetic properties 100, 274
Meisenheimer compound 39, 251, 288
Mercuration, *see* Electrophilic substitution, Iridabenzofuran
Metal-coordinated metallabenzene 94
Metallabenzene
  aromaticity 52, 272
  of iridium, *see* Iridabenzene
  molecular orbital diagram 53, 268
  of osmium, *see* Osmabenzene
  of platinum, *see* Platinabenzene
  of rhenium, *see* Rhenabenzene
  of ruthenium, *see* Ruthenabenzene
Metallabenzofuran
  of iridium, *see* Iridabenzofuran
  of osmium, *see* Osmabenzofuran
  of ruthenium, *see* Ruthenabenzofuran
Metallabenzothiazole
  of osmium, *see* Osmabenzothiazole
Metallabenzothiazolium cation
  of iridium, *see* Iridabenzothiazolium cation
Metallabenzothiophene
  of iridium, *see* Iridabenzothiophene
Metallabenzvalene
  of iridium, *see* Iridabenzvalene
  of rhodium, *see* Rhodabenzvalene
Metallabenzyne
  aromaticity 205
  electronic structure 204
  of osmium, *see* Osmabenzyne
  rearrangement to carbenes 189
  of rhenium, *see* Rhenabenzyne
  ring strain 203
  structural properties 202
Metallacyclopentadiene
  of iridium, *see* Iridacyclopentadiene
Metallanaphthalene
  of iridium, *see* Iridanaphthalene
  of osmium, *see* Osmanaphthalene
Metallanaphthalyne
  of osmium, *see* Osmanaphthalyne
  of ruthenium, *see* Ruthenanaphthalyne
Metallaphenol
  of iridium, *see* Iridaphenol
  of osmium, *see* Osmaphenol
Metallapyridine
  of osmium, *see* Osmapyridine
Metallapyridinium
  of osmium, *see* Osmapyridinium
Metallapyrylium
  of iridium, *see* Iridanapyrylium
Methyl propiolate 11, 14, 17, 19, 81
Mobius aromaticity 52, 273
Molybdenabenzene 94

n

Nickelabenzene 95
Nitration, *see* Electrophilic substitution
Nucleophilic addition
  to coordinated alkynes 219
  to iridabenzenes 38, 167
  to osmabenzenes 38, 233
  to osmabenzynes 193, 228
  to ruthenabenzenes 233
Nucleophilic substitution 38, 251
  theoretical study, 288
Nucleus-independent chemical shift (NICS)
  for metallabenzenes 100, 275
  for metallabenzynes 207

o

$OsCl_2(PPh_3)_3$ 186, 220, 280
$Os(CS)(CO)(PPh_3)$
  reaction with ethyne 2
  reaction with propyne 4
Osmabenzene
  bromination 24
  cine-substitution 237, 290
  electrochemical properties 245
  ligand substitution 232
  mechanism of formation 280

Osmabenzene (*contd.*)
  nitration 24
  nucleophilic addition 234
  nucleophilic substitution 38, 238, 252
  reaction with acetylides 242
  rearrangement to cyclopentadienyl complexes 35
  structural data 4, 26, 229
  spectral data 6, 42, 231
  synthesis 2, 193, 221
Osmabenzofuran
  protonation 12
  structural data 12
  spectral data 13
  synthesis 11
  transesterification 12
Osmabenzothiazole 252
Osmabenzyne
  electronic structure 204
  electrophilic substitution 191
  ligand substitution 189
  migratory insertion 193
  nucleophilic addition 193
  rearrangement to carbenes 193
  rearrangement to cyclopentadienyl complexes 198
  structural data 202
  synthesis 186, 228
Osmabutenethione 11
Osmacyclopropene 11
Osmaisoquinoline 259
Osmaisoquinolyne 259
Osmanaphthalene 223, 282
Osmanaphthalyne
  rearrangement to indenylidene complexes 197
  synthesis 188, 223, 282
Osmaphenol 222, 255
Osmapyridine
  spectral data 231, 249
  structural data 230
  synthesis 226
Osmapyridinium
  cycloaddition 239
  nucleophilic addition 234
  spectral data 231, 249
  structural data 231
  synthesis 227

p

Platinabenzene
  aromaticity 277
  cycloaddition 131
  molecular orbital diagram 269
  spectral data 130
  structural data 131
  stability 129
  synthesis 128

r

Rhenabenzene
  aromaticity 208
  rearrangement to cyclopentadienyl complexes 209, 295
  synthesis 207
Rhenabenzyne
  aromaticity 205
  structural data 202
  synthesis 189
Rhodabenzene
  molecular orbital description 271
Rhodabenzvalene 137
Ring current 274
Ruthenabenzene
  dialdehyde formation 241
  electrochemical properties 244
  ligand substitution 232
  metal-coordinated 95
  nucleophilic addition 233
  nucleophilic substitution 236 288
  optical properties 247
  rearrangement to cyclopentadienyl anions 240
  spectral data 15, 231
  structural data 16, 230
  synthesis 14, 38, 221
Ruthenabenzofuran
  protonation 15, 38
  rearrangement to cyclopentadienyl complexes 38
  spectral data 15
  structural data 16,

synthesis 14
Ruthenacyclohexadiene 233

s

Sandwich complexes
of iridabenzenes 71
of molybdenabenzenes 94
of nickelabenzenes 95
of ruthenabenzenes 95

x

(η^6-*p*-Xylene)$Mo(CO)_3$
coordination to iridabenzenes 71
coordination to iridathiabenzenes 86

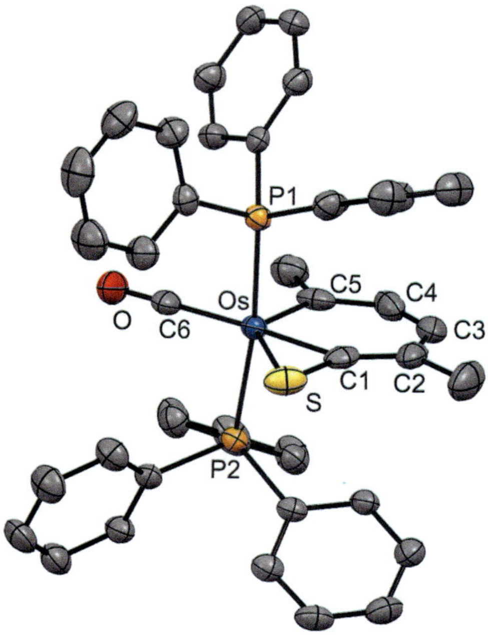

Figure 1.2 Molecular structure of osmabenzene **2** showing 50% probability thermal ellipsoids. Hydrogen atoms have been omitted for clarity. Selected distances [Å]: Os–C1 2.027(4), Os–C5 2.022(5), Os–S 2.4990(12), Os–C6 1.908(4), C1–C2 1.365(6), C2–C3 1.400(7), C3–C4 1.385(7), C4–C5 1.415(6).

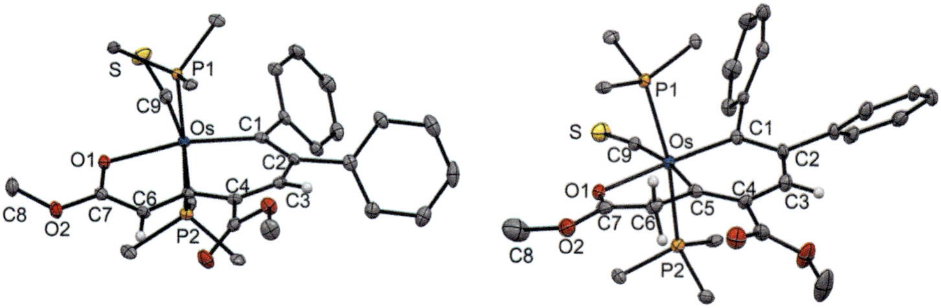

Figure 1.4 Molecular structures of the osmabenzofuran **16** (left) and the cation of the tethered osmabenzene **17** (right) showing 50% probability thermal ellipsoids. Some hydrogen atoms and the phenyl rings of the triphenylphosphine ligands are not shown for clarity. Selected distances [Å] for **16**: Os–C1 2.068(3), Os–C5 2.142(3), Os–O1 2.2178(17), Os–C9 1.846(3), C1–C2 1.376(4), C2–C3 1.442(3), C3–C4 1.359(4), C4–C5 1.436(4), C5–C6 1.371(3), C6–C7 1.416(4), C7–O1 1.250(3). Selected distances [Å] for **17**: Os–C1 1.996(5), Os–C5 2.103(6), Os–O1 2.221(4), Os–C9 1.901(6), C1–C2 1.450(8), C2–C3 1.374(8), C3–C4 1.438(8), C4–C5 1.375(8), C5–C6 1.507(8), C6–C7 1.495(9) C7–O1 1.226(8).

Metallabenzenes: An Expert View, First Edition. Edited by L. James Wright.

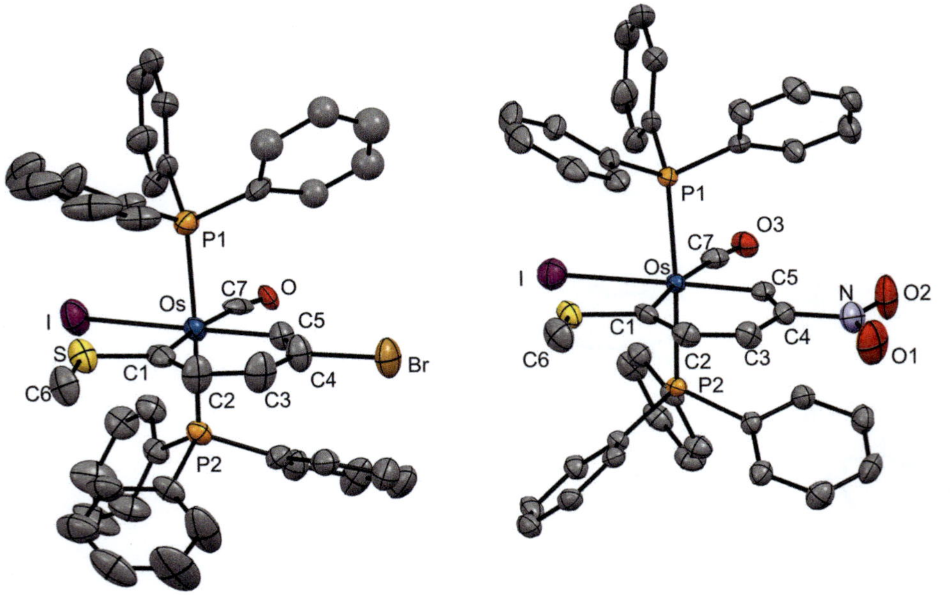

Figure 1.9 Molecular structures of the substituted osmabenzenes **30** (left) and **31** (right) showing 50% probability thermal ellipsoids. Hydrogen atoms are not shown for clarity. Selected distances [Å] for **30**: Os–C1 2.128(9), Os–C5 2.039(9), Os–C7 1.936(13), Os–I 2.8021(9), C1–C2 1.418(13), C2–C3 1.367(14), C3–C4 1.390(14), C4–C5 1.320(13), C1–S 1.702(10), C4–Br 1.968(10). Selected distances [Å] for **31**: Os–C1 2.129(7), Os–C5 2.011(7), Os–C7 1.908(8), Os–I 2.8158(6), C1–C2 1.411(11), C2–C3 1.347(11), C3–C4 1.411(10), C4–C5 1.336(10), C1–S 1.702(7), C4–N 1.476(9).

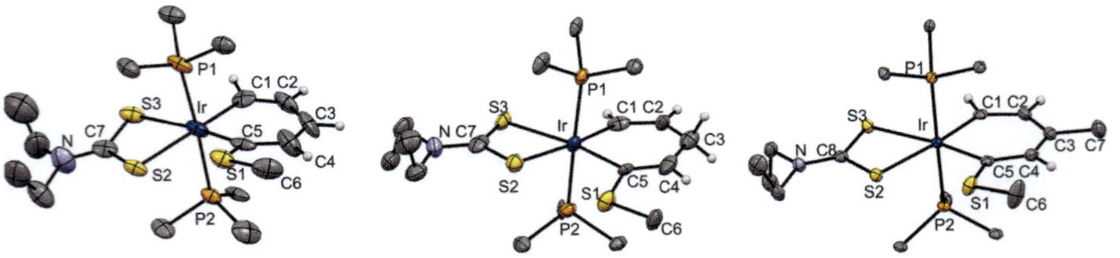

Figure 1.15 Molecular structures of the cation of **52** (left), **54a** (centre) and **56a** (right) showing 50% probability thermal ellipsoids. Phenyl rings on the triphenylphosphine ligand and most hydrogen atoms are not shown for clarity. Selected distances [Å] for **52**: Ir–C1 2.009(16), Ir–C5, 2.009(16), C1–C2, 1.42(2), C2–C3, 1.36(2), C3–C4, 1.44(2), C4–C5, 1.31(2). Selected distances [Å] for **54a**: Ir–C1 2.046(11), Ir–C5 2.073(12), C1–C2 1.321(16), C2–C3 1.460(18), C3–C4 1.465(16), C4–C5 1.340(16). Selected distances [Å] for **56a**: Ir–C1 2.011(3), Ir–C5 2.014(3), C1–C2 1.389(5), C2–C3 1.382(5), C3–C4 1.425(5), C4–C5 1.363(4), C3–C7 1.503(5).

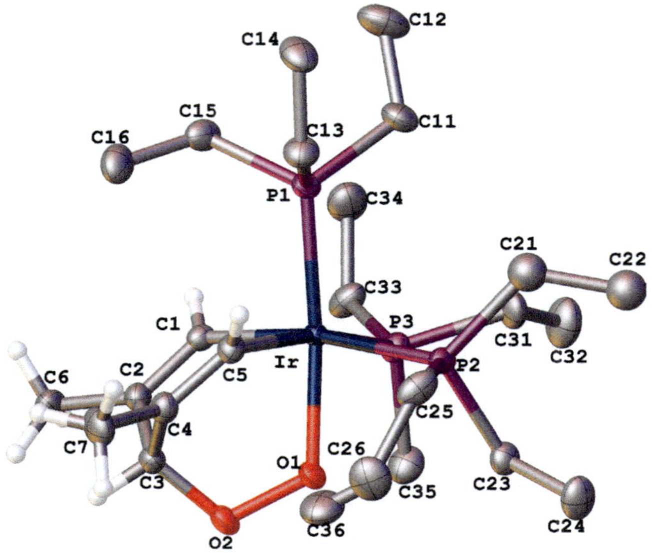

Figure 2.7 X-ray structure of the dioxygen adduct of iridabenzene, **13**, with phosphine H's omitted for clarity. The plot was created using the Olex2 software package [42].

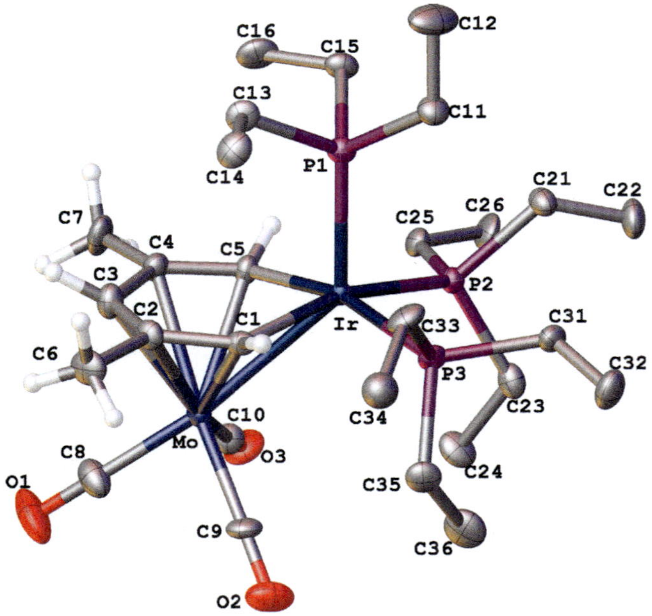

Figure 2.9 X-ray structure of (η^6-iridabenzene)$Mo(CO)_3$, **24**, with phosphine H's omitted for clarity. The plot was created using the Olex2 software package [42].

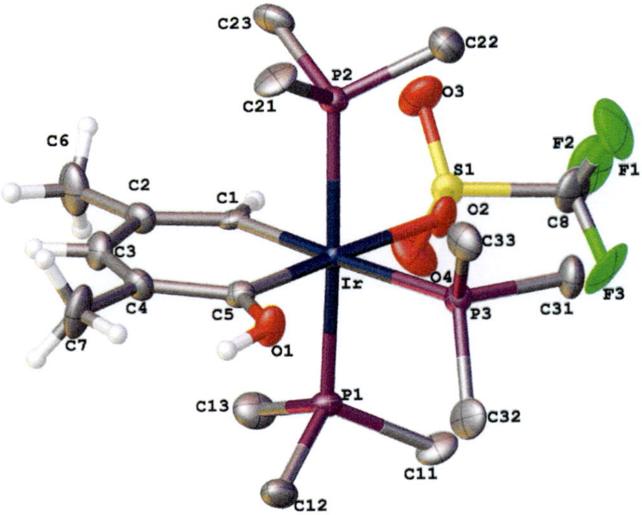

Figure 2.11 X-ray structure of iridaphenol, **31**, with phosphine H's omitted for clarity. The plot was created using the Olex2 software package [42].

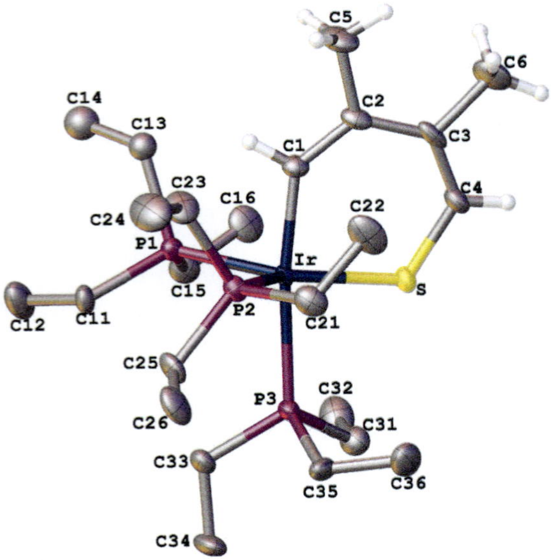

Figure 2.15 X-ray structure of the cation in iridathiabenzene, **50**, with phosphine H's omitted for clarity. The plot was created using the Olex2 software package [42].

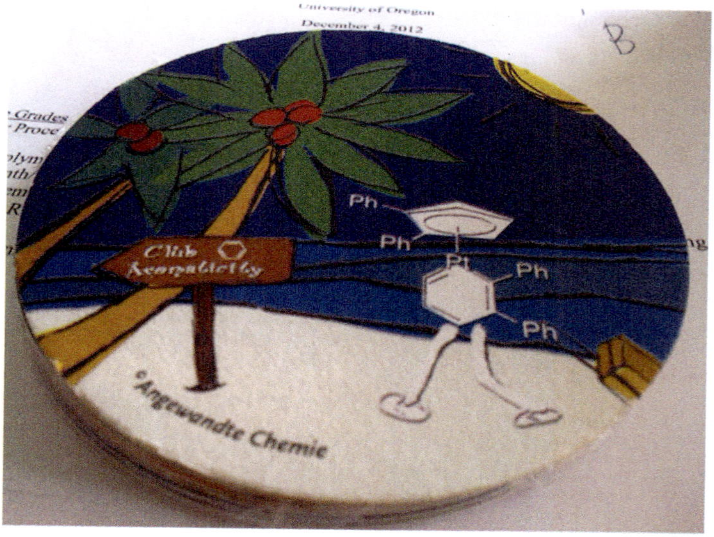

Figure 3.6 The Haley group's platinabenzene beer mat/coaster.

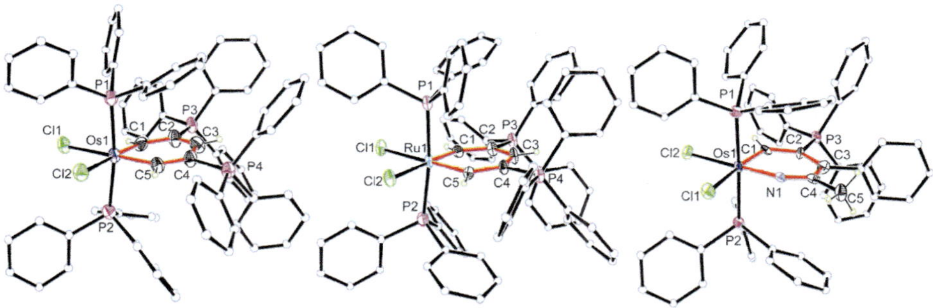

osmabenzene 2 ruthenabenzene 4 osmapyridine 24

Figure 6.1 Molecular structures of osmabenzene **2**, ruthenabenzene **4** and osmapyridine **24**. The hydrogen atoms of PPh_3 are omitted for clarity.

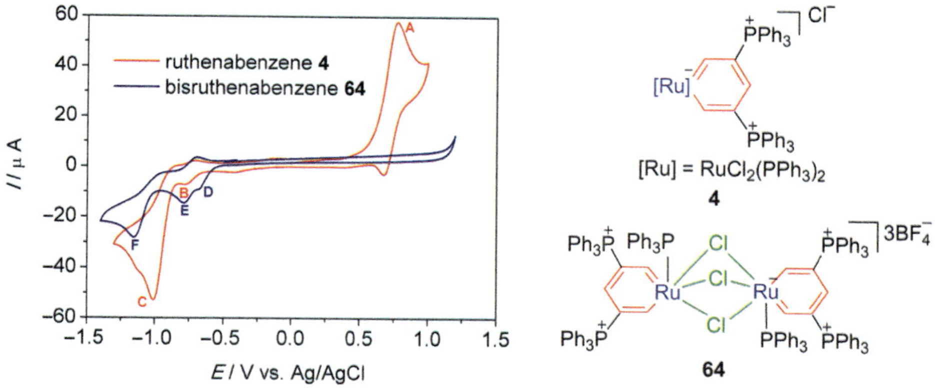

Figure 6.2 Cyclic voltammograms of **4** and **64** measured in DCM with 0.1 M $n\mathrm{Bu_4NClO_4}$ as the supporting electrolyte at a scan rate of $0.10\,\mathrm{V\,s^{-1}}$.

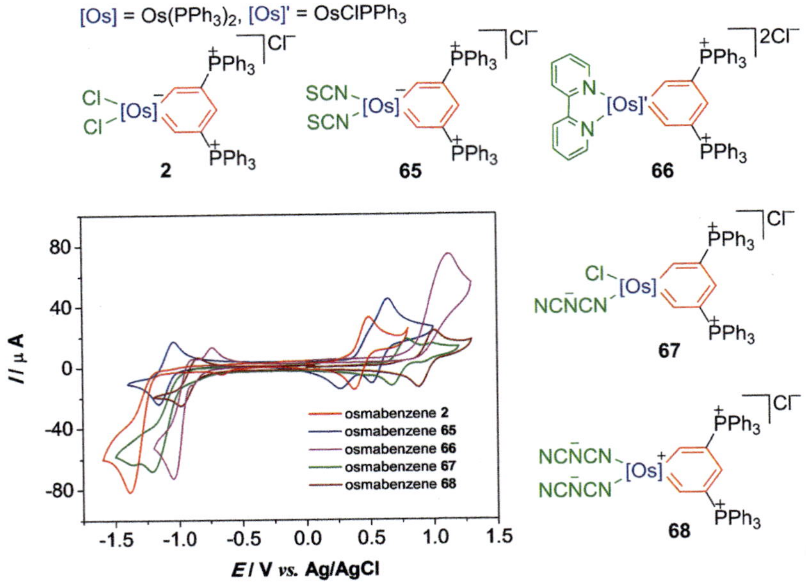

Figure 6.3 Cyclic voltammograms of **2, 65, 66, 67** and **68** measured in DCM with 0.1 M nBu$_4$NClO$_4$ as the supporting electrolyte at a scan rate of 0.10 V s^{-1}.

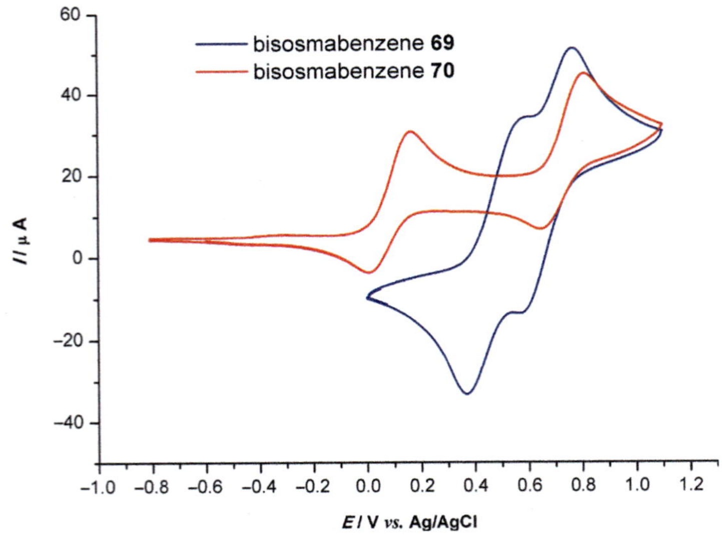

Figure 6.4 Cyclic voltammograms of **69** and **70** measured in DCM with 0.1 M nBu$_4$NClO$_4$ as a supporting electrolyte at a scan rate of 0.10 V s^{-1}.

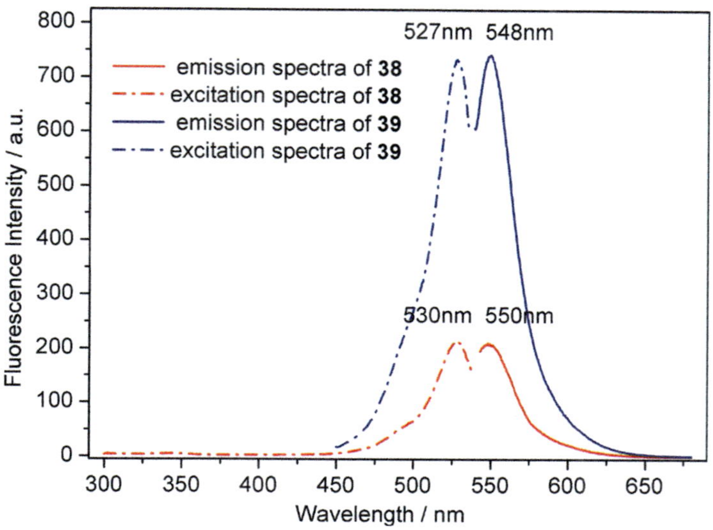

Figure 6.5 Excitation (dot) and emission (solid) spectra of ruthenabenzenes **38** and **39** (1.0×10^{-4} mol L^{-1}) in a methanol-water (1:1, v/v) medium.

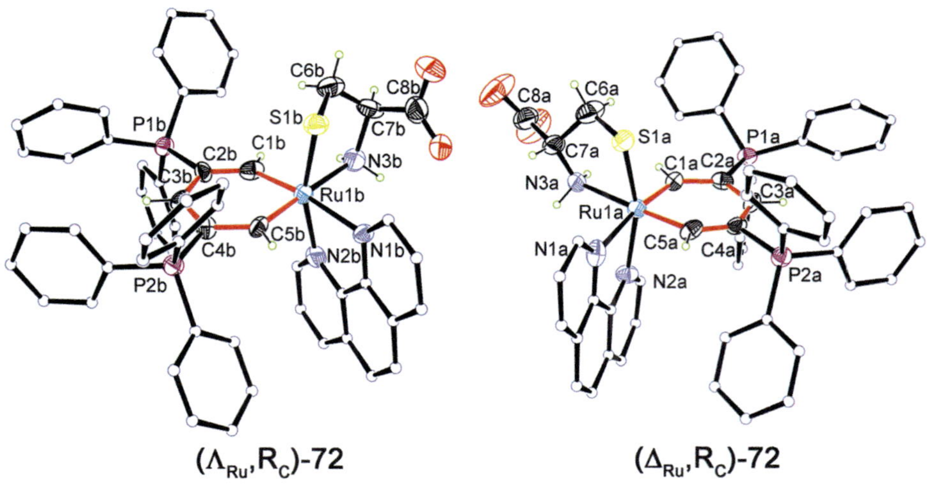

Figure 6.6 Molecular structures of ruthenabenzene (Λ_{Ru}, R_C)-**72** and ruthenabenzene (Δ_{Ru}, R_C)-**72**. The hydrogen atoms of PPh_3 are omitted for clarity.

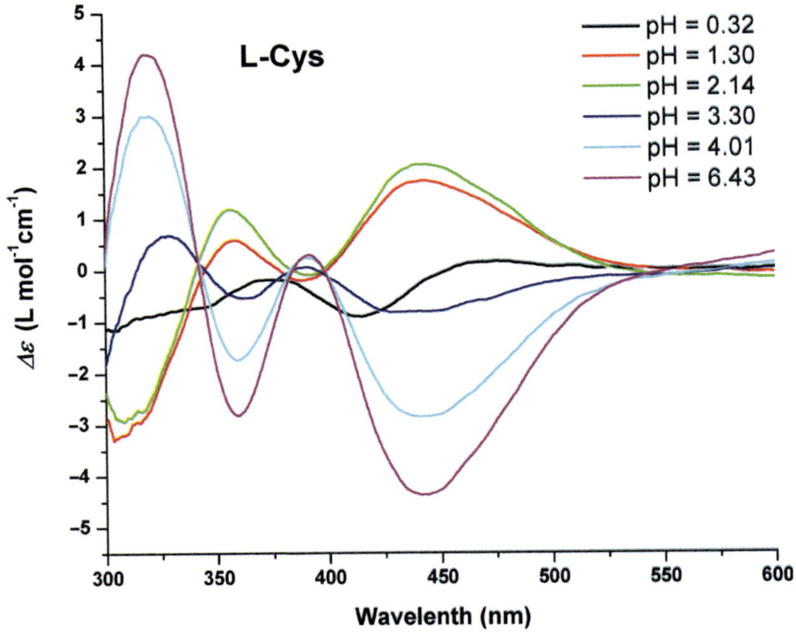

(a)

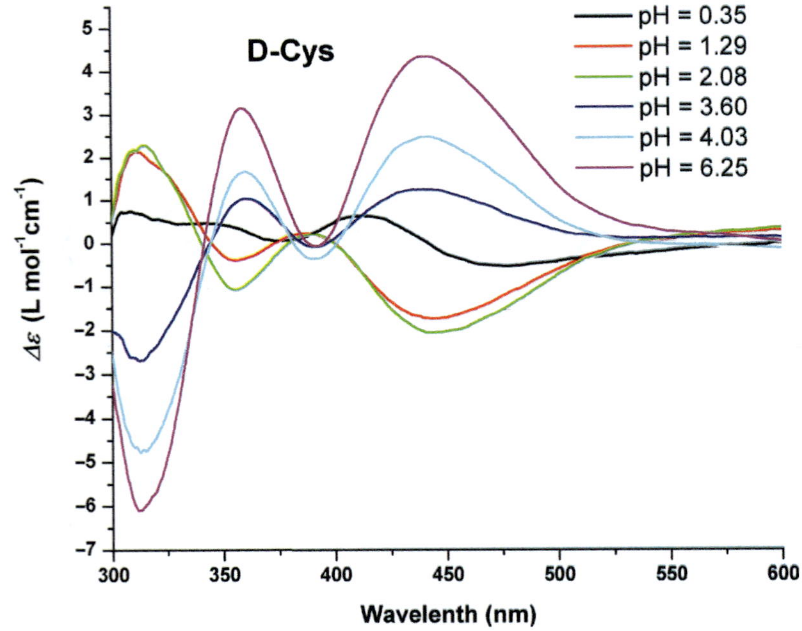

(b)

Figure 6.7 Influence of pH on the end CD responses of **71** towards (a) L- and (b) D-cysteine in 10% ethanol/water. [**71**] = 5.86×10^{-4} M.

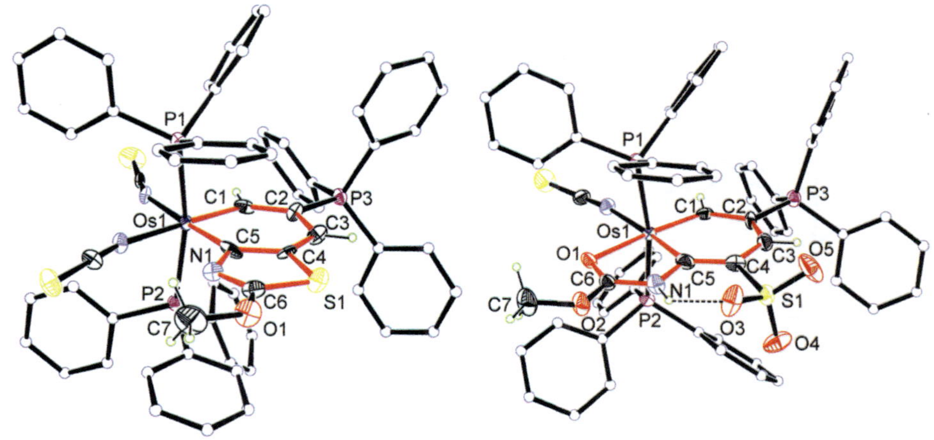

osmabenzothiazole **73** osmabenzoxazole **75**

Figure 6.8 Molecular structures of osmabenzothiazole **73** and osmabenzoxazole **75**. The hydrogen atoms of PPh_3 are omitted for clarity.

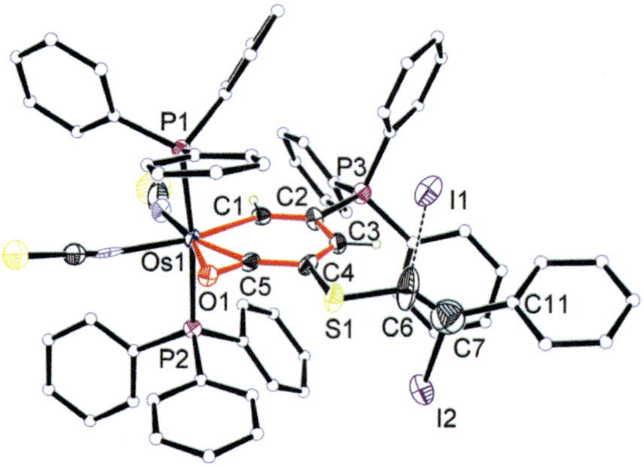

fused osmabenzene **82**

Figure 6.9 Molecular structures of fused osmabenzene **82**. The hydrogen atoms of PPh_3 are omitted for clarity.

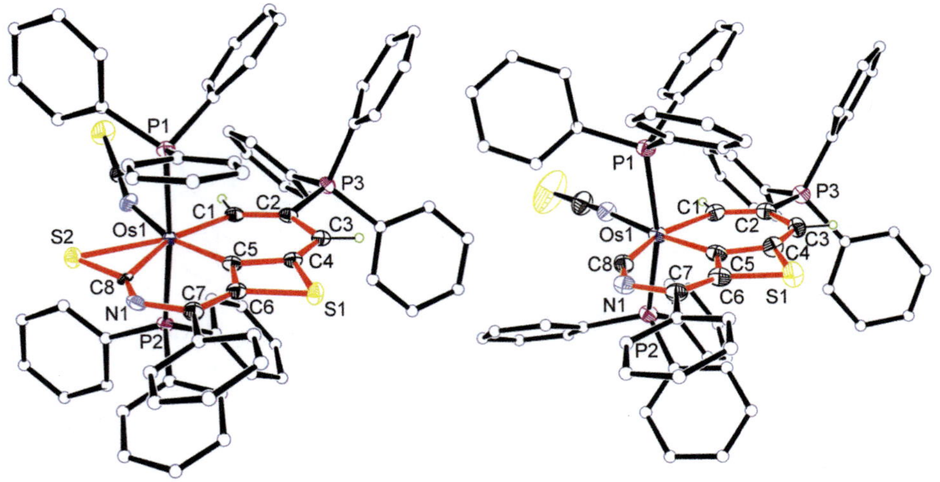

osmaisoquinoline **85** osmaisoquinolyne **86**

Figure 6.10 Molecular structures of osmaisoquinoline **85** and osmaisoquinolyne **86**. The hydrogen atoms of PPh_3 are omitted for clarity.

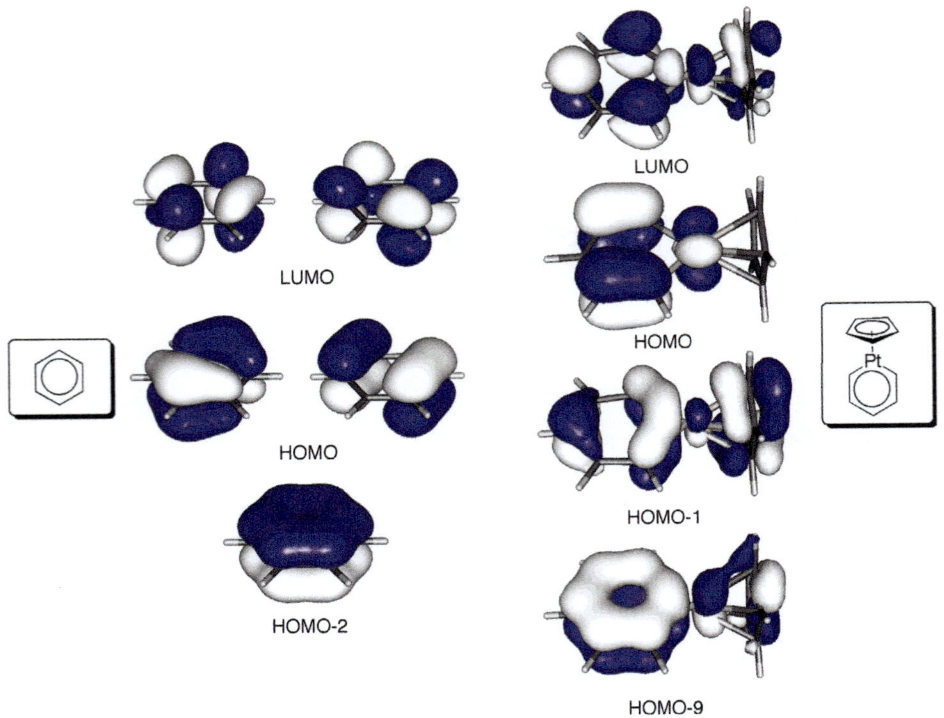

Figure 7.2 Representative MO's computed for benzene (left) and model platinabenzene $Pt[C_5H_5](Cp)$ (right). Figure adapted from reference [17a].

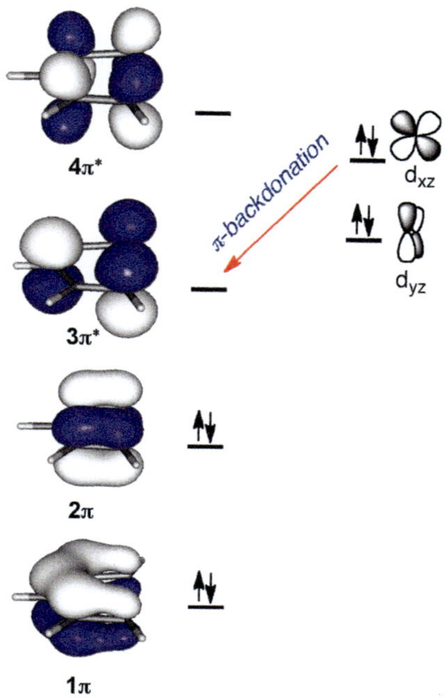

Figure 7.3 Schematic representation of the π-orbital interactions in metallabenzenes, adapted from reference [2a].

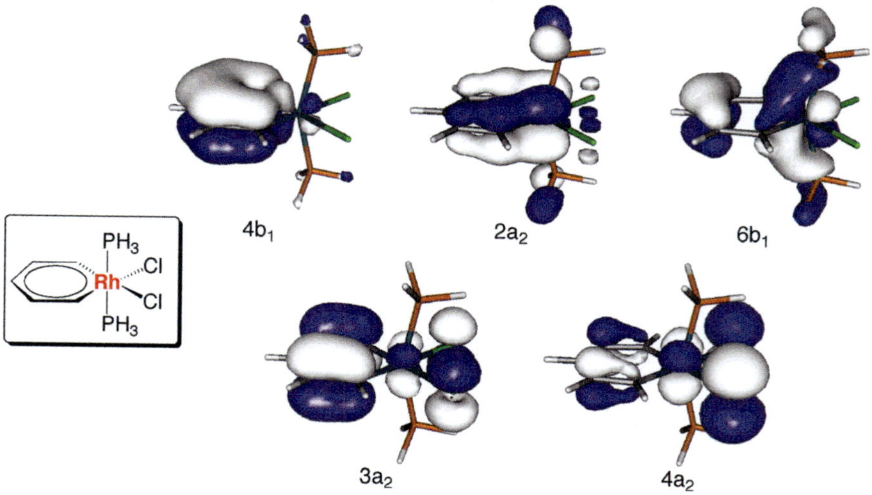

Figure 7.4 π-MO's computed for model rhodabenzene $Rh[C_5H_5](Cl)_2(PH_3)_2$ (figure adapted from reference [20]).

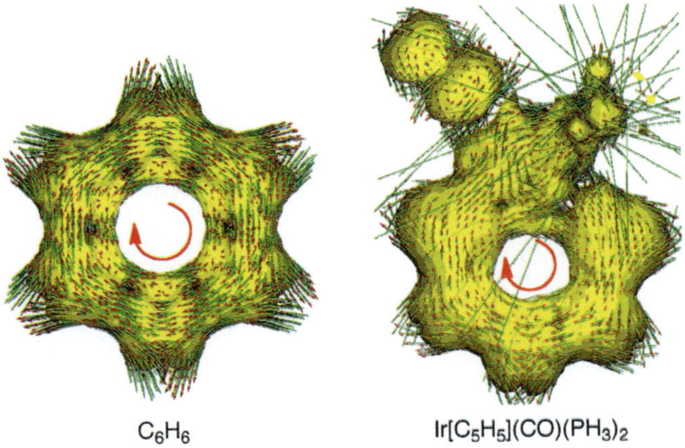

Figure 7.5 ACID plots (isosurface of 0.03 au) computed for benzene and $Ir[C_5H_5](CO)(PH_3)_2$. Aromatic species exhibit clockwise diatropic circulations (arrows). Figure adapted from reference [50].

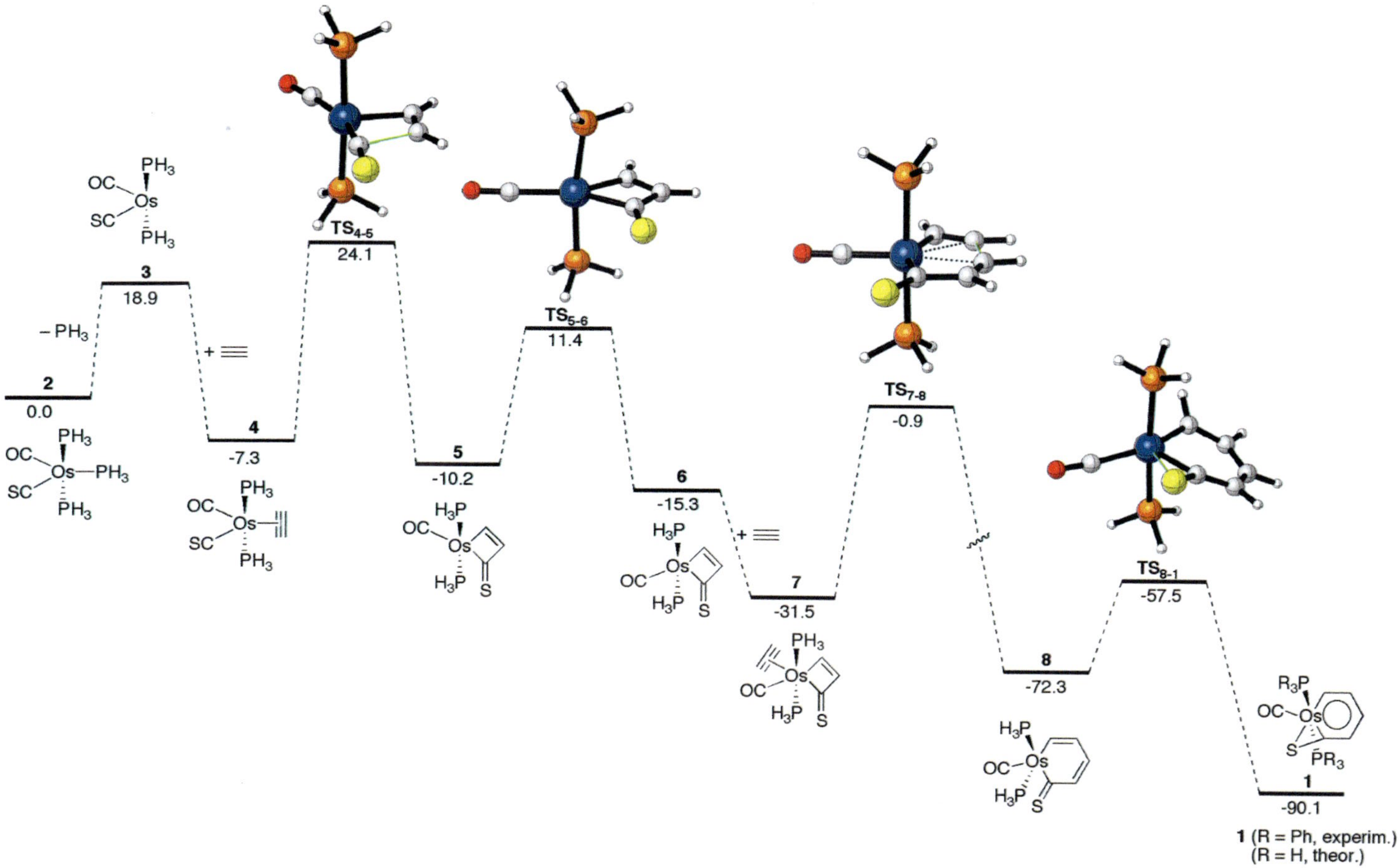

Figure 7.6 Computed reaction profile (PCM(toluene)-mPW1K/SDB-cc-pVDZ//mPW1K/SDD, kcal/mol) for the formation of model osmabenzene **1** (coordinates of the transition states were taken from reference [17a]). Relative free energies (ΔG_{298}, at 298 K) are given in kcal/mol.

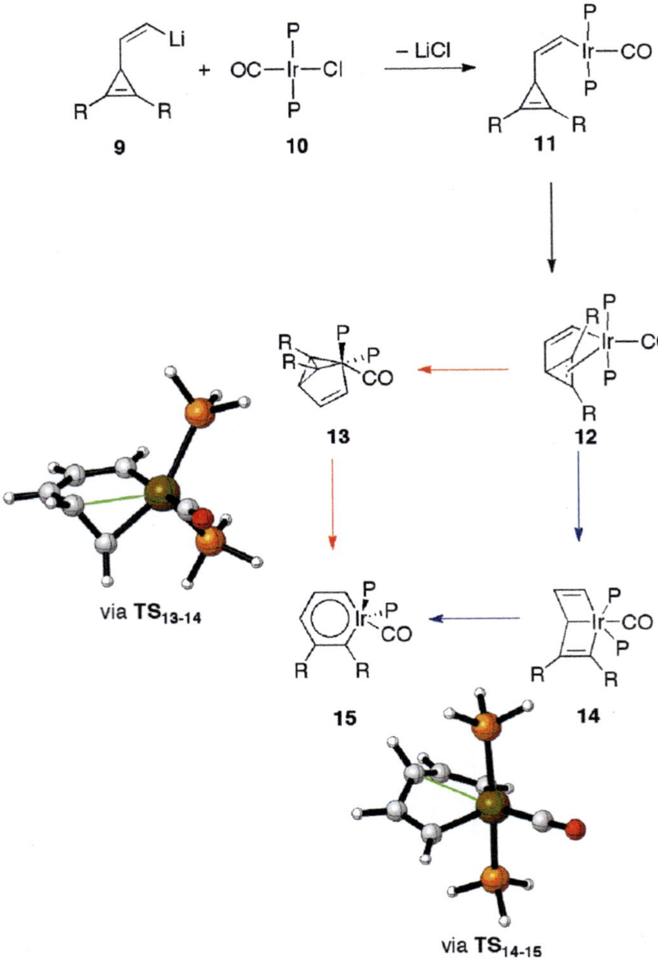

Figure 7.7 Formation of metallabenzenes by reaction of metal halide salts with 2,3-diphenylcyclopropenylvinyllithium **9**.

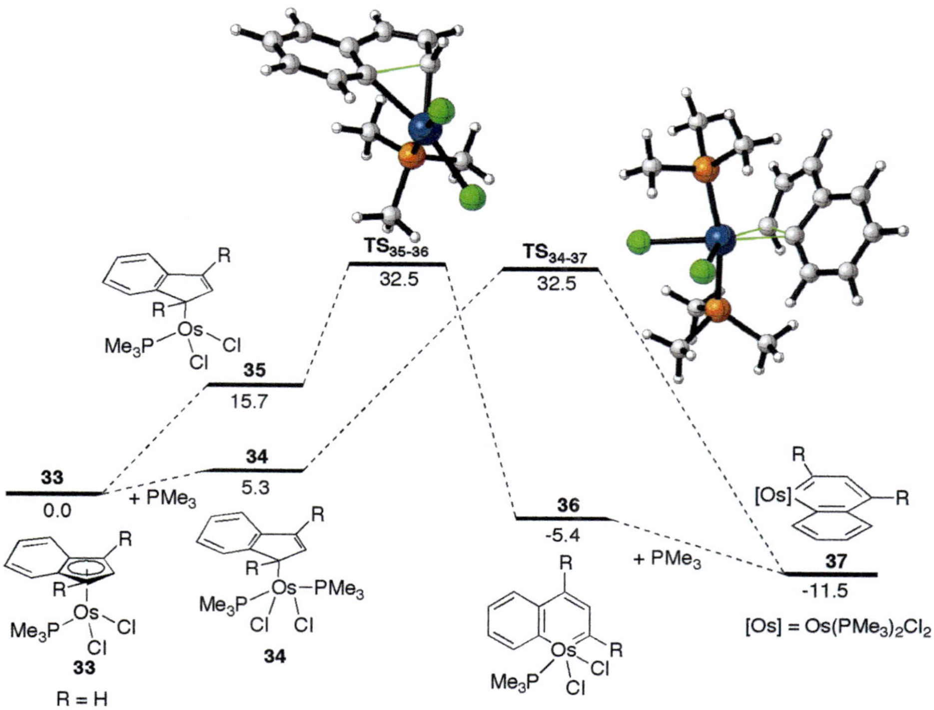

Figure 7.11 Computed (B3LYP/6-31+G(d)&LanL2DZ level) reaction profile for the rearrangement reaction of indenyl-complexes **33** into osmanaphthalenes **37** (coordinates of the transition states were taken from reference [64]). Relative free energies are given in kcal/mol.

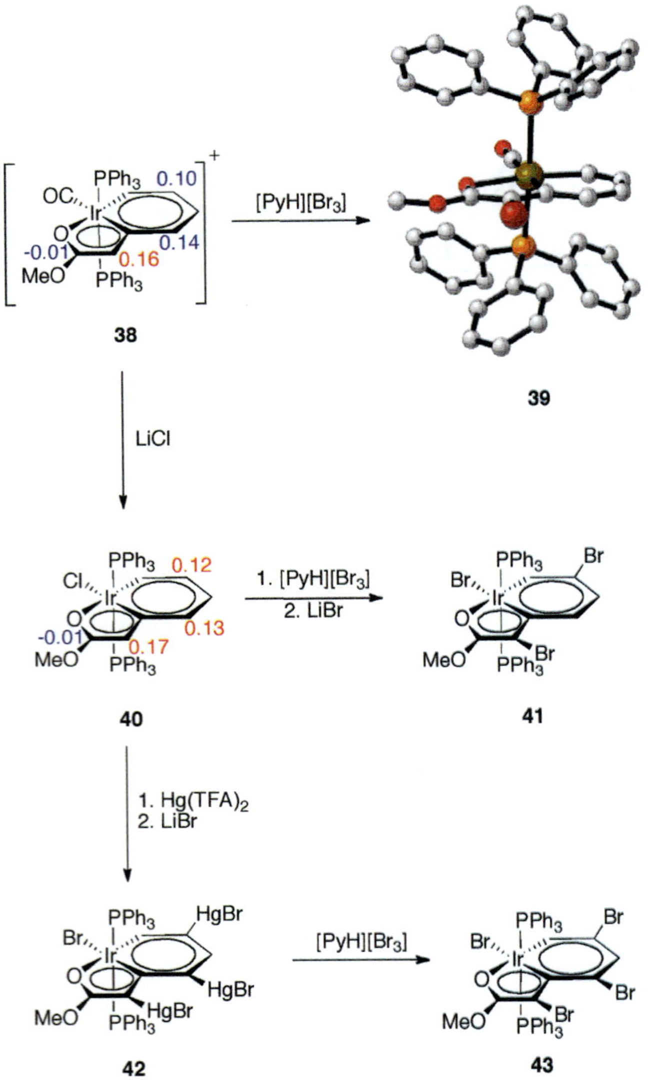

Figure 7.12 Electrophilic substitution reactions on iridabenzofurans. Numbers close to carbon atoms indicate the corresponding computed Fukui functions (f_k^-). Coordinates of complex **39** were taken from reference [66].

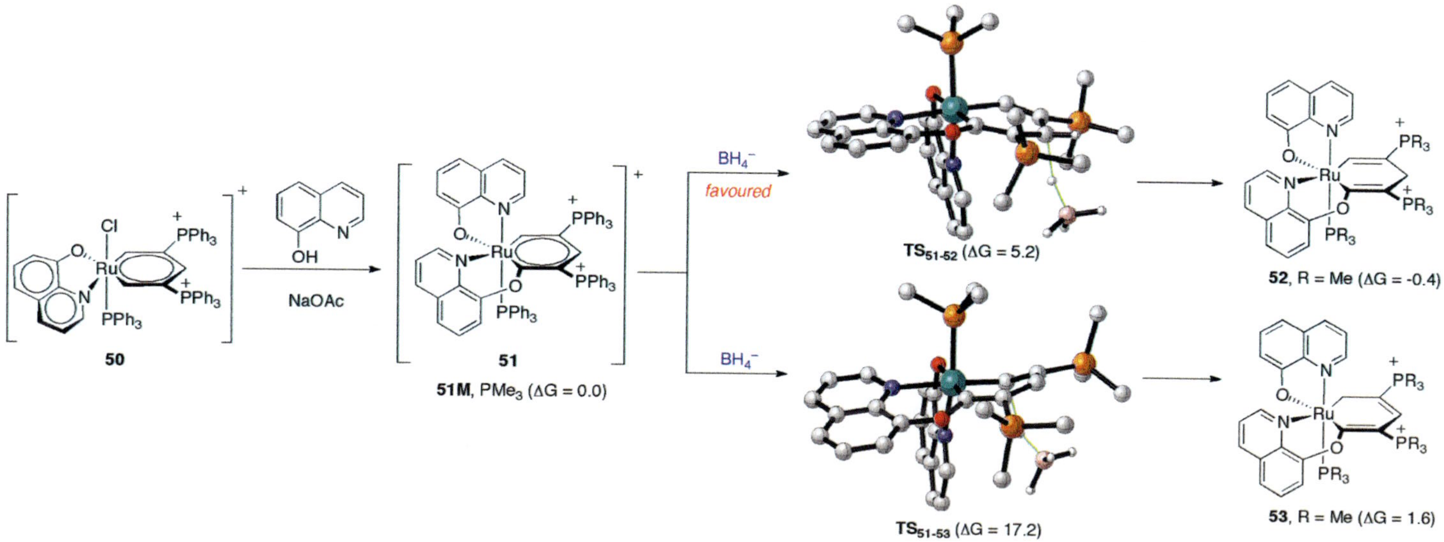

Figure 7.14 Computed (B3LYP/6-31G(d)&LanL2DZ level) reaction profile for the interconversion of ruthenabenzene **51** into ruthenacyclohexa-1,4-diene **52**. Relative free energy values are given in kcal/mol. Coordinates of the transition states were taken from reference [47].

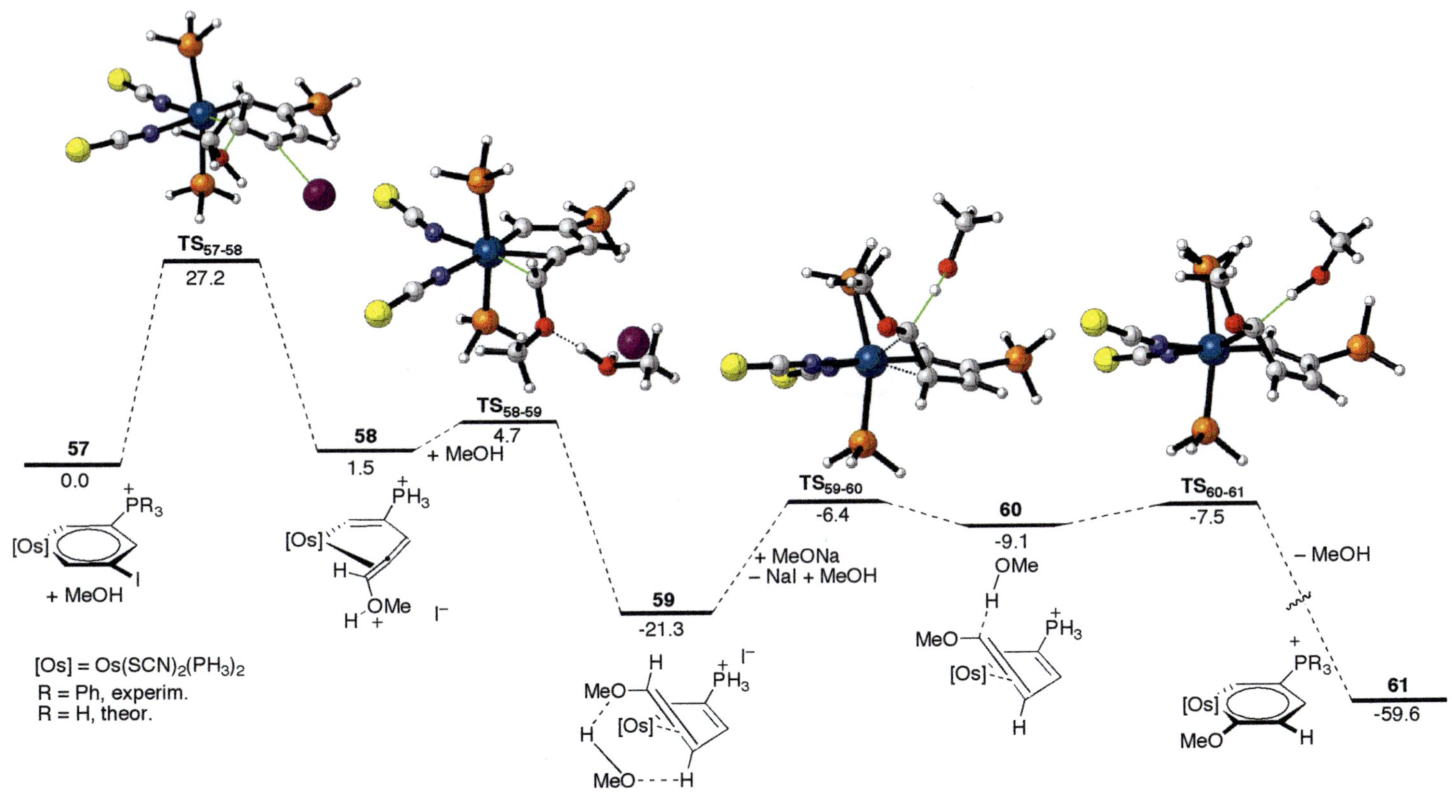

Figure 7.15 Computed reaction profile (B3LYP/6-31G(d, p)&LanL2DZ (plus polarization) level) for the transformation of osmabenzene **57** into **61**. Relative energy values are given in kcal/mol. Coordinates of transition states were taken from reference [70].

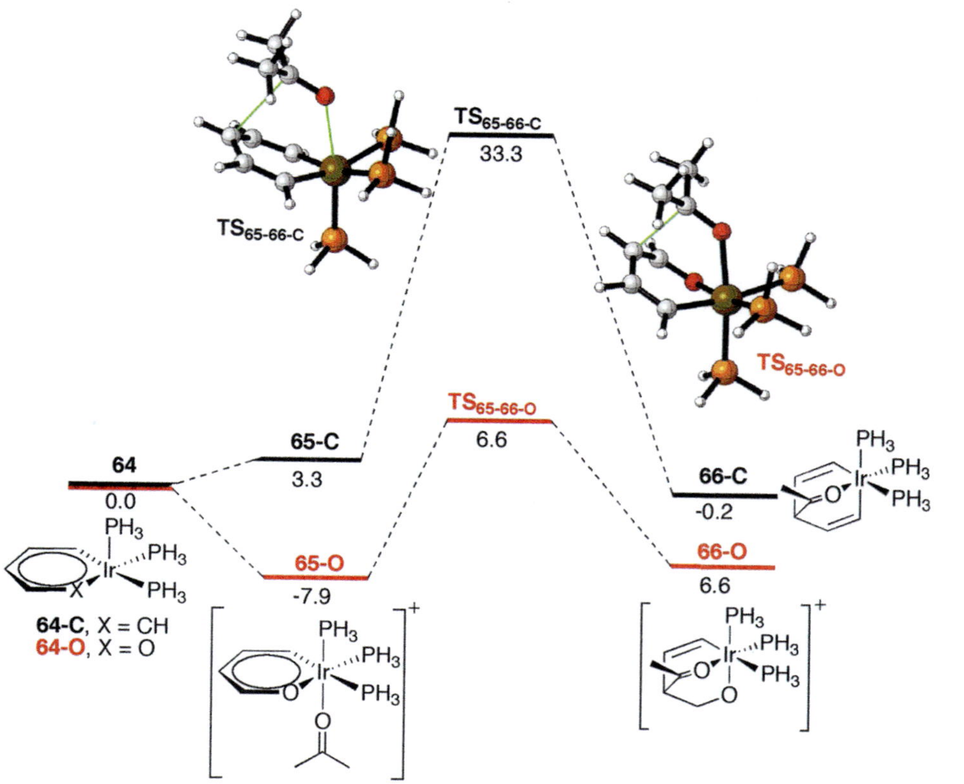

Figure 7.16 Computed reaction profile (mPW1K/SDB-cc-pVDZ//mPW1K/SDD level) for the acetone addition to **64-C** and **64-O**. Relative free energy values are given in kcal/mol. Coordinates of the transition states were taken from reference [73].

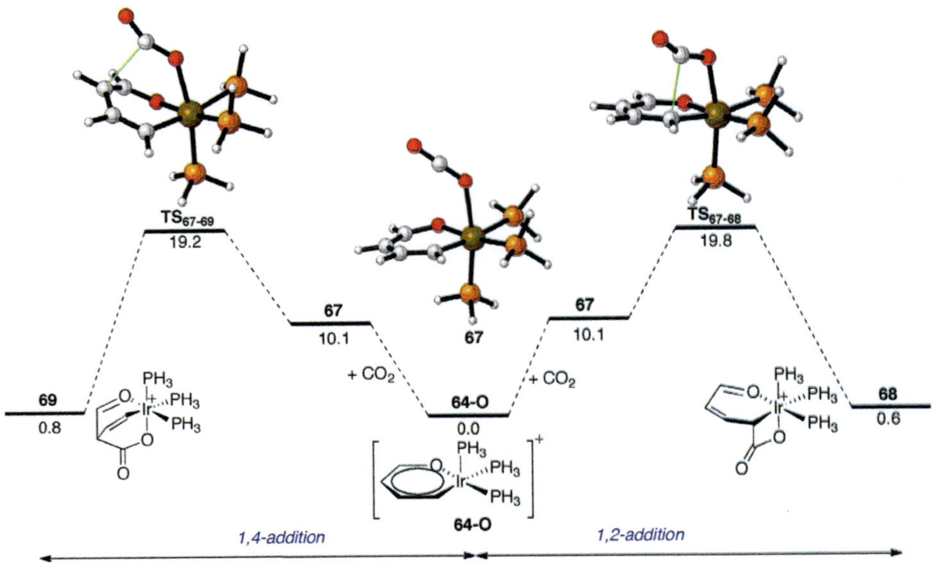

Figure 7.17 Computed reaction profile (PCM/mPW1K/SDB-cc-pVDZ//mPW1K/SDD level) for the cycloaddition reaction of iridapyrylium complex **64-O** and CO_2. Relative free energy values are given in kcal/mol. Coordinates of the transition states were taken from reference [74].

Printed and bound by CPI Group (UK) Ltd, Croydon, CR0 4YY
17/09/2026
14964303-0001